FINITE DIFFERENCES AND NUMERICAL ANALYSIS

H.C. SAXENA

Former Senior Lecturer
Department of Mathematics
Ramjas College
Delhi University, Delhi

S. Chand And Company Limited

(ISO 9001 Certified Company)

S Chand And Company Limited

(ISO 9001 Certified Company)

Head Office: D-92, Sector–2, Noida – 201301, U.P. (India), Ph. 91-120-4682700

Registered Office: A-27, 2nd Floor, Mohan Co-operative Industrial Estate, New Delhi – 110 044 | Phone: 011-49731800

www.**schandpublishing.com**; e-mail: **info@schandpublishing.com**

Marketing Offices:

Chennai	:	Ph: 23632120; chennai@schandpublishing.com
Guwahati	:	Ph: 2738811, 2735640; guwahati@schandpublishing.com
Hyderabad	:	Ph: 40186018; hyderabad@schandpublishing.com
Jalandhar	:	Ph: 4645630; jalandhar@schandpublishing.com
Kolkata	:	Ph: 23357458, 23353914; kolkata@schandpublishing.com
Lucknow	:	Ph: 4003633; lucknow@schandpublishing.com
Mumbai	:	Ph: 25000297; mumbai@schandpublishing.com
Patna	:	Ph: 2260011; patna@schandpublishing.com

First Edition 1961
Subsequent Editions and Reprints 1965, 68, 71, 73, 80, 82, 84, 86, 88 (Twice), 90, 92, 93, 94, 96, 97, 98, 2000, 2006, 2007, 2008 (Twice), 2009, 2011, 2013, 2014 (Twice), 2015, 2017, 2018 (Twice), 2020 (Twice), 2021, 2023

Reprint 2024

ISBN: 978-81-219-0339-4 **Product Code:** H6FDN61MATH10ENAO0XO

PRINTED IN INDIA

By Vikas Publishing House Private Limited, Plot 20/4, Site-IV, Industrial Area Sahibabad, Ghaziabad – 201 010 and Published by S. Chand And Company Limited, A-27, 2nd Floor, Mohan Co-operative Industrial Estate, New Delhi – 110 044.

PREFACE TO THE FIFTEENTH EDITION

A few chapters of the present book have been shifted either earlier or later to have the continuity of the matter. Appendices have been removed and a new chapter on fitting of curves and cubic splines has been added. Also few solved examples have been added at proper places. It is expected that now the misprints have been removed. I am thankful to the M.D. Shri Rajendra Kumar Gupta and Director, Shri Ravindra Kumar Gupta of S. Chand & Co. Ltd. for the fine get up of the book. I shall be grateful for any suggestion leading to the improvement of the book.

67, B.D. Estate, **H.C. SAXENA**
Timarpur, Delhi-54
Phone, 23812585

PREFACE TO THE FIRST EDITION

The present book is based on my experience, extending over a period of about fifteen years of teaching Mathematical Statistics and Calculus of Finite Differences to graduate and postgraduate students of the University of Delhi. During this period, a number of excellent books on the Calculus of Finite Differences have appeared, but most of these books are veritable encyclopaedias and are not at all suited to the needs of the students of the Indian universities. No book, as an introductory text, is available and this is the real reason for the appearance of this book.

The present book completely covers the syllabi in the Calculus of Finite Differences of the various Indian universities. It contains a number of solved examples and a large number of carefully graded examples have been given at the end of each chapter. The examples have been taken from the university papers and standard books and in addition many have been specially constructed. The answers to the questions have been given at the end of the book. As for mathematical pre-requisites, only the elementary knowledge of Calculus up to the B.A. (Pass) standard is assumed, so that the book can be useful for the undergraduate students also, who have Mathematics as one of their subsidiary subjects.

I am much indebted to the authors and publishers of those books which I have often consulted in the course of my teaching.

My thanks are due to the publishers, Shri Shyam Lal Gupta for the fine get up of the book.

I shall be grateful for any suggestion for the improvement of the book.

Delhi

January, 1961 **H.C. SAXENA**

CONTENTS

Part I—CALCULUS OF FINITE DIFFERENCES

Part II—NUMERICAL ANALYSIS

PART - I

CALCULUS OF FINITE DIFFERENCES

1

FINITE DIFFERENCES AND INTERPOLATION WITH EQUAL INTERVALS

1·1. Introduction. The calculus of finite differences, in its broad meaning, deals with the changes that take place in the value of the function, the dependent variable, due to finite changes in the independent variable. It is a study of the relations that exist between the values assumed by the function whenever the independent variable changes by finite jumps whether equal or unequal. In infinitesimal calculus we study, on the other hand, those changes which occur when the independent variable changes continuously in a given interval. In the present chapter, we study the variations in the function when the independent variable changes by equal intervals.

1·2. Differences. *The operator* Δ. Consider the function $y=f(x)$ $=\sin x^{\circ}$. The value of y corresponding to any value of x can be obtained by substituting that value of x on the right hand side of the equation. For example :

x°	$\sin x^{\circ}$
0°	0
10°	·17365
20°	·34202
30°	·50000
40°	·64279
50°	·76604
60°	·86603
70°	·93969
80°	·98481
90°	1·00000

Now suppose that instead of numerical values of x differing by 10°, we have the following consecutive values of x differing by h :

$$a, a+h, a+2h, a+3h, a+4h, \ldots$$

The values of $y=f(x)$ corresponding to these values of x are :

$$f(a), f(a+h), f(a+3h), f(a+4h), \ldots$$

The value of the independent variable is called the *argument* and the value of the dependent variable corresponding to that argument is called the *entry*. The difference between the consecutive values of the independent variable is called the *interval of differencing*.

The differences

$$f(a+h)-f(a), f(a+2h)-f(a+h), f(a+3h)-f(a+2h), \text{etc.}$$

are called the *first differences of the function* $y=f(x)$. *Denoting these differences by* $\Delta f(a), \Delta f(a+h), \Delta f(a+2h)$, *etc., we have*

$$\Delta f(a)=f(a+h)-f(a),$$
$$\Delta f(a+h)=f(a+2h)-f(a+h),$$
$$\Delta f(a+2h)=f(a+3h)-f(a+2h), \text{etc.}$$

The differences of these first differences are called *second differences*. Denoting these by $\Delta^2 f(a), \Delta^2 f(a+h)$, etc., we have

$$\Delta^2 f(a)=\Delta f(a+h)-\Delta f(a)$$
$$=f(a+2h)-2f(a+h)+f(a),$$
$$\Delta^2 f(a+h)=\Delta f(a+2h)-\Delta f(a+h)$$
$$=f(a+3h)-2f(a+2h)+f(a+h) \text{ etc.}$$

The differences of these second differences are called *third differences*. Denoting these by $\Delta^3 f(a)$, etc., we have

$$\Delta^3 f(a)=\Delta^2 f(a+h)-\Delta^2 f(a)$$
$$=f(a+3h)-3f(a+2h)+3f(a+h)-f(a) \quad \ldots(1)$$

and so on. Thus $\Delta^n f(a)=\Delta^{n-1} f(a+h)-\Delta^{n-1} f(a)=\text{etc.}$

These differences are shown in the following table called the *difference table*.

TABLE I. (Finite Differences)

Argument	Entry	First diff.	Second diff.	Third diff.	Fourth diff.
x	$y_x=f(x)$	$\Delta f(x)$	$\Delta^2 f(x)$	$\Delta^3 f(x)$	$\Delta^4 f(x)$
a	$f(a)$				
		$\Delta f(a)$			
$a+h$	$f(a+h)$		$\Delta^2 f(a)$		
		$\Delta f(a+h)$		$\Delta^3 f(a)$	
$a+2h$	$f(a+2h)$		$\Delta^2 f(a+h)$		$\Delta^4 f(a)$
		$\Delta f(a+2h)$		$\Delta^3 f(a+h)$	
$a+3h$	$f(a+3h)$		$\Delta^2 f(a+2h)$		
		$\Delta f(a+3h)$			
$a+4h$	$f(a+4h)$				

Notes 1. $f(a)$, the first entry, is termed as the *leading term* and the differences $\Delta f(a), \Delta^2 f(a), \Delta^3 f(a), \ldots$ are known as the *leading differences*. Also $\Delta^0 = 1$.

2. It should be remembered that Δ is not a quantity but a symbol standing for an "*operation*". Thus Δ^2 does not represent the square of a quantity but denotes the repetition of the operator Δ i.e. $\Delta^2 = \Delta(\Delta)$.

3. Equation (1) shows that the differences of third (any) order can be expressed in terms of the entries.

4. The Operator Δ is called the *forward* or *descending difference operator.*

5. It should be noted that $\Delta f(a) \neq f(a)\,\Delta \Rightarrow \Delta f \neq f \Delta$.

Example 1. *Construct a difference table from the following values of x and y :*

x :	3·0	3·1	3·2	3·3	3·4
$y : (= 1/x)$	·33333	·32258	·31250	·30303	·29412

Sol. The first differences are obtained by subtracting each entry from the next entry, the second differences are obtained by carrying out a similar set of subtractions on the first differences, and so on. The difference table is as follows :

Argument x	Entry $f(x) = 1/x$	First diff. $\Delta f(x)$	Second diff. $\Delta^2 f(x)$	Third diff. $\Delta^3 f(x)$	Fourth diff. $\Delta^4 f(x)$
3·0	·33333				
		– ·01075			
3·1	·32258		·00067		
		– ·01008		– ·00006	
3·2	·31250		·00061		·00001
		– ·00947		– ·00005	
3·3	·30303		·00056		
		– ·00891			
3·4	·29412				

1·3. Theorem. *The nth differences of a rational integral function (polynomial) of the nth degree are constant when the values of the independent variable are at equal intervals.*

Let the rational integral function of the nth degree in x be

$$f(x) = ax^n + bx^{n-1} + cx^{n-2} + \cdots + kx + l$$

where n is a positive integer; $a, b, c, \ldots, l$ are constants and $a \neq 0$.

$$\therefore \quad f(x+h) = a(x+h)^n + b(x+h)^{n-1} + c(x+h)^{n-2} + \cdots + k(x+h) + l$$

Hence $\Delta f(x) = f(x+h) - f(x)$

$$= a[(x+h)^n - x^n] + b[(x+h)^{n-1} - x^{n-1}] + \cdots + kh$$

$$= anhx^{n-1} + b'x^{n-2} + c'x^{n-3} + \cdots + k'x + l' \qquad \ldots(1)$$

where $b', c', \ldots, l'$ are constant coefficients, independent of x.

Thus the first difference of a rational integral function (polynomial) of the nth degree is a rational integral function (polynomial) of degree $(n-1)$.

To find the second difference of $f(x)$, we have

$$\Delta^2 f(x) = \Delta[\Delta f(x)] = \Delta[f(x+h) - f(x)] = \Delta f(x+h) - \Delta f(x)$$
$$= anh\,[(x+h)^{n-1} - x^{n-1}] + b'[(x+h)^{n-2} - x^{n-2}] + \cdots + k'h$$
$$= an(n-1)\,h^2 x^{n-2} + b''\,x^{n-3} + \cdots + k'', \text{ on using (1).}$$

The second differences is thus a polynomial of degree $(n-2)$.

By continuing in this manner, we arrive at the result :

$$\Delta^n f(x) = \Delta^n(ax^n) = a[n(n-1)(n-2)\ldots 1]h^n x^{n-1} = ah^n(n\,!).$$

The nth difference is, therefore, constant, and all higher differences are zero, that is, the $(n+1)$*th and higher differences of a polynomial of the nth degree are zero.*

Note. The converse of the above theorem is also true, namely :

If the nth differences of a tabulated function are constant when the values of the independent variable are taken at equal intervals, the function is a polynomial of degree n.

Example 2. *Evaluate* :

(*i*) $\Delta^3(1-x)(1-2x)(1-3x)$,

(*ii*) $\Delta^n(e^{ax+b})$, *a and b are constants,*

(*iii*) $\Delta(u_x v_x)$,

the interval of differencing being unity and u_x, v_x *are functins of x.*

Sol. (*i*) Here $f(x) = (1-x)(1-2x)(1-3x)$

$$= -6x^3 + 11x^2 - 6x + 1$$

so that $f(x)$ is a polynomial in x of degree 3.

$$\therefore \quad \Delta^3 f(x) = (-6)(3!) = -36.$$

(*ii*) We have $\Delta(e^{ax+b}) = f(x+1) - f(x)$ where $f(x) = e^{ax+b}$

$$= e^{a(x+1)+b} - e^{ax+b} = e^{ax+b}(e^a - 1),$$

$$\Delta^2(e^{ax+b}) = \Delta(\Delta e^{ax+b}) = \Delta[e^{ax+b}(e^a - 1)]$$
$$= e^{a(x+1)+b}(e^a - 1) - e^{ax+b}(e^a - 1)$$
$$= e^{a(x+b}(e^a - 1)^2.$$

Proceeding in this manner, we have

$$\Delta^n(e^{ax+b}) = e^{ax+b}(e^a - 1)^n.$$

(*iii*) Now $\quad \Delta(u_x v_x) = u_{x+1}v_{x+1} - u_x v_x$, by definition

$$= u_{x+1}v_{x+1} - u_{x+1}v_x + u_{x+1}v_x - u_x v_x$$
$$= u_{x+1}\Delta v_x + v_x \Delta u_x.$$

Similarly $\Delta(u_x v_x) = u_x \Delta v_x + v_{x+1} \Delta u_x$.

1·4. Effect of Errors in difference tabular values.

Let $y_0, y_1, \ldots, y_n$ be the true values of a function y_x and suppose that the values y_4 and y_5 are in errors, the errors being ε_1 and ε_2 so that the erroneous values are $y_4 + \varepsilon_1$ and $y_5 + \varepsilon_2$ respectively. The difference table will then be of the following form:

y	Δy	$\Delta^2 y$	$\Delta^3 y$	$\Delta^4 y$	$\Delta^5 y$
y_0					
	Δy_0				
y_1		$\Delta^2 y_0$			
	Δy_1		$\Delta^3 y_0$		
y_2		$\Delta^2 y_1$		$\Delta^4 y_0 + \varepsilon_1$	
	Δy_2		$\Delta^3 y_1 + \varepsilon_1$		
y_3		$\Delta^2 y_2 + \varepsilon_1$		$\Delta^4 y_1 + \varepsilon_1$	
	$\Delta y_3 + \varepsilon_1$		$\Delta^3 y_2 + \varepsilon_2 - 3\varepsilon_1$		
$y_4 + \varepsilon_1$		$\Delta^2 y_3 + \varepsilon_2 - 2\varepsilon_1$		$\Delta^4 y_2 + 6\varepsilon_1 - 4\varepsilon_2$	
	$\Delta y_4 + \varepsilon_2 - \varepsilon_1$		$\Delta^3 y_3 + 3\varepsilon_1 - 3\varepsilon_2$		
$y_5 + \varepsilon_2$		$\Delta^2 y_4 - 2\varepsilon_2 + \varepsilon_1$		$\Delta^4 y_3 + 6\varepsilon_2 - 4\varepsilon_1$	
	$\Delta y_5 - \varepsilon_2$		$\Delta^3 y_4 + 3\varepsilon_2 - \varepsilon_1$		
y_6		$\Delta^2 y_5 + \varepsilon_2$		$\Delta^4 y_4 + \varepsilon_1 - 4\varepsilon_2$	
	Δy_6		$\Delta^3 y_5 - \varepsilon_2$		
y_7		$\Delta^2 y_6$		$\Delta^4 y_5 + \varepsilon_2$	
	Δy_7		$\Delta^3 y_6$		
y_8		$\Delta^2 y_7$			
	Δy_8				
y_9					

Suppose the 3rd differneces are quite irregular and 4th difference are still irregular. Since the algebraic sum of the 4th differences is zero, the 4th differences are *accumulated errors.* The magnitude of the errors $\varepsilon_1, \varepsilon_2$ are determined by writing the entries of the 4th differences column equal to $\varepsilon_1, \varepsilon_2 - 4\varepsilon_1, -4\varepsilon_2 + 6\varepsilon_1, 6\varepsilon_2 - 4\varepsilon_1, -4\varepsilon_2 + \varepsilon_1, \varepsilon_2$ respectively and solving for ε_1 and ε_2.

Looking at the above difference table, we find that

1. The effect of an error increases with the succession.
2. The coefficients of ε's are the binomial coefficients with alternating signs.
3. The algebraic sum of the errors in the various difference columns is zero.
4. The maximum error occurs in the same line as the erroneous tabular value.

With the help of the points noted above, it will be for us to locate the error *i.e.*, its source and so on. However, if there are three or more errors, the successive differences will be irregular and it will not be possible to locate the errors in the tabular values.

Note. If there is error ε_1 only in y_4, then $\varepsilon_2 = 0$. The errors in the column of fourth differences will be $\varepsilon_1 - 4\varepsilon_1, 6\varepsilon_1, -4\varepsilon_1, \varepsilon_1, 0$ while errors in the column of fifth differences will be $\varepsilon, -5\varepsilon, 10\varepsilon, -10\varepsilon, 5\varepsilon, -\varepsilon$.

Example 3. *Locate the error in the second column of the following difference table :*

x	$f(x)$	$\Delta f(x)$	$\Delta^2 f(x)$	$\Delta^3 f(x)$	$\Delta^4 f(x)$	
3·60	0·112046					
		8158				
3·61	0·120204		-12			
		8146		-22		
3·62	0·128350		-34		82	
		8112		60		-4ε
3·63	0·136462		26		-122	
		8138		-62		6ε
3·64	0·144600		-36		82	
		8102		20		-4ε
3·65	0·152702		-16		-21	
		8086		-1		ε
3·66	0·160788		-17		0	
		8069		-1		
3·67	0·168857		-18			
		8051				
3·68	0·176908					

Sol. It is clear from the difference table that irregularity begins in each column on the horizontal lines corresponding to $x = 3{\cdot}63$.

By table

$$-4\varepsilon = 82, -6\varepsilon = 122$$

so the error is nearly equal to $\varepsilon = -{\cdot}000020$

The true value of $f(x)$ at $x = 3{\cdot}63$ is

$$0{\cdot}136462 - (-0{\cdot}000020) = 0{\cdot}136462 + 0{\cdot}000020$$
$$= 0{\cdot}136482$$

Now we replace $f(x)$ at $x = 3{\cdot}63$ by ${\cdot}136482$.

The 3rd difference will be practically constant.

1·5·1. Factorial Notation. The product of factors of which the first factor is x and the successive factors decrease by a constant difference is called a *factorial polynomial* or *function* and is denoted by $x^{(r)}$, r being a positive integer and is read as 'x raised to the power r factorial'. In general, the interval of differencing is 'h'

In particular, we set $x^{(0)} = 1$.

We define $\quad x^{(r)} = x(x-h)(x-2h)\ldots[x-(r-1)h] \quad \ldots(1)$

Also $\Delta x^{(r)} = (x+h)^{(r)} - x^{(r)}$ (Def.)

$$= (x+h)x(x-h)...[x-(r-2)h]$$
$$- x(x-h)(x-2h)...[x-(r-1)h]$$
$$= x(x-h)...(x-\overline{r-2}h)[(x+h)-(x-\overline{r-1}h)]$$
$$= rhx^{(r-1)}. \qquad ...(2)$$

This result is to be remembered. The result of finding difference of $x^{(r)}$ is equal to $h \,.\, [r\, x^{(r-1)}]$.

Similarly, $\Delta^2 x^{(r)} = \Delta(\Delta x^{(r)}) = \Delta[hrx^{(r-1)}] = hr\,\Delta x^{(r-1)}$

$\Rightarrow \qquad \Delta^2 x^{(r)} = h^2 r(r-1)x^{(r-2)}$

and generally,

$$\Delta^m x^{(r)} = h^m r(r-1)...[r-(m-1)]\, x^{(r-m)},\ m \le r.$$
$$= 0,\ m > r,$$
$$\Delta^m x^{(m)} = m\,!\, h^m.$$

Notes 1. If x is an integer greater than $n-1$, then $x^{(n)} = \dfrac{x\,!}{(x-n)!}$.

2. For factorial notation, the operator Δ is analogous to operator D.

3. $x^{(n)} = a_n x^n + a_{n-1}x^{n-1} + \cdots + a_1 x$ where a's are Stirling's numbers of the first kind given by $a_{n+1} = a_n - na_{n-1}$.

4. From (2), page 9, $x^{(r-1)} = \dfrac{1}{rh}\,\Delta x^{(r)}$

5. $x^{(r)} = (x-\overline{r-1}h)\,(x-\overline{r-2}h).....x.$

6. $x^{(0)} = (x+h)^{-1}\,x \Rightarrow x^{(-1)}\,.\,(x+h)^{(-1)} = \dfrac{1}{x+h},\ x^0 \equiv 1.$

7. If h be the interval of differencing, then we define

$$x^{(-n)} = \frac{1}{(x+h)(x+2h)...(x+nh)} = \frac{1}{(x+nh)^{(n)}}$$

In particulars, $x^{(-1)} = \dfrac{1}{x+h},\ x^{(-2)} = \dfrac{1}{(x+2h)^{(2)}}$

or $$x^{(-2)} = \frac{1}{(x+2h)(x+h)} = \frac{1}{(x+h)(x+2h)}$$

Also $\Delta x^{(-n)} = (x+h)^{(-n)} - x^{(-n)}$ [Def.]

1·5·2. Negative Exponent

We define $x^{(-r)} = \dfrac{1}{(x+h)(x+2h)...(x+rh)} = \dfrac{1}{(x+rh)^{(r)}}$

The superscripts are referred to as *pseudo exponents.*

Also $\Delta x^{(-r)} = (x+h)^{(-r)} - x^{(-r)}$

$$= \frac{1}{(x+2h)(x+3h)...(x+\overline{r+1}h)} - \frac{1}{(x+h)(x+2h)...(x+rh)}$$

$$= \frac{(x+h)-(x+\overline{n+1}h)}{(x+h)(x+2h)...(x+nh)(x+\overline{n+1}h)}$$

$$= -nhx^{(-n-1)}.$$

Similarly, $\Delta^2 x^{(-n)} = (-n)(-n-1)h^2 x^{(-n-2)} = n(n+1)h^2 x^{(-n-2)}$

and generally,

$$\Delta^m x^{(-r)} = (-1)^m n(n+1)...(n+m-1)h^m x^{(-n-m)}, m \le n$$

Notes 1. In case of negative exponent, the operator Δ is equivalent to D.

2. Putting $r = -n$ in (1), page 8, we have $x^{(-n)} = \dfrac{1}{(x+nh)^n}$

3. We have $x^{(r)} . (x-r)^{(n)} = x^{(r+n)}$

where r and n are any positive or negative integers. This is referred to as the exponential law for *pseudo exponents*. Putting $n = m - r$, we obtain

$$\frac{x^{(m)}}{x^{(r)}} = (x-r)^{(m-r)} \quad \text{or} \quad x^{(m)} = x^{(r)}(x-r)^{(m-r)}, \quad 0 \le r \le m$$

1·5·3. Factorial function. The function $(x)_n$ defined as follows, is called the *factorial function.*

$$(x)_n = \prod_{r=1}^{n} (x+r-1)$$

$$= x(x+1)(x+2)...(x+n-1), n \ge 1,$$

$$(x)_0 = 1.$$

1·5·4. Representation of a given polynomial in factorial notation.

Example 4. *Represent the function* $f(x)$ *given by*

$$f(x) = 2x^4 - 12x^3 + 24x^2 - 30x + 9$$

and its successive differences in factorial notation.

Sol. Let $f(x) = 2x^4 - 12x^3 + 24x^2 - 30x + 9$

$$= 2x^{(4)} + bx^{(3)} + cx^{(2)} + dx + 9 \text{ (How?)}$$

$$= 2x(x-1)(x-2)(x-3) + bx(x-1)(x-2) + cx(x-1) + dx + 9$$

where b, c and d are constants to be determined.

Putting $x = 1$, we get

$$-7 = 0 + 0 + 0 + d + 9$$

$$\therefore \quad d = -16$$

Similarly, putting $x = 2$ and 3 in turn, we find $c = 4, b = -2$

$$\therefore \quad f(x) = 2x^{(4)} - 2x^{(3)} + 4x^{(2)} - 16x^{(1)} + 9$$

$$\therefore \quad \Delta f(x) = 8x^{(3)} - 6x^{(2)} + 8x^{(1)} - 16 \times 1 + 0 \mid \text{by (2), page 9.}$$

$$\Delta^2 f(x) = 24x^{(2)} - 12x^{(1)} + 8$$

$$\Delta^3 f(x) = 48x^{(1)} - 12$$

$$\Delta^4 f(x) = 48.$$

Alternatively. Method of synthetic division (*see* § **13·2·2**)

Example 5. *Find the function whose first difference is*

$$9x^2 + 11x + 5.$$

Sol. Let $f(x)$ be the required function so that we have

$$\Delta f(x) = 9x^2 + 11x + 5$$

$$\equiv 9x(x-1) + bx + c.$$

Putting $\quad x = 0$ and 1 successively, we find

$$c = 5, b = 20$$

$$\therefore \quad \Delta f(x) = 9x^{(2)} + 20x^{(1)} + 5$$

Hence $\quad f(x) = 3x^{(3)} + 10x^{(2)} + 5x^{(1)} + k$

where k is a constant, on using Note 4 of §1·5·1.

Thus $\quad f(x) = 3x(x-1)(x-2) + 10x(x-1) + 5x + k$

$$= 3x^3 + x^2 + x + k.$$

1·6. The Operators E and Δ. Let $y = f(x)$ be the function of x and let $a, a+h, a+2h, \ldots$ be the consecutive values of x. then

$$\Delta f(x) = f(a+h) - f(a).$$

Here, 1 (unity) is an operator which is such that the function on which it operates leaves the function unaltered. Thus '1' is called *identity operator.*

We define an operator E, called the displacement operator, by the equation

$$Ef(a) = f(a+h).$$

The operator E is also referred to as the *shift* operator, since it results in the next value of the function.

We have already defined,

$$\Delta f(a) = f(a+h) - f(a)$$

$$= Ef(a) - f(a)$$

Hence $\quad Ef(a) = f(a) + \Delta f(a)$

$$= (1 + \Delta) f(a)$$

Since $f(a)$ is arbitrary, this result suggests the following relation between the two operators Δ and E :

$$\boxed{E \equiv 1 + \Delta}$$

the operand $f(a)$ on each side being understood.

Let E^n stand for the operation E being carried n times.

Then

$$E^2f(a) = E.Ef(a) = Ef(a+h) = f(a+2h), \ldots,$$
$$E^nf(a) = E^{n-1}.Ef(a) = E^{n-1}f(a+h) = \cdots = f(a+nh).$$

The inverse operator E^{-1} is defined by

$$E^{-1}f(x) = f(x-h)$$

so that $\quad E^{-1}y_i = y_{i-1}, \quad$ where $y_i = f(x_i)$.

Notes 1. The expressions $E^2f^2(x)$ and $[Ef(x)]^2$ are not identical as is seen by the following example where $f(x) = x$, $f^2(x) = x^2$:

$$E^2(x^2) = E.E\,(x^2) = E\,(x+h)^2 = (x+2h)^2$$

and $\quad (Ex)^2 = (x+h)^2$

Thus $\quad E^2x^2 \neq (Ex)^2$

2. $\quad E^2f(a) = f(a+2h) = f(a) + 2\Delta f(a) + \Delta^2 f(a) = (1+\Delta)^2 f(a)$

$\therefore \quad E^2 \equiv (1+\Delta)^2.$

In general, if x is a positive integer, then

$$E^x \equiv (1+\Delta)^x.$$

3. $E^0 = 1.$

1·6·1. Properties of the two operators Δ and *E*.

The symbols Δ and E obey the following three laws of algebra, namely :

(*i*) *distributive*, (*ii*) *commutative and* (*iii*) *index laws.*

(*i*) $\Delta[f(x) + g(x) + \cdots]$

$$= [f(x+h) + g(x+h) + \cdots] - [f(x) + g(x) + \cdots] \quad \text{[Def.]}$$
$$= [f(x+h) - f(x)] + [g(x+h) - g(x)] + \cdots$$
$$= \Delta f(x) + \Delta g(x) + \cdots$$

Similarly, $E[f(x) + g(x) + \cdots] = Ef(x) + Eg(x) + \cdots$

(*ii*) *The symbols* Δ *and E are commutative with respect to constant only.*

$$\Delta[cf(x)] = cf(x+h) - cf(x)$$
$$= c[f(x+h) - f(x)]$$
$$= c\Delta f(x)$$

and $\quad E[cf(x)] = cf(x+h),$

$$= cEf(x), \text{ where } c \text{ is a constant}$$

(*iii*) $\Delta^p\Delta^q f(x) = \Delta^p\,[\Delta^q f(x)]$

$$= (\Delta\,\Delta\,\Delta \cdots p \text{ times}) \cdot (\Delta\,\Delta\,\Delta \cdots q \text{ times}) f(x)$$
$$= (\Delta\,\Delta\,\Delta \cdots \overline{p+q} \text{ times}) f(x)$$

$= \Delta^{p+q} f(x)$, p and q being positive integers.

Similarly, $E^p E^q f(x) = E^{p+q} f(x)$.

2. If m be a positive integer, then we can define symbol Δ^{-m} as

$\Delta^m [\Delta^{-m} f(x)] = f(x)$.

Similarly, $E^m [E^{-m} f(x)] = f(x)$.

3. $E\Delta = \Delta E$.

For
$$E \Delta f(x) = E[f(x+h) - f(x)]$$
$$= f(x+2h) - f(x+h)$$
$$= \Delta f(x+h)$$
$$= \Delta E f(x)$$

Since $f(x)$ is any arbitrary function, the result follows.

Example 6. *Prove that* $e^x = \left(\dfrac{\Delta^2}{E}\right) e^x \cdot \dfrac{Ee^x}{\Delta^2 e^x}$, *the interval of differencing being h.*

Sol. We have

$$E(e^x) = e^{x+h}$$

$$\Delta e^x = e^{x+h} - e^x = e^x(e^h - 1),\ \Delta^2 e^x = e^x \cdot (e^h - 1)^2$$

$$\left(\frac{\Delta^2}{E}\right) e^x = (\Delta^2 E^{-1}) e^x = \Delta^2 e^{x-h} = e^{-h} \cdot \Delta^2 e^x$$

$$\therefore \quad \text{R.H.S.} = e^{-h} \Delta^2 e^x \cdot \frac{e^{x+h}}{\Delta^2 e^x} = e^x. = \text{L.H.S.}$$

Hence the result is established.

Example 7. *Use the method of separation of symbols to prove the following identities :*

(*i*) $u_x = u_{x-1} + \Delta u_{x-2} + \Delta^2 u_{x-3} + \cdots + \Delta^{n-1} u_{x-n} + \Delta^n u_{x-n}$

(*ii*) $u_0 + {}^x c_1 \Delta u_1 + {}^x c_2 \Delta^2 u_2 + \cdots + \Delta^x u_x$
$$= u_x + {}^x c_1 \Delta^2 u_{x-1} + {}^x c_2 \Delta^4 u_{x-2} + \cdots + \Delta^x u_0$$

(*iii*) $u_1 x + u_2 x^2 + u_3 x^3 + \cdots$
$$= \frac{x}{1-x} u_1 + \frac{x^2}{(1-x)^2} \Delta u_1 + \frac{x^3}{(1-x)^3} \Delta^2 u_1 + \cdots$$

where $y = u_x$ *is any function of x and* $0 < x < 1$.

(*B.Sc. Guwahati, 1996, B.Sc Agra, 1996*)

(*iv*) $$u_0 + u_1 x + u_2 x^2 + = \frac{u_e}{1-x} + \frac{x\Delta u_0}{(1-x)^2} + \frac{x^2 \Delta^2 u_0}{(1-x)^3} + -$$

Sol. (*i*) $u_x - \Delta^n u_{x-n} = u_x - \Delta^n E^{-n} u_x$

$$\Rightarrow \qquad u_x - \Delta^n u_{x-n} = \left\{1 - \frac{\Delta^n}{E^n}\right\} u_x = \frac{E^n - \Delta^n}{E^n} u_x$$

$$= \frac{1}{E^n}\left\{\frac{E^n - \Delta^n}{E - \Delta}\right\} u_x \qquad \because \quad E - \Delta \equiv 1.$$

$$= E^{-n}(E^{n-1} + \Delta E^{n-2} + \Delta^2 E^{n-2} + \cdots + \Delta^{n-1}) u_x$$

$$= (E^{-1} + \Delta E^{-2} + \Delta^2 E^{-3} + \cdots + \Delta^{n-1} E^{-n}) u_x$$

$$= u_{x-1} + \Delta u_{x-2} + \Delta^2 u_{x-3} + \cdots + \Delta^{n-1} u_{x-n}$$

$$\therefore \quad u_x = u_{x-1} + \Delta u_{x-2} + \Delta^2 u_{x-3} + \cdots + \Delta^{n-1} u_{x-n} + \Delta^n u_{x-n}$$

(*ii*) $u_x + {}^x c_1 \Delta^2 u_{x-1} + {}^x c_2 \Delta^4 u_{x-2} + \cdots + \Delta^x u_0$

$$= (1 + {}^x c_1 \Delta^2 E^{-1} + {}^x c_2 \Delta^4 E^{-2} + \cdots + \Delta^x E^{-x}) u_x$$

$$= (1 + \Delta^2 E^{-1})^x u_x$$

$$= \left(\frac{E + \Delta^2}{E}\right)^x u_x = \left(\frac{E^2 - E + 1}{E}\right)^x u_x \qquad [\because \ \Delta = E - 1]$$

$$= \{1 + E(E-1)\}^x \frac{1}{E^x} u_x$$

$$= (1 + \Delta E)^x E^{-x} u_x = (1 + \Delta E)^x u_0$$

$$= (1 + {}^x c_1 \Delta E + {}^x c_2 \Delta^2 E^2 + \cdots + \Delta^x E^x) u_0$$

$$= u_0 + {}^x c_1 \Delta u_1 + {}^x c_2 \Delta^2 u_2 + \cdots + \Delta^x u_x, \qquad [\because \ E^x u_0 = u_x]$$

(*iii*) Right Hand Expression

$$= \frac{x}{1-x} u_1 + \frac{x^2}{(1-x)^2}(E-1) u_1 + \frac{x^3}{(1-x)^3}(E-1)^2 u_1 + \cdots$$

$$= \frac{x}{1-x} u_1 + \frac{x^2}{(1-x)^2} E u_1 - \frac{x^2}{(1+x)^2} u_1$$

$$+ \frac{x^3}{(1-x)^3} E u_1^2 - \frac{2x^3}{(1-x)^3} E u_1 + \frac{x^3}{(1-x)^3} u_1 + \cdots$$

$$= \left\{\frac{x}{1-x} - \frac{x^2}{(1-x)^2} + \frac{x^3}{(1-x)^3} - \cdots\right\} u_1$$

$$+ \left\{\frac{x^2}{(1-x)^2} - \frac{2x^3}{(1-x)^3} + \cdots\right\} E u_1$$

$$+ \left\{\frac{x^3}{(1-x)^3} + \cdots\right\} E^2 u_1 + \cdots$$

$$= \frac{x}{1-x}\left(1 + \frac{x}{1-x}\right)^{-1} u_1 + \frac{x^2}{(1-x)^2}\left(1 + \frac{x}{1-x}\right)^{-2} u_2$$

$$+\frac{x^3}{(1-x)^3}\left(1+\frac{x}{1-x}\right)^{-3}u_3+\cdots$$

$$=u_1x+u_2x^2+u_3x^3+\cdots = \text{Left Hand Expression.}$$

(*iv*) L.H.S.S. $u_0+xEu_0+x^2E^2u_0+\ldots=(1+xE+x^2E^2+)u_0+\dfrac{1}{1-xE}u_0$

$$=\frac{1}{1-x(1+D)}u_0$$

$$=\frac{1}{1-x}\left[1-\frac{x\Delta}{1-x}\right]u_0=\frac{1}{1-x}[1+\frac{x\Delta}{1-x}+\frac{x^2\Delta^2}{(1-x)^2}+]u_0$$

$$=\frac{u_0}{1-x}+\frac{x}{(1-x)^2}\Delta u_0+\ldots.\ \text{Hence the result.}$$

1·6·2. The operators *D* and ∇. Relations between the operators Δ, *E*, ∇, and *D*. If $D^n y_x$ represents the *n*th derivative of y_x. Taylor's series is given by

$$y_{x+h}=y_x+hDy_x+\frac{h^2}{2!}D^2y_x+\cdots+\frac{h^n}{n!}D^ny_x+\cdots$$

or
$$Ey_x\equiv\left(1+hD+\frac{h^2D^2}{2!}+\cdots+\frac{h^nD^n}{n!}+\cdots\right)y_x$$

$$=e^{hD}y_x$$

where h is the interval of differencing.

Thus we have the symbolic equivalence,

$$E\equiv e^{hD}=e^U,\quad \text{say where } U\equiv hD.$$

Hence the operators E, Δ and D are connected by the relations

$$\boxed{E\equiv 1+\Delta\equiv e^{hD}.}$$

From these, we deduce that

$$hD\equiv\log_e E\equiv\log_e(1+\Delta)\equiv\Delta-\frac{1}{2}\Delta^2+\frac{1}{3}\Delta^3-\cdots$$

Also $\quad \Delta\equiv e^{hD}-1\equiv E-1.$

The operator ∇ is called the **backward or ascending operator** or ascending difference operator and is defined by

$$\nabla f(a+h)=f(a+h)-f(a)$$

so that $\quad \nabla f(x)=f(x)-f(x-h).$

Also $\quad \nabla f(x)=f(x)-E^{-1}f(x),\qquad \therefore\ \nabla\equiv 1-E^{-1}$

Also $e^{-hD}=1-\nabla\Rightarrow\nabla=1-e^{-hD}=1-E^{-1}$

$$\therefore\quad hD\equiv-\log_e(1-\nabla)=\nabla+\frac{1}{2}\nabla^2+\frac{1}{3}\nabla^3+\cdots \quad ...(1)$$

The second backward differences are denoted by $\nabla^2 f(x)$ and defined by

$$\nabla^2 f(a+2h) = \nabla f(a+2h) - \nabla f(a+h)$$
$$= [f(a+2h) - f(a+h)] - [f(a+h) - f(a)]$$
$$\therefore \quad = f(a+2h) - 2f(a+h) + f(a). \quad ...(2)$$

In general $\nabla^n f(x) = \nabla^{n-1} [\nabla f(x)] = \nabla^{n-1} f(x) - \nabla^{n-1} f(x-h) = \text{etc.}$

1·7·1. Interpolation. Extrapolation. *Interpolation* means insertion or filling up intermediate terms of a series. It is the technique of estimating the value of a function under certain assumptions, for any intermediate value of the independent variable when the values of the function corresponding to a number of the values of the variable are given. The process of computing the value of a function outside the range of given values of the variable is called *extrapolation.*

Assumptions for extrapolation

1. The function can be approximated by a polynomial of degree n which coincides with it for the $(n+1)$ points $(x_0, y_0), (x_1, y_1), ..., (x_n, y_n)$ or can be represented to any desired degree of approximation or accuracy by a polynomial.
2. The rate of change of figures from one period to another is uniform.

1·7·2. Methods of Interpolation

The various methods of interpolation are :

(*a*) Graphic method

(*b*) Method of Curve Fitting

(*c*) Use of Finite Difference Formulae.

We shall describe the first two methods briefly and deal with third method in details in this and other chapters.

(*a*) **Graphic method.** We are given the n values of x and the corresponding values of y for a given function $y = f(x)$. We plot these n observed points (x_i, y_i), $i = 1, 2, ...$ and draw a free hand curve passing through these plotted points. From the graph so obtained, we can find out the value of y for any intermediate value of x. There is one *drawback* in the graphic method which states that the value of y obtained is the estimated value of y. The estimated value of y differs from the actual value of y.

(*b*) **Method of curve fitting.** This method can be applied only when the form of the function is known. Knowing the form of the function, we use the method of least squares and thus we are in a position to fit the curve to the set of n observed points. From the fitting curve, we can calculate the unknown value of y for an intermediate value of x, y being a function of x. This method has also drawbacks which are :

(1) This method gives an estimate of y for a given x.

(2) It is useful only when the number of observed points is sufficiently large.

(3) It gives better approximation than the graphic method.

We shall describe this method including the Spline Cubic Method later on. (See Chapter 19)

(*c*) **Use of finite differences formulae.** We shall use the techniques of interpolation (*i*) with equal intervals in this chapter and (*ii*) with unequal intervals and (*iii*) Central difference in the next two chapters.

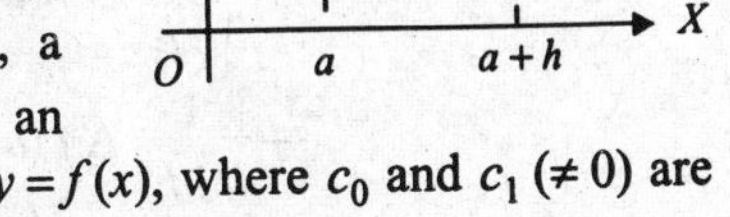

1·8. Interpolation with equal intervals

Linear interpolation. Suppose $[a, f(a)]$, $[a+h, f(a+h)]$ are the points A and B given to us. Then $y = c_0 + c_1 x$, a linear expression in x, is assumed to be an adequate representation to the function $y = f(x)$, where c_0 and c_1 ($\neq 0$) are unknown constants. Then

$$\Delta f(a) = [c_0 + c_1 (a+h)] - [(c_0 + c_1 a)]$$
$$= c_1 h$$
$$\therefore \quad = (c_0 + c_1 a) + \theta c_1 h f(a + \theta h) = c_0 + c_1 (a + \theta h)$$
$$\Rightarrow f.(a + \theta h) = f(a) + \theta \, \Delta f(a)$$
$$= (1 - \theta) f(a) + \theta f(a+h),$$

which is the required linear interpolation formula.

1·8·1. Newton-Gregory Formula for Forward Interpolation. Let $y = f(x)$ denote a function which takes the values $f(a), f(a+h), f(a+2h), \ldots, f(a+nh)$ for $(n+1)$ equidistant values $a, a+h, a+2h, \ldots, a+nh$ of the independent variable x and let $P_n(x)$ be a polynomial in x of degree n. Then

$$P_n(x) = A_0 + A_1 (x-a) + A_2 (x-a)(x-a-h)$$
$$+ A_3 (x-a)(x-a-h)(x-a-2h)$$
$$+ \cdots + A_n (x-a)(x-a-h) \ldots (x - a - \overline{n-1}h) \qquad \ldots(1)$$

We choose the coefficients $A_0, A_1, \ldots, A_n$ so that

$$P_n(a) = f(a), P_n(a+h) = f(a+h), \ldots, P_n(a+nh) = f(a+nh).$$

Putting successive in (1), $x = a, a+h, \ldots, a+nh$ and writing the values of $P_n(a), P_n(a+h), \ldots, P_n(a+nh)$, we obtain

$$f(a) = A_0 \quad \text{or} \quad A_0 = f(a),$$

$$f(a+h) = A_0 + hA_1 \quad \Rightarrow A_1 = \frac{f(a+h) - f(a)}{h} = \frac{1}{h} \Delta f(a)$$

$$f(a+2h) = A_0 + 2hA_1 + 2h.h.A_2$$

$$\Rightarrow \quad A_2 = [f(a+2h) - 2\{f(a+h) - f(a)\} - f(a)]/2h^2$$

$$= [f(a+2h) - 2f(a+h) + f(a)]/2h^2$$

$$= \frac{1}{2\,!}\, h^2 \Delta^2 f(a)$$

Similarly, $A_3 = \dfrac{1}{3\,!\,h^3} \cdot \Delta^3 f(a), \ldots, \quad A_n = \dfrac{1}{n\,!\,h^n}\, \Delta^n f(a).$

Substituting the values of $A_0, A_1, \ldots, A_n$ in (1), page 17, we get

$$P_n(x) = f(a) + \frac{1}{h}\Delta f(a)\,(x-a) + \frac{1}{2\,!\,h^2}\Delta^2 f(a)\,(x-a)\,(x-a-h)$$

$$+ \frac{1}{3\,!\,h^3}\Delta^3 f(a)\,(x-a)\,(x-a-h)\,(x-a-2h) + \cdots$$

$$+ \frac{1}{n\,!\,h^n}\Delta^n f(a)\,(x-a)\,(x-a-h)\ldots(x-a-\overline{n-1}h).$$

This is Newton-Gregory formula for *forward* interpolation.

Another form. Setting $(x-a)/h = u$ or $x = a + hu$, the formula takes the simple form

$$P_n(x) = P_n(a+hu) = f(a) + u\,\Delta f(a) + \frac{u\,(u-1)}{2\,!}\,\Delta^2 f(a)$$

$$+ \frac{u(u-1)\,(u-2)}{3\,!}\,\Delta^3 f(a) + \cdots$$

$$+ \frac{u\,(u-1)(u-2)\ldots(u-n+1)}{n\,!}\,\Delta^n f(a) \qquad \ldots(1)$$

or $\quad P_n(x) = f(a) + u^{(1)}\,\Delta f(a) + \dfrac{1}{2\,!}\,u^{(2)}\,\Delta^2 f(a) + \cdots + \dfrac{1}{n\,!}\,u^{(n)}\,\Delta^{(n)} f(a)$

where $\quad u^{(n)} = u(u-1)(u-2)\ldots(u-n+1),\ x = a + hu.$

This is the form which Newton-Gregory formula takes.

Aliter. We have

$$\nabla f(a) = f(a) - f(a-h)$$

$$\Rightarrow f(a-h) = (1-\nabla)\,f(a)$$

Similarly $f(a-2h) = (1-\nabla)\,f(a-h)$

$$= (1-\nabla)\,(1-\nabla)\,f(a)$$

$$= (1-\nabla)^2 f(a)$$

Proceeding this way, we obtain

$$f(a-nh) = (1-\nabla)^n f(a)$$

$$= \left[\,1 - n\nabla + \frac{n(n-1)}{2\,!}\,\nabla^2 - \frac{n(n-1)(n-2)}{3\,!}\,\nabla^3 + \cdots\right] f(a)$$

$$= f(a) - n\nabla f(a) + \frac{n(n-1)}{2!} \nabla^2 f(a) - \frac{n(n-1)(n-2)}{3!} \nabla^3 f(a) + \cdots \quad ...(2)$$

Working table for Backward Formula

x	y	∇	∇^2	∇^3	∇^4
$a-4h$	y_4				
		∇y_3			
			$\nabla^2 y_2$		
$a-3h$	y_3			$\nabla^3 y_1$	
		∇y_2			
$a-2h$	y_2		$\nabla^2 y_1$		$\nabla^4 y_0$
		∇y_1		$\nabla^3 y_0$	
$a-h$	y_1		$\nabla^2 y_0$		
		∇y_0			
a	y_0				

Example 8. *Find the cubic polynomial interpolation which takes on the values $f_0 = 5$, $f_1 = 1$, $f_2 = 9$, $f_3 = 25$, $f_4 = 55$.* (*A.M.I.E. 1993*)

Sol. Hence find f_5

x	f	∇	∇^2	∇^3	∇^4
0	5				
		-4			
1	1		12		
		8		-4	
2	9		8		10
		16		6	
3	25		14		
		30			
4	55				

Here $a = 4, h = 1, a - nh = 5 \Rightarrow n = -1$.

using formula (2), page 18 we obtain

$$f(5) = f(4) - (-1)\nabla f(4) + \frac{(-1)(-2)}{2!} \nabla^2 f(4) - \frac{(-1)(-2)(-3)}{3!} \nabla^2 f(4) + \cdots$$

$$= 55 + 30 + 14 + 6 + 10 = 115 \text{ approx.}$$

Example 9. *Write down Newton's Backward differences formula. Use it to obtain the numerical value of the first derivative at $x = 0{\cdot}4$ of the function $f(x)$ defined as under:*

x :	0·1	0·2	0·3	0·4
$f(x)$:	1·10517	1·22140	1·34986	1·49182

(*A.M.I.E. 1991*)

Sol. Newton's backward difference formula is

$$f(a - nh) = (1 - \nabla)^n f(a)$$

$$= f(a) - n\nabla f(a) + \frac{n(n-1)}{2!} \nabla^2 f(a) - \frac{n(n-1)(n-2)}{3!} \nabla^3 f(a) + \cdots$$

$\therefore$ Differentiating both sides w.r.t. 'n', we obtain

$$f'(a-nh)=\frac{1}{h}\left[\nabla f(a)-\frac{2n-1}{2\,!}\nabla^2 f(a)+\frac{3n^2-6n+2}{3\,!}\nabla^3 f(a)-\cdots\right] \quad ...(1)$$

For the given question, we have

$$a=0{\cdot}4,\ h=0{\cdot}1,\ a-nh=0{\cdot}4 \Rightarrow n=0.$$

Substituting these values of a, n and h in formula (2), we obtain

$$f'(0{\cdot}4)=\frac{1}{0{\cdot}1}\left[0{\cdot}14196+\frac{1}{2}(0{\cdot}013150)+\frac{1}{3}(0{\cdot}00127)\right]$$

$$=10\,[0{\cdot}14196+0{\cdot}00675+0{\cdot}000427]$$

$$=10\times 0{\cdot}149133=1{\cdot}49133.$$

x	$f(x)$	$\nabla f(x)$	$\nabla^2 f(x)$	$\nabla^3 f(x)$
0·1	1·10517			
		0·11623		
0·2	1·22140		0·01223	
		0·12846		0·00127
0·3	1·34986		0·01350	
		0·14196		
0·4	1·49182			

Note. The reason for the name "forward" interpolation formula lies in the fact that the formula contains values of the tabulated function from $f(a)$ onward to the right [forward from $f(a)$] and none to the left of this value. This formula is used mainly for interpolating the values of $f(x)$ near the *beginning* of a set of tabulated values.

Cor. Put $h=1, a=0$ so that x takes the values 0, 1, 2, ..., n we get

$$P_n(x)=f(0)+{}^xc_1\,\Delta f(0)+{}^xc_2\,\Delta^2 f(0)+\cdots+{}^xc_n\,\Delta^n f(0). \quad ...(1)$$

This is Newton-Gregory *advancing difference formula.*

Another form. Let $a_1=1-\frac{1}{2}(x+1),\ a_2=1-\frac{1}{3}(x+1)$; etc. so that

$$a_1\,x=\frac{(1-x)\,x}{2}=-{}^xc_2$$

$$a_1, a_2 . x=a_2\,.\,a_1\,x=-\frac{(x-2)}{3}\cdot\frac{(1-x)\,x}{2}={}^xc_3,\ \text{etc.}$$

Thus the formula (1) can be written as

$$P_n(x)=f(0)+x\Delta f(0)-xa_1\,\Delta^2 f(0)+xa_1\,a_2\,\Delta^3 f(0)-\cdots$$

1·8·2. Newton-Gregory formula for Backward Interpolation. The formula of the preceding section cannot be used for interpolating a value of $f(x)$ *near* the *end* of the tabular values. To derive a formula for this case, we write

$$P_n(x)=A_0+A_1(x-a-nh)+A_2(x-a-nh)(x-a-nh+h)$$
$$+A_3(x-a-nh)(x-a-nh+h)(x-a-nh+2h)+\cdots$$

$$+ A_n (x-a-nh)(x-a-nh+h)\ldots(x-a-h) \qquad \ldots(1)$$

We determine the coefficients $A_0, A_1, \ldots, A_n$ so as to make

$$P_n(a+nh) = f(a+nh), \ldots, P_n(a) = f(a).$$

Putting successively in (1), $x = a+nh, a+nh-h, \ldots$ and also putting $P_n(a+nh) = f(a+nh), \ldots$ we get

$$f(a+nh) = A_0 \text{ or } A_0 = f(a+nh)$$

$$f(a+nh-h) = A_0 + A_1(-h)$$

$$\Rightarrow \quad A_1 = \frac{f(a+nh) - f(a+nh-h)}{h}$$

$$\Rightarrow \quad A_1 = \frac{1}{h} \nabla f(a+nh).$$

where ∇ is called the *backward operator* and is defined by the relation

$$\nabla f(x) = f(x) - f(x-h).$$

Also $f(a+nh-2h) = A_0 + A_1(-2h) + A_2(-2h)(-h)$

or $$A_2 = \frac{f(a+nh) - 2f(a+\overline{n-1}h) + f(a+\overline{n-2}h)}{2h^2}$$

$$= \frac{1}{2!\,h^2} \nabla^2 f(a+nh)$$

Similarly, $$A_3 = \frac{1}{3!\,h^3} \nabla^3 f(a+nh).$$

$$\ldots \quad \ldots \quad \ldots$$

$$A_n = \frac{1}{n!\,h^n} \nabla^n f(a+nh).$$

Substituting the values of $A_1, A_2, \ldots, A_n$ in (1), get

$$P_n(x) = f(a+nh) + \frac{\nabla f(a+nh)}{h}(x - \overline{a+nh})$$

$$+ \frac{\nabla^2 f(a+nh)}{2!\,h^2}(x-\overline{a+nh})(x-\overline{a+nh-h}) + \cdots$$

$$+ \frac{\nabla^n f(a+nh)}{n!\,h^n}(x-\overline{a+nh})(x-\overline{a+nh-h})\ldots(x-\overline{a+h})$$

This is Newton-Gregory formula for backward interpolation,

Setting $$u = \frac{x-(a+nh)}{h} \quad \text{or} \quad x = a+nh+hu,$$

we find

$$P_n(x) = P_n(a+nh+hu)$$

$$= f(a+nh) + u\nabla f(a+nh) + \frac{u(u+1)}{2!} \nabla^2 f(a+nh)$$

$$+\frac{u(u+1)(u+2)}{3\,!}\nabla^3 f(a+nh)+\cdots$$

$$+\frac{u(u+1)\ldots(u+n-1)}{n\,!}\nabla^n f(a+nh) \qquad \ldots\text{(II)}$$

This is the form in which Newton-Gregory formula for backward interpolation is usually written.

This is the backward interpolation formula and is so called because it contains values of the tabulated function from $f(a+nh)$ backward to the left and none to the right of $f(a+nh)$. This formula is used mainly for interpolating the value of $f(x)$ near the end of a set of tabular values.

1·8·3. Equidistant Terms with One or More Missing Terms. When one or more of the values of $f(x)$ corresponding to the equidistant values of x are missing, either of the following methods may be used for finding the missing value(s) :

First Method.

We have $\Delta^n f(a)=(E-1)^n f(a)$

$$=[E^n-{}^nc_1\,E^{n-1}+{}^nc_2\,E^{n-2}-\cdots+(-1)^n]f(a)$$

$$\Rightarrow \quad \Delta^n f(a)=E^n f(a)-{}^nc_1\,E^{n-1}f(a)+{}^nc_2\,E^{n-2}f(a)-\cdots+(-1)^n f(a)$$

$$=f(a+nh)-{}^nc_1\,f(a+\overline{n-1}h)+{}^nc_2 f(a+\overline{n-2}h)-\cdots+(-1)^n f(a).$$

Assume that $f(x)$ is a polynomial in x of degree $(n-1)$ so that $\Delta^n f(x)=0$ for all x, we get

$$f(a+nh)-{}^nc_1\,f(a+\overline{n-1}h)+\cdots+(-1)^n f(a)=0.$$

This equation enables us to find out the missing value of $y=f(x)$ when only n values of $f(x)$ are known.

Second Method.

Let n values out of $(n+1)$ values of $y=f(x)$ corresponding to $(n+1)$ equidistant values of x be known and let it be required to find out the remaining value of y corresponding to some known value of x. Denote the missing value of y by u and form the difference table. Since only n values of y are known, we assume that $y=f(x)$ is a polynomial in x of degree $(n-1)$. Equating to zero the nth difference, we obtain the value of u.

Example 10. *The population of a country in the decennial census were as under. Estimate the population for the year 1975.*

Year	x :	*1941*	*1951*	*1961*	*1971*	*1981*
Population (in lakhs)	y :	*46*	*67*	*83*	*95*	*102*

(Delhi B.A. Hons. 1997)

Sol. The difference table is :

x	y	∇y	$\nabla^2 y$	$\nabla^3 y$	$\nabla^4 y$
1941	46				
		21			
1951	67		-5		
		16		1	
1961	83		-4		-2
		12		-1	
1971	95		-5		
		7			
1981	102				

Since five values of x and y are given, we assume that fourth degree polynomial can be fitted. We require the entry for $x = 1975$, $h = 10$, $x_n = 1981$

We shall apply backward formula II, page 22

Hence $u = \dfrac{x - 1981}{h} = \dfrac{1975 - 1981}{10} = -0{\cdot}6$

$$\therefore \quad y = 102 + (-0{\cdot}6) \times 7 + \frac{(-0{\cdot}6)(0{\cdot}4) \times (-5)}{2\,!}$$

$$+ \frac{(-0{\cdot}6)(0{\cdot}4)(1{\cdot}4) \times (-1)}{3\,!} + \frac{(-0{\cdot}6)(0{\cdot}4)(1{\cdot}4)(2{\cdot}4) \times (-2)}{4\,!} - \ldots$$

$$= 102 - 4{\cdot}2 + 0{\cdot}6 + 0{\cdot}0112 - 0{\cdot}1008 \text{ approx.}$$

$$= 98{\cdot}3104$$

Hence the population for 1975 is estimated to be 96·64 lakhs.

Example 11. *Given* $\log_{10} 100 = 2$, $\log_{10} 101 = 2{\cdot}0043$, $\log_{10} 103 = 2{\cdot}0128$, $\log_{10} 104 = 2{\cdot}0170$, *find* $\log_{10} 102$.

Sol. First Method.

The data can be put as under :

x :	100	101	102	103	104
$f(x) = \log_{10} x$:	2	2·0043	...	2·0128	2·0170

Since four values are given, we assume that a polynomial of degree 3 can be fitted, which will have fourth differences as zero.

$$\Delta^4 f(x) = 0 \;\forall\; x.$$

Putting $x = 100$, we have $\Delta^4 f(100) = 0$

$$\Rightarrow \quad (E - 1)^4 f(100) = 0$$

$$\Rightarrow \quad (E^4 - 4E^3 + 6E^2 - 4E + 1) f(100) = 0$$

$$\Rightarrow \quad f(104) - 4f(103) + 6f(102) - 4f(101) + f(100) = 0$$

$$\Rightarrow \quad 2{\cdot}0170 - 4 \times 2{\cdot}0128 + 6 \log(102) - 4 \times 2{\cdot}0043 + 2 = 0$$

$$\Rightarrow \quad \log_{10} 102 = \frac{12{\cdot}0514}{6} = 2{\cdot}0086 \text{ approx.}$$

Second Method.

Let the missing value be y_2, then the difference table is :

x	$y = \log_{10} x$	Δy	$\Delta^2 y$	$\Delta^3 y$	$\Delta^4 y$
100	2·0000				
		0·0043			
101	2·0043		$y_2 - 2·0086$		
		$y_2 - 2·0043$		$6·0257 - 3y_2$	
102	y_2		$4·0171 - 2y_2$		$6y_2 - 12·0514$
		$2·0128 - y_2$		$3y_2 - 6·0257$	
103	2·0128		$y_2 - 2·0086$		
		0·0042			
104	2·0170				

Since fourth differences must be zero by assumption we have

$$6y_2 = 12·0514$$

$$\Rightarrow \qquad \log_{10} 102 = \frac{12·0514}{6} = 2·0086 \text{ approx}$$

Example 12. *Obtain the estimates of the missing figures in the following table :*

x :	*2·0*	*2·1*	*2·2*	*2·3*	*2·4*	*2·5*	*2·6*
y_x :	*0·135*	—	*0·111*	*0·100*	—	*0·082*	*0·074*

Sol. Since five values are given, we assume that a polynomial of fourth degree can be fitted, so that fifth differences are zero.

Hence $\qquad \Delta^5 y_x = 0 \quad \text{for all } x$

$\Rightarrow \qquad (E - 1)^5 y_x = 0 \quad \text{for all } x \quad (\forall\, x)$

$\therefore \qquad (E - 1)^5 y_{2·0} = 0 \quad \text{when } x = 2·0$

and $\qquad (E - 1)^5 y_{2·1} = 0 \quad \text{when } x = 2·1.$

Taking the interval of differencing to be 0·1, these equations give

$$y_{2·5} - 5y_{2·4} + 10y_{2·3} - 10y_{2·2} + 5y_{2·1} - y_{2·0} = 0$$

and $$y_{2·6} - 5y_{2·5} + 10y_{2·4} - 10y_{2·3} + 5y_{2·2} - y_{2·1} = 0$$

Substituting the values of $y_{2·0}, y_{2·2}, y_{2·3}, y_{2·5}$ and $y_{2·6}$ in these equations and solving, we get

$$y_{2·1} = 0·123 \quad \text{and} \quad y_{2·4} = 0·090.$$

Note. The function in this equation is $y_x = e^{-x}$ and the tabular value of $y_{2·4} = 0·091$. The difference in the observed and expected values is due to the fact that our *assumption* that y_x is a rational integral function of x of the fourth degree is only approximately true.

Example 13. *From the following table, find the number of students who obtained less than 45 marks ;*

Marks	*No. of students*	*Cumulative frequency (No. of students below)*
30—40	*31*	*31*
40—50	*42*	*31 + 42 = 73*
50—60	*51*	*31 + 42 + 51 = 124*
60—70	*35*	*31 + 42 + 51 + 35 = 159*
70—80	*31*	*31 + 42 + 51 + 135 + 31 = 190*

Sol. The difference table is as under :

Marks x	*No. of students* y	*First diff.* Δy	*Second diff.* $\Delta^2 y$	*Third diff.* $\Delta^3 y$	*Fourth diff.* $\Delta^4 y$
Below 40	31				
		42			
,, 50	73		9		
		51		– 25	
,, 60	124		– 16		37
		35		12	
,, 70	159		– 4		
		31			
,, 80	190				

Here $h = 10, y_{45} = E^{\frac{1}{2}} y_{40} = (1 + \Delta)^{\frac{1}{2}} y_{40}$

or $$y_{45} = y_{40} + \frac{1}{2} \Delta y_{40} - \frac{1}{8} \Delta^2 y_{40} + \frac{1}{16} \Delta^3 y_{40} - \frac{5}{128} \Delta^4 y_{40} \text{ approx.}$$

$$= 31 + \frac{1}{2}(42) - \frac{1}{3}(9) + \frac{1}{16}(-25) - \frac{5}{128}(37)$$

$$= 47{\cdot}868 \text{ approx.}$$

Hence the number of students getting marks less than 45 is 48.

1·9. Sub-division of intervals. Suppose we are given every nth value of a function and it is required to find out the values of the function at individual points. For example, the quinquennial values $u_0, u_5, u_{10}, u_{15}, \ldots$ or decennial values $u_0, u_{10}, u_{20}, u_{30}, \ldots$ are known and it is required to complete the series $u_0, u_1, u_2, u_3, \ldots$

Let δu_x denote the difference for unit interval of x. Then

$$u_{x+n} = u_x + \Delta u_x = (1 + \delta)^n u_x$$

Thus we have the identity:

$$(1 + \delta)^n \equiv 1 + \Delta$$

i.e., $$1 + \delta = (1 + \Delta)^{1/n}$$

$\Rightarrow$ $$\delta = (1 + \Delta)^{1/n} - 1$$

$\Rightarrow$ $$\delta = \frac{1}{n}\Delta + \frac{\frac{1}{n}\left(\frac{1}{n} - 1\right)}{2\,!}\Delta^2 + \frac{\frac{1}{n}\left(\frac{1}{n} - 1\right)\left(\frac{1}{n} - 2\right)}{3\,!}\Delta^3 + \cdots \qquad \ldots(1')$$

Thus we have

$$\left.\begin{aligned}\delta u_x &= \left(\frac{1}{n}\Delta - \frac{n-1}{2!\,n^2}\Delta^2 + \frac{(n-1)(2n-1)}{3!\,n^3}\Delta^3 - \cdots\right)u_x \\ \text{Similarly squaring (1}'\text{),}\quad \delta^2 u_x &= \left(\frac{1}{n^2}\Delta^2 - \frac{n-1}{n^3}\Delta^3 + \cdots\right)u_x \\ \text{Also}\quad \delta^3 u_x &= \left(\frac{1}{n^3}\Delta^3 - \cdots\right)u_x\end{aligned}\right\} \quad \ldots(1)$$

and so on.

Example 14. *Given*

$$\sum_{1}^{10} f(x) = 500426,\ \sum_{4}^{10} f(x) = 329240,\ \sum_{7}^{10} f(x) = 175212$$

and $f(10) = 40365,$

find $f(1)$.

Sol. Here we have

$$u_1 = f(10) = 40365,\quad u_4 = \sum_{7}^{10} f(x) = 175212,$$

$$u_7 = \sum_{4}^{10} f(x) = 329240,\quad u_{10} = \sum_{1}^{10} f(x) = 500426.$$

We want to determine δu_9.

since $f(x) = u_{10} - u_9 = \delta u_9.$

Thus we are given the values u_1, u_4, u_7, u_{10} and we want to complete the series $u_1, u_2, u_3, \ldots$

We find that

$$\Delta u_1 = 134847,\quad \Delta^2 u_1 = 19181,\quad \Delta^3 u_1 = -2023,$$

the third differences being constant since only four values u_1, u_4, u_7 and u_{10} are known.

Putting $n = 3$ in (1), page 26, we get

$$\begin{aligned}\delta u_x &= \frac{1}{3}\Delta u_x - \frac{1}{9}\Delta^2 u_x + \frac{5}{81}\Delta^3 u_x \\ &= \frac{1}{3}(134847) - \frac{1}{9}(19181) + \frac{5}{81}(-2023) \\ &= 42692{\cdot}90\end{aligned}$$

$$\delta^2 u_x = \frac{1}{9}\,\Delta^2 u_x - \frac{2}{27}\,\Delta^3\, u_x = \frac{1}{9}\,(19181) - \frac{2}{27}\,(-\,2023) = 2281{\cdot}07$$

$$\Delta^3 u_x = \frac{1}{27}\,\Delta^3 u_x = -\,74{\cdot}92, \text{ up to two places of decimals.}$$

Difference Table

x	u_x	δu_x	$\delta^2 u_x$	$\delta^2 u_x$
1	40365			
		42692·90		
2			2281·07	
		44973·97		74·92
3			2206·15	
		47180·12		-74·92
4			2131·23	
		49311·35		74·92
5			2056·31	
		51367·66		74·92
6			1981·39	
		53349·05		74·92
7			1906·47	
		55255·52		74·92
8			1831·55	
		57087·07		74·92
9			1756·63	
		58843·70		
10				

Thus $f(1) = 58,843{\cdot}70$ upto 2 places of decimals.

Aliter. Let $\quad u_1 = \sum_{1}^{10} f(x) = 500426$

$$u_4 = \sum_{4}^{10} f(x) = 329240$$

$$u_7 = \sum_{7}^{10} f(x) = 175212$$

and $\quad u_{10} = f(10) = 40365$

Difference table for u_x

x	u_x	Δ	Δ^2	Δ^3
1	500426			
		– 171186		
4	329240		17158	
		– 154028		2023
7	175212		19181	
		– 134847		
10	40365			

Now $$f(1) = \sum_{1}^{10} f(x) - \sum_{2}^{10} f(x)$$

where $$u_2 = \sum_{2}^{10} f(x)$$

Now $$a + xh = 2 \text{ so that } x = \frac{1}{3}$$

and $$u_2 = u_1 + x\Delta u_1 + \frac{x(x-1)}{2\,!}\Delta^2 u_1 + \frac{x(x-1)(x-2)}{3\,!}\Delta^3 u_1 + \cdots$$

$$= 500426 + \frac{1}{3}(-171186) + \frac{\frac{1}{3}\left(\frac{-2}{3}\right)}{2\,!}17158$$

$$+ \frac{\frac{1}{3}\left(\frac{-2}{3}\right)\left(\frac{-5}{3}\right)}{3\,!}202{\cdot}3$$

$$= 441582{\cdot}432.$$

Hence $f(1) = 500426 - 441582{\cdot}432 = 58843{\cdot}568.$

1·10·1. Differences of zero. If n and m be positive integers we have, on using $E^n x^m = (x+n)^m$,

$$\Delta^n x^m = (E-1)^n x^m$$

$$= [E^n - {}^n c_1 E^{n-1} + {}^n c_2 E^{n-2} - \cdots + (-1)^{n-1}\,{}^n c_1 E + (-1)^n]\, x^m$$

$$= (x+n)^m - {}^n c_1 (x+n-1)^m + {}^n c_2 (x+n-2)^m - \cdots$$

$$+ {}^n c_1 (-1)^{n-1} (x+1)^m + (-1)^n x^m$$

Putting $x = 0$, we get

$$[\Delta^n x^m]_{x=0} = n^m - {}^n c_1 (n-1)^m + {}^n c_2 (n-2)^m - \cdots + {}^n c_1 (-1)^{n-1} \quad \ldots(1)$$

The expression $[\Delta^n x^m]_{x=0}$ is frequently written as $\Delta^n O^m$. The quantities $\Delta^n O^m$ are known as *differences of zero* because the leading term is always zero.

We can obtain the values of $\Delta^n O^m$ for various integral values of n and m.

For example, if $n = 3$, $m = 4$, we have form (1)

$$\Delta^3 0^4 = 3^4 - 3{\cdot}2^4 + 3{\cdot}1^4 = 36$$

If $n = 1,\ m = 3,\ \Delta 0^3 = 1^3 = 1,$

$n = 2,\ m = 3,\ \Delta^2 0^3 = 2^3 - 2{\cdot}1^3 = 6,$

$n = 3,\ m = 3,\ \Delta^3 0^3 = 3^3 - 3{\cdot}2^3 + 3{\cdot}1^2 = 6,$ etc.

1·10·2. To express x^m in factorial notation. If y_x be a polynomial in x of degree m, then Newton-Gregory formula,

$$y_x = y_0 + {}^x c_1 \Delta y_0 + {}^x c_2 \Delta^2 y_0 + \cdots + {}^x c_m \Delta^m y_0$$

In particular, if $y_x = x^m$, we have

$$x^m = 0^m + {}^x c_1 \Delta 0^m + {}^x c_2 \Delta^2 0^m + \cdots + {}^x c_m \Delta^m 0^m$$

or
$$x^m = \sum_{i=0}^{m} {}^x c_i \Delta^i 0^m = \sum_{i=0}^{m} \left[\frac{x^{(i)}}{i\,!} \Delta^i 0^m. \right]$$

1·10·3. Relation between Δ^n and Δ^{n-1}. *To show that*

$$\Delta^n 0^m = n(\Delta^{n-1} 0^{m-1} + \Delta^n 0^{m-1}),$$

where n and m are positive integers.

From (1) of §1·10·1, we have

$$\Delta^n 0^m = n^m - {}^n c_1 (n-1)^m + {}^n c_2 (n-2)^m - \cdots + {}^n c_{n-1} (-1)^{m-1}.$$

$$= n[n^{m-1} - {}^{n-1}c_1 (n-1)^{m-1} + {}^{n-1}c_2 (n-2)^{m-1} - \cdots]$$

$$= n[(1+n-1)^{m-1} - {}^{n-1}c_1 (1+n-2)^{m-1} + {}^{n-1}c_2 (1+n-3)^{m-1} - \cdots]$$

$$= n[E^{n-1} x^{m-1} - {}^{n-1}c_1 E^{n-2} x^{m-1} + {}^{n-1}c_2 E^{n-3} x^{m-1} - \cdots]_{x=1}$$

$$= n[(E-1)^{n-1} x^{m-1}]_{x=1} = n \Delta^{n-1} 1^{m-1}$$

$$= n \Delta^{n-1} E 0^{m-1} = n \Delta^{n-1} (1+\Delta) 0^{m-1}$$

$$= n(\Delta^{n-1} 0^{m-1} + \Delta^n 0^{m-1})$$

i.e.
$$\frac{\Delta^n 0^m}{n\,!} = \frac{n\Delta^n 0^{m-1}}{n\,!} + \frac{\Delta^{n-1} 0^{m-1}}{(n-1)\,!}. \qquad \ldots(1)$$

This recurrence relation is helpful in calculating $\Delta^n 0^m$, when $\Delta^{n-1} 0^{m-1}$ and $\Delta^n 0^{m-1}$ are known. Further a table of the coefficients in the expansion of x^m in terms of successive values of the factorial $x^{(m)}$ can be written down. We have

$$\Delta^n 0^n = n\,!,$$

$$\Delta^{n+r} 0^n = 0, \quad r > 0$$

i.e.,
$$\Delta^m 0^n = 0, \quad n < m.$$

A table of differences of zero by continued application of equation (1), page 29 is :

m	$\Delta^1 0^m$	$\Delta^2 0^m$	$\Delta^3 0^m$	$\Delta^4 0^m$
1	1	0		
2	1	2	0	0
3	1	6	6	24
4	1	14	36	240
5	1	30	150	1560
6	1	62	540	

and so on.

1·11·1. Herschel's Theorem

If $\phi(\lambda)$ *be a polynomial with constant coefficients and if* $E^n 0^m = n^m h^m$, *then*

$$\phi(e^{ht}) = \phi(1) + t\,\phi(E)0 + \frac{t^2}{2\,!}\,\phi(E)\,0^2 + \cdots$$

Proof. Since $\phi(\lambda)$ is a polynomial in λ with constant coefficients,

$$\phi(e^{ht}) = \sum_{n=0} A_n\, e^{nht},$$

where A's are constants.

Now
$$e^{(x+nh)\,t} = 1 + t(x+nh) + \frac{t^2}{2\,!}(x+nh)^2 + \cdots$$

$$= 1 + t\,E^n x + \frac{t^2}{2\,!} E^n x^2 + \cdots$$

Putting $x = 0$, we have

$$e^{nth} = 1 + t\,E^n\,0 + \frac{t^2}{2\,!} E^n\,0^2 + \cdots$$

Hence
$$\phi(e^{ht}) = \sum_{n=0} A_n\, e^{nht}$$

$$= \sum_{n=0} A_n \left(1 + tE^n\,0 + \frac{t^2}{2\,!} E^n\,0^2 + \cdots\right)$$

$$= \Sigma A_n + t\,\Sigma A_n\, E^n 0 + \frac{t^2}{2\,!}\Sigma A_n\, E^n\,0^2 + \cdots$$

Thus
$$\phi(e^{ht}) = \phi(1) + t\phi(E)0 + \frac{t^2}{2\,!}\,\phi(E)0^2 + \cdots \qquad \text{...(1)}$$

1·11·2. Difference Quotients in terms of Derivatives. We shall by Herschel's theorem, express *differences* of a function in terms of its *derivatives.*

We have, on using (1), page 30 and taking $\phi(E) = (E-1)^n$,

$$(e^{ht}-1)^n = t(E-1)^n\,0 + \frac{t^2}{2\,!}(E-1)^n\,0^2 + \cdots$$

$$\Rightarrow \qquad (e^{ht}-1)^n = t\,\Delta^n\,0 + \frac{t^2}{2\,!}\Delta^n\,0^2 + \cdots$$

where $\Delta^n\,0^m = 0 \quad \text{if} \quad m < n$

$\qquad = n\,! \quad \text{if} \quad m = n$

Hence $\quad (e^{ht}-1)^n = t^n + \dfrac{t^{n+1}}{(n+1)\,!}\Delta^n\,0^{n+1} + \dfrac{t^{n+2}}{(n+2)\,!}\Delta^n\,0^{n+2} + \cdots$

But $\quad \Delta^n = (e^{hD}-1)^n$

Thus
$$\Delta^n u_x = D^n u_x + \frac{\Delta^n 0^{n+1}}{(n+1)\,!} D^{n+1}\,u_x + \frac{\Delta^n\,0^{n+2}}{(n+2)\,!} D^{n+2}\,u_x + \cdots$$

We shall close this chapter with the solution of two miscellaneous examples.

Example 15. *Show that* $B(m+1, n) = (-1)^m\,\Delta^m\left(\frac{1}{n}\right)$ *where m is a positive integer,* $n > 0$.

Sol. We have

$$\int_0^\infty e^{-nx}\,dx = \frac{1}{n}$$

$$\therefore \qquad \Delta^m \int_0^\infty e^{-nx}\,dx = \Delta^m\left(\frac{1}{n}\right)$$

$$\Rightarrow \qquad \int_0^\infty \Delta^m\,e^{-nx}\,dx = \Delta^m\left(\frac{1}{n}\right)$$

But
$$\Delta^m\,e^{-nx} = \Delta^{m-1}\,(e^{-(n+1)x} - e^{-nx})$$
$$= \Delta^{m-1}\,e^{-nx}\,(e^{-x}-1) = \cdots$$
$$= e^{-nx}(e^{-x}-1)^m.$$

Hence
$$\int_0^\infty e^{-nx}(e^{-x}-1)^m\,dx = \Delta^m\left(\frac{1}{n}\right).$$

Putting $z = e^{-x}$, we obtain

$$(-1)^m \int_0^\infty z^{n-1}(1-z)^m\,dz = \Delta^m\left(\frac{1}{n}\right)$$

$$\Rightarrow \qquad B(m+1, n) = (-1)^m \Delta^m \left(\frac{1}{n}\right).$$

Example 16. *Prove that if n is an integer, then*

$$(x\,\Delta)^{(n)} u_x = (x+n-1)^{(n)}\,\Delta^n u_x.$$

Sol. We shall prove this result with the help of Mathematical Induction.

When $n = 1$, we have

$$(x\,\Delta)\,u_x = x\,\Delta\,u_x$$

which is clearly true.

When $n = 2$, the left side

$$\begin{aligned}(x\,\Delta)^{(2)} u_x &= (x\,\Delta)(x\,\Delta - 1)\,u_x \\ &= (x\,\Delta)(x\,\Delta\,u_x) - x\,\Delta\,u_x \\ &= x\,\Delta\,(x\,\Delta\,u_x) - x\,\Delta\,u_x\end{aligned}$$

Using $\Delta\,(u_x\,v_x) = u_{x+1}\,\Delta v_x + v_x\,\Delta\,u_x$, page 6, we have

$$\begin{aligned}(x\,\Delta)^{(2)} u_x &= x\,[(x+1)\,\Delta^2 u_x + 1\cdot\Delta u_x] - x\,\Delta\,u_x \\ &= (x+1)^{(2)}\,\Delta^2\,u_x.\end{aligned}$$

Thus the theorem holds good for $n = 2$.

When $n = 3$, the left side

$$\begin{aligned}(x\,\Delta)^{(3)} u_x &= (x\,\Delta)(x\,\Delta - 1)(x\,\Delta - 2)\,u_x \\ &= (x\,\Delta - 2)\cdot x\,\Delta\,(x\,\Delta - 1)\,u_x \\ &= (x\,\Delta - 2)\cdot(x+1)^{(2)}\,\Delta^2 u_x \\ &= x\,\Delta\,[(x+1)^{(2)}\,\Delta^2 u_x] - 2(x+1)^{(2)}\,\Delta^2 u_x \\ &= (x+2)^{(3)}\,\Delta^3 u_x + 2x(x+1)\,\Delta^2\,u_x - 2(x+1)^{(2)}\,\Delta^2 u_x \\ &= (x+2)^{(3)}\cdot\Delta^3 u_x.\end{aligned}$$

Hence the theorem is true for $n = 3$.

Suppose the theorem holds good for n. We shall show that it is true for the index $n + 1$.

$$\begin{aligned}(x\,\Delta)^{(n+1)}\,u_x &= (x\,\Delta)(x\,\Delta - 1) - \cdots\,(x\,\Delta - \overline{n-1})(x\,\Delta - n)\,u_x \\ &= (x\,\Delta - n)\cdot(x\,\Delta)^n\,u_x \\ &= (x\,\Delta - n)\,[(x+n-1)^{(n)}\,\Delta^n\,u_x] \\ &= x\,\Delta\,[(x+n-1)^{(1)}\,\Delta^n u_x] - n(x+n-1)^{(n)}\,\Delta^n u_x \\ &= x\,[(x+n)^{(n)}\,\Delta^{n+1} u_x + n(x+n-1)^{(n-1)}\,\Delta^n u_x] \\ &\qquad - n(x+n-1)^{(n)}\,\Delta^n u_x\end{aligned}$$

$$= (x+n)^{(n+1)} \Delta^{n+1} u_x + n(x+n-1)^{(n)} \Delta^n u_x - n(x+n-1)^{(n)} \Delta^n u_x$$
$$= (x+n)^{(n+1)} \Delta^{n+1} u_x$$

which is of the same form as the theorem. Hence the result holds good universally.

EXERCISE 1

Take interval of differencing to be unity.

1. Evaluate : $(a > 0)$ a, b, c are constants.

(*i*) Δa^x, (*ii*) $\Delta (ab^{cx})$, $(b > 0, c > 0)$

(*iii*) $\Delta [2^x/(x+1)!]$, (*iv*) $\Delta \cot 2^x$,

(*v*) $\Delta \sinh (a + bx)$, (*vi*) $\Delta \cosh (a + bx)$,

(*vii*) $\Delta \tan ax$, (*viii*) $\Delta \tan^{-1} ax$.

(*ix*) $\Delta \log x$. (*B.Sc. Maths. Hons. Delhi, 1990*)

(*x*) $\Delta \log f(x)$, h being the interval of differencing

(*AMIE 1991, 1993*, B.Sc. Agra, 1995)

2. Evaluate : [$a, b, c, d, a_0, a_1, \ldots, a_r$ and e are constants]

(*i*) $\Delta^2 x^3$, (*ii*) $\Delta^4 (ae^x)$,

(*iii*) $\Delta^3[(1-x)(1-2x)(1-3x)]$, and in general,

$\Delta^n[(a_0 + a_1 x)(a_0 + a_2 x)\ldots(a_0 + a_1 x)]$

(*iv*) $\Delta^{10} [(1-ax)(1-bx^2)(1-cx^3)(1-dx^4)]$,

(*v*) $\Delta^2(ab^{cx})$. (*vi*) $\Delta^2 \sin x$ (*vii*) $\Delta^2(x^3 - x^2)$

(*viii*) $(\nabla + \nabla)^2 (x^2 + x)$, given that $h = 1$ (*AMIE 1993*)

3. (*a*) Evaluate :

(*i*) $\dfrac{\Delta^2}{E} x^3$, (*ii*) $\dfrac{\Delta^2 x^3}{Ex^2}$.

(*iii*) $\left(\dfrac{\Delta}{E}\right)^2 x^3$ (*iv*) $\dfrac{\Delta^2 x^3}{E^2 x^3}$

(*v*) $\dfrac{\Delta^2}{E} \sin (x+h) + \dfrac{\Delta^2 \sin (x+h)}{E \sin (x+h)}$,

(*vi*) $\dfrac{\Delta^2}{E} e^x \cdot \dfrac{E e^x}{\Delta^2 e^x}$, (Agra, 1995, 97, *AMIE 1993*)

h being the interval of differencing in (*iii*) to (*vi*).

(*b*) Explain the difference between $(\nabla^2/E)\, u_x$ and $\dfrac{\Delta^2 u_x}{Eu_x}$ (*Ranchi 1990*)

4. (*a*) Show that if n is a positive integer, a and b are constants then

(*i*) $\Delta^n \sin (a + bx)$.

$$= \left(2 \sin \frac{b}{2}\right)^n \times \sin\left[a + bx + \frac{n}{2}(b + \pi)\right],$$ 1 is the interval of differencing

(*ii*) $\Delta^n \cos(a + bx)$

$$= \left(2 \sin \frac{b}{2}\right)^n \times \cos\left[a + bx + \frac{n}{2}(b + \pi)\right], \text{ 1 is the interval of differencing}$$

(*iii*) $\Delta^n \left(\frac{1}{n}\right) = (-1)^n n!\, h^n (x - h)^{-(n+1)}$, h is the interval of differencing,

(*iv*) $\Delta^n x^{(n)} = n!$

[**Hint.** Use Mathematical Induction to prove (*i*), (*ii*) and (*iii*)]

(*v*) $\Delta^n (e^{ax+b})$ where a and b are constants. (*AMIE 1996, 1998*)

(*B.Sc. Guwahati, 1996*)

(*b*) If Δ denotes the forward difference operator with respect to k and m is an integer ≥ 0, then prove that

$$\Delta^m (1 + j - k)_r = (-r)_m (1 + j - k)_{r-m}$$

(*c*) Show that if n is a positive integer, then

(*i*) $E^n y_x = \sum_{k=0}^{n} \binom{n}{k} \Delta^k y_x$ and thus $E^3 y_x = y_x + 3\Delta y_x + 3\Delta^2 y_x + \Delta^3 y$

(*ii*) $\Delta^n y_n = \sum_{r=0}^{n} \binom{n}{r} (-1)^r y_{2n-r}$

(*iii*) $\Delta^n y_n = 0$ where $y_n = (-1)^{n-1} \left\{ \binom{\frac{1}{2}}{n+1} + \binom{-\frac{1}{2}}{n+1} \right\}$,

where $\binom{\frac{1}{2}}{n+1} = {}^{\frac{1}{2}}C_{n+1}$.

Sol. (*c*) (*iii*) $\Delta^n y_n = (E - 1)^n y_n = \sum_{r=0}^{n} \binom{n}{r} (-1)^r E^{n-r} y_n$

$$= \sum_{r=0}^{n} \binom{n}{r} (-1)^r y_{2n-r} \text{ where } \binom{n}{r} = {}^n c_r$$

$$= \sum_{r=0}^{n} \binom{n}{r} (-1)^{2n-r-1+r} \left\{ \binom{\frac{1}{2}}{2n-r+1} + \binom{-\frac{1}{2}}{2n-r+1} \right\}$$

$$= -\sum_{r=0}^{n} \left[\binom{n}{r} \binom{-\frac{1}{2}}{2n-r+1} + \binom{n}{r} \binom{\frac{1}{2}}{2n-r+1} \right]$$

$$= -\text{ Coeff. of } S^{2n+1} \text{ in } [(1+S)^n (1+S)^{1/2} + (1+S)^n (1+S)^{1/2}]$$

$$= -\text{ Coeff. of } S^{2n+1} \text{ in } [(1+S)^{n+1/2} + (1+S)^{n-1/2}]$$

$$= -\left[\binom{n+\frac{1}{2}}{2n+1} + \binom{n-\frac{1}{2}}{2n+1} \right]$$

$$= -\left[\frac{\left(n+\frac{1}{2}\right)\left(n-\frac{1}{2}\right)\cdots\left(n+\frac{1}{2}-2n\right)+\left(n-\frac{1}{2}\right)\cdots\left(n-\frac{1}{2}-2n\right)}{(2n+1)\,!}\right]$$

$$= -\left[\frac{\left(n+\frac{1}{2}\right)\cdots\left(-n+\frac{1}{2}\right)\left\{\left(n+\frac{1}{2}\right)+\left(-n-\frac{1}{2}\right)\right\}}{(2n+1)\,!}\right]$$

$$= 0.$$

Hence the result.

5. (*a*) If $f(x) = e^{ax}$. show that $f(0)$ and its leading differences form a geometrical progression, a being a constant.

(*b*) Given $u_0 = 3,\ u_1 = 12,\ u_2 = 81,\ u_3 = 200,\ u_4 = 100, u_4 = 8$ find $\Delta^5\, u_0$.

6. (*a*) Prove that if $f(x)$ and $g(x)$ are any functions of x, a being a constant.

(*i*) $\Delta\,[f(x) \pm g(x)] = \Delta f(x) \pm \Delta\, g(x)$.

(*ii*) $\Delta\,[af(x)] = a\,\Delta f(x)$, a being a constant,

(*iii*) $\Delta\,[f(x)g(x)] = f(x+1)\,\Delta\, g(x) + g(x)\,\Delta f(x)$
$= f(x)\,\Delta\, g(x) + g(x+1)\,\Delta f(x)$,

(*iv*) $\Delta\left[\dfrac{f(x)}{g(x)}\right] = \dfrac{g(x)\,\Delta f(x) - f(x)\,\Delta\, g(x)}{g(x)\, g(x+1)}$,

(*v*) $\Delta\left[\dfrac{1}{f(x)}\right] = \dfrac{-\Delta f(x)}{f(x) f(x+1)}$.

(*vi*) $\Delta\,[f(x)]^2 = [f(x) + f(x+1)].\,\Delta f(x)$

(*vii*) $\Delta\,[u(x)v(x)] = Eu(x)\,\Delta\, v(x) + v(x)\,\Delta\, u(x)$

(*viii*) $(E-a)\,(xa^x) = a.a^x$, a being a constant.

(*ix*) $(E-a)^x\,(xd) = 0$

(*b*) Show that if $f(E)$ is a polynomial in E then
$f(E)a^x = a^x f(a)$, a being a constant.

(*c*) Prove the operator relations :

(*i*) $(1+\Delta)(1-\nabla) = 1$ (*ii*) $\nabla = 1 - E^{-1}$.

(*d*) Prove $(x)_{2n} = 2^{2n}\left(\dfrac{x}{2}\right)_n\left(\dfrac{x+1}{2}\right)_n$

In general, $(x)_{kn} = k^{nk}\left(\dfrac{x}{k}\right)_n\left(\dfrac{x+1}{k}\right)_n \cdots \left(\dfrac{x+k-1}{k}\right)_n$

[**Hint.** $(x)_{2n} = x(x+1)(x+2)(x+3)\ldots(x+2n-1)$.
Group alternate factors and take 2 common from each factor on the right.]

7. (*a*) A third degree polynomial passes through the points $(0,-1)$, $(1, 1)$, $(2, 1)$, and $(3,-2)$. Find the polynomial.

(*b*) Find the lowest degree polynomial u_x which takes the following values :

(*i*) $u_1 = 1,\ u_2 = 9,\ u_3 = 25,\ u_4 = 55,\ u_5 = 105$.

(*ii*)

$x =$	0	1	2	3	4	5
$f(x) =$	0	3	8	15	24	35

(*c*) Obtain a function whose first differences is

(*i*) e^x, (*ii*) $x^3 + 3x^2 + 5x + 12$. (*iii*) a, a being constant (*iv*) $x^{(2)}$

8. (*a*) Express the function $2x^3 - 3x^2 + 3x - 10$ and its differences in factorial notation.

[**Hint.** *First Method.* Proceed as in solved Ex. 4, page 10.

Second Method. This problem can be solved by Newton-Gregory theorem also since what we need is y_0 and its leading differences which on constructing a difference table are found to be – 10, 2, 6, 12.

Hence $y = -10 + 2x^{(1)} + 6\dfrac{x^{(2)}}{2!} + 12\dfrac{x^{(3)}}{3!}$]

(*b*) Represent the function $x^4 - 12x^2 + 42x^2 - 30x + 9$ in the factorial notation. Show that $\Delta^3 y = 18$ where $y = 3x^3 + x^2 + x - 1$ (*AMIE 1989*)

(*c*) Find the relation between α, β, γ in order that $\alpha + \beta x + \gamma x^2$ may be expressible in one term in the factorial notation.

[**Hint.** $\gamma x^2 + \beta x + \alpha \equiv \gamma(x+h)(x+h-1) = \gamma[x^2 + (2h-1)x + h(h-1)]$. Equate coefficients and eliminate h.]

(*d*) Show that if Δ operates on n, then

$$\Delta\binom{n}{x+1} = \binom{n}{x} \text{ where } \binom{n}{x} = {}^nC_x$$

and hence that

$$\sum_{n=1}^{N}\binom{n}{x} = \binom{N+1}{x+1} - \binom{1}{x+1}$$

(*e*) Find $\nabla n^{(m)}$ and $\nabla n^{(-m)}$. What happens when $n \lesseqgtr m$?

(*B.Sc. Ranchi 1991*)

(*f*) Form the difference table to locate and correct the error in the following tables :

(*i*) 20736, 28561, 38416, 50625, 65540, 83521, 104976, 130321, 160000

(*ii*) 0·09983, 0·14944, 0·19867, 0·24740, 0·29552, 0·34290, 0·38945, 0·43497, 0·47943, 0·52269, 0·56464, 0·60519, 0·64422.

(*g*) Suppose a table contains two errors in successive entries. Show how the difference pattern will overlap in the second and subsequent differences? How would you determine the magnitude of each error?

9. Prove that if m is positive integer, then

(*a*) $\Delta^2 x^{(m)} = m(m-1)x^{(m-2)}$, $x^{(m)} = x(x-1)(x-2)...(x-m+1)$

(*b*) $\Delta^2 x^{(-m)} = m(m+1)\, x^{(-m-2)}$

(*c*) $\dfrac{(x+1)^{(m)}}{m!} = \dfrac{x^{(m)}}{m!} + \dfrac{x^{(m-1)}}{(m-1)!}$

(*d*) $x^{(m+1)} + mx^{(m)} = x.x^{(m)}$.

10. (*a*) Show that Newton-Gregory formula can be written as

$P_{n(x)} = f(0) + x\,\Delta f(0) - xa_1\,\Delta^2 f(0) + xa_1 a_2\,\Delta^3 f(0) - \cdots$ where

$a_1 = 1 - \dfrac{1}{2}(x+1)$, $a_2 = 1 - \dfrac{1}{3}(x+1)$, etc.

(*b*) Prove that $y_4 = y_3 + \Delta y_2 + \Delta^2 y_1 = \Delta^3 y_1$

(*c*) Show that $y_4 = y_0 + 4\Delta y_0 + 6\Delta^2 y_{-1} + 10\Delta^3 y_{-1}$

$$= y_0 + 4\nabla y_0 + 6\nabla^2 y_{-1} + 10\nabla^3 y_{-1}$$

as far as third differences. (*B.Sc. Ranchi 91*)

[**Hint.** (*a*) $y_4 - y_3 = \Delta y_3 = \Delta(y_2 + \Delta y_2) = \Delta y_2 + \Delta^2 y_2$

$$= \Delta y_2 + \Delta^2(y_1 + \Delta y_1) = \text{etc.}$$

(*b*) $y_4 = E^5 y_{-1} = (1+\Delta)^5 y_{-1} = \text{etc.}$]

Use the method of separation of symbols to prove the identities (11-17).

11. (*i*) $u_1 + u_2 + \cdots + u_n = {}^nc_1 u_1 + {}^nc_2 \Delta u_1 + {}^nc_3 + \cdots \ \Delta u_1 \quad + \Delta^{n-1} u_1$

(*ii*) $u_0 - u_1 + u_2 - \cdots = \frac{1}{2} u_0 - \frac{1}{4}\Delta u_0 + \frac{1}{8}\Delta^2 u_0 - \frac{1}{16}\Delta^3 u_0 + \cdots$ (*B.Sc. Agra, 1997*)

Deduce from (i), the value of $1^2 + 2^2 + 3^2 + \cdots + n^2$

[**Hint.** (*i*) Left side $= (1 + E + E^2 + \cdots + E^{n-1})\, u_0$

$$= \frac{E^n - 1}{E - 1} u_0 = \frac{(1+\Delta)^n - 1}{\Delta} u_0 = \text{etc.}]$$

For deduction, take $u_n = n^2$. The values are 1, 4, 9, 16, ..., first diff: 3,5,7,.....

12. $u_0 + \frac{u_1 x}{1!} + \frac{u_2 x^2}{2!} + \cdots = e^x\left[u_0 + x\Delta u_0 + \frac{x^2}{2!}\Delta^2 u_0 + \cdots\right]$

13. (*a*) $\Delta^n y_x = y_{x+n} - {}^nc_1 y_{x+n-1} + {}^nc_2 y_{x+n-2} - \cdots + (-1)^n y_x,\ \Delta \equiv E - 1.$

(*b*) $u_0 + {}^nc_1 u_1 x + {}^nc_2 u_2 x^2 + {}^mc_3 u_3 x^3 + \cdots + u_n x^n$

$$= (1+x)^n u_0 + {}^nc_1 (1+x)^{n-1} x\Delta u_0 + {}^nc_2 (1+x)^{n-2} x^2 \Delta^2 u_0 + \cdots + x^n \Delta^n u_0$$

14. (*a*) $u_x = u_{x-1} + \Delta u_{x-2} + \Delta^2 u_{x-3} + \cdots + \Delta^{n-1} u_{x-n} + \Delta^n u_{x-n}$

(*b*) $\sum_{r=0}^{n-1} u_r x^r = \frac{u_0 - x^n u_n}{1-x} + \frac{x}{(1-x)^2}(\Delta u_0 - x^n \Delta u_n) + \cdots$

15. $u_{x+n} = u_n + {}^xc_1 \Delta u_{n-1} + {}^{x+1}c_2 \Delta^2 u_{n-2} + {}^{x+2}c_3 \Delta^3 u_{n-3} + \cdots$

[**Hint.** $u_{x+n} = E^x u_n = (E^{-1})^{-x} u_n = \text{etc},\ E^{-1} = 1/E = \frac{E - \Delta}{E} = 1 - \Delta E^{-1}$]

16. $u_x - u_{x+1} + u_{x+2} - u_{x+3} \cdots$

$$= \frac{1}{2}\left[u_{x-1/2} - \frac{1}{8}\Delta^2 u_{x-3/2} + \frac{1\cdot 3}{2!\,8^2}\Delta^4 u_{x-5/2} = \cdots\right]$$

(*AMIE 1995, B.Sc. Ranchi 1990*)

[**Hint.** Right side $= \frac{1}{2}E^{-1/2}\left(1 + \frac{\Delta^2}{4E}\right)^{-1/2} u_x$

$$= (4E + \Delta^2)^{-1/2} u_x = (2+\Delta)^{-1} = u_x,\ E \equiv 1 + \Delta$$

$$= (1+E)^{-1} u_x = \text{etc.}]$$

17. (*a*) $u_x - \frac{1}{8}\Delta^2 u_{x-1} + \frac{1\cdot 3}{8\cdot 16}\Delta^4 u_{x-2} - \frac{1\cdot 3\cdot 5}{8\cdot 16\cdot 24}\Delta^6 u_{x-3} + \cdots$

$$= u_{x+1/2} - \frac{1}{2}\Delta u_{x+1/2} + \frac{1}{4}\Delta^2 u_{x+1/2} - \frac{1}{8}\Delta^2 u_{x+1/2} + \cdots$$

(*b*) $\sum_{x=0}^{\infty} u_{2x} = \frac{1}{2}\sum_{x=0}^{\infty} u_x + \frac{1}{2}\left(1 - \frac{1}{2}\Delta + \frac{1}{4}\Delta^2 - \cdots\right)u_0$

[**Hint.** (*a*) Left side $= \left(1 + \frac{1}{4}\Delta^2 E^{-1}\right)^{-1/2} u_x$

$$= 2E^{1/2}(2 + \Delta)^{-1} u_x = \text{etc.}]$$

18. Sum to n terms the series :

$1\cdot 2\,\Delta x^n - 2\cdot 3\,\Delta^2 x^n + 3\cdot 4\,\Delta^3 x^n - 4\cdot 5\,\Delta^4 x^n + \cdots$ (*Statistics Sub. Ranchi 1990*)

[**Hint.** Since $\Delta^{n+1} x^n$ and higher differences of x^n are zero, the sum of the series to n terms is the same as the sum to infinity. Hence the given series

$$= 2\Delta[1 - 3\Delta + 3\cdot 2\,\Delta^2 - \cdots]\,x^n$$
$$= 2\Delta(1 + \Delta)^{-3} x^n = 2(E - 1)E^{-3} x^n$$
$$= 2(E^{-2} - E^{-3})\,x^n = 2[(x-2)^{-n} - (x-3)^{-n}].$$

19. Show that

$$u_{2n} - {}^n c_1 \cdot 2^1 u_{2n-1} + {}^n c_2\, 2^2 u_{2n-2} - \cdots + (-2^n)\,u_n = (-1)^n (c - 2\,an)$$

where $u_x = ax^2 + bx + c$. (*B.A. Math. Hons., Delhi, 1997*)

20. Given the following table, construct a difference table and from it estimate y when $x = 0\cdot 7$.

x:	0	0·1	0·2	0·3	0·4
y:	1	1·095	1·179	1·251	1·310

21. If p, q, r, s be the successive entries corresponding to equidistant arguments in a table, show that when third differences are taken into account, the entry corresponding to the argument half-way between the arguments of q and r is $A + \frac{1}{24}B$, where A is the arithmetic mean of q and r and B is the arithmetic mean of q and r and B is the arithmetic mean of $3q - 2p - s$ and $3r - 2s - p$.

[**Hint.** If the arguments are $a, a+h, a+2h$ and $a+3h$, then we are to find $y_{a+3h/2}$. Now $y_{a+3h/2} = E^{3/2} y_a = (1+\Delta)^{3/2} y_a = \text{etc.}$]

22. (*a*) Find the missing term in the following table :

x :	0	1	2	3	4
y :	1	3	9	X	81

Explain why the resulting value differs from 3^3 or 27. (*B.Sc. Agra, 1997*)

(*b*) x:	16	18	20	22	24	26
y:	39	85	...	151	264	388

23. (*a*) If l_x represents the number of persons living at age x in a life table, find as accurately as the data will permit, l_x for values of $x = 35$, 42 and 47. Given

$$l_{20} = 512, \ l_{30} = 439, \ l_{40} = 346, \ l_{50} = 243.$$

(*b*) Given sin 45° = ·7071, sin 50° = ·7660, sin 55° = ·8192, sin 60° = ·8660, find sin 52°. (*B.Sc. Agra, 1995*)

24. The following data gives the melting points (m.p.) of an alloy of lead and zinc.

Percentage in alloy	(*P*) :	50	60	70	80
Temperature	(*Q* °C) :	2·5	225	248	274

Find the m.p. of the alloy containing 54% of lead, using appropriate interpolation formula. (*AMIE 1992*)

25. (*a*) The following table gives the values of the function

$$y = \frac{1}{\sqrt{2x}} \int_{-\infty}^{x} e^{-\frac{1}{2}t^2}\, dt$$

for $x = 0{\cdot}30(0{\cdot}02)0{\cdot}40$. Evaluate y for $x = 0{\cdot}316$.

x :	0·30	0·32	0·34	0·36	0·38	0·40
y :	0·6179114	0·6255158	0·6330717	0·6405764	0·6480273	0·6554217

Note. 0·30(0·02)0·40 means the values of x ranging from 0·30 to 0·40 with increment 0·02, i.e., 0·30, 0·32, 0·34, 0·36, 0·38, 0·40.

(*b*) Estimate the missing term in the following table :

x :	1	2	3	4	5	6	7
y :	2	4	8	—	32	64	128

Explain why the result differs from 2^4.

(*c*) Apply Gregory-Newton backward formula to the following table to find sun's declination on Feb. 12 :

Date	1	3	5
Declination	– 17°0′ 19·0″	– 16°25′ 22·9″	– 15°49′ 18·8″

7	9	11	13
– 15°12′ 09·8″	– 14°33′ 59·1″	– 13°54′ 49·8″	– 13°14′ 45·0″

26. (*a*) By constructing a difference table, find the 7th term as well as the general term of the sequence

$$0, 0, 2, 6, 12, 20.$$

(*b*) Find the first term of the series whose second and subsequent terms are 8, 3, 0, – 1, 0.

[**Hint.** $f(1) = E^{-1} f(2) = (1 - \Delta + \Delta^2 \ldots) f(2)$, etc.]

27. From the following data, estimate the number of persons earning wages between 60 and 70 rupees.

Wage (in Rs.)	No. of persons (in thousands)
Below 40	250
,, 40—60	120
,, 60—80	100
,, 80—100	70
,, 100—120	50

(*Guwahati 1994*)

28. In an examination the number of candidates who obtained marks between certain limits were as follows :

Marks	No. of candidates
0—19	41
20—39	62
40—59	65
60—79	50
80—99	17

Estimate the number of candidates who obtained fewer than 70 marks.

(AMIE 1987, B.Sc. Agra, 1997)

29. (*a*) If u_x is a rational function of second degree, prove that

$$\left(\frac{d}{dx}u_x\right) = \sum \frac{b+c}{(a-b)(c-a)} u_a.$$

Hence evaluate $\Sigma a^2 (b+c)(b-c)$.

(*b*) Given that u_x is a polynomial of second degree and $u_0 = 1,\ u_1 + u_2 = 10,\ u_3 + u_4 + u_5 = 65$, find the value of u_{10}.

(*c*) Given $f(-1) = 1{\cdot}25, f(0) = 1{\cdot}75$ and $f(1) = 1{\cdot}50$, locate approximately the value of x for which $f(x)$ attains a maximum in the interval $(-1, 1)$.

(*d*) Given that $f(1) + f(2) + f(3) = 25$,

$$f(4) = 29,$$

and $$f(5) + f(6) = 113,$$

estimate the value of $f(10)$.

30. Given $u_0 + u_8 = 1{\cdot}9243$, $\quad u_1 + u_7 = 1{\cdot}9590$,

$u_2 + u_6 = 1{\cdot}9823$, and $\quad u_3 + u_5 = 1{\cdot}9956$,

find u_4, stating the assumptions made.

[Use $\Delta^8 u_0 = 0$ or $(E-1)^8 u_0 = 0$]

31. (*a*) A function $f(x)$ is given in equidistant points $x_0, x_1, ..., x_n$ where $x_k = x_0 + kh$, and the corresponding differences of different orders are denoted by $\Delta f, \Delta^2 f, ...$ By also taking the points mid-way between the first points, we obtain $x_0, x_{1/2}, x_1, x_{3/2}, ...$ and the corresponding differences are denoted by $\Delta_1 f, \Delta_1^2 f, ...$

Show that

$$\Delta_1{}^r = 2^{-r}[\Delta^r - a\,\Delta^{r+1} + b\,\Delta^{r+2} - \cdots]$$

and compute the constants a and b.

[**Hint.** $\Delta_1 x_0 = x_{1/2} - x_0 = (E^{1/2} - 1)\, x_0 = [(1+\Delta)^{1/2} - 1]\, x_0$

thus $\quad \Delta_1 = (1+\Delta)^{1/2} - 1 =$ etc.]

(*b*) If u_x be a function whose differences when the increment of x is unity are denoted by $\delta u_x, \delta^2 u_x, ...$ and by $\Delta u_x, \Delta^2 u_x ...$ when the increment of x is n, then if $\delta^2 u_x, \delta^2 u_{x+1}, \delta^2 u_{x+2}, ...$ are in G. P. with common ratio q, show that

$$\frac{\Delta u_x - n\,\delta u_x}{(q^n - 1) - n(q-1)} = \frac{\delta^2 u_x}{(q-1)^2}. \quad \text{(B.A. Stat. Hons. Delhi 1995)}$$

[**Hint.** $\delta^2 u_{x+1} = q\,\delta^2 u_x$ (Given)

$\Rightarrow \quad \delta^2 u_{x+1} - \delta^2 u_x = (q-1)\,\delta^2 u_x$ [Subtract $\delta^2 u_x$ from both sides]

$\Rightarrow \quad \delta^3 u_x = (q-1)\,\delta^2 u_x,\ \delta^4 u_x = (q-1)\,\delta^3 u_x = (q-1)^2\,\delta^2 u_x$, etc.

Further, $\Delta^n u_x - {}^nC_1\,\delta\,u_x = \dfrac{1}{q-1}(-{}^nC_1\,\delta^2 u_x + {}^nC_1\,\delta^2 u_{x-1})$

$$= \frac{-n}{q-1} + \frac{1}{(q-1)^2}[{}^nC_1\,(q-1) + {}^nC_2(q-1)^2 + \cdots]$$

$$= \frac{-n}{q-1} + \frac{1}{(q-1)^2}[\{(1+q-1)\}^n - 1]$$

$$= \frac{-n}{q-1} + \frac{q^n - 1}{(q-1)^2} = \frac{q^n - 1 - n(q-1)}{(q-1)^2}.$$

Hence $\dfrac{\Delta u_x - n\,\delta\,u_x}{\delta^2 u_x} =$ etc.]

32. Prove that

$$\frac{\Delta^n 0^m}{n\,!} = \frac{n\,\Delta^n 0^{m-1}}{n\,!} + \frac{\Delta^{n-1} 0^{m-1}}{(n-1)\,!}$$

and form a table of the coefficients in the expansion of x^m in terms of the factorials $x^{(k)}$, illustrating its use by an example.

33. Calculate the values of

(*i*) $\Delta^2\,0^5$, (*ii*) $\Delta^3\,0^6$ (*iii*) $\Delta^4\,0^5$, (*iv*) $\Delta^5\,0^5$, (*v*) $\Delta\,0^n$.

34. Prove that

$$\Delta^n\,0^m = n^m - {}^nC_1\,(n-1)^m + {}^nC_2\,(n-2)^m - \cdots$$

and deduce that

$$n\,! = n^n - {}^nC_1\,(n-1)^n + {}^nC_2\,(n-2)^n - \cdots$$

35. Show that

(*i*) $\Delta\,0^m - \frac{1}{2}\Delta^2\,0^m + \frac{1}{3}\Delta^3 0^m - \cdots = 0$, if $m > 1$.

(*ii*) $(1 + \log E)^r\,0^n = \dfrac{r\,!}{(r - n\,!)}$, $r > n$.

(*iii*) $(e^t - 1)^k = \displaystyle\sum_{v=0}^{\infty} \frac{(\Delta^k\,0^v)\,t^v}{v\,!}$

(*iv*) $(2x+5)^{(3)} = (2x+5)(2x+3)(2x+1)$.

(*v*) $(3x-4)^{-2} = 1/(3x-1)(3x+2)$

[**Hint.** (*iii*) $(e^t - 1)^k = \sum_{m=0}^{k} (-1)^{k-m}\, {}^mC_k\, e^{mt}$

where $$e^{mt} = \sum_{v=0}^{\infty} \frac{m^v t^v}{v!}$$

so that

$$(e^t - 1)^k = \sum_{v=0}^{\infty} \frac{t^v}{v!} \left\{ \sum_{m=0}^{k} (-1)^{k-m}\, {}^mC_k\, m^v \right\},$$

where $$\Delta^k 0^0 = [\Delta^k x^0]_{x=0} = \Delta^k . 1 = 0.]$$

36. Prove that

$$\Delta^n 0^{n+1} = \frac{1}{2} n (n+1) \Delta^n 0^n = \frac{1}{2} n [(n+1)!]$$ (*B.Sc. Hons. Statistics, 1996*)

[**Hint.** Let $x^{n+1} \equiv x^{(n+1)} + A_1 x^{(n)} + \cdots + A_n x^{(1)} + A_{n+1}$ where on comparing the coefficients of x^n on both sides, we find

$$A_1 = 1 + 2 + \cdots + n = \frac{1}{2} n (n+1)].$$

37. Show that

$$(n+1) \Delta^n 0^n = 2 [\Delta^{n-1} 0^n + \Delta^n 0^n].$$

[**Hint.** Use Example 36]

38. Prove the identity :

$$\Delta^m 0^n + {}^nC_1 \Delta^m 0^{n-1} + {}^nC_2 \Delta^m 0^{n-2} + \cdots + \frac{n!}{(n-m)!}$$

$$= \frac{1}{m+1} \Delta^{m+1} 0^{n+1}.$$

[**Hint.** L.H.S. $= [\Delta^m (x+1)^n]_{x=0} = [\Delta^m E x^n]_{x=0}$

$$= (\Delta^m + \Delta^{m+1}) 0^n = \Delta^m 0^n + \Delta^{m+1} 0^n$$

$$= \frac{1}{m+1} \Delta^{m+1} 0^{n+1}$$ by §1·10·1, page 29.]

39. Show that, if $E^n 0^m = n^m$, then

$$\phi(E^n) 0^m = n^m \phi(E) 0^m$$

where $\phi(E)$ is a rational integral function of E.

40. (*i*) If $f(E)$ is a polynomial in E, show that

$$f(E) 0^n = Ef'(E) 0^{n-1}.$$

Hence deduce that $\Delta^m 0^n = m \Delta^{m-1} 0^{n-1} + m \Delta^m 0^{n-1}$.

[**Hint.** (*i*) Let $f(E) = \sum_{r=0} A_r E^r$ so that

$$f(E) 0^n = \Sigma A_r r^n$$

Also $Ef'(E) = \Sigma A_r . rE^r$ so that

$$Ef'(E)\,0^{n-1} = \Sigma r\,.\,A_r r^{n-1} = \Sigma A_r r^n = f(E)\,0^n].$$

(*ii*) Let $f(E) = \Delta^m = E^m - {}^mc_1 E^{m-1} + \cdots + (-1)^m$, a polynomial in E.

Show that $\Delta^m 0^n = mE\,\Delta^{m-1} 0^{n-1} = m(\Delta^{m-1} 0^{n-1} + \Delta^m 0^{n-1})$.]

(*iii*) Prove that $f(E)a^x \phi(x) = a^x f[a^h E]\,\phi(x)$, a being a constant where $f(x)$ and $\phi(x)$ are polynomials.

(*iv*) If $\phi(x)$ is a polynomial, prove $\phi(E)\,u_x = a^x \phi(a^h E)\,[a^{-x} u_x]$,

a is a constant.

41. Show that, a being a constant,

(*i*) $\Delta^{m-1} 2^n = (\Delta^m 1^{n+1} + \Delta^{m-1} 1^n)/(m+1)$. (*B.A. Stat. Hons. Delhi 1995*)

(*ii*) $\Delta^m 1^{n+1} = (m+1)\,\Delta^m 1^n + m\Delta^{m-1} 1^n$.

42. Prove that, if $0 < x < 1$,

$$1^n + 2^n x + 3^n x^2 + \cdots$$

$$= \frac{1}{(1-x)^2}\left\{\Delta 0^n + \frac{x}{1-x}\Delta^2 0^n + \left(\frac{x}{1-x}\right)^2 \Delta^3 0^n + \cdots\right\}$$

(*B.A. Hons Stat. 1997*)

$$\left[\textbf{Hint. } \text{R.H.S.} = \frac{1}{x(1-x)}\cdot\left[\frac{x}{1-x}\Delta + \left(\frac{x}{1-x}\right)^2 \Delta^2 + \left(\frac{x}{1-x}\right)^3 \Delta^3 + \cdots\right]0^n\right.$$

$$= \frac{1}{x(1-x)}\left[\left(1 - \frac{x}{1-x}\Delta\right)^{-1} - 1\right]0^n$$

$$= \left[\left\{\frac{1}{x}(1-xE)^{-1} - \frac{1}{x(1-x)}\right\}z^n\right]_{z=0}$$

$$= \left[\left(\frac{1}{x} + E + xE^2 + \cdots\right)z^n - \frac{1}{x}(1-x)^{\frac{1}{2}} z_n\right]_{z=0}$$

$$= [(1 + E + xE^2 + \cdot)\,z^n]_{z=0} = \text{etc.}$$

43. Show that

$$e^{xt} = e\left\{1 + e^{\Delta}\frac{0t}{1\,!} + e^{\Delta}\frac{0^2 t^2}{2\,!} + \cdots\right\}$$

and calculate the first four terms of the expression.

[**Hint.** R.H.S. $= e\{1 + e^{\Delta}(e^{xt} - 1)_{x=0}\}$

$$= e\{1 + e^{\Delta} e^{xt} - e^{\Delta}.1\}_{x=0}$$

since $\quad e^{\Delta} = 1 + \dfrac{\Delta}{1\,!} + \dfrac{\Delta^2}{2\,!} + \cdots$

and $\quad \Delta e^{xt} = e^{(x+1)t} - e^{xt} = e^{xt}(e^t - 1)$

$\therefore \quad$ R.H.S. $= e[1 + e^{xt}(e^{xt} - 1) - 1]_{x=0} = e[1 + (e^{xt} - 1)]$

$$= e^{xt}.$$

44. Sum the series :

$$(n-1)^2\,{}^nC_1 + (n-3)^2\,{}^nC_3 + (n-5)^2\,{}^nC_5 + \cdots$$

[**Hint.** Given expression $= [({}^nC_1 E^{-1} + {}^nC_3 E^{-3} + \cdots)\, x^2]_{x=n}$

$$= [\tfrac{1}{2}\, [(1+E^{-1})^n - (1-E^{-1})^n]\, x^2]_{x=n}$$

$$= [\tfrac{1}{2}\, [(\Delta+2)^n - \Delta^n]\, E^{-n} x^2]_{x=n}$$

$$= \tfrac{1}{2}\, [(\Delta+2)^n - \Delta^n]\, (x-n)^2]_{x=n}$$

$$= \tfrac{1}{2}\, [(\Delta+2)^n - \Delta^n)\, 0^2]$$

$$= \tfrac{1}{2}\, [2^n 0^2 + n\, 2^{n-1}\, \Delta\, 0^2 + \tfrac{1}{2}\, n(n-1)\, 2^{n-2}\, \Delta^2\, 0^2 + \cdots]$$

$$= \tfrac{1}{2}\, [n 2^{n-1} + n(n-1)\, 2^{n-2}] = \text{etc.}]$$

45. (*a*) If $D \equiv x\dfrac{d}{dx}$, prove that

$$e^D f(x) = f(e^x)$$

(*b*) Show that

$$\left(\frac{d}{d\theta}\right)^n = \frac{\Delta\, 0^n}{1} D + \frac{\Delta^2\, 0^n}{2\,!} D^2 + \frac{\Delta^3\, 0^n}{3\,!} D^3 + \cdots$$

where $x = e^{\theta}$ and $D^n \equiv x^n \dfrac{d^n}{dx^n} = \left(\dfrac{d}{d\theta}\right)^n$

[Use §1·11·2]

46. Prove that if m and n are any integers, then

(*i*) $(x\,\Delta)^m\, (xE)^m = (xE)^m\, (x\,\Delta + m)^m$.

(*ii*) $(x\,\Delta)^m (xE)^m = (xE)^m (x\,\Delta + m)^m$, $(x\Delta)\, (xE)^m = (xE)^m\, (n\Delta + m)$.

(*iii*) $(xE)^m = (x+m-1)^{(m)}\, E^m$.

(*iv*) $(E-a)^2\, (xa^x) = 0$

(*v*) $(x\,\Delta)^{(n)} u_x = (x+n-1)^{(n)}\, \Delta^n u_x$

[**Hint.** (*ii*) R.H.S. $= (xE)^m (x\,\Delta + m) = (xE)^m (xE - x + m)$

$= (x+m-1)^{(m)}\, E^m (xE - x + m)$ by (*iii*)

$= (x+m)^{(m+1)}\, E^{m+1} - (x+m)(x+m-1)^{(m)}\, E^m + m(x+m-1)^{(m)}\, E^m$

$= (x+m)^{(m+1)}\, E^{m+1} - x(x+m-1)^{(m)}\, E^m$, on simplification

$= (xE)^{m+1} - x(xE)^m$ by (*iii*)

$= (xE - x)(xE)^m$

$= x\,\Delta\, (xE)^m =$ L.H.S.

[**Hint.** (*i*) Use the method of induction

Also $(xE)^m (x\,\Delta + m)^{m+1} = (xE)^m (x\,\Delta + m)^m [xE - (x-m)]$

$= (xE)^m (x\,\Delta + m)^m xE - (xE)^m (x\,\Delta + m)^m (x-m)$

$= (x\,\Delta)^m (xE)^m xE - (x\,\Delta + m)^m (xE)^m (x-m)$

$= (x\,\Delta)^m (xE)^m [xE - (x-m)]$

$= (x\,\Delta)^m (xE)^m (x\,\Delta + m) = (x\,\Delta)^m x\,\Delta\,(xE)^m$

$= (x\,\Delta)^{m+1} (xE)^m.$

(*v*) **[Hint.** $(x\,\Delta)^{(n+1)} u_x = (x\,\Delta)(x\,\Delta - 1)\ldots(x\,\Delta - n)\, u_x$

$= (x\,\Delta - n)(x\,\Delta)^{(n)}\, u_x = [x\,\Delta - n]\,[x + n - 1)^{(n)}\,\Delta^n u_x]$

$= x\,\Delta\,[(x + n - 1)^{(n)}\,\Delta^n u_x] - n(x + n - 1)^{(n)}\,\Delta^n u_x$

$= x[(x + n)^{(n)}\,\Delta^{n+1} v_x + n(x + n - 1)^{(n-1)}\,\Delta^n u_x] - n(x + n - 1)^{(n)}\,\Delta^n u_x$

$= (x + n)^{(n+1)}\,\Delta^{n+1} u_x + n(x + n - 1)^{(n)}\,\Delta^n u_x - n(x + n - 1)^{(n)}\,\Delta^n u_x$

= etc.]

47. If $\sum_0^{\infty} t^n\, u_x = f(e^t)$, prove that

$$u_x = f(E)\, 0^n \div n\,!$$

[See 1.11.1, page 30]

48. Show that

$$\int_0^{\infty} e^{-nx}\,(e^{-x} - 1)^m x^{n-1}\, dx = \Gamma\,(a)\,\Delta^m n^{-a},$$

m, *n* and *a* being positive quantities and Δ operating on *n* alone.

[Hint. L.H.S. $= \int_0^{\infty} e^{-nx}\,[e^{-mx} - {}^m c_1\, e^{-(m-1)\,.\,x} + {}^m c_2\, e^{-(m-2)\,x} - \cdots$

$+ (-1)^m]\, x^{a-1}\, dx$

$$= \int_0^{\infty} e^{-(n+m)\,x}\, x^{a-1}\, dx - {}^m c_1 \int_0^{\infty} e^{-(n+m-1)\,x}\, x^{a-1}\, dx + \cdots$$

$$+ (-1)^m \int_0^{\infty} e^{-nx}\, x^{a-1}\, dx$$

$$= \frac{\Gamma\, a}{(n+m)^a} - {}^m c_1\, \frac{\Gamma\, a}{(n+m-1)^a} + \cdots + \frac{(-1)^m\,\Gamma\, a}{n^a}$$

$= \Gamma\, a\,[(n+m)^{-a} - {}^m c_1\,(n+m-1)^{-a} + \cdots + (-1)^m\, n^{-a}]$

$= \Gamma\, a\,[E^m n^{-a} - {}^m c_1\, E^{m-1} n^{-a} + \cdots + (-1)^m n^{-a}]$

$= \Gamma\,(a)\,(E - 1)^m n^{-a} = \Gamma\,(a)\,\Delta^m n^{-a}].$

49. Prove that

$$\int_0^{\infty} \Delta^m\, \frac{x^{a+1}}{x^2 + n^2}\, dx = \frac{\pi}{2}\,(-1)^{k+1}\,(\sin \tfrac{1}{2}\, a\,\pi)^{-1}.\ \Delta^m n^a,$$

m being a positive integer greater than a and $2k$ the even integer next greater than $a+1$ and the operator Δ acts on n alone.

[**Hint.** Let $2k$ be the even integer next greater than $a+1$, then

$$\frac{x^{a+1}}{x^2+n^2} = x^{a-2k+1}\left[\frac{x^{2k}-n^{2k}}{x^2+n^2} + \frac{n^{2k}}{x^2+n^2}\right]$$

Now the first member of the R.H.S. is a rational integral function of x an order lower than m. It therefore vanishes, when Δ^m operates on it.

$$\therefore \quad 1 = \int_0^\infty \Delta^m \frac{n^{2k}}{x^2+n^2} x^{a-2k+1}\, dx = \Delta^m \int_0^\infty \frac{n^{2k}}{x^2+n^2} x^{a-2k}\, dx$$

$$= \frac{1}{2}\Delta^m n^a \cdot \int_0^\infty \frac{y^{1/2-k}}{y+1}\, dy, \qquad \text{when } x^2 = n^2 y]$$

$$= \frac{1}{2}\frac{\pi}{\sin\left(\frac{a}{2}-k+1\right)\pi}\Delta^m n^a = \text{etc.}]$$

50. Show that if $2n > m > a > 0$, then

$$\int_0^\infty \frac{\sin 2x \sin^m x}{x^{a+1}}\, dx = \frac{\pi\, \Delta^m (2n-m)^a}{2^{m+1}\,\Gamma(a+1)\cos\frac{a+m}{2}\pi}$$

51. The length of the day was 12 hours on March 19th, 14 hours on April 18th and 15 hours 40 minutes on May 18th. Compute (*a*) the length of the day on May 3rd, (*b*) the mean length of the day during the period, March 19th to May 18th.

[**Hint.** (*a*) We have $y_0 = 12,\ y_{30} = 14,\ y_{60} = 47/3$

We want to evaluate y_{45} which is $E^{3/2} y_0 = (1+\Delta)^{3/2} y_0 = \text{etc.}$

(*b*) We have $12 = \Sigma x_1/30 \ \Rightarrow\ \Sigma x_1 = 360$

Also $\Sigma x_2 = 30 \times 14 = 420.\ \Sigma x_3 = 30 \times \frac{47}{3} = 4{\cdot}70$

$$\text{Mean length} = \frac{x_1+x_2+x_3}{3} = 13\frac{8}{9}\text{ hrs.}]$$

52. Show that *(B.Sc. Agra, 1996)*

(*i*) $\sum_{k=0}^{n-1} \Delta^2 f_k = \Delta f_n - \Delta f_0$

(*ii*) $\sum_{k=0}^{n-1} (E-1)^2\ f_k = \sum_{k=0}^{n-1} (f_{k+2} - 2f_{k+1} + f_k)$

[**Hint.** (*i*) L.H.S. $= f_{n+1} - f_n - (f_1 - f_0)$, on simplification $= \Delta f_n - \Delta f_0$]

2

INTERPOLATION WITH UNEQUAL INTERVALS OF THE ARGUMENT

2·1. Introduction. The formulae which we have obtained in the preceding chapter hold good only when the given values are at equidistant intervals of the independent variable or argument. Before we develop formulae which are true in case the arguments are unequally spaced, we define what are known as divided differences.

2·2. Divided Differences with Unequal Arguments

Let $f(x_0), f(x_1), ..., f(x_n)$ be the entries corresponding to the arguments $x_0, x_1, ..., x_n$ where the intervals $x_1 - x_0, x_2 - x_1, ...\ x_n - x_{n-1}$ may not be equal, *i.e.,* not necessarily equally spaced. Then

The *first divided differences* of $f(x)$ for the two arguments x_0, x_1 ; x_1, x_2, etc., are defined as

$$\left.\begin{aligned} \underset{x_1}{\Delta'} f(x_0) &= \underset{x_1}{\Delta'} f(x_0) = f(x_0, x_1) = \frac{f(x_0) - f(x_1)}{x_0 - x_1}, \\ \underset{x_2}{\Delta'} f(x_1) &= \underset{x_2}{\Delta'} f(x_1) = f(x_1, x_2) = \frac{f(x_1) - f(x_2)}{x_1 - x_2}, \\ &\text{etc.} \qquad \text{etc.} \end{aligned}\right\} \qquad ...(2\cdot1)$$

The *second divided difference* of $f(x)$ for the three arguments x_0, x_1 and x_2 defined as

$$\underset{x_0, x_1}{\Delta^{12}} f(x_0) = \underset{x_1, x_2}{\Delta^2} f(x_0) = f(x_0, x_1, x_2) = \frac{f(x_0, x_1) - f(x_1, x_2)}{x_0 - x_2} \qquad ...(2\cdot2)$$

and the *n*th *divided difference* for the $n + 1$ arguments $x_0, x_1, ..., x_n$ as

$$\underset{x_1, x_2, ..., x_{n-1}}{\Delta'^n} f(x_0) = \underset{x_1, x_2, ..., x_{n-1}}{\Delta'^n} f(x_0) = f(x_0, x_1, x_2, ..., x_n)$$

$$= \frac{f(x_0, x_1, x_2, ..., x_{n-1}) - f(x_1, x_2, ...x_n)}{x_0 - x_n} \qquad ...(2\cdot3)$$

A divided difference table can be constructed for any set of tabulated values of $f(x)$; it has the appearance of Table 2·1.

Table 2·1. Divided Difference Table

Argument x	*Entry* $f(x)$	*First Div. diff.* $\Delta' f(x)$	*Second Div. diff.* $\Delta'^2 f(x)$	*Third Div. diff.* $\Delta'^3 f(x)$
x_0	$f(x_0)$			
		$f(x_0, x_1)$		
x_1	$f(x_1)$		$f(x_0, x_1, x_2)$	
		$f(x_1, x_2)$		$f(x_0, x_1, x_2, x_3)$
x_2	$f(x_2)$		$f(x_1, x_2, x_3)$	
		$f(x_2, x_3)$		
x_3	$f(x_3)$			

2·2·1. Divided Differences when two or more arguments are same or coincident. If two of the arguments coincide, the divided difference can be given a meaning. The meaning assigned is obtained by taking the limit. Thus

$$f(x_0, x_0) = \lim_{\varepsilon \to 0} f(x_0 + \varepsilon, x_0)$$

$$= \lim_{\varepsilon \to 0} \frac{f(x_0 + \varepsilon) - f(x_0)}{(x_0 + \varepsilon) - x_0} = f'(x_0) \qquad ...(2\cdot4)$$

if $f(x)$ is differentiable at x_0.

Again, $f(x_0, x_0, x_0) = \lim\limits_{\substack{\varepsilon \to 0 \\ \varepsilon' \to 0}} f(x_0, x_0 + \varepsilon, x_0 + \varepsilon')$ $\qquad ...(2\cdot5)$

$$= \lim_{\substack{\varepsilon \to 0 \\ \varepsilon' \to 0}} \frac{f(x_0, x_0 + \varepsilon) - f(x_0 + \varepsilon, x_0 + \varepsilon')}{x_0 - (x_0 + \varepsilon')}$$

$$= \lim_{\varepsilon' \to 0} \frac{f'(x_0) - f(x_0, x_0 + \varepsilon')}{-\varepsilon'}$$

$$= \lim_{\varepsilon' \to 0} \frac{f'(x_0) - \dfrac{f(x_0 + \varepsilon') - f(x_0)}{\varepsilon'}}{-\varepsilon'}$$

$$= \lim_{\varepsilon' \to 0} \frac{\varepsilon' f'(x_0) - f(x_0 + \varepsilon') + f(x_0)}{-\varepsilon'^2}$$

$$= \lim_{\varepsilon' \to 0} \frac{f'(x_0) + \varepsilon' f''(x_{0)} - f'(x_0 + \varepsilon')}{-2\varepsilon'}$$

$$= \frac{1}{2!} f''(x_0).$$

by applying D' Hospital's ule twice.

Proceeding in this way, we find that

$$f(x_0, x_0, \ldots x_0) = \frac{1}{r!} f^r(x_0). \qquad \ldots(2\cdot5')$$

where the $(r+1)$ arguments are each equal to x_0.

Thus if all the $(r+1)$ arguments are equal to x_0, then rth divided difference is equal to the rth derivative of $f(x)$ at x_0 divided by r!

Finally, we also get

$$f(x, x, x_0, x_1, \ldots x_n) = \frac{d}{dx} f(x, x_0, x_1, \ldots, x_n).$$

Note. The limiting value of a divided difference which arises when two or more of the arguments coincide, is called a *confluent* divided difference arising from the confluence of the arguments in question.

2·2·2. Properties of the Divided Difference

Theorem 1. *The divided difference are symmetrical in all their arguments, that is, the value of any divided difference is independent of the order of the arguments.*

Proof. $f(x_0, x_1) = \dfrac{f(x_0) - f(x_1)}{x_0 - x_1}$

$$= \frac{f(x_0)}{x_0 - x_1} + \frac{f(x_1)}{x_1 - x_0} = \sum \frac{f(x_0)}{x_0 - x_1}$$

Thus $f(x_0, x_1)$ is symmetrical in x_0, x_1.

Again, $f(x_0, x_1, x_2) = \dfrac{f(x_0, x_1) - f(x_1, x_2)}{x_0 - x_2}$

$$= \frac{\dfrac{f(x_0)}{x_0 - x_1} + \dfrac{f(x_1)}{x_1 - x_0}}{x_0 - x_2} - \frac{\dfrac{f(x_1)}{x_1 - x_2} + \dfrac{f(x_2)}{x_2 - x_1}}{x_0 - x_2}$$

$$= \frac{f(x_0)}{(x_0 - x_1)(x_0 - x_2)} + \frac{f(x_1)}{(x_1 - x_0)(x_1 - x_2)} + \frac{f(x_2)}{(x_2 - x_0)(x_2 - x_1)}$$

or $$f(x_0, x_1, x_2) = \sum_{x_0}^{x_2} \frac{f(x_0)}{(x_0 - x_1)(x_0 - x_2)},$$

which is symmetrical in x_0, x_1, x_2.

We shall now show that

$$f(x_0, x_1, \ldots x_n) = \frac{f(x_0)}{(x_0 - x_1)(x_0 - x_2)\ldots(x_0 - x_n)} + \frac{f(x_1)}{(x_1 - x_0)(x_1 - x_2)\ldots(x_1 - x_n)} + \cdots$$

$$+\frac{f(x_i)}{(x_i-x_0)\dots(x_i-x_{n-1})(x_i-x_n)}+\cdots$$

$$+\frac{f(x_n)}{(x_n-x_0)(x_n-x_1)\dots(x_n-x_i)\dots(x_n-x_{n-1})} \qquad \text{...(2·6)}$$

Here the coefficient of $f(x_i)$ is

$\alpha_i^{(n)}=\dfrac{1}{(x_i-x_0)(x_i-x_1)\dots(x_i-x_n)}$. $i=0, 1, 2, ..., n$ where the zero factor x_i-x_i is to be omitted in the denominator.

We assume that the result holds good for index n.

Now $f(x_0, x_1, x_2, ..., x_{n+1})$

$$=\frac{1}{x_{n+1}-x_0}[(f(x_1, x_2, ..., x_{n+1})-f(x_0, x_1, ..., x_n)]$$

$\therefore$ Coefficient of $f(x_i)$

$$=\frac{1}{x_{n+1}-x_0}\left[\frac{1}{(x_i-x_1)\dots(x_i-x_{n+1})}-\frac{1}{(x_i-x_0)\dots(x_i-x_n)}\right]$$

$$=\frac{1}{(x_i-x_0)(x_i-x_1)\dots(x_i-x_{n+1})}=\alpha_i^{(n+1)}$$

Therefore, $\alpha_i^{(n+1)}$ is of the same form as $\alpha_i^{(n)}$ but has $(n+1)$ factors.

Thus if (2·6) is valid for n; it is valid for $n+1$. Since it has been established for $n=1, n=2$, it is therefore valid for any positive integer n, as was to be shown.

Note. Thus we see that

(*i*) a divided difference is a symmetrical function of all the arguments involved and

(*ii*) for any function $f(x)$ the value of a divided difference remains unaltered when any of the arguments involved are interchanged *i.e.*, value of a divided difference depends only on the value of arguments involved and not on the order in which they are taken. Thus

$$f(x_0, x_1)=f(x_1, x_0)$$

$$f(x_0, x_1, x_2)=f(x_1, x_0, x_2)=f(x_2, x_1, x_0) \text{ and so on.}$$

Theorem 2. *The nth divided differences of a polynomial of the nth degree are constant.*

Proof. Consider first that function $f(x)=x^n$, we shall show later that it will not matter if we omit the other terms of degree less than or equal to $(n-1)$.

The first divided differences are given by

$$f(x_r, x_{r+1})=\frac{f(x_{r+1})-f(x_r)}{x_{r+1}-x_r}=\frac{x_{r+1}^n-x_r^n}{x_{r+1}-x_r}$$

$$= x_{r+1}^{n-1} + x_r x_{r+1}^{n-2} + \cdots + x_r^{n-2} x_{r+1} + x_r^{n-1},$$

a homogeneous polynomial of degree $(n-1)$ in x_r, x_{r+1}.

The second divided differences are given by

$$f(x_r, x_{r+1}, x_{r+2}) = \frac{f(x_r, x_{r+1}) - f(x_{r+1}, x_{r+2})}{x_r - x_{r+2}}$$

$$= \frac{f(x_{r+1}, x_{r+2}) - f(x_r, x_{r+1})}{x_{r+2} - x_r}$$

$$= [(x_{r+2}^{n-1} + x_{r+1} x_{r+2}^{n-2} + \cdots + x_{r+1}^{n-2} x_{r+2} + x_{r+1}^{n-1})$$
$$- (x_{r+1}^{n-1} + x_r x_{r+1}^{n-2} + \cdots + x_r^{n-2} x_{r+1} + x_r^{n-1})]/(x_{r+2} - x_r)$$

$$= \frac{x_{r+2}^{n-1} - x_r^{n-1}}{x_{r+2} - x_r} + x_{r+1} \frac{(x_{r+2}^{n-2} - x_r^{n-2})}{x_{r+2} - x_r} + \cdots$$
$$+ \frac{x_{r+1}^{n-1}(x_{r+2} - x_r)}{x_{r+2} - x_r}$$

$$= (x_{r+2}^{n-2} + \cdots + x_r^{n-2}) + x_{r+1}(x_{r+2}^{n-3} + \cdots + x_r^{n-3})$$
$$+ \cdots + x_{r+1}^{n-2},$$

a homogeneous polynomial of degree $n-2$ in x_r, x_{r+1} and x_{r+2}.

By induction it can be shown that

$f(x_r, x_{r+1} \ldots, x_{r+m})$ is a polynomial of degrees $n-m$.

In particular, the nth divided difference of $f(x) = x^n$ is a polynomial of degree zero, that is a constant, and is, therefore, independent of the values of $x_r, x_{r+1}, \ldots, x_{r+n}$.

For a polynomial of the nth degree with leading term $a_0 x^n$, the nth divided difference of all terms but the leading terms are zero, so that the nth divided differences of this polynomial are constant and of value a_0.

If $f(x) = a_0 x^{n-1} + \cdots + a_{n-1} x + a_n$ be a polynomial of degree n, then the nth divided difference of all the terms except as x^n will be zero.

Note. In the above proof, we have assumed the following two results which can be easily proved :—

1. The divided difference of the algebraic sum of any number of functions is the algebraic sum of their separate divided differences

i.e., $\Delta' [f(x) \pm g(x) \pm \ldots] = \Delta' f(x) \pm \Delta' g(x) \pm \ldots$

2. The divided difference of the product of a constant and a function is the product of the constant and the divided difference of the functions, *i.e.*

$$\Delta' [af(x)] = a\, \Delta' f(x).$$

Theorem 3. *The nth divided difference can be expressed as the quotient of two determinants, each of order n + 1.*

Proof. We shall prove the result for third difference and then generalise the result. Consider the third divided difference

$$f(x_0, x_1, x_2, x_3) = \Sigma \frac{f(x_0)}{(x_0 - x_1)(x_0 - x_2)(x_0 - x_3)}$$

$$= \Sigma \left[\frac{f(x_0).\ (\text{difference-product of } x_1, x_2, x_3)}{(\text{difference-product of } x_0, x_1, x_2, x_3)} \right]$$

By an important theorem of determinants due to Vandermonde, we have

$$\text{difference product of } x_1, x_2, x_3 = \begin{vmatrix} x_1^2 & x_2^2 & x_3^2 \\ x_1 & x_2 & x_3 \\ 1 & 1 & 1 \end{vmatrix}$$

$$f(x_0, x_1, x_2, x_3) = \Sigma \left[f(x_0) \begin{vmatrix} x_1^2 & x_2^2 & x_3^2 \\ x_1 & x_2 & x_3 \\ 1 & 1 & 1 \end{vmatrix} \div \begin{vmatrix} x_0^3 & x_1^3 & x_2^3 & x_3^3 \\ x_0^2 & x_1^2 & x_2^2 & x_3^2 \\ x_0 & x_1 & x_2 & x_3 \\ 1 & 1 & 1 & 1 \end{vmatrix} \right]$$

$$= \begin{vmatrix} f(x_0) & f(x_1) & f(x_2) & f(x_3) \\ x_0^2 & x_1^2 & x_2^2 & x_3^2 \\ x_0 & x_1 & x_2 & x_3 \\ 1 & 1 & 1 & 1 \end{vmatrix} \div \begin{vmatrix} x_0^3 & x_1^3 & x_2^3 & x_3^3 \\ x_0^2 & x_1^2 & x_2^2 & x_3^2 \\ x_0 & x_1 & x_2 & x_3 \\ 1 & 1 & 1 & 1 \end{vmatrix}$$

Proceeding in this way, we can express the higher order differences as the quotient of two determinants each of order one more than the order of the difference.

Theorem 4. *The $(n-1)$th divided difference can be expressed as the product of multiple integrals, i.e.,*

$$f(x_1, x_2, \ldots, x_n) = \int_0^1 dt_1 \int_0^{t_1} dt_2 \int_0^{t_2} dt_3 \ldots \int_0^{t_{n-2}} f^{n-1}(u_n)\, dt_{n-1}$$

where $u_n = (1 - t_1)x_1 + (t_1 - t_2)x_2 + \cdots + (t_{n-2} - t_{n-1})x_{n-1} + t_{n-1}x_n$, $t_0 = 1$, $t_1, t_2 \ldots, t_n$ are independent variables, and f^{n-1} means the $(n-1)$th derivative of $f \equiv f(x_1, x_2, \ldots, x_n)$ with respect to t's.

Proof. We first prove the result for $n = 2$ and $n = 3$.

When $n = 2$,

$$\text{R.H.S.} = \int_0^1 f'(u_2)\, dt_1 = \int_2^1 f'\{(1 - t_1)x_1 + t_1 x_2\}\, dt_1 = \left[\frac{f\{x_1 + t_1(x_2 - x_1)\}}{x_2 - x_1} \right]_0^1$$

$$= \frac{f(x_2) - f(x_1)}{x_2 - x_1} = f(x_1, x_2).$$

When $\quad n = 3,$

$$\text{R.H.S.} = \int_0^1 dt_1 \int_0^{t_1} f''(u_3)\, dt_2$$

$$= \int_0^1 dt_1 \int_0^{t_1} f''[(1-t_1)x_1 + (t_1 - t_2)x_2 + t_2 x_3]\, dt_2$$

$$= \int_0^1 dt_1 \Big[f'(1-t_1)x_1 + (t_1 - t_2)x_2 + t_2 \,.\, x_3 \Big]_0^{t_1} / (x_3 - x_2)$$

$$= \frac{1}{x_3 - x_2} \left[\int_0^1 f'\{(1-t_1)x_1 + t_1 x_3\}\, dt_1 - \int_0^1 f'\{(1-t_1)\,.\,x_1 + t_1 x_2\}\, dt_1 \right]$$

$$= \frac{1}{x_3 - x_2} \cdot \frac{1}{x_3 - x_1} [f(x_3) - f(x_1)] - \frac{1}{(x_3 - x_2)(x_2 - x_1)} [f(x_2) - f(x_1)]$$

$$= \frac{f(x_1)}{(x_1 - x_2)(x_1 - x_3)} + \frac{f(x_2)}{(x_2 - x_1)(x_2 - x_3)} + \frac{f(x_3)}{(x_3 - x_1)(x_3 - x_2)}$$

$$= f(x_1, x_2, x_3)$$

Suppose that the result is valid for n arguments. We shall show that it is valid for $n+1$ arguments.

We have

$$= \int_0^{t_{n-1}} f^n \{(1-t_1)x_1 + \cdots + (t_{n-1} - t_n)x_n + t_n x_{n+1}\}\, dt_n$$

$$= \frac{1}{x_{n+1} - x_n} \Big[f^{n-1} \{(1-t_1)x_1 + \cdots + (t_{n-1} - t_n)x_n + t_n x_{n+1}\} \Big]_0^{t_{n-1}}$$

$$= \Big[f^{n-1} [(1-t_1)x_1 + \cdots + (t_{n-2} - t_{n-1})x_{n-1} + t_{n-1} x_n] - f^{n-1} [(1-t_1)x_1 + \cdots + (t_{n-2} - t_{n-1})x_{n-1} + t_{n-1} x_n] \Big] \div (x_n - x_{n+1})$$

Hence

$$\int_0^1 dt_1 \int_0^{t_1} dt_2 \ldots \int_0^{t_{n-1}} f^n(u_{n+1})\, dt_n$$

$$= \frac{f(x_1, x_2, ..., x_n) - f(x_2, ..., x_{n-1}, x_n, x_{n+1})}{x_1 - x_{n+1}}$$

$$= \frac{f(x_n, x_2, ..., x_1) - f(x_n, x_2, \ldots x_{n-1}, x_{n+1})}{x_1 - x_{n+1}}$$

on interchanging x_1 and x_n in numerator only.

$$= \frac{f(x_1, x_2, ..., x_n) - f(x_2, x_3 \ldots x_n, x_{n+1})}{x_1 - x_{n+1}}$$

$$= f(x_1, x_2, ..., x_{n+1})$$

so that the result holds good for $n + 1$ arguments. The result holds good for $n = 2$ and $n = 3$ arguments and hence it holds for n arguments.

2·3. Newton's formula for Unequal Intervals

Let $f(x_0), f(x_1), ..., f(x_n)$ be the values of $f(x)$ corresponding to the arguments $x_0, x_1, ..., x_n$, not necessarily equally spaced. From the definition of divided differences given in §2·1.

$$f(x, x_0) = \frac{f(x) - f(x_0)}{x - x_0}$$

$$\Rightarrow \qquad f(x) = f(x_0) + (x - x_0) f(x, x_0) \qquad \ldots(2·7)$$

Also $$f(x, x_0, x_1) = \frac{f(x, x_0) - f(x_0, x_1)}{x - x_1}$$

$$\Rightarrow \qquad f(x, x_0) = f(x_0, x_1) + (x - x_1) f(x, x_0, x_1) \qquad \ldots(2·7')$$

Similarly, $f(x, x_0, x_1) = f(x_0, x_1, x_2) + (x - x_2) f(x, x_0, x_1, x_2)$ ...(2·7″)

....

$$f(x, x_0, x_1, ..., x_{n-1}) = f(x_0, x_1, ..., x_n) + (x - x_n) f(x, x_0, x_1, ..., x_n) \qquad \ldots(2·7''')$$

Multiplying equation (2·7′) by $x - x_0$, equation (2.7″) by $(x - x_0)$. $(x - x_1)$, ..., equation (2.7″) by $(x - x_0)(x - x_1) \ldots (x - x_{n-1})$ and adding to equation (2·7) we have

$$f(x) = f(x_0) + (x - x_0) f(x_0, x_1) + (x - x_0)(x - x_1) f(x_0, x_1, x_2) + \cdots + (x - x_0)(x - x_1) \ldots (x - x_{n-1}) f(x_0, x_1, ..., x_n) + R_n,$$

where the remainder R_n is given by

$$R_n = (x - x_0)(x - x_1) \ldots (x - x_n) f(x, x_0, x_1, ..., x_n)$$

Assuming that $f(x)$ is a polynomial of degree n, so that

$f(x, x_0, x_1, ..., x_n)$ vanishes.

$$\therefore\ f(x) = f(x_0) + (x - x_0) f(x_0, x_1) + (x - x_0)(x - x_0)(x - x_1) f(x_0, x_1, x_2) + \cdots$$
$$+ (x - x_0)(x - x_1) \ldots (x - x_{n-1}) f(x_0, x_1, \ldots, x_n) \qquad \text{...(III)}$$

This formula is called *Newton's divided difference interpolation formula*

Note. Putting $x - x_i = X_i$, $i = 0, 1, \ldots, n-1$; III reduces to

$$f(x) = f(x_0) + X_0 \underset{x_1}{\Delta} f(x_0) + X_0 X_1^2 \underset{x_1 x_2}{\Delta} f(x_0) + \cdots + X_0 X_1 \ldots X_{n-1}^n \underset{x_1 \ldots x_n}{\Delta} f(x_0)$$

We note that in each term of R.H.S., the number of small letters is one more than the number of capital letters.

Newton-Gregory Formula as a particular case of Newton's Formula III. Let the arguments $x_0, x_1, x_2, \ldots, x_n$ be equally spaced, that is

$$h = x_1 - x_0 = x_2 - x_1 = \ldots = x_n - x_{n-1}$$

and let $x = x_0 + nh$. Then

$$f(x_0, x_1) = \frac{f(x_1) - f(x_0)}{x_1 - x_0} = \frac{1}{h} . \Delta f(x_0)$$

$$f(x_0, x_1, x_2) = \frac{f(x_0, x_1) - f(x_1, x_2)}{x_0 - x_2}$$

$$= \frac{\frac{1}{h} \Delta f(x_0) - \frac{1}{h} \Delta f(x_1)}{-2h}$$

$$= \frac{1}{2!\, h^2} . \Delta^2 f(x_0)$$

Similarly

$$f(x_0, x_1, \ldots x_n) = \frac{1}{(n!)\, h^n} \Delta^n f(x_0)$$

Substituting the values of the divided differences in Newton's formula (III), we get

$$f(x_0 + nh) = f(x_0) + \frac{nh}{1!\, h} \Delta f(x_0) + \frac{nh\,(nh - h)}{2!\, h^2} \Delta^2 f(x_0) + \cdots$$
$$+ \frac{nh\,(nh - h)(nh - 2h) \ldots (nh - \overline{n-1}h)}{n!\, h^n} \Delta^n f(x_0)$$
$$= f(x_0) + nc_1 \Delta f(x_0) + {}^n c_2 \Delta^2 f(x_0) + \cdots + \Delta^n f(x_0), \qquad \text{...(2·8)}$$

which is *Newton's formula 1.*

Example 1. *(a) Find the third divided difference with arguments 2, 4, 9, 10 of the function* $f(x) = x^3 - 2x$.

(b) Find a polynomial satisfied by (– 4, 1245), (– 1, 33), (0, 5), (2, 9) and (5, 1335).

Sol. (*a*)

x	$f(x)$	$\Delta' f(x)$	$\Delta'^2 f(x)$	$\Delta'^3 f(x)$
2	4			
		$\frac{56-4}{4-2}=26$		
4	56		$\frac{131-26}{9-2}=15$	
		$\frac{711-56}{9-4}=131$		$\frac{23-15}{10-2}=1$
9	711		$\frac{269-131}{10-4}=23$	
		$\frac{980-711}{10-9}=269$		
10	980			

Hence the third divided difference is 1.

(*b*) The divided difference table is :

x	$f(x)$	$\Delta' f(x)$	$\Delta'^2 f(x)$	$\Delta'^3 f(x)$	$\Delta'^4 f(x)$
− 4	1245				
		− 404			
− 1	33		94		
		− 28		− 14	
0	5		10		3
		2		13	
2	9		88		
		442			
5	1335				

$$\therefore \quad f(x) = 1245 - 404(x+4) + 94(x+4)(x+1)$$
$$- 14(x+4)(x+1)x + 3(x+4)(x+1)x(x-2),$$
$$= 3x^4 - 5x^2 + 6x^2 - 14x + 5.$$

Example 2. *By means of Newton's divided difference formula, find the values of f (2), f (8) and f (15) from the following table :*

x :	*4*	*5*	*7*	*10*	*11*	*13*
$f(x)$:	*48*	*100*	*294*	*900*	*1210*	*2028*

Sol. The divided difference table is :

x	$f(x)$	$\Delta' f(x)$	$\Delta'^2 f(x)$	$\Delta'^3 f(x)$	$\Delta'^4 f(x)$
4	48				
		$\frac{100-48}{5-4}=52$			
5	100		$\frac{97-52}{7-4}=15$		
		$\frac{294-100}{7-5}=97$		$\frac{21-15}{10-4}=1$	
7	294		$\frac{202-97}{10-5}=21$		0
		$\frac{900-294}{10-7}=202$		$\frac{27-21}{11-5}=1$	
10	900		$\frac{310-202}{11-7}=27$		0
		$\frac{1210-900}{11-10}=310$		$\frac{33-27}{13-7}=1$	
11	1210		$\frac{409-310}{13-10}=33$		
		$\frac{2028-1210}{13-11}=409$			
13	2028				

Using the result (III), we have

$$f(x) = 48 + 52\,(x-4) + 15\,(x-4)\,(x-5) + 1\,(x-4)\,(x-5)\,(x-7)$$
$$= x^2\,(x-1)$$
$$f(8) = 48 + (8-4) \times 52 + (8-4)\,(8-5) \times 15 + (8-4)\,(8-5)\,(8-7) \times 1$$
$$= 8^2\,(8-1) = 448.$$
$$f(15) = 48 + (15-4) \times 52 + (15-4)\,(15-5) \times 15 + (15-4)\,(15-5)\,(15-7) \times 1$$
$$= 3150$$
$$f(2) = 2^2\,(2-1) = 4.$$

Example 3. *Using the following table, find f (x) as a polynomial in powers of (x − 6) :*

x	-1	0	2	3	7	10
$f(x)$	-11	1	1	1	141	561

Also find f′ (6), f″ (6) and f‴ (6).

Sol. The divided difference table is as follows :

x	$f(x)$	$\Delta' f(x)$	$\Delta'^2 f(x)$	$\Delta'^3 f(x)$
-1	-11			
		$\frac{1+11}{0+1} = 12$		
0	1		$-\frac{0-12}{2+1} = -4$	
		$\frac{1-1}{2-0} = 0$		$\frac{0+4}{3+1} = 1$
			$0 = 0$	
2	1			$\frac{7-0}{7-0} = 1$
		$\frac{1-1}{3-2} = 0$	$\frac{35-0}{7-2} = 7$	
				$\frac{15-7}{10-2} = 1$
3	1	$\frac{141-1}{7-3} = 35$	$\frac{140-35}{10-3} = 15$	
7	141	$\frac{561-141}{10-7} = 140$		
10	561	....	....	....
6	c	b	a	1
6	c	b'	a'	1
6	c	b''	a''	1

Since the third difference are constant, $f(x)$ is a polynomial of third degree and hence the argument 6 should appear three times successively in the difference table. The extended table occurs below the dotted line shown in the above table.

Now from column V, $\dfrac{a-15}{6-3} = 1 \Rightarrow a = 18$

Also $\qquad \dfrac{a'-a}{6-7} = 1$ and $\dfrac{a''-a'}{6-10} = 1$

Thus $\qquad a' = 17, \qquad a'' = 13.$

Again, $\dfrac{b-140}{6-7}=a, \quad \dfrac{b'-b}{6-10}=a', \quad \dfrac{b''-b'}{6-6}=a''$

$\Rightarrow \qquad b=122, \quad b'=54, \quad b''=b'=54.$

Finally, $\dfrac{c-561}{6-10}=b=122$

$\therefore \qquad c=73.$

Newton's formula with arguments x_0, x_1, x_2, x_3 is

$$f(x)=f(x_0)+(x-x_0)f(x_0,x_1)+(x-x_0)(x-x_1)f(x_0,x_1,x_2)$$
$$+(x-x_0)(x-x_1)(x-x_2)f(x_0,x_1,x_2,x_3)$$

Putting $x_0=x_1=x_2=x_3=6$, we get

$$f(x)=f(6)+(x-6)\,b'+(x-6)^2\,a''+(x-6)^3\times 1$$
$$=73+54\,(x-6)+13\,(x-6)^2+(x-6)^3.$$

From (2·5′), page 49 we have

$$f'(6)=f(6,6)=b'=54,$$
$$f''(6)=(2\,!)\,f(6,6,6)=2\times a''=2\times 13=26,$$

and $\qquad f'''(6)=(3\,!)\,f(6,6,6,6)=6\times 1=6.$

Example 4. *Show that the nth divided difference of x^n is 1.*

Sol. We have

$$\frac{x^{n+1}}{(x-x_0)(x-x_1)\ldots(x-x_n)}=1+\sum{}'\frac{x_0^{n+1}}{(x_0-x_1)\ldots(x_0-x_n)}\cdot\frac{1}{x-x_0}$$

dash indicating that the term x_i-x_i is to be excluded from denominator.

Put $\quad x=0$ on both sides, we have

$\therefore \qquad 0=1-f(x_0,x_1,\ldots,x_n)$

$\Rightarrow \qquad f(x_0,x_1,\ldots,x_n)=1.$

2·4·1. Lagrange's Interpolation Formula for Unequal Intervals.

Let $f(x_0), f(x_1), f(x_2), \ldots, f(x_n)$ be the values of the function $y=f(x)$ corresponding to the $(n+1)$ arguments $x_0, x_1, x_2, \ldots, x_n$, **not necessarily equally spaced.**

Suppose that $P_n(x)$ is a polynomial in x of degree n. Then

$$P_n(x)=A_0(x-x_1)(x-x_2)\ldots(x-x_n)$$
$$+A_1(x-x_0)(x-x_2)\ldots(x-x_n)+\cdots$$
$$+A_n(x-x_0)(x-x_1)\ldots(x-x_{n-1}) \qquad \text{...(2·9)}$$

where A's are constants. We determine the $(n+1)$ constants $A_0, A_1, \ldots, A_n$ so as to make

$$P_n(x_0)=f(x_0),\ P_n(x_1)=f(x_1),\ldots,P_n(x_n)=f(x_n).$$

To determine A_0, put $x=x_0$ and $P_n(x_0)=f(x_0)$, so that

$$f(x_0) = A_0 (x_0 - x_1)(x_0 - x_2)\dots(x_0 - x_n).$$

$$\therefore \quad A_0 = \frac{f(x_0)}{(x_0 - x_1)(x_0 - x_2)\dots(x_0 - x_n)}$$

Similarly

$$A_1 = \frac{f(x_1)}{(x_1 - x_0)(x_1 - x_2)\dots(x_1 - x_n)}.$$

$$\dots\dots\dots\dots\dots\dots\dots\dots\dots$$

$$A_n = \frac{f(x_n)}{(x_n - x_0)(x_n - x_1)\dots(x_n - x_{n-1})}.$$

Substituting the values of A's in (2·9), we get

$$P_n(x) = \frac{(x - x_1)(x - x_2)\dots(x - x_n)}{(x_0 - x_1)(x_0 - x_2)\dots(x_0 - x_n)} f(x_0)$$

$$+ \frac{(x - x_0)(x - x_2)\dots(x - x_n)}{(x_1 - x_0)(x_1 - x_2)\dots(x_1 - x_n)} f(x_1)$$

$$+ \frac{(x - x_0)(x - x_1)(x - x_3)\dots(x - x_n)}{(x_2 - x_0)(x_2 - x_1)(x_2 - x_3)\dots(x_2 - x_n)} f(x_2)$$

$$+ \dots + \frac{(x - x_0)(x - x_1)\dots(x - x_{n-1})}{(x_n - x_0)(x_n - x_1)\dots(x_n - x_{n-1})} f(x_n) \qquad \dots(2\cdot10)$$

which is Lagrange's interpolation formula.

Note. (2·10) can also be written as

$$P_n(x) = \sum_{r=0}^{n'} \frac{(x - x_0)(x - x_1)\dots(x - x_r)}{(x_r - x_0)(x_r - x_1)\dots(x_r - x_n)} f(x_r)$$

$$= \sum_{r=0}^{n'} \left[\frac{\pi_n(x) f(x_r)}{x - x_r}\right]\left[\frac{1}{(x_r - x_0)(x_r - x_1)\dots(x_r - x_r)}\right]$$

where $\pi_n(x) = \prod_{r=0}^{n} (x - x_r)$ and, therefore,

$$\pi_n'(x_r) = \left[-\frac{d}{dx}\pi_n(x)\right]_{x = x_r}$$

$$= (x_r - x_0)(x_r - x_1)\dots(x_r - x_n),$$

Hence

The notation $\sum^{n'}$ means that the factor $x_r - x_r$ is excluded from the product for $r = 1, 2\dots, n$.

$$P_n(x) = \sum_{r=0}^{n'} \frac{\pi_n(x) f(x_r)}{(x - x_r)\, \pi_n'(x_r)} \qquad ...(2\cdot11)$$

Aliter. Let $\pi(x) = (x - x_0)(x - x_1)...(x - x_n)$ so that

$\pi'(x_0) = (x_0 - x_1)(x_0 - x_2)...(x_0 - x_n)$, $\pi'(x_1) = (x_1 - x_0)(x_1 - x_2) \ldots (x_1 - x_n)$

Hence (2·10) can be written as

$$P(x) = \left[\frac{\dfrac{\pi(x)}{x - x_0}}{\pi'(x_0)} \right] f(x_0) + \cdots$$

Second Method. Suppose that $P_n(x)$ is a polynomial of degree n, then by Cor., Theorem 2·2 the $(n-1)$ the divided differences are zero so that if x be any other argument, we have

$$P_n(x_0, x_1, x_2, \ldots x_n, x) = 0$$

or
$$\frac{f(x_0)}{(x_0 - x_1) \ldots (x_0 - x_n)(x_0 - x)} + \frac{f(x_1)}{(x_1 - x_0)\ldots(x_1 - x_n)(x_1 - x)} + \cdots$$

$$+ \frac{f(x_n)}{(x_n - x_0) \ldots (x_n - x_{n-1})(x_n - x)}$$

$$+ \frac{f(x)}{(x - x_0)\ldots(x - x_{n-1})(x - x_n)} = 0,$$

on putting $P_n(x_0) = f(x_0)$, $P_n(x_1) = f(x_1)$, ..., $P_n(x_n) = f(x_n)$.

Multiplying both sides by $(x - x_0)(x - x_1)...(x - x_n)$, we get

$$f(x) = \frac{(x - x_1)(x - x_2)\ldots(x - x_n)}{(x_0 - x_1)(x_0 - x_2)\ldots(x_0 - x_n)} f(x_0)$$

$$+ \frac{(x - x_0)(x - x_2)\ldots(x - x_n)}{(x_1 - x_0)(x_1 - x_2)\ldots(x_1 - x_n)} f(x_1) + \cdots$$

$$+ \frac{(x - x_0)(x - x_1)\ldots(x - x_{n-1})}{(x_n - x_0)(x_n - x_1)\ldots(x_n - x_{n-1})} f(x_n)$$

which is Lagrange's interpolation formula.

Third Method. Let the polynomial passing through the $(n+1)$ points

$$[x_0, f(x_0)], [x_1, f(x_1)], \ldots, [x_n, f(x_n)]$$

be
$$P_n(x) = A_0 + A_1 x + A_2 x^2 + \cdots + A_n x^n.$$

Using the condition,
$$f(x_0) = A_0 + A_1 x_0 + A_2 x_0^2 + \cdots + A_n x_0^n,$$

$$f(x_1) = A_0 + A_1 x_1 + A_2 x_1^2 + \cdots + A_n x_1^n$$

$$\cdots\cdots\cdots\cdots\cdots\cdots\cdots\cdots\cdots\cdots$$

$$f(x_n) = A_0 + A_1 x_n + A_2 x_n^2 + \cdots + A_n x_n^n$$

Eliminating $A_0, A_1, ..., A_n$, we have

$$\begin{vmatrix} P_n(x) & 1 & x & x^2 & ....x^n \\ f(x_0) & 1 & x_0 & x_0^2 & ....x_0^n \\ f(x_1) & 1 & x_1 & x_1^2 & ....x_1^n \\ ... & ... & ... & ... & ..... \\ f(x_n) & 1 & x_n & x_n^2 & ....x_n^n \end{vmatrix} = 0$$

Expanding in terms of the elements of the first column and using the value of Vandermonde's determinant, we get the required result.

Cor. For $n = 1, P(x) = \dfrac{x - x_1}{x_0 - x_1} f(x_0) + \dfrac{x - x_0}{x_1 - x_0} f(x_1) = a_0 f(x_0) + a_1 f(x_1)$

where $$a_0 = \frac{x - x_1}{x_0 - x_1}, \; a_1 = \frac{x - x_0}{x_1 - x_0}$$

a's are called Lagrange's fundamental polynomials, each of degree 1. They satisfy the condition $a_0 = a_1 = 1$,

2·4·2. Lagrange's interpolation formula for equal intervals

In case the argumnets are equally spaced, then letting

$$x_s = x_0 + sh, x = x_0 + hu, s = 1, 2, ..., n$$

we have

$$P_n(x) = (-1)^n \frac{(u-1)(u-2)...(u-n)}{n!} f(x_0) + \cdots$$

Notes: 1. Lagrange's formula is merely a relation between two variables either of which may be taken as the independent variable. Regarding y as the independent variable, we get the following formula, on interchanging x and y in (2·9) giving x as a function of y :

$$\psi(y) = \frac{(y-y_1)(y-y_2)...(y-y_n)}{(y_0-y_1)(y_0-y_2)...(y_0-y_n)} x_0 + \frac{(y-y_0)(y-y_2)...(y-y_n)}{(y_1-y_0)(y_1-y_2)...(y_1-y_n)} x_1$$
$$+ \frac{(y-y_0)(y-y_1)...(y-y_n)}{(y_2-y_0)(y_2-y_1)...(y_2-y_n)} x_2 + \cdots$$
$$+ \frac{(y-y_0)(y-y_1)...(y-y_{n-1})}{(y_n-y_0)(y_n-y_1)...(y_n-y_{n-1})} x_n \qquad ...(2·12)$$

2. It can be proved that the sum of the Lagrangian coefficients is unity. See Q. 8 (*b*), page 70.

Example 6. *Given* $\log_{10} 654 = 2.8156$, $\log_{10} 658 = 2.8182$, $\log_{10} 659 = 2.8189$, $\log_{10} 661 = 2.8202$; *find* $\log_{10} 656$. *(B.Sc. Guwahati 1996)*

Sol. We have

x :	654	658	659	661
$\log_{10} x$:	2·8156	2·8182	2·8189	2·8202

Let $y = f(x) = \log_{10} x$.

Here $x_0 = 654, x_1 = 658, x_2 = 659, x_3 = 661, x = 656$.

Substituting these values in Lagrange's formula, we get

$$\log_{10} 656 = \frac{(656-658)(656-659)(656-661)}{(654-658)(654-659)(654-661)} \times 2.8156$$
$$+ \frac{(656-654)(656-659)(656-661)}{(658-654)(658-659)(658-661)} \times 2.8182$$
$$+ \frac{(656-654)(656-658)(656-661)}{(659-654)(659-658)(659-661)} \times 2.8189$$
$$+ \frac{(656-654)(656-658)(656-659)}{(661-654)(661-658)(661-659)} \times 2.8202$$

$$\log_{10} 656 = \frac{(-2)(-3)(-5)}{(-4)(-5)(-7)} \times 2.8156 + \frac{(2)(-3)(-5)}{(4)(-1)(-3)} \times 2.8182$$
$$+ \frac{(2)(-2)(-5)}{(5)(1)(-2)} \times 2.8189 + \frac{(2)(-2)(-3)}{(7)(3)(2)} \times 2.8202$$
$$= 0.6033 + 7.0455 - 5.6378 + 0.8059 = 2.8169$$

Thus the estimated value of $\log_{10} 656 = 2.8169$.

Example 7. *The mode of a certain frequency curve $y = f(x)$ is very near $x = 9$ and the value of the frequency density $f(x)$ for $x = 8.9$, 9.0 and 9.3 are respectively equal to 0.30, 0.35 and 0.25. Calculate the approximate value of the mode.*

Sol. We have

x	8·9	9·0	9·3
$f(x)$	0·30	0·35	0·25

We shall first find the form of the frequency density $f(x)$. By Lagrange's formula, we get

$$f(x) = \frac{(x-9)(x-9.3)}{(8.9-9)(8.9-9.3)} \times 0.30 + \frac{(x-8.9)(x-9.3)}{(9-8.9)(9-9.3)} \times 0.35$$
$$+ \frac{(x-8.9)(x-9)}{(9.3-8.9)(9.3-9)} \times 0.25$$

or
$$f(x) = \frac{-25}{12} x^2 + \frac{453.5}{12} x - \frac{2052.3}{12}$$

This is the form of the frequency density. It will be maximum for the value of x lying in the range (8·9, 9·3) for which $f'(x) = 0$ and $f''(x)$ is negative.

Now $f'(x) = 0$ gives

$$\frac{-5}{12}(10x - 90.7) = 0 \quad \text{or} \quad x = 9.07$$

and $$f''(x) = \frac{-25}{6} \text{ which is a negative quantity.}$$

Hence the mode of the frequency curve $y = f(x)$ is at $x = 9{\cdot}07$.

Example 8. *Using Lagrange's interpolation formula, prove that* $y_1 = y_3 - 0{\cdot}3\,(y_5 - y_{-3}) + 0{\cdot}2\,(y_{-3} - y_{-5})$ *approximately.*

Sol. If we shift the origin to 5, then we are given y_0, y_2, y_3 and y_{10}. We are to find y_6.

Applying Lagrange's formula, we have

$$y_6 = \frac{(6-2)(6-8)(6-10)}{(0-2)(0-8)(0-10)} y_0 + \frac{(6-0)(6-8)(6-10)}{(2-0)(2-8)(2-10)} y_2 + \frac{(6-0)(6-2)(6-10)}{(8-0)(8-2)(8-10)} y_8 + \frac{(6-0)(6-2)(6-8)}{(10-0)(10-2)(10-8)} y_{10}$$

or $$y_6 = -0{\cdot}2\,y_0 + 0{\cdot}5\,y_2 + y_8 - 0{\cdot}3\,y_{10}$$

Shifting back the origin, we get

$$y_1 = y_3 - 0{\cdot}3\,(y_5 - y_{-3}) + 0{\cdot}2\,(y_{-3} - y_{-5}).$$

Example 9. *Four equidistant values u_{-1}, u_0, u_1 and u_2 being given, a value is interpolated by Lagrange's interpolation formula. Show that it may be written in the form*

$$u_x = yu_0 + xu_1 + \frac{y(y^2-1)}{3!}\Delta^2 u_{-1} + \frac{x(x^2-1)}{3!}\Delta^2 u_0,$$

where $x + y = 1$.

Sol. We have $\Delta^2 u_{-1} = \Delta u_0 - \Delta u_{-1} = (u_1 - u_0) - (u_0 - u_{-1})$.

$$= u_1 - 2u_0 + u_{-1}$$

Similarly, $$\Delta^2 u_0 = u_2 - 2u_1 + u_0.$$

$\therefore$ Right hand side

$$= (1-x)\,u_0 + xu_1 + \frac{(1-x)\,x\,(x-2)}{3}(u_1 - 2u_0 + u_{-1}) + \frac{(x-1)\,x\,(x+1)}{3!}(u_2 - 2u_1 + u_0)$$

$$= -\frac{x(x-1)(x-2)}{6} u_{-1} + \left[-(x-1) + \frac{x(x-1)(x-2)}{3} + \frac{(x-1)\,x\,(x+1)}{6}\right] u_0 + \left[x - \frac{x(x-1)(x-2)}{6} - \frac{(x-1)\,x\,(x+1)}{3}\right] u_1 + \frac{(x-1)\,x\,(x+1)}{6} u_2$$

$$= -\frac{x(x-1)(x-2)}{6}u_{-1} + \frac{(x-2)(x-1)(x+1)}{6}u_0$$

$$-\frac{(x-2)x(x+1)}{2}u_1 + \frac{(x-1)x(x+1)}{6}u_2$$

$$= (x+1)x(x-1)(x-2)\left[\frac{-u_{-1}}{6(x+1)} + \frac{u_0}{2x} - \frac{u_1}{2(x-1)} + \frac{u_2}{6(x-2)}\right]$$

$$= (x+1)x(x-1)(x-2)$$

$$\times\left[u_{-1}\frac{1}{(-1-0)(-1-1)(-1-2)(x+1)}\right.$$

$$+u_0\frac{1}{(0+1)(0-1)(0-2)x} + u_1 \cdot \frac{1}{(1+1)(1-0)(1-2)(x+1)}$$

$$\left.+u_2 \cdot \frac{1}{(2+1)(2-0)(2-1)(x-2)}\right]$$

$= u_x$ by Lagrange's formula.

Example 10. *Solve Example 14, page 26 by Lagranges's method.*

Sol. Let

$$S_{10} = \sum_{1}^{10} f(n) = 500426$$

$$S_7 = \sum_{4}^{10} f(n) = 329240$$

$$S_4 = \sum_{7}^{10} f(n) = 175212$$

$$S_1 = f(10) = 40365$$

Let us denote sum in terms of S. In other words we are given cumulative function S_n for $n = 1, 4, 7, 10$.

In order to obtain $u_9, u_8, ..., u_1$, we proceed as follows :

x	S_n
1	40365
4	175212
7	329240
10	500426

By Lagrange's formula :

$$S_2 = \frac{(9-4)(9-7)(9-10)}{(1-4)(1-7)(1-10)} \times 40365$$

$$+\frac{(9-1)(9-7)(9-10)}{(4-1)(4-7)(4-10)} \times 175212$$

$$+\frac{(9-1)(9-4)(9-10)}{(7-1)(7-4)(7-10)}\times 329240$$

$$+\frac{(9-1)(9-4)(9-7)}{(10-1)(10-4)(10-7)}\times 500426$$

or $$S_9=\frac{5\times 2\times 1}{3\times 6\times 9}\times 40365-\frac{8\times 2\times 1}{3\times 7\times 3}\times 525636$$

$$+\frac{8\times 5\times 1}{6\times 3\times 3}\times 329240+\frac{8\times 5\times 2}{9\times 3\times 3}\times 500426$$

$$=\frac{201825}{81}-\frac{4205088}{81}+\frac{19754400}{81}+\frac{20017040}{81}$$

$$=\frac{201825+19754400+20017040-4205088}{81}$$

$$=\frac{35768177}{81}=441582{\cdot}93$$

Now $U_1=S_{10}-S_9$

$$=500426{\cdot}00-441582{\cdot}93$$
$$=58843{\cdot}07$$

Example 11. *The following are the mean temperatures (Fahrenheit) on three days, 30 days apart, round the periods of summer. Estimate the approximate date and value of the maximum temperature.*

Summer

Day	*Date*	*Temperature*
0	15th June	58·8
63	15th July	63·4
60	14th Aug.	62·5

Sol. Difference Table for Summer (30 days = 1 unit)

x	Temp. u_x	Δu_x	$\Delta^2 u_x$
0	58·8	4·6	
1	63·4		
		−0·9	−2·75
2	63·5		

$$u_x=u_0+x\,\Delta u_0+(x-0)(x-1)\,\Delta^2 u_0$$
$$=58{\cdot}8+x(4{\cdot}6)+x(x-1)\cdot(-2{\cdot}75)$$
$$=-2{\cdot}75x^2+7{\cdot}35x+58{\cdot}8.$$

For maximum or minimum value of u_x, we have

$$0=u_x'=-5{\cdot}5x+7{\cdot}35 \text{ or } x=\frac{7{\cdot}35}{5{\cdot}5}=1{\cdot}342,\ (0<x<2)$$

and $$u_x''=-5{\cdot}5<0.$$

Hence u_x is maximum at $x=1{\cdot}342$

The maximum temperature $-2\cdot75\,(1\cdot342)^2+7\cdot35\,(1\cdot342)+58\cdot8$ = etc. (in Fahrenheit)

2.5. Hermite's Interpolation Formula

So far we have considered the interpolation formulae which make use only of a certain number of functional values. We now derive an interpolations formula in which both the function and its derivatives are to be assigned to each point of interpolation.

Theorem. *If f' is continuous on $[a, b]$ and $x_0, x_1, ..., x_n \in [a, b]$ are distinct, then the unique polynomial of least degree agreeing with f and f' at $x_0, x_1, ..., x_n$ is a polynomial of degree at most $2n+1$ given by*

$$P(x)=\sum_{i=0}^{n}[1-2(x-x_i)]\,l_i\,(x)^2 f(x_i)$$

$$+\sum_{i=0}^{n}(x-x_i)\,l_i\,(x)^2 f'(x_i),\ l_i\,(x)^2=[l_i\,(x)]^2$$

The Hermite interpolating polynomial interpolation are not only the function $f(x)$ but also its derivatives at a given set of tabular points *i.e.*,

$$p(x_i)=f(x_i),\quad p'(x_i)=f'(x_i),\quad i=0,1,...,n$$

Since there are $(2n+2)$ conditions to be satisfied, $p(x)$ must be a polynomial of degree $\le 2n+1$. The required polynomial may be written as

$$p(x)=\sum_{i=0}^{n}A_i\,(x)\,f(x_i)+\sum_{i=0}^{n}B_i\,(x)\,f'(x_i) \qquad ...(1)$$

where $A_i\,(x)$ and $B_i\,(x)$ are polynomials of degree $\le 2n+1$ and satisfy

$$\left.\begin{array}{ll}(i) & A_i\,(x_j)=\begin{cases}0, & i\neq j\\ 1, & i=j\end{cases}\\ (ii) & A_i'\,(x_j)=0 \text{ for all } i \text{ and } j\\ & B_i\,(x_j)=0 \text{ fo all } i \text{ and } j\\ & B_i\,(x_j)=0 \text{ for all } i \text{ and } j\\ (iii) & B_i'\,(x_j)=\begin{cases}0, & i\neq j\\ 1, & i=j\end{cases}\end{array}\right] \qquad ...(2)$$

using the Lagrange's theorem, $l_i\,(x)$ where

$$l_i\,(x)=\frac{(x-x_0)(x-x_1)\ldots(x-x_{i-1})(x-x_{i+1})\ldots(x-x_n)}{(x_i-x_0)(x_i-x_1)\ldots(x_i-x_{i-1})(x_i-x_{i+1})\ldots(x_i-x_n)}$$

and

$$\begin{cases}A_i\,(x)=r_i\,(x)\,l_i^2\,(x), & l_i\,(x)=\{l_i\,(x)\}^2\\ B_i\,(x)=\delta_i\,(x)\,l_i^2\,(x)\end{cases} \qquad ...(3)$$

Since $l_i^2(x)$ is a polynomial of degree $2n, \gamma_i(x)$ and $\delta_i(x)$ are both the linear polynomials.

Thus let $\quad \gamma_i(x) = a_i x + b_i \quad$ and $\quad \delta_i(x) = c_i x + d_i$ ...(4)

where a_i, b_i, c_i and d_i are constants to be determined.

Using the conditions (2) we obtain

$$a_i = -2l\, l_i(x_i), \quad b_i = l + 2x_i \cdot l_i(x_i)$$
$$c_i = 1, \quad d_i = -x_i \qquad \text{...(5)}$$

Substituting (5) into (4) we obtain $\gamma_i(x)$ Now using (3), the eq. (1) and $\delta_i(x) = x - x_i$. the eq. (1) now becomes

$$P(x) = \sum_{i=0}^{n}\left[1 - 2(x - x_i)\, l'_i(x)\, l_i^2(x_i) f(x_i)\right]$$

$$+ \sum_{i=0}^{n}[(x - x_i)\, l_i^2(x) f'(x_i)]$$

$$\sum h_i f(x_i) + \sum \bar{h}'_i f(x_i)$$

where $h_i(x) = 1 - 2\, x - x_i)\;\; l'(x_i)\;\; h_i(x) = (x - x_i)\, l_j^2(x)$

which is the required Hermite Interpolation form.

Example 12. *For the following tabular values of $f(x)$ and $f'(x)$, estimate the values of $f(x)$ at*

(i) $x = 0{\cdot}5$ and (ii) $x = -0{\cdot}5$,

Using Hermite interpolating polynomial :

x :	-1	0	1
$f(x)$:	1	1	3
$f'(x)$:	-5	1	7

(*B.Sc. Hons. Math, Delhi, 1995*)

Sol. By Lagrange's formula for arguments $-1, 0, 1$, the function $f(x)$ is given by

$$f(x) = \frac{(x-0)(x-1)}{(-1-0)(-1-1)} \cdot 1 + \frac{(x+1)(x-1)}{(0+1)(0-1)} \cdot 1 + \frac{(x+1)\,x}{(1+1)\cdot 1} \cdot 3$$

$$= \tfrac{1}{2}(x^2 - x) - (x^2 - 1) + \tfrac{3}{2}(x^2 + x)$$

$$l_0 = \tfrac{1}{2}(x^2 - x),\; l_1 = -(x^2 - 1),\; l_2 = \tfrac{3}{2}(x^2 + x),\; x_0 = -1,\; x_1 = 0,\; x_2 = 1$$

Here $n = 2$, $x_0 = -1$, $x_1 = 0$ and $x_2 = 1$

The Hermite interpolation is

$$P(x) = \sum_{i=0}^{2} A_i(x) f(x_i) + \sum_{i=0}^{2} B_i(x) f'(x_i) \text{...(1)}$$

where $A_0(x) = [1 - 2(x - x_0)\, l_0'(x_0)]\, l_0^2(x)$,

$$A_1(x) = [1 - 2(x - x_1)\, l_1'(x_1)]\, l_1^2(x)$$

$$A_2(x) = [1 - 2(x - x_2)\, l_2'(x_2)]\, l_2^2(x)]$$

$$B_0(x) = (x - x_0)\, l_0^2(x)$$

$$B_1(x) = (x - x_1)\, l_1^2(x)$$

$$B_2(x) = (x - x_2)\, l_2^2(x).$$

$$l_0(x) = \frac{(x-0)(x-1)}{(-1-0)(-1-1)} = \tfrac{1}{2}x(x-1),\ l_0'(-1) = \frac{-3}{2}$$

$$l_1(x) = \frac{(x+1)(x-0)}{(0+1)(0-1)} = -(x^2-1),\ l_1'(0) = 0$$

$$l_2(x) = \frac{(x+1)(x-0)}{(1+1)(1-0)} = \frac{x(x+1)}{2},\ l_2'(1) = \frac{-3}{2}$$

$$A_0(x) = [1 + 3(x+1)]\left[\frac{x^2(x-1)^2}{4}\right] = \frac{1}{4}[3x^5 - 2x^4 - 5x^3 + 4x^2]$$

$$A_1(x) = [1 - 2(x-0) \times (0)]\,(x^2-1)^2 = x^4 - 2x^2 + 1$$

$$A_2(x) = 1 - 3(x-1)\,\frac{x^2(x+1)^2}{4} = \frac{1}{4}(-3x^5 - 2x^4 + 5x^3 + 4x^2)$$

$$B_0(x) = \frac{(x-1)\,x^2(x+1)^2}{4} = \frac{1}{4}(x^5 - x^4 - x^3 + x^2)$$

$$B_1(x) = x(x^2-1)^2 = x^5 - 2x^3 + x$$

$$B_2(x) = \frac{1}{4}(x^5 + x^4 - x^3 - x^2)$$

Substituting these values in (1) we obtain

$$\begin{aligned} P(x) &= \frac{1}{4}(3x^5 - 2x^4 - 5x^3 + 4x^2)(1) + (x^4 - 2x^2 + 1)(1) \\ &\quad + \frac{1}{4}(-3x^5 - 2x^4 + 5x^3 + 4x^2)(3) \\ &\quad + \frac{1}{4}(x^5 - x^4 - x^3 + x^2)(-5) + (x^5 - 2x^3 + x)(1) \\ &\quad + \frac{1}{4}(x^5 + x^4 - x^3 - x^2)(7) \\ &= 2x^4 - x^2 + x + 1 \end{aligned}$$

Putting $x = -0{\cdot}5$ and $0{\cdot}5$, we get

$$f(-0{\cdot}5) \simeq \frac{3}{8},\quad \text{exact value} = \frac{33}{64}$$

$$f(0{\cdot}5) \simeq \frac{11}{8},\quad \text{exact value} = \frac{97}{64}.$$

Ex. *Construct the polynomial* $P_3(x)$ *for which* $P(0) = 1$, $P(1) = 2$, $P(-1) = 5$, $P''(0) = 0$.

EXERCISE 2

1. (*a*) Prove that the third divided differences with arguments *a, b, c, d* of the function $1/x$ is equal to $-1/abcd$. (*B.Sc. Ranchi 1990*)

In general, show that if $f(x)=\frac{1}{x}$, the divided difference

$$f(x_1, x_2, ..., x_{p+1})=\frac{(-1)^p}{x_1 x_2 \ldots x_p x_{p+1}}$$ (*B.A. Hons Maths. Delhi, 1997*)

(*b*) Find the first three divided differences of the function $1/x^2$ for the arguments *a, b, c, d.*

(*c*) Show that (*i*) $\underset{y,z}{\Delta^2} x^2 = 1$. (*ii*) $\underset{y,2}{\Delta^2} (ax+b)(cx+d) = a.c$

(*iii*) $\Delta^2_{x,z} x^3 = x+y+z.$ (*B.Sc. Ranchi 1990*)

In general, show that

$$f(x_0, x_1, ..., x_n) = x_0 + x_1 + \cdots + x_n.$$

(*d*) Show that the divided difference of x^n is unity, no matter which of the arguments $x_0, x_1, ..., x_n$ are used.

2. (*a*) Show that the divided differences of $f(x)+\phi(x)$ are the sums of corresponding divided differences of $f(x)$ and of $\phi(x)$.

(*b*) Show that the divided differences of $kf(x)$ where *k* is a constant are *k* times the corresponding divided differences of $f(x)$.

(*c*) If $f(x)=u(x)v(x)$, show that

$$f(x_0, x_1) = u(x_0)v(x_0, x_1) + u(x_0, x_1)v(x_1).$$

(*d*) Show that $f(c, x, x, ..., x) = f^r(c, x)/(r\,!)$, *x* occurs in the left. function $(r+1)$ times.

3. (*a*) Show that the divided difference $f(x_0, x_1, x_2, ..., x_n)$ of $f(x)$ is the symmetrical function of the arguments $x_0, x_1, ..., x_n$. Express it as the ratio of two determinants. (*M.A. Patna 1996*)

(*b*) Values of y_x are given for all integral values of *x* from 0 to $n-1$. Show that y_x is capable of expression in the form

$$\frac{x\,!}{(x-n)\,!\,(n-1)\,!}\left[\frac{y_{n-1}}{x-n+1} - {}^{n-1}C_1\frac{y_{n-2}}{x-n+2} + {}^{n-1}C_2\frac{y_{n-3}}{x-n+3} + \cdots \pm {}^{(n-1)}C_{n-1}\frac{y_0}{x}\right].$$

[**Hint.** Apply Lagrange's formula].

4. Find out the divided difference of y_x, given that

x :	1	2	3	4	5
y_x :	22	30	82	106	216

5. Given the following table, find log 656

x :	654	658	659	661
$\log x$:	2·8156	2·8182	2·8189	2·8202

6. Given $y_{35\cdot0}=1175$, $y_{35\cdot5}=1280$, $y_{39\cdot5}=2180$, $y_{40\cdot5}=2420$, obtain a value for y_{40}.

7. (*a*) Given the following data, find $f(x)$ as a polynomial in powers of $(x-5)$.

x :	0	2	3	4	7	9
$f(x)$:	4	26	58	112	466	922

(*b*) From the following table, find $f(x)$ in powers of $(x-3)$

x :	5	11	27	34	42
$f(x)$:	23	899	17315	35606	68510

(*c*) Assuming that the third divided differences of $f(x)$ is constant for all x, fill in the spaces in the following divided difference table, from right, and hence evaluate (*i*) $f'(8)$ and $f''(8)$. (*ii*) $f'(6)$ and $f''(10)$.

(*i*)

x	$f(x)$	1st	2nd	3rd
0	3	1	4	1
1	4	13	10	1
3	30	63	17	—
6	219	148	—	—
8	515	—	—	
8	515			

(*ii*)

x	$f(x)$	1st	2nd	3rd
0	− 1	1	14	2
2	1	71	26	2
5	—	—	—	—
6	—	—	—	—
6	—	—	—	
10	—	—		
10	—			

8. (*a*) Obtain Lagrange's interpolation formula in the form

$$f(x)=\sum_{r=1}^{n}\frac{L(x)f(x_r)}{(x-x_r)\,L'(x_r)}=\sum_{r=1}^{n}L_r(x)f(x_r) \quad \text{where}$$

$$L(x)=(x-x_1)\ldots(x-x_n)=\prod_{r=1}^{n}(x-x_r).$$

(*b*) Show that $\sum_{r=1}^{n}L_r(x)=1$,

In general, $\sum_{r=1}^{n}r^k L_r(x)=x^k,\ k<n$ and $\sum_{r=1}^{n}r\,L_r(x)=x$

(*c*) What relation does this give when

$f(x)=c$ (a constant)?

(*d*) Apply Lagrange's formula to find $f(5)$ and $f(6)$ given that $f(1)=2$, $f(2)=4$, $f(3)=8$, $f(4)=16$, $f(7)=128$ and explain why the results differ from those obtained by completing the series of powers of 2.

[Hint. (*b*) $\dfrac{1}{L(x)} = \dfrac{1}{(x-x_1)\ldots(x-x_n)} \equiv \dfrac{A_1}{x-x_1} + \dfrac{A_2}{x-x_2} + \cdots + \dfrac{A_n}{x-x_n}$

$$= \sum_{r=1}^{n} \frac{A_r}{x - x_r}$$

where $A_r = \left[\dfrac{x-x_r}{L(x)}\right]_{x=x_r} = \dfrac{1}{L'(x_r)} \quad \therefore \quad \dfrac{1}{L(x)} = \sum_{r=1}^{n} \dfrac{1}{(x-x_r)\,L'(x_r)}$

$$\Rightarrow \quad 1 = \sum_{r=1}^{n} \frac{L(x)}{(x-x_r)\,L'(x_r)} = \sum_{r=1}^{n} L_r(x).$$

Thus sum of Lagrangian coefficients is unity].

9. The observed values of function are respectively 168, 120, 72 and 63 at the four positions 3, 7, 9 and 10 of the independent variable. What is the best estimate you can give for the value of the function at the position 6 of the independent variable?

10. (*a*) The following table is given. What is the form of the function?

(*i*)

x :	0	1	2	5
$f(x)$:	2	3	12	147

(*AMIE 1991*)

(*ii*)

x :	0	1	4	5
$f(x)$:	8	11	78	123

(*iii*)

x :	– 2	– 1	1	2
$f(x)$:	– 5	– 2	0	7

(*B.Sc. Delhi 1992*)

(*iv*) Find the value of y for $x = 4$, given that

x :	0	1	2	5
y :	2	5	7	8

(*B.Sc. Agra, 1995*)

(*b*) The function $y = f(x)$ is given at the points (7, 3), (8, 1), (9, 1), and (10, 9), Find the value of y for $x = 9{\cdot}5$, using Lagrange's interpolation formula.

11. By means of Lagrange's interpolation formula, prove that approximately

(*a*) $y_0 = \frac{1}{2}(y_1 + y_{-1}) - \frac{1}{3}\left[\frac{1}{2}(y_3 - y_1) - \frac{1}{2}(y_{-1} - y_{-3}\right]$,

(*b*) $y_3 = {\cdot}05(y_0 + y_6) - {\cdot}3(y_1 + y_5) + {\cdot}75(y_2 + y_4)$,

(*c*) $f(1) = f(3) - 0{\cdot}3\,[f(-3)] + 0{\cdot}2\,[f(-2) - f(-5)]$.

12. (*a*) Establish Lagrange's interpolation formula and prove that sum of Langrangian Coefficients is always unity.

(*b*) Using Lagrange's interpolation formula, estimate u_4 given :

$$u_2 = -12;\ u_3 = -3,\ u_5 = 75,\ u_6 = 156.$$

polynomial of degree n, then $\Delta^{n+1} P(x) = 0$.

13. (*a*) Find the polynomial of the lowest possible degree which assumes the values 3, 12, 15, – 21 when x has the values 3, 2, 1, –1 respectively.

(*b*) Find the polynomial of the fifth degree from the following data :

$u_0 = -18, u_1 = 0, u_3 = 0, u_5 = -248, u_6 = 0, u_9 = 13104$.

14. If all the terms except y_5, of the sequence $y_1, y_2, y_3, ..., y_9$ be given, show that the value of y_5, is approximately

$[56 (y_4 + y_6) - 28 (y_3 + y_7) + 8 (y_2 + y_8) - (y_1 + y_9)] \div 70$

15. (*a*) Find approximately the real root of the equation

$$y^3 - 2y - 5 = 0.$$

[**Hint.** (*a*) Let $x = y^3 - 2y - 5$ so that we get $y = f(x)$. We want the value of $f(0)$. Give the values 1·9, 2·0, 2·1, 2·2 and 2·3 of y and form difference table. Then apply Newton's divided difference formula.]

(*b*) Apply Lagrange's formula to find a root of the equation $f(x) = 0$, given that $f(30) = -30$, $f(34) = -13$, $f(38) = 3$ and $f(42) = 18$.

16. (*a*) Given that $f(0) = 8$, $f(1) = 68$ and $f(5) = 123$, construct a table of divided differences. Use the table to determine the value of $f(2)$.

(*b*) The function $y = \int_x^\infty \frac{e^{-t}}{t}\, dt$

is given in the table below. Find y for $x = 0·0378$

x :	0·00	0·01	0·02	0·03	0·04	0·05	0·06
y :	∞	4·0379	3·3547	2·9591	2·6813	2·4679	2·2953

(*c*) Given the table below, find e^x.

x	e^x	x	e^x
3·10	22·197951	3·13	22·873980
3·11	22·421044	3·14	23·103867
3·12	22·640380	3·15	23·336065

17. If the data are $u_0, u_3, u_4, u_7, u_{11}$ and the interpolation formula is

$u_x = u_4 + c_1 \Delta'_3\, u_4 + c_2 \Delta'^2_{4,7}\, u_3 + c_3 \Delta'^3_{3,4,7}\, u_0 + c_4 \Delta'^4_{3,4,7,11}\, u_0$,

find the values of c_1, c_2, c_3, and c_4 where $\Delta'^3_{3,4,7}\, u_0$ is the third order difference for the arguments, 3, 4, 7, 0.

18. Given $u_1 = 1, u_1 + u_3 = 5·41, u_4 + u_5 + u_6 = 18·47$,

$$u_7 + u_8 + u_9 + u_{10} + u_{11} + u_{12} = 90·36,$$

find the values of u_x for $x = 1, 2, ..., 12$.

[**Hint.** Obtain s_1, s_3, s_6, s_{12}

where $s_3 = u_1 + u_2 + u_3, s_6 = u_1 + u_2 + \cdots + u_6, s_{12} = u_1 + \cdots + u_{12}$].

19. Show that if $P(x) = A_0 + A_1 \cos 2x + A_2 \sin 2x$ agrees with $f(x)$ when $x = x_0, x_1$ and x_2, then

$$P(x) = \frac{\sin(x - x_1)\sin(x - x_2)}{\sin(x_0 - x_1)\sin(x_0 - x_2)} f(x_0) + \text{two similar terms.}$$

[**Hint.** Use the identity :

$$\begin{vmatrix} 1 & \cos 2a_1 & \sin 2a_1 \\ 1 & \cos 2a_2 & \sin 2a_2 \\ 1 & \cos 2a_3 & \sin 2a_3 \end{vmatrix} = 4\sin(a_2 - a_1)\sin(a_3 - a_2)\sin(a_1 - a_3)]$$

20. (*a*) Show that the limit of Lagrange's interpolation formula for arguments x_0, x_1 and $x_0 + \varepsilon$ as $\varepsilon \to 0$, is

$$f(x) = \frac{(x_1 - x)(x + x_1 - 2x_0)}{(x_1 - x_0)^2} f(x_0) + \frac{(x - x_0)(x_1 - x)}{x_1 - x_0} f'(x_0) + \frac{(x - x_0)^2}{(x_1 - x_0)^2} f(x_1)$$

(*b*) A function $f(x)$ is known in three points x_1, x_2 and x_3 in the vicinity of an extreme point x_0, Show that

$$x_0 \simeq \frac{x_1 + 2x_2 + x_3}{4} - \frac{f(x_1, x_3) + f(x_2, x_3)}{4f(x_1, x_2, x_3)}$$

[**Hint.** (*b*) Differentiate with respect to x the formula :

$$f(x) = f(x_1) + (x - x_1) f(x_1, x_2) + (x - x_1)(x - x_2) f(x_1, x_2, x_3)$$

and get $x = \frac{1}{2}(x_1 + x_2) - \frac{1}{2} f(x_1, x_2)/f(x_1, x_2, x_3)$

Similarly $x = \frac{1}{2}(x_2 + x_3) - \frac{1}{2} f(x_2, x_3)/f(x_1, x_2, x_3)$ by changing $x_1 \to x_2$, $x_2 \to x_3$, $x_3 \to x_1$ and therefore, find the average value of x].

21. The value of $f(x)$ are given at *a, b* and *c*. Show that under certain conditions to be mentioned, the maximum or minimum is attained at

$$x = \Sigma(b^2 - c^2) f(a)/[2\Sigma(b - c) f(a)].$$

[**Hint.** Find $f(x)$ by Lagrange's formula and then put $f(x) = 0$ to obtain x.

22. Given that u_x has the fifth differences constant show that

$$u_{2½} = \frac{1}{2} c + [25(c - b) + 3(a - c)]/256$$

where $a = u_0 + u_5, b = u_1 + u_4, c = u_2 + u_3.$ (*B.Sc. Agra, 1996*)

[**Hint.** Clearly u_x should be polynomial in x of degree 5. Put $x = 2\frac{1}{2}$ in Lagrange's formula :

$$u_x = \frac{(x-1)(x-2)(x-3)(x-4)(x-5)}{-120} u_0 + \cdots \text{etc.}$$

23. The following data are given for certain function

x	$f(x)$	$f'(x)$
0·4	1·554284	0·243031
0·5	1·561136	– 0·089618

In the interval $0·4 < x < 0·5$, $f(x)$ has a maximum at $x = a$. Find $[a, f(a)]$.

[**Hint.** By Lagrange's formula,

$$f'(a) = \frac{(a - ·5)}{(·4 - ·5)} \times ·243031 + \frac{(a - ·4)}{(·5 - ·4)} \times -0·089618$$

24. Show that for the interpolation of $f(x)$ relative to $0, \alpha, 1$. Lagrange's formula gives

$$f(x) \simeq \left[1 - \frac{x(x-\alpha)}{1-\alpha}\right] f(0) + \frac{x(1-x)}{1-\alpha} \frac{f(\alpha)-f(0)}{\alpha} + \frac{x(x-\alpha)}{1-\alpha} f(1)$$

Also show that if $a \to 0$, it reduces to

$$f(x) = (1-x^2) f(0) + x(1-x) f'(0) + x^2 f(1).$$

25. Express the confluent divided difference $f(a, x, b, c)$ in the form

$$\begin{vmatrix} f(a) & f'(a) & f(b) & f(c) \\ a^2 & 2a & b^2 & c^2 \\ a & 1 & b & c \\ 1 & 0 & 1 & 1 \end{vmatrix} \div \begin{vmatrix} a^3 & 3a^2 & b^3 & c^3 \\ a^2 & 2a & b^2 & c^2 \\ a & 1 & b & c \\ 1 & 0 & 1 & 1 \end{vmatrix}$$

[**Hint.** $g(x) = f(a, x, b, c)$

$$= \begin{vmatrix} f(a) & f(x) & f(b) & f(c) \\ a^2 & x^2 & b^2 & c^2 \\ a & x & b & c \\ 1 & 1 & 1 & 1 \end{vmatrix} \div \begin{vmatrix} a^3 & x^3 & b^3 & c^3 \\ a^2 & x^2 & b^2 & c^2 \\ a & x & b & c \\ 1 & 1 & 1 & 1 \end{vmatrix}$$

Now $f(a, a, b, c) = \lim_{x \to a} g(x) =$ the reqd. result.]

26. Denoting the interpolation of $f(x)$ on the set of points $x_0, x_1, \ldots, x_n$ by $\sum_{k=0}^{n} l_k(x) f(x_k)$, find an expression for $\sum_{k=0}^{n} l_k(0) x^n$.

27. For what value of x is y a minimum ?

x :	3	4	5	6	7	8
y :	– 205	– 240	– 259	– 262	– 252	– 224

[**Hint.** Find y using Lagrange's formula and then $\frac{dy}{dx} = 0$ gives $x = 3{\cdot}5$.]

28. Show that $f(x_2, x_1, x_0) = f(x_{\alpha_0}, x_{\alpha_1}, x_{\alpha_2})$

where $\alpha_0, \alpha_1, \alpha_2$ is any permutation of integers 2, 1, 0 and $f(x_2, x_1, x_0)$ is finite divided difference of order 2.

29. Give the table below of the natural logarithm and its derivative

x	logx	$\frac{1}{x}$
.40	–.916291	2.50
.50	– .693147	2.00
.70	– .356675	1.43
. 80	– .223144	1.25

Estimate the value of ln .60 using the hermite interpolation formula,

$h_1(.60) = {}^{11}/_{54}$, $\bar{h}_1(.60) = {}^{1}/_{180}$

$h_2(.60) = {}^{-8}/_{27}$, $\bar{h}_2(.60) = {}^{2}/_{45}$

$h_3(-.60) = {}^{8}/_{27}$, $\bar{h}_3(.60) = {}^{-2}/_{45}$

$h_4(.60) = {}^{11}/_{54}$, $\bar{h}_4(.60) = {}^{-1}/_{180}$

3

CENTRAL DIFFERENCE INTERPOLATION FORMULAE

3·1. Introduction. We have earlier obtained Newton's and Lagrange's interpolation formulae. Lagrange's formula is tedious to apply and involves a great deal of computation and does not give accurate results if the values of the independent variable are quite apart. Newton's formulae are fundamental and are applicable to nearly all cases of interpolation, but in general they do not converge as rapidly as central difference formulae. Further, sometimes it is useful to shift the origin to some convenient point so that the values $y_0, y_1, y_2, ...,$ of the entry y become

$$y_{-3}, y_{-2}, y_{-1}, y_0, y_1, y_2, y_3, \dots$$

and then we need a formula which involves these values and the differences $\Delta y_0, \Delta^2 y_{-1}$, etc.

The central difference formulae are particularly suited for interpolating values of the function near the middle of a tabulated set. The central difference formulae are likely to converge more rapidly. There are various central difference formulae, the important onces, being due to Gauss, Stirling, Bessel and Laplace-Everett. Before we derive these formulae, we introduce Sheppard's Central Difference notation.

3·2. Central Difference Operators : δ, ∇, σ and μ. The first central difference $\delta f(x)$ of $f(x)$ is defined as

$$\delta f(x) = f\left(x + \frac{h}{2}\right) - f\left(x - \frac{h}{2}\right)$$

$$= (E^{1/2} - E^{-\frac{1}{2}}) f(x), \text{ where } E = 1 + \Delta$$

Thus $\delta \equiv E^{1/2} - E^{-1/2}$ and is called the **central difference operator.**

Also $\delta = E^{-1/2}(E - 1) = E^{-1/2}\Delta = \Delta E^{-1/2}$

The lower case delta is used for central differences, the normal capital delta for forward differences and the inverted capital delta for backward differences.

$$\therefore \qquad \delta E^{1/2} f(x) = \Delta f(x)$$

so that

$$\delta f(x + \tfrac{1}{2} h) = f(x + h) - f(x) = \Delta f(x) \qquad \ldots(3\cdot1)$$

The first backward or ascending difference $\nabla f(x)$ of $f(x)$ is defined as

$$\nabla f(x) = f(x) - f(x - h)$$

$$= (1 - E^{-1}) f(x)$$

$$\therefore \qquad \nabla \equiv 1 - E^{-1} \equiv 1 - e^{-hD} \qquad \ldots(3\cdot2)$$

∇ is called **backward difference** (the nabla) operator.

From (3·1) and (3·2), we have the symbolic equivalence

$$\delta = \Delta E^{-1/2} = \nabla E^{1/2} \qquad \ldots(3\cdot3)$$

The nth power of these operators applied to y_x gives

$$\delta^n y_x = \Delta^n E^{-n/2} y_x = \nabla^n E^{n/2} y_x$$

or

$$\delta^n y_x = \Delta^n y_{x - n/2} = \nabla^n y_{x + n/2} \qquad \ldots(3\cdot4)$$

The operator μ, called the **averager** or **mean operator** is defined by the operator equation

$$\mu = \tfrac{1}{2} (E^{1/2} + E^{-1/2})$$

$$\therefore \qquad \mu\, y_x = \tfrac{1}{2} (E^{1/2} + E^{1/2})\, y_x$$

$$= \tfrac{1}{2} (y_{x+1/2} + y_{x-1/2}) \qquad \ldots(3\cdot5)$$

and

$$\mu\, \delta^n y_x = \tfrac{1}{2} (\delta^n y_{x+1/2} + \delta^n y_{x-1/2})$$

Also

$$\mu^2 y_x = \tfrac{1}{4} (E^{1/2} + E^{-1/2})^2\, y_x$$

$$= \tfrac{1}{4} [(E^{1/2} - E^{-1/2})^2 + 4]\, y_x$$

$$= \tfrac{1}{4} (\delta^2 + 4)\, y_x$$

$$= (1 + \tfrac{1}{4} \delta^2)\, y_x$$

$$\therefore \qquad \mu^2 \equiv 1 + \tfrac{1}{4} \delta^2 \qquad \ldots(3\cdot6)$$

Again, since

$$\delta\, y_x = y_{x+1/2} - y_{x-1/2}$$

and

$$\mu\, y_x = \tfrac{1}{2} (y_{x+1/2} + y_{x-1/2})$$

$$\therefore \qquad 2\mu\, y_x + \delta\, y_x = 2y_{x+1/2}$$

or

$$(2\mu + \delta)\, y_x = 2E^{1/2} y_x$$

$$2\mu + \delta \equiv 2E^{1/2}$$

or

$$E^{1/2} = \mu + \tfrac{1}{2} \delta \qquad \ldots(3\cdot7)$$

$E = e^U$ where $hD = U$, $h = 1$, by $1·6·2, page 15

$$\therefore \quad \left.\begin{aligned} \delta &= E^{1/2} - E^{-1/2} = 2\sinh\frac{U}{2} \\ \text{and} \quad \mu &= \tfrac{1}{2}(E^{1/2} + E^{-1/2}) = \cosh\frac{U}{2} \end{aligned}\right\} \qquad ...(3.8)$$

The operator inverse to δ is the *central-sum operator* $\sigma\,(=\delta^{-1})$ such that $\delta\,\sigma f(x) = f(x)$ and is defined by

$$\sigma f(x) = \sigma f(x-h) + f(x - \tfrac{1}{2}h) \qquad ...(3.9)$$

$$\Rightarrow \quad f(x) = \sigma f\left(x + \frac{h}{2}\right) - \sigma f\left(x - \frac{h}{2}\right)$$

Also $(\sigma - \sigma E^{-1}) f(x) = E^{-1/2} f(x)$, on using (3.9)

$$\therefore \quad \sigma(1 - E^{-1}) \equiv E^{-1/2}$$

or

$$\sigma = \frac{E^{-1/2}}{1 - E^{-1}} \quad \text{or} \quad \sigma = \frac{E^{1/2}}{E - 1} \qquad ...(3.9')$$

With the operator δ defined by (3.1), the appearance of a central difference table is :

$$\begin{array}{ccccccccc}
y_{-2} & & & & \delta^2 y_{-2} & & & & \delta^4 y_{-2} \\
 & & \delta\, y_{-3/2} & & & & \delta^3 y_{-3/2} & & \\
y_{-1} & & & & \delta^2 y_{-1} & & & & \delta^4 y_{-1} \\
 & & \delta\, y_{-1/2} & & & & \delta^3 y_{-1/2} & & \\
\underline{y_0} & & & & \underline{\delta^2 y_0} & & & & \underline{\delta^4 y_0} \\
 & & \delta\, y_{1/2} & & & & \delta^3 y_{1/2} & & \\
y_1 & & & & \delta^2 y_1 & & & & \delta^4 y_1 \\
 & & \delta\, y_{3/2} & & & & \delta^3 y_{3/2} & & \\
y_2 & & & & \delta^2 y_2 & & & & \delta^4 y_2.
\end{array}$$

Note. (*i*) $\delta\,\sigma \neq \sigma\,\delta$.

(*ii*) $\delta\, y_{1/2} = \Delta\, y_0$, $\delta^2 y_0 = \Delta^2 y_{-1}$, $\delta^3 y_{1/2} = \Delta^3 y_{-1}$, $\delta^4 y_0 = \Delta^4 y_{-2}$, etc.

3.3. Gauss's Central Difference Interpolation Formulae. Let y_x denote a function which takes the value $y_{-3}, y_{-2}, y_{-1}, y_0, y_1, y_2, \ldots$ for equally spaced values and with unit intervals of u *viz.*, $-3, -2, -1, 0, 1, 2, \ldots$ where $u = (x - x_0)/h$.

Newton's advancing difference formula is

$$y_u = y_0 + {}^uC_1\,\Delta\, y_0 + {}^uC_2\,\Delta^2 y_0 + {}^uC_3\,\Delta^3 y_0 + \cdots$$

But $\Delta^2 y_0 = \Delta^2 y_{-1} + \Delta^3 y_{-1}$; $\Delta^3 y_0 = \Delta^3 y_{-1} + \Delta^4 y_{-1}$,

$\Delta^3 y_{-1} = \Delta^3 y_{-2} + \Delta^4 y_{-2}$, $\Delta^4 y_{-1} = \Delta^4 y_{-2} + \Delta^5 y_{-2}$, etc.

$$\begin{aligned}
\therefore \quad y_u &= y_0 + {}^uC_1\,\Delta\, y_0 + {}^uC_2\,(\Delta^2 y_{-1} + \Delta^3 y_{-1}) \\
&\quad + {}^uC_3\,(\Delta^3 y_{-1} + \Delta^4 y_{-1}) + {}^uC_4\,(\Delta^4 y_{-1} + \Delta^5 y_{-1}) + \cdots \\
&= y_0 + {}^uC_1\,\Delta\, y_0 + {}^uC_2\,\Delta^2 y_{-1} + ({}^uC_2 + {}^uC_3)\,\Delta^3 y_{-1} \\
&\quad + {}^uC_3\,(\Delta^4 y_{-2} + \Delta^5 y_{-2}) + {}^uC_4\,(\Delta^4 y_{-2} + \Delta^5 y_{-2} + \Delta^5 y_{-1}) + \cdots
\end{aligned}$$

or $$y_u = y_0 + {}^uC_1 \Delta y_0 + {}^uC_2 \Delta^2 y_{-1} + {}^{u+1}C_3 \Delta^3 y_{-1} + {}^{u+1}C_4 \Delta^4 y_{-2} + \cdots + {}^{u+k-1}C_{2k-1} \Delta^{2k-1} y_{-k+1} + {}^{u+k-1}C_{2k} \Delta^{2k} y_{-k} + \cdots$$

$$= y_0 + u\,\Delta y + \frac{u(u-1)}{2!} \Delta^2 y_1 + \frac{u(u^2-1)}{3!} \Delta^3 y_1 + \frac{u(u^2-1)(u-2)}{4!} \Delta^4 y_2 + \cdots \qquad \ldots(3{\cdot}10)$$

The result is known as Gauss's forward formula for equal intervals. It employs the odd differences just below the central line from y_0 and the even differences on the central line.

Another formula due to Gauss may be derived in a similar manner.

$$\Delta y_0 = \Delta y_{-1} + \Delta^2 y_{-1},$$

$$\Delta^3 y_{-1} = \Delta^3 y_{-2} + \Delta^4 y_{-2}.$$

and so on

Substituting these in (3·10) and grouping the terms in a way slightly different from the previous one, (3·10) takes the form

$$y_{a+hu} = y_u = y_0 + {}^uC_1 \Delta y_{-1} + {}^{u+1}C_2 \Delta^2 y_{-1} + {}^{u+1}C_3 \Delta^3 y_{-2} + {}^{u+2}C_4 \Delta^4 y_{-2} + \cdots + {}^{u+k-1}C_{2k-1} \Delta^{2k-1} y_{-k} + {}^{u+k}C_{2k} \Delta^{2k} y_{-k} + \cdots \qquad \ldots(3{\cdot}11)$$

This is known as **Gauss's backward formula for equal intervals.** This formula employs the odd differences just *above* the central line through y_0 and even differences on the central line.

Another form

$$y_{a+hu} = y_0 + {}^uC_1 \delta y_{-1/2} + {}^{u+1}C_2 \delta^2 y_0 + {}^{u+1}C_3 \delta^3 y_{-1/2} + {}^{u+2}C_4 \delta^4 y_0 + \cdots$$

using note (*ii*)

3·4. Stirling's interpolation Formula. Taking the average of the two Gauss's formulae, we find

$$y_u = y_0 + u\,\frac{\Delta y_0 + \Delta y_{-1}}{2} + \frac{u^2}{2!} \Delta^2 y_{-1} + \frac{u(u^2-1^2)}{3!} \cdot \frac{\Delta^3 y_{-1} + \Delta^3 y_{-2}}{2} + \frac{u^2(u^2-1^2)}{4!} \Delta^4 y_{-2} + \cdots$$

$$+ \frac{u(u^2-1^2)(u^2-2^2)\ldots[u^2-(k-1)^2]}{(2k-1)!} \times \frac{\Delta^{2k-1} y_{-k+1} + \Delta^{2k-1} y_{-k}}{2}$$

$$+ \frac{u^2(u^2-1^2)(u^2-2^2)\ldots[u^2-(k-1)^2]}{(2k)!} \times \Delta^{2k} y_{-k} + \ldots\ldots \qquad \ldots(3{\cdot}12)$$

where $u = (x - x_0)/h.$

$$\therefore \quad y_u = y_0 + u\,\mu\,\delta y_0 + \frac{u^2}{2!}\delta^2 y_0 + {}^{u+1}C_3\,\mu\,\delta^3 y_0$$

$$+ \frac{u^2(u^2-1^2)}{4!}\delta^4 y_0 + {}^{u+2}C_5\,u\,\delta^5 y_0 + \frac{u^2(u^2-1^2)(u^2-2^2)}{6!}\delta^6 y_0 + \cdots$$

This is Stirling's formula. It employs the mean of the odd differences above and below the central line and even differences on the central line. In this formula there are $2k+1$ terms, and the polynomial coincides with the given function at the $2k+1$ points.

$$u = -k, -(k-1), \ldots, -2, -1, 0, 1, 2, \ldots, k-1, k.$$

Note: Stirling formula may be used when $-\frac{1}{4} < u < \frac{1}{4}$. For convenience take u = 0.

3·5. Bessel's Interpolation Formula. Gauss's backward formula with origin transferred to unity takes the form

$$y_u = y_1 + {}^{u-1}C_1\,\Delta y_0 + {}^{u}C_2\,\Delta^2 y_0 + {}^{u}C_3\,\Delta^3 y_{-1} + \cdots$$

Taking the mean of this and Gauss's forward formula we obtain

$$y_u = \tfrac{1}{2}(y_0 + y_1) + (u - \tfrac{1}{2})\,\Delta y_0 + \frac{u(u-1)}{2!} \times \frac{\Delta^2 y_{-1} + \Delta^2 y_0}{2}$$

$$+ \frac{(u-\frac{1}{2})\,u(u-1)}{3!}\Delta^3 y_{-1} + \cdots$$

$$+ \frac{(u+1)\,u(u-1)\,u(u-2)}{4!} \times \frac{\Delta^4 y_{-1} + \Delta^4 y_{-2}}{2} + \cdots$$

$$+ \frac{(u+k-1)(u+k-2)\ldots(u-k)}{2k!} \times \frac{\Delta^{2k} y_{-k} + \Delta^{2k} y_{-k+1}}{2}$$

$$+ \frac{(u-\frac{1}{2})(u+k-1)(u+k-2)\ldots(u-k)}{(2k+1)!} \times \Delta^{2k+1} y_{-k} + \cdots \quad \ldots(3\cdot 13)$$

This is Bessel's interpolation formula. In this formula, the coefficients of the odd order difference are all zero at $u + \frac{1}{2}$ as well as at $u = 0$ and 1. This is an advantage of Bessel's formula over most other interpolation formulae which involve all orders of differences.

(3·13) can be written more symmetrically by introducing the variable $v = u - \frac{1}{2}$

$$\therefore \quad y_u = y_{v+1/2} = \tfrac{1}{2}(y_0 + y_1) + v\,\Delta y_0 + \frac{v^2 - \frac{1}{2}}{2!} \cdot \frac{\Delta y^2_{-1} + \Delta^2 y_0}{2}$$

$$+ \frac{v\left(v^2 - \frac{1}{4}\right)}{3!}\Delta^3 y_{-1} + \frac{\left(v^2 - \frac{1}{4}\right)\left(v^2 - \frac{9}{4}\right)}{4!} \cdot \frac{\Delta^4 y_{-2} + \Delta^4 y_{-1}}{2} + \cdots$$

$$+ \frac{\left(v^2 - \frac{1}{4}\right)\left(v^2 - \frac{3}{4}\right)\ldots\left[v^2 - \frac{1}{4}(2k-1)^2\right]}{(2k)!} \cdot \frac{\Delta^{2k} y_{-k} + \Delta^{2k} y_{-k+1}}{2}$$

$$+\frac{v\left(v^2-\frac{1}{4}\right)\left(v^2-\frac{9}{4}\right)\ldots\left[v^2-\frac{1}{4}(2k-1)^2\right]}{(2k+1)!}\Delta^{2k+1}y_{-k}+\cdots \qquad \ldots(3\cdot14)$$

Bessel's formula gives better result for $1/4 < u < 3/4$. For convenience table $u = 1/2$

Cor. Putting $u=\frac{1}{2}$ or $v=0$ in (3·14), we get

$$y_{1/2}=\frac{y_0+y_1}{2}-\frac{1}{8}\cdot\frac{\Delta^2y_{-1}+\Delta^2y_0}{2}+\frac{3}{128}\cdot\frac{\Delta^4y_{-2}+\Delta^4y_{-1}}{2}$$

$$-\frac{5}{1024}\cdot\frac{\Delta^6y_{-3}+\Delta^6y_{-2}}{2}+\cdots+(-1)^k\frac{[1\cdot3\cdot5\ldots(2k-1)]^2}{2^{2k}(2k)!}\cdot\frac{\Delta^{2k}y_{-k}+\Delta^{2k}y_{-k+1}}{2}+$$

This is called the *formula for interpolating to halves* or formula for halving an interval. It is used for computing values of the function mid-way between any two given values. In the notation of operator μ, we have

$$y_{1/2}=\mu y_{1/2}-\tfrac{1}{8}\mu\delta^2y_{1/2}+\tfrac{3}{128}\mu\delta^4y_{1/2}-\cdots$$

3·6. Laplace-Everett Interpolation Formula. Eliminating the odd differences in Gauss's formula (3·10) and using the relations

$$\Delta^{2k+1}y_{-k}=\Delta^{2k}y_{1-k},-\Delta^{2k}y_{-k},\ k=0,1,2,\ldots$$

we arrive at the following expansion in factorial notation :

$$y=y_0+u(y_1-y_0)+\frac{u^{(2)}}{2!}\Delta^2y_{-1}+\frac{(u+1)^{(3)}}{3!}(\Delta^2y_0-\Delta^2y_{-1})$$

$$+\frac{(u+1)^{(4)}}{4!}\Delta^4y_{-2}+\cdots+\frac{(u+k-1)^{(2k)}}{2k!}\Delta^{2k}y_{-k}$$

$$+\frac{(u+k)^{(2k+1)}}{(2k+1)!}(\Delta^{2k}y_{1-k}-\Delta^{2k}y_{-k})+\cdots$$

$$\Rightarrow y_x=(1-u)\,y_0+uy_1+\left[\frac{u^{(2)}}{2!}-\frac{(u+1)^{(3)}}{3!}\right]\Delta^2y_{-1}$$

$$+\frac{(u+1)^{(3)}}{3!}\cdot\Delta^2y_0+\left[\frac{(u+1)^{(4)}}{4!}-\frac{(u+2)^{(5)}}{5!}\right]\Delta^4y_{-2}$$

$$+\frac{(u+2)^{(5)}}{5!}\Delta^4y_{-1}+\cdots+\left[\frac{(u+k-1)^{(2k)}}{(2k)!}-\frac{(u+k)^{(2k+1)}}{(2k+1)!}\right]\Delta^{2k}y_{-k}$$

$$+\frac{(u+k)^{(2k+1)}}{2k+1)!}\Delta^{2k}y_{1-k}+\cdots \qquad \ldots(3\cdot16)$$

In order to obtain a convenient form of this formula known as Laplace-Everett formula, we introduce the variable $v=1-u$.

Now $$\frac{(u+k-1)^{(2k)}}{(2k)!}-\frac{(u+k)^{(2k+1)}}{(2k+1)!}=[(2k+1)-(u+k)]\frac{(u+k-1)^{(2k)}}{(2k+1)!}$$

$$= -\frac{(u+k-1)^{(2k+1)}}{(2k+1)\,!} = -\frac{(k-v)^{(2k+1)}}{(2k+1)\,!}$$

$$= \frac{v(v^2-k^2)^{(k)}}{(2k+1)\,!}$$

$$= \frac{v(v^2-1^2)(v^2-2^2)\ldots(v^2-k^2)}{(2k+1)\,!}$$

Therefore (3·16) takes the elegant form :

$$y_u = vy_0 + \frac{v(v^2-1)}{3\,!}\Delta^2 y_{-1} + \frac{v(v^2-1)(v^2-4)}{5\,!}\Delta^4 y_{-2} + \cdots$$

$$+ \frac{v(v^2-1)(v^2-4)\ldots(v^2-k^2)}{(2k+1)\,!}\Delta^{2k} y_{-k} + \cdots$$

$$+ uy_1 + \frac{u(u^2-1^2)}{3\,!}\Delta^2 y_0 + \frac{u(u^2-1)(u^2-4)}{5\,!}\Delta^4 y_{-1} + \cdots$$

$$+ \frac{u(u^2-1^2)(u^2-4)\ldots(u^2-k^2)}{(2k+1)\,!}\Delta^{2k} y_{1-k} + \cdots$$

$$= vy_0 + {}^{v+1}C_1\delta^2 y_0 + {}^{v+2}C_2\delta^4 y_0 + \cdots$$

$$+ uy_1 + {}^{u+1}C_3\delta^2 y_1 + {}^{u+2}C_3\delta^4 y_1 + \cdots$$

This is *Laplace-Everett's formula* in the most common form. This formula is convenient especially when using tables in which only differences of even order are tabulated. It can be used when $0 < u < 1$.

Table of Central Differences

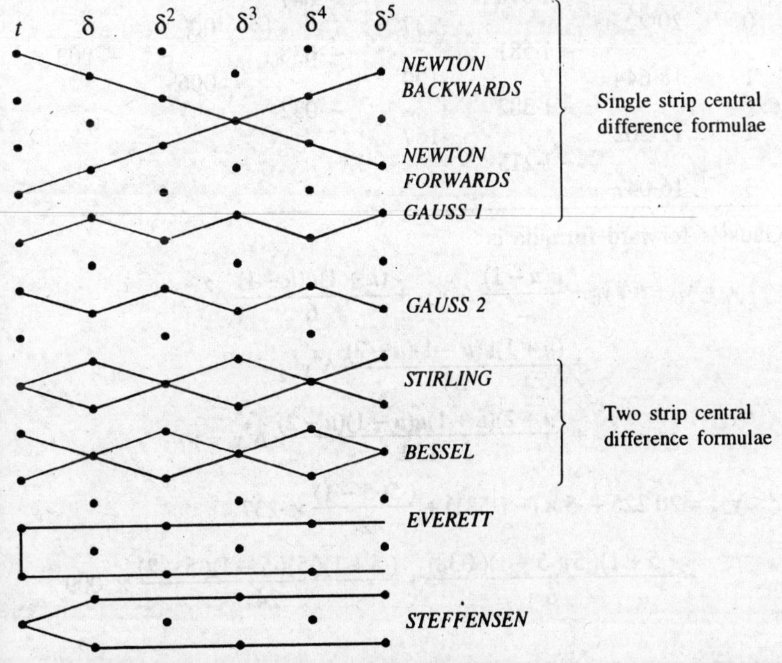

Let φ and ψ (with or without indices) denote even and oldd functions of δ, respectively. The best known formulae have the following structure :

1. $y_x = \varphi(\delta)y_0 + \psi(\delta)y_{-\frac{1}{2}}$ (Gauss 1)

2. $y_x = \varphi(\delta)y_0 + \psi(\delta)y_{-\frac{1}{2}}$ (Gauss 2)

3. $y_x = \varphi(\delta)y_0 + \mu\psi(\delta)y_0$ (Stirling)

4. $y_x = \mu\varphi(\delta)y_{\frac{1}{2}} + \psi(\delta)y_{\frac{1}{2}}$ (Bessel)

5. $y_x = \varphi_0(\delta)y_0 + \varphi_1(\delta)y_1$ (Everett)

Example 1. *Use Gauss's forward formula to find the value of y when $x = 3{\cdot}75$ given the following table :*

x :	2·5	3·0	3·5	4·0	4·5	5·0
y_x :	24·145	22·043	20·225	18·644	17·262	16·047

Sol. Take 3·5 as the origin and 0·5 as the unit, the value of y required will be for $u = \dfrac{3{\cdot}75 - 3{\cdot}50}{0{\cdot}5} = 0{\cdot}5$

The difference table is :

u	y_u	$\Delta^2 y_u$	$\Delta^3 y_u$	$\Delta^4 y_u$	$\Delta^5 y_u$	
− 2	24·145					
		− 2·102				
− 1	22·043		·284			
		− 1·818		− ·047		
0	20·225		·237		·009	
		− 1·581		− ·038		− ·003
1	18·644		·199		·006	
		− 1·382		− ·032		
2	17·262		·167			
		− 1·215				
3	16·047					

Gauss's forward formula is

$$y_u = y_0 + u\Delta y_0 + \frac{u(u-1)}{2}\Delta^2 y_{-1} + \frac{(u+1)u(u-1)}{6}\Delta^2 y_{-1}$$

$$+ \frac{(u+1)u(u-1)(u-2)}{24}\Delta^4 y_{-2}$$

$$+ \frac{(u+2)(u+1)u(u-1)(u-2)}{120}\Delta^5 y_{-2} + \cdots$$

$$\therefore \quad y_{\cdot 5} = 20{\cdot}225 + {\cdot}5 \times (-1{\cdot}581) + \frac{({\cdot}5({\cdot}5 - 1)}{2} \times {\cdot}237$$

$$+ \frac{({\cdot}5+1)({\cdot}5)({\cdot}5-1)({\cdot}038)}{6} + \frac{({\cdot}5+1)({\cdot}5)({\cdot}5-1)({\cdot}5-2)}{24} \times {\cdot}009$$

$$+\frac{(\cdot 5+2)(\cdot 5+1)(\cdot 5)(\cdot 5-1)(\cdot 5-2)}{120}\times(-\cdot 003)$$

$$= 22\cdot 225 - 0\cdot 7905 - 0\cdot 029625 + \cdot 00238 + 0\cdot 0023750 + 0\cdot 0002106$$

$$= 19\cdot 41 \text{ approx.}$$

Hence the estimated value of y when $x = 3\cdot 75$ is 19·41 approx.

Example 2. *Interpolate by means of Gauss's backward formula the population for the year 1936, given the following table :*

Year	:	1901	1911	1921	1931	1941	1951
Population ('000)	:	12	15	20	27	39	52

Sol. Take 1931 as the origin and 10 years as the unit, then the population is to be estimated for $u = \frac{1936-1931}{10} = 0\cdot 5.$

The difference table is :

u	y_u	Δy_u	$\Delta^2 y_u$	$\Delta^3 y_u$	$\Delta^4 y_u$	$\Delta^5 y_u$
−3	12					
		3				
−2	15		2			
		5		0		
−1	20		2		3	
		7		3		−10
0	27		5		−7	
		12		−4		
1	39		1			
		13				
2	52					

Gauss's backward formula is :

$$y_u = y_0 + u\Delta y_{-1} + \frac{(u+1)u}{2}\Delta^2 y_{-1} + \frac{(u+1)u(u-1)}{6}\Delta^3 y_{-2}$$

$$+\frac{(u+2)(u+1)u(u-1)}{24}\Delta^4 y_{-2} + \cdots$$

$$\therefore\ y_{\cdot 5} = 27 + \cdot 5\times 7 + \frac{1\cdot 5\times\cdot 5}{2}\times 5 + \frac{1\cdot 5\times\cdot 5\times -\cdot 5}{6}\times 3$$

$$+\frac{2\cdot 5\times 1\cdot 5\times\cdot 5\times -\cdot 5}{24}\times 7 + \frac{2\cdot 5\times 1\cdot 5\times\cdot 5\times -\cdot 5\times -1\cdot 5}{120}\times -10$$

$$= 27 + 3\cdot 5 + 1\cdot 875 - 0\cdot 1875 + 0\cdot 2734 - 0\cdot 1172$$

$$= 32\cdot 6484 - 0\cdot 3047 = 32\cdot 3427 \text{ thousands.}$$

Hence the estimated population for 1936 is 32·3437 thousands.

Example 3. *Use Stirling's formula to find y_{35}, given*

$$y_{20} = 512,\ y_{30} = 439,\ y_{40} = 346,\ y_{50} = 243,$$

where y_x represents the number of persons at age x years in a life table.

Sol. Take 30 years as the origin and 10 years as the unit, then we are to find y_x for value of $u = \dfrac{35-30}{10} = \cdot 5$. The difference table is :

u	y_u	Δy_u	$\Delta^2 y_u$	$\Delta^3 y_u$
−1	512			
		−73		
0	439		−20	
		−93		+10
1	346		−10	
		−103		
2	243			

Stirling's formula is

$$y_u = y_0 + u\frac{\Delta y_0 + \Delta y_{-1}}{2} + \frac{u^2}{2!}\Delta^2 y_{-1}$$

$$+\frac{u(u^2-1^2)}{3!}\left(\frac{\Delta^3 y_{-2} + \Delta^3 y_{-1}}{2}\right) + \cdots$$

$$\therefore \qquad y_{\cdot 5} = 439 + \cdot 5\frac{(-93-73)}{2} + \frac{\cdot 25}{2}\times(-20)$$

$$= 439 - 41{\cdot}5 - 2{\cdot}5$$

$$= 395.$$

Thus the estimated value of $y_{35} = 395$.

Example 4. *Apply Bessel's formula to obtain y_{25}, given*

$$y_{20} = 2854,\ y_{24} = 3162,\ y_{28} = 3544,\ y_{32} = 3992.$$

Sol. Take 24 as origin and 4 as unit, then we are to find the value of $y_{1/4}$. The difference table is :

u	y_u	Δy_u	$\Delta^2 y_u$	$\Delta^3 y_u$
−1	2854			
		308		
0	3162		74	
		382		−8
1	3544		66	
		448		
2	3992			

Bessel's formula is

$$y_u = \tfrac{1}{2}(y_0 + y_1) + \left(u - \tfrac{1}{2}\right)\Delta y_0 + \frac{u(u-1)}{2!}\left(\frac{\Delta^2 y_{-1} + \Delta^2 y_0}{2}\right)$$

$$+\frac{\left(u-\tfrac{1}{2}\right)u(u-1)}{3!}\Delta^3 y_{-1} + \cdots$$

$$\therefore \qquad y_{1/4} = \tfrac{1}{2}(3162 + 3544) + \left(\tfrac{1}{4} - \tfrac{1}{2}\right)\times 382 + \left(\tfrac{1}{4}\right)\left(\tfrac{1}{4} - 1\right)\times\frac{(74+66)}{3!}$$

$$+\frac{\left(\frac{1}{4}-\frac{1}{2}\right)\left(\frac{1}{4}\right)\left(\frac{1}{4}-1\right)}{3\,!}\times -8$$

$$= 3353 - 95{\cdot}5 - 6{\cdot}5625 - 0{\cdot}0625$$
$$= 3250{\cdot}875.$$

Thus the value of $y_{25} = 3250{\cdot}875$.

Example 5. *Given the table :*

x	$\log_{10} x$
310	2·4913617
320	2·5051500
330	2·5185139
340	2·5314789
350	2·5440680
360	2·5563025

Find the value of $\log_{10} 337{\cdot}5$ *by Laplace-Everett formula.* *(B.Sc. Agra, 1996)*

Sol. Taking 330 as the origin and 10 as the unit, we are to find $\log_{10} x$ when

$$u = \frac{337{\cdot}5 - 330}{10} = {\cdot}75$$

$$\therefore \quad v = 1 - u = 1 - {\cdot}75 = {\cdot}25 = \tfrac{1}{4}.$$

The difference table is :

u	y_u	Δy_u	$\Delta^2 y_u$	$\Delta^3 y_u$	$\Delta^4 y_u$	$\Delta^5 y_u$
– 2	2·4913617					
		·0137883				
– 1	2·5051500		– 4244			
		·0133639		255		
0	2·5185139		– 3989		– 25	
		·0129650		230		
1	2·5314789		– 3759		– 17	
		·0125891		213		
2	2·5440680		– 3546			
		·0122345				
3	2·5563025					

Laplace-Everett's formula where $v = 1 - u = \frac{1}{4}$ is

$$y_u = vy_0 + \frac{v(v^2-1)}{3\,!}\Delta^2 y_{-1} + \frac{v(v^2-1)(v^2-4)}{5\,!}\Delta^4 y_{-2} + \cdots$$
$$+ uy_1 + \frac{u(u^2-1)}{3\,!}\Delta^2 y_0 + \frac{u(u^2-1)(u^2-4)}{5\,!}\Delta^4 y_{-1} + \cdots$$

$$\therefore \quad y_{.75} = \tfrac{1}{4}(2{\cdot}5185139) + \left(-\frac{5}{128}\right)(-{\cdot}0003989)$$
$$+ \frac{63}{8192}({\cdot}000255) + \tfrac{3}{4}(2{\cdot}5314789)$$

$$+\left(-\frac{7}{123}\right)(\cdot0003759)+\frac{77}{8192}(-\cdot0000017)$$

$$= 0\cdot629628473 + \cdot0000155 - \cdot00002 + 1\cdot898609175$$

$$+ \cdot0000206 - \cdot0000001$$

$$= 2\cdot5282537 \text{ approx.}$$

$\therefore \quad \log 337\cdot5 = 2\cdot5282537.$

Example 6. *If third differences are constant, prove that*

$$y_{x+1/2} = \frac{1}{2}(y_x + y_{x+1}) - \frac{1}{16}(\Delta^2 y_{x-1} + \Delta^2 y_x).$$

Sol. Bessel's formula upto third differences, is

$$y_x = \frac{y_0 + y_1}{2} + \left(x - \frac{1}{2}\right)\Delta y_0 + \frac{x(x-1)}{2\,!}\ \frac{\Delta^2 y_{-1} + \Delta^2 y_0}{2}$$

$$+ \frac{(x-\frac{1}{2})x(x-1)}{3\,!}\Delta^3 y_{-1}$$

Put $\quad x = \frac{1}{2}$.

$$\therefore \quad y_{1/2} = \frac{y_0 + y_1}{2} - \frac{1}{16}(\Delta^2 y_{-1} + \Delta^2 y_0)$$

Changing the origin to x, we have the approximation

$$y_{x+1/2} = \frac{1}{2}(y_x + y_{x+1}) - \frac{1}{16}(\Delta^2 y_{x-1} + \Delta^2 y_x).$$

Example 7. *Given $y_0, y_1, y_2, y_3, y_4, y_5$ (fifth differences constant), prove that*

$$y_{2\frac{1}{2}} = \frac{1}{2}c + \frac{25(c-b) + 3(a-c)}{256}$$

where $a = y_0 + y_5$; $b = y_1 + y_4$; $c = y_2 + y_3$.

Sol. Put $x = \frac{1}{3}$ in Bessel's formula.

$$y_{1/2} = \frac{1}{2}(y_0 + y_1) - \frac{1}{8}\left(\frac{\Delta^2 y_{-1} + \Delta^2 y_0}{2}\right) + \frac{3}{128}\left(\frac{\Delta^4 y_{-2} + \Delta^4 y_{-1}}{2}\right)$$

Shift the origin to 2,

$$y_{2\frac{1}{2}} = \frac{1}{2}(y_2 + y_3) - \frac{1}{16}(\Delta^2 y_1 + \Delta^2 y_2) + \frac{3}{128}\left(\frac{\Delta^4 y_0 + \Delta^4 y_1}{2}\right) \quad \ldots(1)$$

Now

$$\Delta^2 y_1 = y_3 - 2y_2 + y_1,\ \Delta^4 y_0 = y_4 - 4y_3 + 6y_2 - 4y_1 + y_0, \text{ etc.}$$

Substituting the values in (1), we get the required result.

Aliter. $u_{5/2} = E^{5/2}\, u_0 = (1 + \Delta)^{5/2} u_0$.

$$= \left(1 + \frac{5}{2}\Delta + \frac{\frac{5}{2}\cdot\frac{3}{2}}{2\,!}\Delta^2 + \frac{\frac{5}{2}\cdot\frac{3}{2}\cdot\frac{1}{2}}{3\,!}\Delta^3 + \frac{\frac{5}{2}\cdot\frac{3}{2}\frac{1}{2}\cdot\frac{-1}{2}}{4\,!}\Delta^4 + \frac{\frac{5}{2}\cdot\frac{3}{2}\cdot\frac{1}{2}\cdot\frac{-1}{2}\cdot\frac{-3}{2}}{5\,!}\Delta^5 + \cdots \right) u_0$$

$$= u_0 + \frac{5}{2}\Delta u_0 + \frac{15}{8}\nabla^2 u_0 + \frac{5}{16}\Delta^3 u_0 - \frac{5}{128}\Delta^4 u_0 + \frac{3}{256}\Delta^5 u_0$$

Now $\Delta^5 u_0 = (E-1)^5 u_0 = u_5 - 5u_4 + 10u_3 - 10u_2 + 5u_1 - u_0$, etc.

$$u_{5/2} = \frac{3}{256}(u_0 + u_5) - \frac{25}{256}(u_1 + u_4) + \frac{75}{128}(u_2 + u_3) \text{ approx.}$$

$$= \frac{3}{256}a - \frac{25}{256}b + \frac{75}{128}c$$

$$= \frac{1}{2}c + \frac{3(a-c) + 25(c-b)}{256}.$$

EXERCISE 3

1. (*a*) Define the operators $E, \Delta, \mu, \delta,$ and ∇ used in Calculus of Finite Differences and show that these all commute with one another.

(*b*) Prove :

(*i*) $\delta \equiv \Delta E^{-1/2} \equiv \nabla E^{1/2}$ (*B.Sc. Hons., Delhi 1995*)

(*ii*) $\mu \equiv \frac{1}{2}(E^{1/2} + E^{-1/2}) \equiv \frac{2+\Delta}{2\sqrt{1+\Delta}} \equiv \frac{2-\nabla}{2\sqrt{1-\nabla}} \equiv \sqrt{1 + \frac{1}{4}\delta^2}$

(*iii*) $E^{\pm 1/2} = \mu \pm \frac{1}{2}\delta.$

(*iv*) $\Delta \equiv \frac{1}{2}\delta^2 + \delta\sqrt{1 + \frac{1}{4}\delta^2}.$ (*Delhi B.Sc. Math. Hons, 1997*)

(*v*) $\sqrt{1 + \delta^2\mu^2} \equiv 1 + \frac{1}{2}\delta^2,$

(*c*) Establish the relations :

(*i*) $\Delta\nabla \equiv \nabla\Delta \equiv \Delta - \nabla \equiv \delta^2,$ (*AMIE, 1985, Agra, 1990*)

(*ii*) $\mu\delta = \frac{1}{2}(\Delta + \nabla) \equiv \frac{1}{2}\Delta E^{-1} + \frac{1}{2}\Delta = \sinh(hD),\ E = e^{hD}$ (*AMIE, 1995*)

(*iii*) $\Delta + \nabla \equiv \Delta/\nabla - \nabla/\Delta.$

(*d*) Show that :

$$\Delta[f(x-1)\Delta g(x-1)] \equiv \nabla[f(x)\,\Delta g(x)] \equiv \Delta\,[f(x-1)\nabla g(x)]$$

(*e*) Show that :

(*i*) $\delta\,[f(x)g(x)] = \mu f(x)\,\delta g(x) + \mu g(x)\delta f(x)$

(*ii*) $\delta\left(\frac{f(x)}{g(x)}\right) = \frac{\mu g(x)\delta f(x) - \mu f(x)\delta g(x)}{g\left(x - \frac{1}{2}\right)g\left(x + \frac{1}{2}\right)}.$

(*iii*) $\mu\,[f(x)g(x)] = \mu f(x)\mu g(x) + \frac{1}{4}\delta f(x)\delta g(x).$

(iv) $\mu\left(\dfrac{f(x)}{g(x)}\right) = \dfrac{\mu f(x)\mu g(x) - \frac{1}{4}\delta f(x)\delta g(x)}{g\left(x-\frac{1}{2}\right)g\left(x+\frac{1}{2}\right)}$.

(v) $\Delta[f(x)g(x)] = f(x)\nabla g(x) + g(x-1)\nabla f(x)$.

(vi) $\nabla\left[\dfrac{f(x)}{g(x)}\right] = \dfrac{f(x)\nabla g(x) - g(x)\nabla f(x)}{g(x-1)g(x)}$

(f) Show that $\displaystyle\sum_{k=0}^{n-1} \delta^2 f_{2k+1} = (f_{2n} - f_0)\tanh\frac{U}{2}$

[**Hint.** $\displaystyle\sum_{k=0}^{n-1} \delta^2 f_{2k+1} = \sum_{k=0}^{n-1}(E + E^{-1} - 2)(f_{2k+1}) = \sum_{k=0}^{n-1}(f_{2k+2} + f_{2k} - 2f_{2k+1})$

$$= \sum_{k=0}^{n-1}(f_{2k+2} - f_{2k}) - 2\sum_{k=0}^{n-1}(f_{2k+1} - f_{2k})$$

$$= (f_{2n} - f_0) + 2(f_0 - f_1 + f_2 - \cdots - f_{2n-1})$$

$$\left(\because \sum_{k=0}^{n-1}(f_{2k+2} - f_{2k}\right)$$

$$= f - f_0 + f_4 - f_2 + \ldots\ldots + f_{2n} - f_{2n-2}$$

$$= f_{2n} - f_0$$

$$= (f_{2n} - f_0) + 2\left(\frac{1 - E^{2n}}{1 - E}\right) f_0 = \text{etc.}$$]

2. If D, E, δ and μ be the operators with usual meaning and if $hD = U$, where h is the interval of differencing, prove the following relations between the operators :

(a) $E = e^U \quad \delta = 2\sinh\dfrac{U}{2} \quad \mu = \cosh\dfrac{U}{2}, \quad e^{-U} = 1 - \nabla.$

$(E + 1)\,\delta = 2\,(E - 1)\,\mu.$ *(AMIE 1993, 1996)*

(b) $\dfrac{U}{\delta} = \dfrac{2}{\delta}\sinh^{-1}\dfrac{\delta}{2} = 1 - \dfrac{\delta^2}{24} + \dfrac{3}{640}\delta^4 - \dfrac{5}{7168}\delta^6 + 0(\delta)^8,$

$$\left(\frac{U}{\delta}\right)^2 = 1 - \frac{\delta^2}{12} + \frac{\delta^4}{90} - \frac{\delta^6}{560} + 0(\delta)^8$$

$$\left(\frac{U}{\delta}\right)^4 = 1 - \frac{\delta^2}{6} + \frac{7}{240}\delta^4 + 0(\delta)^8$$

$$\left(\frac{U}{\delta}\right)^6 = 1 - \frac{\delta^2}{4} + 0(\delta)^4,$$

(c) $\dfrac{U}{\delta\mu} = \dfrac{U}{\delta}\left(1 + \dfrac{\delta^2}{4}\right)^{-1/2} = 1 - \dfrac{\delta^2}{6} + \dfrac{\delta^4}{30} - \dfrac{\delta^6}{140} + 0(\delta)^8.$

(*d*) $\left(\frac{\delta}{U}\right)^2 = 2\,(\cosh U - 1)/U^2.$

(*e*) $\frac{\delta}{\mu U} = \tanh \frac{U}{2} / \frac{U}{2}.$

(*f*) $\frac{\mu\delta}{U} = \sinh U/U.$

(*g*) $\nabla^2 = U^2 - U^3 + \frac{7}{12} U^4 - \cdots$

(*h*) $U^2 = \nabla^2 + \nabla^3 + \frac{11}{12} \nabla^4 + \cdots$

3. (*a*) Show that backward difference operator ∇ satisfies the following identities :

(*i*) $\nabla(y_i + y_j) = \nabla y_i + \nabla y_j,$

(*ii*) $\nabla(ay_i) = a\nabla y_i$, a being a constant

(*iii*) $\nabla^m(\nabla^n y_i) = \nabla^{m+n} y_i = \nabla^n(\nabla^m y_i).$

(*b*) Prove that :

$$\delta^2 y_0 = y_1 - 2y_0 + y_{-1},$$

$$\delta^3 y_{1/2} = y_2 - 3y_1 + 3y_0 - y_{-1},$$

and, in general,

$$\delta^n y_x = \sum_{k=0}^{n} (-1)^k \frac{n!}{k!\,(n-k)!} y_{x+n/2-k} = \nabla^n y_{x-n/2}.$$

[**Hint.** $\delta^n = E^{-n/2}(E-1)^n$

$\therefore$ $\delta^n y_x = (E-1)^n\, y_{x-n/2} = \text{etc.}$]

(*c*) Let $\delta'^2 y_0 = y_n - 2y_0 + y_{-n}$ and $\delta^2 y_0 = y_1 - 2y_0 + y_{-1}$.

Obtain δ'^2 as a power series in δ^2.

(*d*) Prove :

$$y_{-2} = y_0 - 2\nabla y_0 + \nabla^2 y_0,$$

$$y_{-3} = y_0 - 3\nabla y_0 + 3\nabla^2 y_0 - \nabla^3 y_0$$

and, in general,

$$y_{-n} = \sum_{k=0}^{n} (-1)^k \binom{n}{k} \nabla^k y_0$$

(*e*) If $\nabla f(x) = f(x) - f(x-1),$

show that (*i*) $\nabla^n f(x) = \nabla^{n-1} f(x) - \nabla^{n-1} f(x-h)$ (*B.Sc. Agra, 1996*)

(*ii*) $\nabla^n f(x) = f(x) - {}^nC_1\, f(x-h) + {}^nC_2\, f(x-2h) - \cdots + (-1)^n f(x-n)h.$

(*f*) Establish the following relations :

$$\Delta^n f_i = \delta^n f_{i+n/2} = \nabla^n f_{i+n} = (n!)\, h^n f[x_i, x_{i+1}, \ldots, x_{i+n}]$$

where $f_i = f(x_i) \equiv y_i.$

[**Hint. Prove by induction**].

$$\Delta^{n+1}f_i = \Delta(\Delta^n f_i) = \Delta\delta^n(f_{i+n/2}) = \delta^n(\Delta f_{i+n/2}) = \delta^n[f_{i+n/2+1} - f_{i+n/2}]$$
$$= \delta^n(\delta f_{i+\frac{n+1}{2}} = \delta^{n+1} f_{i+\frac{n+1}{2}}$$
$$\nabla^n y_{i+n} = (1 - E^{-1})^n y_{i+n} = (E-1)^n E^{-n} y_{i+n} = (E-1)^n y_i$$

4. Prove that if third differences are assumed to be constant

$$y_x = xy_1 + \frac{x(x^2-1)}{3!}\Delta^2 y_0 + uy_0 + \frac{u(u^2-1)}{3!}\Delta^2 y_{-1},$$

where $u = 1 - x$.

Apply this formula to find the values of y_{11} and y_{16} given that

$y_0 = 3010, y_5 = 2710, y_{10} = 2285, y_{15} = 1860, y_{20} = 1560, y_{25} = 1510$ and $y_{30} = 1835$.

5. (*a*) Use Gauss's forward formula to find y_{30}, given that

$y_{21} = 18{\cdot}4708$; $y_{25} = 17{\cdot}8144$; $y_{29} = 17{\cdot}1070$; $y_{33} = 16{\cdot}3432$; $y_{37} = 15{\cdot}5154$.

(*b*) Given $y_2 = 10, y_1 = 8, y_0 = 5, y_{-1} = 10$, find $y_{1/2}$ by Gauss's forward formula.

6. Given that

$\sqrt{12500} = 111{\cdot}803399$; $\sqrt{12510} = 111{\cdot}848111$;
$\sqrt{12520} = 111{\cdot}892806$; $\sqrt{12530} = 111{\cdot}937483$;

show by Gauss's backward formula that

$\sqrt{12516} = 111{\cdot}874930$.

7. Apply a central difference formula to obtain

(*i*) $f(25)$, given that $f(20) = 14$, $f(24) = 32$, $f(28) = 35$, $f(32) = 40$.

(*ii*) $f(32)$, given that $f(25) = 0{\cdot}2707$, $f(30) = 0{\cdot}3027$, $f(35) = 0{\cdot}3386$, $f(40) = 0{\cdot}3794$.

8. The values of e^{-x} for certain equidistant values of x are given in the following table. Find the value of e^{-x} when $x = 1{\cdot}7489$ by Bessel as well as Stirling's formulae.

x	e^{-x}
1·72	0·1790661429
1·73	0·1772844100
1·74	0·1755204100
1·75	0·1737739435
1·76	0·1720458636
1·77	0·1703329888
1·78	0·1686381473

9. From the following table, find the value of $\log_{10} 337{\cdot}5$ by Gauss, Sitrling, Bessel and Everett formulae.

x	$\log_{10} x$
310	2·4913617
320	2·5051500
330	2·5185139

340	2·5314789
350	2·5440680
360	2·5563025

10. Apply (*a*) Sitrling's interpolation formula, (*b*) Bessel's interpolation formula, to find the value of $f(1{\cdot}22)$ from the following table which gives the value of

$$f(t)=\frac{1}{\sqrt{2\pi}}\int_0^t \exp\left(-\frac{t^2}{2}\right)dt$$

at intervals of $t = 0{\cdot}5$ from $t = 0$ to $t = 3$.

t	$f(t)$
0·0	0·0
0·5	0·19146
1·0	0·34134
1·5	0·43319
2·0	0·47725
2·5	0·49379
3·0	0·49865.

Show also that Newton's forward formula gives $f(1{\cdot}22) = 0{\cdot}38867$. Find also the error in each case when the true values of $f(1{\cdot}22) = 0{\cdot}338877$.

11. Use Stirling's formula :

(a) to find $f(1{\cdot}22)$ from the following table

x	1·0	1·1	1·2	1·3	1·4
$f(x)$	0·84147	0·89121	0·93204	0·96356	0·98545
x	1·5	1·6	1·7	1·8	
$f(x)$	0·99749	0·99957	0·97385	0· 97385	

(*b*) to show that tan 16° = 0·2867, given

θ :	0°	5°	10°	15°	20°	25°	30°
tan θ :	0·0000	0·0875	0·1763	0·2679	0·3640	0·4663	0·5774

$$\left[\textbf{Hint. } \theta = 15°,\ u = \frac{(16° - 15°)}{5°} = 0{\cdot}2°\right]$$

(*c*) to find u_{28}, given

$u_{10} = 49225, u_{25} = 48316, u_{30} = 47236, u_{35} = 45926$ and $u_{40} = 44306$

$$\left[\textbf{Hint. } \text{Take } x = 30 \text{ as origin, } h = 5 \text{ and } u = \frac{28-30}{5} = -0{\cdot}4\right]$$

(*d*) to find cos 0·806595, given

cos 0·8050 = 0·6931111235
cos 0·8055 = 0·692750733
cos 0·8060 = 0·692390058
cos 0·8065 = 0·692029210
cos 0·8070 = 0·691668188
cos 0·8075 = 0·691306994
cos 0·8080 = 0·690945627

(*e*) to find f(35) given that

$f(20) = 519, f(30) = 439, f(40) = 456, f(50) = 243$ (*B.Sc. Agra 1995*)

12. (*a*) From the following table, find $f(0{\cdot}5437)$, using a suitable central difference formula :

x	0·51	0·52	0·53	0·54
$f(x)$	0·5292437	0·52378987	0·5464641	0·5549392
x	0·55	0·56	0·57	
$f(x)$	0·5633233	0·5716157	0·5798158	

(*b*) Solve the equation $x = 10 \log_{10} x$, using the following table :

x :	13·5	13·6	13·7	13·8
$\log_{10} x$:	1·1303	1·1335	1·1367	1·1399

13. y_x is a polynomial of degree n and $x_0, x_1, x_2, ..., x_{n-1}$ are a set of values of x. If y_x is written as

$$y_x = A_0 + A_1(x - x_0) + A_2(x - x_0)(x - x_1) + \cdots + A_n(x - x_0)(x - x_1)...(x - x_{n-1})$$

determine $A_0, A_1, ..., A_n$.

Hence or otherwise derive Gauss's forward and backward difference formulae.

[**Hint.** (*i*) Replace $x_0, x_1, x_2, \ldots$ by $x_0, x_1, x_{-1}, x_{-2}, \ldots$ and make use of the result

$$f(x_i, x_{i+1}, x_{i+2}, ..., x_{i+r}) = \frac{1}{r!\,h^r}\,\delta^r f_{i-r/2}$$

(*ii*) Replace $x_0, x_1, x_2 \ldots$ by $x_0, x_{-1}, x_1, x_{-2}, x_2, ...$]

14. (*a*) Eliminate odd differences from Gauss's forward formula to derive Everett formula

$$y_n = (1 - \alpha) f_0 + \alpha f_1 - \frac{\alpha(\alpha - 1)(\alpha - 2)}{3\,!}\,\delta^2 f_0 + \frac{(\alpha + 1)\alpha(\alpha - 1)}{3\,!}\,\delta^2 f_1 + \cdots$$

where $y = f(x_0 + \alpha h), \alpha = (x - x_0)/h$

(*b*) Given the following table, construct a difference table and from it estimate y when (*i*) $x = 0{\cdot}15$ and (*ii*) $x = 0{\cdot}35$

x :	0	0·1	0·2	0·3	0·4
y :	1	1·095	1·129	1·251	1·310

(*B.Sc. Math. Hons., Delhi 1998*)

[**Hint.** Form diff. table and then $f(x) = 1 + x\,\dfrac{0{\cdot}095}{0{\cdot}1} + \dfrac{x(x - 0{\cdot}1)}{2(0{\cdot}1)^2}(-\,0{\cdot}011) + \dfrac{x(x - 0{\cdot}1)(x - 0{\cdot}2)}{6\,(0{\cdot}1)^3}(-\,{\cdot}001)$

$= \dfrac{-1}{6}x^3 - 0{\cdot}5x^2 + 1{\cdot}0016x,\; f(0{\cdot}15) = 1149,\; f(0{\cdot}35) = -\,1{\cdot}282)$]

4

INVERSE INTERPOLATION

4·1. Introduction. The problem of direct interpolation is to find the value of $y = f(x)$ for a value of x lying between two tabulated values of argument. The problem in inverse interpolation is to find the value of x corresponding to the value of y lying between two tabulated values of y.

Definition. *Inverse interpolation is the process of finding the value of the argument corresponding to a given value of the function when the latter is intermediate between two tabulated values.*

For certain functions, the problem of inverse interpolation does not offer any difficulty.

For example, if $y = \sin x$, then $x = \sin^{-1} y$, if $y = \log_e x$, then $x = e^y$; and the required values of x can be calculated easily on substituting the value of y. In other cases, specially when the corresponding numerical values of x and y are given, the problem of inverse interpolation offers difficulty and can be solved by several methods, the important ones being :

(1) *Use of Lagrange's Formula.*

(2) *Method of Successive Approximations or Iteration.*

(3) *Method of Reversion of Series.*

4·2. Methods of solving the problem of inverse interpolation.

4·2·1. By Making Use of Lagrange's Interpolation formula. The problem of inverse interpolation can be solved by the use of Lagrange's formula in the form (2·11) in which x is expressed as a function of y.

Example 1. *Apply Lagrange's formula inversely to find to one decimal place, the age for which the annuity value is 13·6, given the following table :*

Age	*Annuity value of* $4\frac{1}{2}$ *per cent*
x	$y = u_x$
30	15·9
35	14·9

40	14·1
45	13·3
50	12·5

Sol. Here $x_0 = 30$, $x_1 = 35$, $x_2 = 40$, $x_3 = 45$, $x_4 = 50$, $y_0 = 15{\cdot}9$, etc., $y = 13{\cdot}6$.

Applying Lagrange's formula giving x as a function of y,

$$x = \frac{(13{\cdot}6 - 14{\cdot}9)(13{\cdot}6 - 14{\cdot}1)(13{\cdot}6 - 13{\cdot}3)(13{\cdot}6 - 12{\cdot}5)}{(15{\cdot}9 - 14{\cdot}9)(15{\cdot}9 - 14{\cdot}1)(15{\cdot}9 - 13{\cdot}3)(15{\cdot}9 - 12{\cdot}5)} \times 30$$

$$+ \frac{(13{\cdot}6 - 15{\cdot}9)(13{\cdot}6 - 14{\cdot}1)(13{\cdot}6 - 13{\cdot}3)(13{\cdot}6 - 12{\cdot}5)}{(14{\cdot}9 - 15{\cdot}9)(14{\cdot}9 - 14{\cdot}1)(14{\cdot}9 - 13{\cdot}3)(14{\cdot}9 - 12{\cdot}5)} \times 35$$

$$+ \cdots + \frac{(13{\cdot}6 - 15{\cdot}9)(13{\cdot}6 - 14{\cdot}9)(13{\cdot}6 - 14{\cdot}1)(13{\cdot}6 - 13{\cdot}3)}{(12{\cdot}5 - 15{\cdot}9)(12{\cdot}5 - 14{\cdot}9)(12{\cdot}5 - 14{\cdot}1)(12{\cdot}5 - 13{\cdot}3)} \times 50$$

$$= 40{\cdot}1.$$

4·2·2. By using the Technique of Successive Approximations or Iteration. Let us suppose that the function $y = f(x)$ is a polynomial of the nth degree. We use any of the interpolation formulae of chapters II and III to find the value of x corresponding to the given value of y. Consider Newton formula (1) :

$$y = y_0 + x\,\Delta y_0 + \frac{x(x-1)}{2\,!}\Delta^2 y_0 + \frac{x(x-1)(x-2)}{3\,!}\Delta^3 y_0 + \frac{x(x-1)(x-2)(x-3)}{4\,!}\Delta^4 y_0 + \cdots \qquad ...(4{\cdot}1)$$

Since y is given, one needs to solve for a real root of this equation. This equation can be solved by successive approximations.

Transposing and dividing throughout by Δy_0, we have

$$x = \frac{y - y_0}{\Delta y_0} - \frac{x(x-1)}{2\,\Delta y_0}\Delta^2 y_0 - \frac{x(x-1)(x-2)}{3\,!\,\Delta y_0}\Delta^3 y_0 - \frac{x(x-1)(x-2)(x-3)}{4\,!\,\Delta y_0}\Delta^4 y_0 \qquad ...(4{\cdot}2)$$

neglecting higher differences.

To get a *first approximation for* x, we neglect all differences higher than the first and therefore

$$x = x_1 = \frac{y - y_0}{\Delta y_0}$$

The *second approximation for* x is obtained by substituting x_1 in the right side of (4·2). We then have

$$x_2 = \frac{y - y_0}{\Delta y_0} - \frac{x_1(x_1 - 1)}{2\,!}\cdot\frac{\Delta^2 y_0}{\Delta y_0} - \frac{x_1(x_1-1)(x_1-2)}{3\,!}\cdot\frac{\Delta^3 y_0}{\Delta y_0}$$

$$-\frac{x_1(x_1-1)(x_1-2)(x_1-3)}{4\,!}\cdot\frac{\Delta^4 y_0}{\Delta\, y_0}$$

The *third approximation for x* obtained by substituting x_2 in the right side of (4·2), is

$$x_3=\frac{y-y_0}{\Delta\, y_0}-\frac{x_2(x_2-1)}{2\,!}\cdot\frac{\Delta^2 y_0}{\Delta\, y_0}-\frac{x_2(x_2-1)(x_2-2)}{3\,!}\cdot\frac{\Delta^2 y_0}{\Delta\, y_0}$$

$$-\frac{x_2(x_2-1)(x_2-2)(x_2-3)}{4\,!}\cdot\frac{\Delta^4 y_0}{\Delta\, y_0}$$

and so on further approximations till we reach stability in the value of x.

If the differences $\Delta^2 y_0, \Delta^3 y_0, \Delta^4 y_0$ decrease rapidly, then the sequence of successive approximations *viz.,* $x_1, x_2, x_3 \ldots$ converges quickly to the desired root of the equation (4·1). Thus we see that the iterative process is one way of inverse interpolation based on forward interpolation formula. This method is specially useful for finding a real root of an equation.

Example 2. *Given coshx = 1·285 find x by inverse interpolation, using the data in columns II and III of the following table.*

u	x	$y = \cosh x$	$\Delta\, y$	$\Delta^2 y$	$\Delta^3 y$	$\Delta^4 y$	$\Delta^5 y$
– 2	0·736	1·2832974					
			8049				
– 1	0·737	1·2841023		13			
			8062		– 1		
0	0·738	1·2849085		12		3	
			8074		2		– 6
1	0·739	1·2857179		14		– 3	
			8088		– 1		
2	0·740	1·2865247		13			
			8101				
3	0·741	1·2873348					

Sol. The above table shows that the desired value of x lies between 0·738 and 0·739. We take $x_0 = 0{\cdot}738$, $h = 0{\cdot}001$, $y = 1{\cdot}285$ and use Bessel's formula *viz.,*

$$y=\frac{y_0+y_1}{2}+u\,\Delta\, y_0+\frac{\left(u^2-\frac{1}{4}\right)}{2\,!}\cdot\frac{\Delta^2 y_{-1}+\Delta^2 y_0}{2}+\frac{u\left(u^2-\frac{1}{4}\right)}{3\,!}\Delta^3 y_{-1}+\cdots$$

or $$1{\cdot}285 = 1{\cdot}2853122 + u\times 0{\cdot}0008074+\frac{u^2-0{\cdot}25}{2}\times{\cdot}0000013$$

$$+\frac{u(u^2-0{\cdot}25)}{6}({\cdot}0000002)$$

Transposing and dividing throughout by ·0008074, we get

$$u = -\cdot 3867327217 - (u^2 - 0\cdot 25)(\cdot 00000065)$$
$$- u(u^2 - 0\cdot 25)(\cdot 000000033)(\cdot 0008074)$$

A first approximation for u is obtained by neglecting all terms beyond the first in right side of (2). Hence

$$u_{(1)} = -\cdot 3867327217$$

Substituting this for u in right side of (2), we get as second approximation

$$u_{(2)} = -\cdot 3867239591.$$

Similarly,

$$u_{(3)} = -\cdot 3867238231.$$

The value differs only slightly from $u_{(2)}$ and we therefore make no further approximations.

$$\therefore \qquad x = 0\cdot 738 + \cdot 001\,(-\cdot 3867239591 + \tfrac{1}{2})$$
$$= 0\cdot 73811330.$$

4·2·3. To find the Roots of an Algebraic Equation by Inverse Interpolation. Suppose that a real root of $f(x) = 0$ has been isolated, then $f(x)$ may be tabulated using a finite interval in the neighbourhood of the root and represented closely by a central difference formula. This produces a polynomial equation in which the terms of higher degree have very small influence; consequently an accurate approximation to the required root can be found, and this approximation can be improved by an iterative procedure.

Example 3. *Find by the method of inverse interpolation the real root of the equation $x^3 + x - 3 = 0$ which lies between 1·2 and 1·3.*

(B.A. Math. Hons. Delhi, 1982)

Sol. The difference table is :

x	X	$y (= x^3 + x - 3)$	Δy	$\Delta^2 y$	$\Delta^3 y$	$\Delta^4 y$
1	·2	−1				
			·431			
1·1	·1	−0·569		·066		
			·497		·006	
1·2	0	0·072		·072		0
			·569		·006	
1·3	·1	+0·497		·078		
			·647			
1·4	·2	+1·144				

The root of the equation is 1·2 + 0·1 x_0 where we are to find x_0.

Taking the origin at 1·2, interval of difference is 0·1, and using Stirling's formula *viz.*,

$$y = y_0 + x\,\frac{\Delta y_0 + \Delta y_{-1}}{2} + \frac{x^2}{2}\,\Delta^2 y_{-1} + \frac{x(x^2-1)}{6}\cdot\frac{\Delta^3 y_{-1} + \Delta^3 y_{-2}}{2}$$

we get

$$0 = -\cdot 072 + x\,\frac{\cdot 569 + \cdot 497}{2} + \frac{x^2}{2}(\cdot 072) + \frac{x(x^2-1)}{3}(\cdot 006)$$

$$= -\cdot 072 + \cdot 531\,x + \cdot 0036\,x^2 + \cdot 002\,x^3 \qquad ...(1)$$

This cubic equation can be solved by successive approximations.

(1) can be written as

$$x = \frac{\cdot 072}{\cdot 531} - \frac{\cdot 0036}{\cdot 531}x^2 - \frac{\cdot 001}{\cdot 531}x^3$$

∴ First approximation of x is

$$x_1 = \frac{\cdot 072}{\cdot 531} = 0\cdot 1353$$

For second approximation, substitute this value of x in (1), we get

$$x_2 = \cdot 1353 - \cdot 067\,(\cdot 1353)^2 - \frac{\cdot 001}{\cdot 531}(\cdot 1353)^3$$

$$= \cdot 1353 - \cdot 001211075 - \cdot 0003428675 = \cdot 1338539425$$

$$= \cdot 134 \text{ approx.}$$

∴ Real root $= 1\cdot 2 + \cdot 1 \times 1\cdot 34 = 1\cdot 2134$ approx.

Example 4. *Find by the method of iteration a real root of* $2x - \log_{10} x = 7$, using *graphical method.*

Sol. We find from the intersection of the graphs of $y = 2x - 7$ and $y = \log_{10} x$ or otherwise that an approximate value of the root of the given equation is 3·8, *i.e.*, $x_0 = 3\cdot 8$.

The given equation can be written as

$$x = \tfrac{1}{2}(\log_{10} x + 7)$$

∴ As a first approximation,

$$x_{(1)} = \tfrac{1}{2}(\log_{10} 3\cdot 8 + 7) = 3\cdot 79$$

Second approximation,

$$x_{(2)} = \tfrac{1}{2}(\log 3\cdot 79 + 7) = 3\cdot 7893$$

Third approximation,

$$x_{(3)} = \tfrac{1}{2}(\log 3\cdot 7989 + 7) = 3\cdot 7983.$$

Since $x_{(3)}$ is same as $x_{(2)}$ we do not repeat the process but take 3·7893 as the correct value of the root to four places of decimals.

Example 5. *Solve the equation* $x = 10 \log_{10} x$, *given the following data* :

x	:	*1·35*	*1·36*	*1·37*	*1·38*
log x	:	*0·1303*	*0·1335*	*0·1367*	*0·1399*

Sol. The difference table is :

x	$y = 10 \log_{10} x$	Δy	$\Delta^2 y$
1·35	1·303		
		·032	
1·36	1·335		0
		·032	
1·37	1·367		0
		·032	
1·38	1·399		

We find that all differences above the first are zero. Take $x_0 = 1{\cdot}37$ because x and $10 \log_{10} x$ are nearest to each other for this value of x; $h = {\cdot}01$.

By Stirling's formula,

$$y = y_0 + x \frac{(\Delta y_0 + \Delta y_{-1})}{2} + \cdots$$

or
$$1{\cdot}37 = 1{\cdot}367 + x \frac{({\cdot}032 + {\cdot}032)}{2}$$

This gives

$$x = \frac{{\cdot}003}{{\cdot}022} = 0{\cdot}136$$

$\therefore$ Required root $= x_0 + hx = 1{\cdot}37 + {\cdot}01 \times {\cdot}136$
$= 1{\cdot}37136$ approx.

Example 6. *The following table gives the values of sinh x for certain equidistant values of x. If sinh x = 62, find x.*

x	y	$y - \sinh x$	Δy	Δy^2	$\Delta^3 y$
4·80	-2	*60·7511*			
			·6106		
4·81	-1	*61·3617*		*·0062*	
			·6168		*0*
4·82	*0*	*61·9785*		*·0062*	
			·6230		*0*
4·83	*1*	*62·6015*		*·0062*	
			·6292		
4·84	*2*	*63·2307*			

Sol. We notice that all differences higher than the second vanish. Also the desired value of x lies between 4·82 and 4·83 and therefore we take $x_0 = 4{\cdot}82$, $h = {\cdot}01$. By Stirling's formula,

$$y = y_0 + x \frac{(\Delta y_0 + \Delta y_{-1})}{2} + \frac{x^2}{2\,!} \Delta^2 y_{-1} + \cdots$$

or
$$62 = 61{\cdot}9785 + x \frac{({\cdot}6230 + {\cdot}6168)}{2} + \frac{x^2}{2\,!} \times {\cdot}0062$$

This is a quadratic equation in x and we therefore do not use successive approximations. We have

$$31x^2 + 6199x - 215 = 0$$

Solving, $x = 0{\cdot}0347$, neglecting negative root.

$\therefore \qquad x_1 = x_0 + hx = 4{\cdot}82 + {\cdot}01 \times {\cdot}0347 = 4{\cdot}8203$ approx.

Thus the value of x for which $\sinh x = 62$ is $4{\cdot}8203$.

4·2·4. By the Method of Reversion of Series. The interpolation formulae developed so far express y in the form of a power series in x so that

$$y = a_0 + a_1 x + a_2 x^2 + a_3 x^3 + a_4 x^4 + \cdots \qquad \text{...(4·3)}$$

Let $\qquad w = \dfrac{y - a_0}{a_1}$

and assume that the given power series is convergent so that x can be expanded as a power series in w,

$$x = c_1 w + c_2 w^2 + c_3 w^3 + c_4 w^4 + \cdots$$

Substituting this expression for x and $a_1 w$ for $y - a_0$ in (4·3) and equating the coefficients of like powers of w, we find

$$\left.\begin{aligned} c_1 &= 1 \\ c_2 &= -\frac{a_2}{a_1} \\ c_3 &= -\frac{a_3}{a_1} + 2\left(\frac{a_2}{a_1}\right)^2 \\ c_4 &= -\frac{a_4}{a_1} + 5\,\frac{a_2 a_3}{a_1^2} - 4\left(\frac{a_2}{a_1}\right)^3, \end{aligned}\right\} \qquad \text{...(4·4)}$$

Note. Gregory-Newton's formula with fourth differences regarded as constant is :

$$\begin{aligned} y &= y_0 + x\Delta\, y_0 + \frac{x\,(x-1)}{2\,!}\,\Delta^2 y_0 + \frac{x\,(x-1)(x-2)}{3\,!}\,\Delta^3 y_0 \\ &\quad + \frac{x\,(x-1)(x-2)(x-3)}{4\,!}\,\Delta^4 y_0 \\ &\equiv a_0 + a_1 x + a_2 x^2 + a_3 x^3 + a_4 x^4. \end{aligned} \qquad \text{...(4·5)}$$

Equating the coefficient of like powers of x, we get

$$\left.\begin{aligned} a_0 &= y_0 \\ a_1 &= \Delta\, y_0 - \frac{\Delta^2 y_0}{2} + \frac{\Delta^3 y_0}{3} - \frac{\Delta^4 y_0}{4} \\ a_2 &= \frac{\Delta^2 y_0}{2} - \frac{\Delta^3 y_0}{2} + \frac{11\,\Delta^4 y_0}{24}, \\ a_3 &= \frac{\Delta^3 y_0}{6} - \frac{\Delta^4 y_0}{4}, \\ a_4 &= \frac{\Delta^4 y_0}{24}. \end{aligned}\right\} \qquad \text{...(4·6)}$$

Similarly, we can obtain the values of a's for Stirling and Bessel's formulae.

Example 7. *Solve Example 6, page 98, by the method of Reversion of Series.*

Sol. Stirling's formula, neglecting Δ^5 and higher order differences, is

$$y = y_0 + x\left(\frac{\Delta y_0 + \Delta y_{-1}}{2}\right) + \frac{x^2}{2!}\Delta^2 y_{-1} + \frac{x(x^2-1)}{3!}\left(\frac{\Delta^3 y_{-1} + \Delta^3 y_{-2}}{2}\right)$$

$$+ \frac{x^2(x^2-1)}{2!}\Delta^4 y_{-2}$$

$$= y_0 + \left(\frac{\Delta y_0 + \Delta y_{-1}}{2} - \frac{\Delta^3 y_{-1} + \Delta^3 y_{-2}}{12}\right)x + \left(\frac{\Delta^2 y_{-1}}{2!} - \frac{\Delta^4 y_{-2}}{24}\right)x^2$$

$$+ \frac{\Delta^3 y_{-1} + \Delta^3 y_{-2}}{12}x^3 + \frac{\Delta^4 y_{-2}}{24}x^4$$

$$\equiv a_0 + a_1 x + a_2 x^2 + a_3 x^3 + a_4 x^4$$

$$\therefore \quad a_0 = y_0,\ a_1 = \frac{\Delta y_0 + \Delta y_{-1}}{2} - \frac{\Delta^3 y_{-1} + \Delta^3 y_{-2}}{12},$$

$$a_2 = \frac{\Delta^2 y_{-1}}{2} - \frac{\Delta^4 y_{-1}}{24},\quad a_3 = \frac{\Delta^3 y_{-1} + \Delta^3 y_{-2}}{12};\quad a_4 = \frac{\Delta^3 y_{-2}}{24}$$

In this case,

$$a_6 = 61{\cdot}9785,\ a_1 = 0{\cdot}6192,\ a_2 = 0{\cdot}0031,\ a_3 = 0,\ a_4 = 0$$

$$\therefore \quad w = \frac{y - a_0}{a_1} = \frac{62 - 61{\cdot}9785}{0{\cdot}6192} = 0{\cdot}034683,$$

$$\therefore \quad c_2 = -\frac{a_2}{a_1} = -0{\cdot}005001.$$

$$c_3 = -\frac{a_2}{a_1} + 2\left(\frac{a_2}{a_1}\right)^2$$

$$= 0 + 2\,({\cdot}005001) = {\cdot}00005002.$$

$$c_4 = 0 \text{ practically}$$

$$\therefore \quad X = w + c_2 w^2 + c_3 w^3 + c_4 w^4$$

$$= {\cdot}034683 - {\cdot}005001\,({\cdot}034683)^2 + {\cdot}00005002\,({\cdot}034683)^3$$

$$= {\cdot}0347 \text{ approx.}$$

$$\therefore \quad x = x_0 + wX = 4{\cdot}82 + 0{\cdot}1 \times {\cdot}0347 = 4{\cdot}8203 \text{ approx.}$$

as previously obtained by the method of successive approximations.

EXERCISE 4

1. (*a*) Write a short note on Inverse Interpolation and convergence of interative process.

(*b*) Explain how you will solve the problem of inverse interpolation by the method of iteration.

(*c*) Explain the method of reversion of series.

(*d*) Obtain a value of x when $y_x = 19$, given the following values :

x :	0	1	2
y_x :	0	1	20

(*B.Sc. B.R. Nov. 94*)

2. (*a*) Explain the use of Largrange's interpolation formula for inverse interpolation.

(*b*) Find the value of x for which y is 18,600 given

x :	52	53	54	55	56
y :	19,231	18,868	18,519	18,182	17,855

3. (*a*) Find x when y_x is 0·163 given

x :	80	82	84	86	88
y_x :	0·134	0·154	0·176	0·200	0·227

(*b*) Find the age for which the annuity value is 13·6 when following table is available

Age (x) :	30	35	40	45	50
Annuity value (y_x) :	15·9	14·9	14·1	13·3	12·5

(*B.Sc., BR. Nov. 94*)

4. $f(0) = 16{\cdot}35$, $f(5) = 14{\cdot}88$, $f(10) = 13{\cdot}59$, $f(15) = 12{\cdot}46$, find x when $f(x) = 14{\cdot}00$.

5. $f(44) = 13{\cdot}40$, $f(45) = 13{\cdot}16$, $f(46) = 12{\cdot}93$, $f(47) = 12{\cdot}68$ find x when $f(x) = 13{\cdot}00$.

6. Apply Lagrange's formula (inversely) to find a root of the equation $f(x) = 0$, when $f(30) = -30$, $f(34) = -13$, $f(38) = 3$, $f(42) = 18$.

7. (*a*) The equation $f(x) = x^3 - 6x - 11 = 0$ has a root between 3 and 4. Obtain it by inverse interpolation correct to three places of decimals.

[Obtain $f(0), f(1), ..., f(5)$ and use Newton's advancing difference formula]

(*b*) The equation $f(x) = x^3 + x^2 - 7x + 1 = 0$ has a root which is approximately 2. Find it.

[**Hint.** For $x = 1{\cdot}9, 2{\cdot}0, 2{\cdot}1, 2{\cdot}2, 2{\cdot}3$, $f(x)$ is $-1{\cdot}831, -1{\cdot}000, -{\cdot}029$, $1{\cdot}088, 2{\cdot}357$. Take 2·1 as origin and apply Sitrling's formula. Since the root lies in the range $|t| < \frac{1}{2}|$. Bessel's formula is preferable if the root lies in the range $|t - \frac{1}{2}| < \frac{1}{4}|$.]

8. (*a*) Find the root of the equation (*i*) $x^3 - 9x - 14 = 0$ (*ii*) $x^3 - 6x - 11 = 0$ between 3 and 4 up to three places of decimals.

[**Hint.** (*ii*) Form the difference table with values $(1, -16), (2, -15)$

(3, – 2), (4, 29).]

(*b*) Find the root of the equation $10x^3 - 15x + 3 = 0$ which lies between 1 and 2 up to three places of decimals.

(*c*) The equation $x^3 - 2x - 5 = 0$ has a root lying between 2 and 3. Find it up to three places of decimals.

9. Find to two places of decimals the smallest positive root of the equation $x^3 - 5x + 3 = 0$ by making a difference table when $x = 0, 1, 2$ and 3.

10. Find the real root of the equation by the method of iteration, correct to four places of decimals.

(*i*) $x + \log_{10} x = 0{\cdot}5$,

(*ii*) $2 + \log_{10} x = 2e^{-x}$

(*iii*) $x \log_e x - 1{\cdot}15 = 0$, use $\log_e 10 = 2{\cdot}30$

(*iv*) $x \log_{10} x - 1{\cdot}2 = 0$

[**Hint.** (*i*) Root lies between 0·6 and 0·7,

(*ii*) Root lies between 0·3 and 0·4,

(*iii*) Root lies between 1·5 and 2·0,

(*iv*) Root lies between 2 and 3]

11. Find the value of x correct to five decimal places lying between 15 and 20 satisfying the equation

$$x^2 + 250 \log_{10} x = 635.$$

12. Apply the method of iteration to find to four places of decimals the root of the equation $xe^x - 2 = 0$ which lies between 0·8 and 0·9.

13. The equation $x^3 - 15x + 4 = 0$ has a root close to 0·3. Obtain this root up to 6 decimal places, using inverse interpolation with Bessel's interpolation formula.

14. Explain how Newton-Gregory formula can be used to carry out inverse interpolation of second order. Show that a good approximation to the interpolating factor x will be given by

$$x = \frac{f_x - f_0}{\Delta f_0}\left[1 - \left\{\frac{f_x - f_0}{2(\Delta f_0)^2}\right\}\Delta^2 f_0\right]$$

15. Find to two decimal places, the root of the equation $x^3 + x^2 + x - 100 = 0$ which is approximately 4.

16. Use the method of iteration to solve the equation $x = \exp(-x)$, starting with $x = 1{\cdot}00$. Perform four iterations taking the reading upto four decimal places. (*A.M.I.E., 1988*)

[**Hint.** $x = e^{-x} \therefore \phi(x) = e^{-x}, \phi'(x) = -e^{-x}, |\phi(x)| = \left|e^{-x}\right| = \frac{1}{e} = {\cdot}36 < 1$

Also $x_1 = e^{-1} = 0{\cdot}3679$,

$$x_2 = e^{-0{\cdot}3679} = 0{\cdot}692,\ x_3 = e^{-0{\cdot}692} = 0{\cdot}5,\ x_4 = e^{-0{\cdot}5} = 0{\cdot}6065.]$$

5

NUMERICAL DIFFERENTIATION

5·1. Introduction. The numerical differentiation is the process of finding or evaluating the derivative or derivatives of a function at some particular value of the independent variable when the values of the function corresponding to the given value of the independent variable are known. The problem of differentiation is solved by first approximating the function by a polynomial by means of an interpolation formula and then differentiating this formula as many times as desired. If the values of the argument are equally spaced, we represent the function by Newton-Gregory formula. Further, if it is required to find the derivative of the function at a point near the beginning (end) of a set of tabular values, we use Newton-Gregory forward (backward) formula. When it is desired to find derivative at a point near the middle of the table, we should use one of central difference formulae. If the entries are given for unequal intervals of the argument, Newton's divided difference formula should be used to represent the function.

To find the maximum or minimum value of a tabulated function we compute the *necessary differences* from the given table and substitute them in an appropriate interpolation formula. Next we find the first' derivative of the function obtained from the interpolation formula and put the derivative to zero to obtain the value of the critical value(s) and then the maximum or minimum value by substituting the value of the critical point in the function already obtained.

5.2. Use of central difference interpolatin formulae in obtaining the derivative(s) of an interpolating polynominal.

(*a*) Use of Newton-Gregory forward difference formula for obtaining $f'(x), f''(x), \ldots$ or $\frac{dy}{dx}, \frac{d^2y}{dx^2}, \ldots$

Suppose we are given $(n+1)$ ordered pairs (x_i, y_i), $i = 0, 1, \ldots n$ having equidistant values of x *i.e.*, $x_1 - x_0 = x_2 - x_1 = \ldots\ldots = h$, say. Then at 0 point nearer the starting point x_0, using eq. (*i*), page 18, we have

$$y_x = y_0 + u\,\Delta y_0 + \frac{u(u-1}{2\,!}\,\Delta^2 y_0 + \frac{u(u-1)\,(u-2)}{3\,!}\,\Delta^3 y_0$$
$$+ \frac{u(u-1)\,(u-2)\,(u-3)}{4\,!}\,\Delta^4 y_0 + \ldots(1)$$

where $u = (x - x_0)/h$ and $y_x = y(x_0 + uh)$ is a polynomial in x of degree n

Also $\dfrac{du}{dx} = \dfrac{1}{h}, \dfrac{dy_x}{dx} = \dfrac{dy_u}{du} \cdot \dfrac{du}{dx} = \dfrac{1}{h}\dfrac{dy_u}{du}, \dfrac{d^2y_x}{dx^2} = \dfrac{d}{dx}\left(\dfrac{1}{h}\dfrac{dy_u}{du}\right) = \dfrac{1}{h^2}\dfrac{d^2y}{du^2}$

Similarly $\dfrac{d^3y_x}{dx^3} = \dfrac{1}{h^3}\dfrac{d^3y}{du^3}, \ldots$

Differentiating both sides of (1) w.r.f. x, we obtain

$$\frac{dy_x}{dx} = 0 + \frac{1}{h}\,.\,\Delta y_0 + \frac{1}{h}\,.\,\frac{2u-1}{2}\,\Delta^2 y_0 + \frac{3u^2-6u+2}{6}\,.\,\frac{1}{h}\,\Delta^3 y_0$$
$$+ \frac{4u^3-18u^2+22u-6}{24}\,.\,\frac{1}{h}\,\Delta^4 y_0 +$$
$$= \frac{1}{h}\left[\Delta y_0 + \frac{2u-1}{2}\,\Delta^2 y_0 + \frac{3u^2-6u+2}{6}\,\Delta^3 y_0 + \frac{4u^3-18u^2+22u-6}{24}\,\Delta^4 y_0 + \ldots\right] \quad \ldots(2)$$

The eq. (2) gives the value of the first derivative of y_x at any point x

For a particular values of x say at $x = x_0$, we have $u = 0$ and eq. (2) reduces to

$$\left(\frac{dy_x}{dx}\right)_{x=x_0} = \frac{1}{h}\left[\Delta y_0 - \frac{1}{2}\,\Delta^2 y_0 + \frac{1}{3}\,\Delta^3 y_0 - \frac{1}{4}\,\Delta^4 y_0 + \ldots\right] \quad \ldots(2')$$

Differentiating (2) again w.r.t. x both sides, we obtain

$$\frac{d^2y_x}{dx^2} = \frac{1}{h^2}\left[0 + 1\,.\,\Delta^2 y_0 + \frac{6u-6}{6}\,\Delta^3 y_0 + \frac{6u^2-18u+11}{12}\,\Delta^4 y_0 + \ldots\right]$$
$$= \frac{1}{h^2}\left[\Delta^2 y_0 + (u-1)\,\Delta^2 y_0 + \frac{1}{12}\,(6u^2-18u+11)\Delta^4 y_0 + \ldots\right] \quad \ldots(3)$$

Hence $\left(\dfrac{d^2y_x}{dx^2}\right)_{x=x_o} = \dfrac{1}{h^2}\left[\Delta^2 y_0 - \Delta^3 y_0 + \dfrac{11}{12}\,\Delta^4 y_0 \ldots\right], u = \dfrac{x_0 - x_0}{h} = 0. \quad \ldots(3')$

Similarly it can be easily established that

$$\frac{d^3y_x}{dx^3} = \frac{1}{h^3}\left[\Delta^3 y_0 + \frac{12u-18}{12}\,\Delta^4 y_0 + \ldots\right]$$

so that $\left(\dfrac{d^3y_x}{dx^3}\right) = \dfrac{1}{h^3}\left[\Delta^3 y_0 - \dfrac{3}{2}\,\Delta^4 y_0 + \ldots\right]$

(b) Use of Newton-Gregory backward difference formula for obtaining

$$\frac{dy}{dx}, \frac{d^2y}{dx^2}, \ldots.$$

From II, Page 2.1, we have, at a point nearer the end point with $a = 0$,

$$y_x = y(x_n + uh) = y_n + u\,\nabla y_n + \frac{u(u+1)}{2\,!}\nabla^2 y_n + \frac{u(u+1)(u+2)}{3\,!}\nabla^3 y_n$$

$$+ \frac{u(u+1)(u+2)}{4\,!}\nabla^4 y_n + \ldots,$$

Differentiating both sides w.r.t.x, remembering $\frac{dy_x}{dx} = \frac{1}{h}\frac{dy}{du}$, we have

$$\frac{dy_x}{dx} = \frac{1}{h}[\nabla y_n + \frac{2u+1}{2}\nabla^2 y_n + \frac{3u^2+6u+2}{6}\nabla^3 y_n + \frac{4u^3+18u^2+22u+6}{24}\nabla^4 y_n + \ldots]$$

$$\frac{d^2y_x}{dx^2} = \frac{1}{h^2}\left[\nabla^2 y_n + (u+1)\,\nabla^3 y_n + \frac{6u^2+18u+1}{12}\nabla^4 y_n + \ldots.\right]$$

and so on.

The last three equations give the values of $\frac{dy}{dx}, \frac{d^2y_x}{dx^2}, \frac{d^3y_x}{dx^3}$, at any general x.

For the particular value of $x = x_n$ or $u = 0$, we find that

$$\left(\frac{dy}{dx}\right) = \frac{1}{h}\left[\nabla y_n + \frac{1}{2}\nabla^2 y_n + \frac{1}{3}\nabla^3 y_n + \frac{1}{4}\nabla^4 y_n \ldots\right]$$

$$\left(\frac{d^2y}{dx^2}\right)_{x=x_n} = \frac{1}{h^2}\left[\nabla^2 y_n + \nabla^3 y_n + \frac{1}{12}\nabla^4 y_h + \right]$$

$$\left(\frac{d^3y}{dx_3}\right)_{x=x_n} = \frac{1}{h^3}\left[\nabla^3 y_n + \frac{3}{2}\nabla^4 y_n + -\right].$$

(*c*) Use of Bessel' interpolation formula for obtaining $\frac{dy_x}{dx}, \frac{d^2y_x}{dx^2}$,

From (3.14), page 79, we have

$$y_x = \frac{1}{2}(y_0 + y_1) + \left(u - \frac{1}{2}\right)\Delta y_0 + \frac{u(u-1)}{2\,!}\cdot\frac{\Delta^2 y_{-1} + \Delta^2 y_0}{2}$$

$$+ \frac{\left(u - \frac{1}{2}\right)u(u-1)}{3\,!}\Delta^3 y_{-1}$$

$$+ \frac{(u+1)\,u\,(u-1)(u-2)}{4\,!}\cdot\frac{\Delta^4 y_{-1} + \Delta^4 y_{-2}}{2} + \ldots,\quad u = (x - x_o)/h,\ \frac{du}{dx} = \frac{1}{h}$$

$$\frac{dy_x}{dx} = \frac{1}{h}\left[\ \Delta y_0 + \left(u - \frac{1}{y}\right)^2 . \frac{\Delta^2 y_{-1} + \Delta^2 y_0}{2} + \frac{(3u^2 - 3u + ½)}{6}\Delta^3 y_{-1}\right.$$

$$+ \frac{4u^3 - 6u^2 - 2u + 2}{24} . \frac{\Delta^4 y_{-1} + \Delta^4 y_{-2}}{2} t + ...$$

$$\therefore \left(\frac{dy_x}{dx}\right) = \frac{1}{h}\left[\Delta y_0 - \frac{1}{24}\Delta^3 y_{-1} - \frac{5}{96}\frac{\Delta^4 y_{-1} + \Delta^4 y_{-2}}{2}\right.$$

$u = \frac{1}{2}$ or $x = x_0 + \frac{1}{2}h$

$$\left(\frac{d^2 y_x}{dx}\right) = \frac{1}{h^2}\left[\frac{\Delta^2 y_{-1} + \Delta^2 y_0}{2} - \frac{5}{24} . \frac{\Delta^4 y_{-1} + \Delta^4 y_{-2}}{2} \quad ...\right]$$

$$\left(\frac{d^3 y_2}{dx^3}\right)_{u = \frac{1}{2}} = \frac{1}{h^3}\left[\Delta^3 y_{-1} \quad ...\right]$$

(d) Use of stirling's formula for obtaining derivative (s)

From (2), page 78, we have

$$y_x = y_u = y_0 + u\frac{(\Delta y_0 + \Delta y_{-1})}{2} + \frac{u^2}{2}\Delta^2 y_{-1} + \frac{u(u^2 - 1)}{6}\left(\frac{\Delta^3 y_{-1} + \Delta^3 y_{-2}}{2}\right)$$

$$+ \frac{u^2(u^2 - 1)}{41}\Delta^4 y_{-2} + ...$$

$$\therefore \frac{dy_x}{dx} = \frac{1}{h}\left[\frac{1}{2}(\Delta y_0 + \Delta y_{-1}) + u\Delta^2 y_{-1}) + \frac{3u^2 - 1}{12}(\Delta^3 y_{-1} + \Delta^3 y_{-2})\right.$$

$$+ \frac{1}{12}\left(2u^3 - u\right)\Delta^4 y_{-2} + ..$$

Again differentiating both sides w.r.t. x, we hve

$$\frac{d^2 y_x}{dx^2} = \frac{1}{h^2}\left[\Delta^2 y_{-1} + \frac{u}{2}\left(\Delta^3 y_{-1} + \Delta^3 y_{-2}\right) + \frac{6u^2 - 1}{12}\Delta^4 y_{-2} + ...\right],$$

$$\frac{d^3 y}{dx^3} = \frac{1}{h^3}\left[\frac{1}{2}\left(\Delta^3 y_{-1} + \Delta^3 y_{-2}\right) + u\Delta^4 y_{-2} + ... - \right], ...$$

Letting $x = x_0$ *i.e.* $u = 0$, we obtain

$$\left(\frac{dy_x}{dx}\right)_{x = x_0} = \frac{1}{h}\left[\frac{1}{2}(\Delta y_0 + \Delta y_1) - \frac{1}{12}(\Delta^3 y_{-1} + \Delta^3 y_{-2}) + \right]$$

$$\left(\frac{d^2 y_x}{dx^2}\right)_{x = x_0} = \frac{1}{h^2}(\Delta^2 y_{-1} - \frac{1}{12}\Delta^4 y_{-2} + ...$$

$$\left(\frac{d^3y}{dx^3}\right)_{x=x_0} = \frac{1}{h^3}\left[\frac{1}{2}(\Delta^3 y_{-1} + \Delta^3 y_{-2}) + ...\right]$$

(*e*) Use of Laplace Everett' formula for obtaining derivate(s).

Proceed as in (*a*) or (*b*) or (*c*) in order to obtain the derivative(s) when $u = \frac{1}{2}$ or $v = 1 - u = \frac{1}{2}$

In the previous article, we have seen how the derivative(s) of a function can be expressed in terms of differences of various order and thus by solving $\frac{dy}{dx} = 0$ we can determine two values after ignoring higher differences above third and hence the value(s) of x for which there is the function y or (are) extreme values of $f(x)$. Thus we shall be able to determine the maximum and minimum value(s) if any, of y or $f(x)$.

Example. Find the minimum value of y from the table :

x :	0	1	2	3	4	5
y :	0	0.25	0	2.25	16.00	56.25

(*B.Sc. B.R. Nov. 1989*)

Sol. Here the values of x are equidistant. Also clearly the minimum value occurs at $x = 0$ or $x = 2$. We apply Newton's forward formula.

Difference Table

x	y	Δy	Δ^2	Δ^3
0	0			
		0.25		
1	0.25		–0.50	
		–0.25		3
2	0		2.50	
		2.25		9
3	2.25		11.50	
		13.75		15
4	16.00		26.50	
		40.25		
5	56.25			

$$\left(\frac{dy}{dx}\right) = y_0 + u\,\Delta y_0 + \frac{u\,(u-1)}{2}\Delta^2 y_0 + ...$$

$\therefore$ For minimum of y at x, we have

$$0 = 0 + u(\tfrac{1}{4}) + \frac{u^2 - u}{2}\left(-\frac{1}{2}\right)$$

$\Rightarrow \quad u - u^2 + u = 0 \quad \Rightarrow u(u-2) = 0$

$\therefore \quad u = 0, \quad u = 2$

$\therefore \; x = x_0 + hu = 0 + 1.u = 0 + 1.0 = 0$

Also $\quad x = x_0 = hu = 0 = 1 \,.\, 2 = 0 + 2 = 2$

Also $$\left(\frac{d^2y}{dx^2}\right)_{x=0} = \frac{1}{h^2}\left[\Delta^2 y_0 - \Delta^3 y_0 + \quad ...\right]$$

$$= [-0.50 - 3] = -3.50 < 0$$

and $$\left(\frac{d^2y}{dx^2}\right)_{x=2} = \frac{1}{h^2}\left[\Delta^2 y_0 - \Delta^3 y_0 + \quad ....\right]$$

$$= [2.25 - 9] = -6.75 < 0$$

Hence there is minimum at $x = 0, 2$

Minimum value of y in each case is 0.

Example. *From the following given table of values, find the value of x for which y is a maximum and the maximum value.*

Sol. Looking at the given table of values, we conclude that in the neighbourhood of $x = 65$ there will be a maximum value. Hence we use Stirling's formula as the values of x are equally spaced.

x	$y = f(x)$	Δy	$\Delta^2 y$	$\Delta^3 y$	$\Delta^4 y$
45	26.2	10			
55	36.2	$6(= \Delta y_{-1})$	$-4\ (= \Delta^2 y_{-2})$		
$65\ (=x_0)$	$42.2 = y_0$	$-1.4\ (=\Delta y_0)$	$-7.4\ (= \Delta^2 y_{-1})$	$6.6\ (= \Delta^3 y_{-2})$	$10(=\Delta^4 y_1)$
75	40.8	-2.1	$-0.8\ (=\Delta^2 y_0)$		
85	38.6	2			

Here $h = 10$, $y_0 = 42.2$, $\Delta y_0 = -1.4$, $\Delta^2 y_0 = -0.8$, $\Delta y_{-1} = 6$, $\Delta^2 y_{-1} = -7.4$, etc.

Stirling formula is

$$y(x) = y_0 + \frac{u}{2}(\Delta y_0 + \Delta y_{-1}) + \frac{u^2}{2}\Delta^2 y_{-1} + \frac{u(u^2 - 1^2)}{12}\left(\Delta^3 y_{-1} + \Delta^3 y_{-2}\right)$$
$$+ \frac{u\,(u^2 - 1^2)}{24}\Delta^4 y_{-2} + ...$$

$$= 42.2 + \frac{u}{2}(-1.4 + 6) + \frac{u^2}{2}(-7.4) + \frac{(u^3 - u)}{12}(-3.4 + 6.6)$$

$$= 42.2 + 2.3u - 3.7u^2 + (u^3 - u)\,(0.272)$$

$$\Rightarrow \quad y = 0.272\,u^3 - 3.7u^2 + 2.03u + 42.2 \qquad ...\ (1)$$

For an exterme value, we have $\frac{dy}{dx} = 0$

$$0.816u^2 - 7.4u + 2.03 = 0$$

$$\therefore \quad u = 7.4 \pm \frac{\sqrt{(7.4)^2 - 4(2.03)(0.816)}}{1.632} = \frac{7.4 \pm \sqrt{48.335}}{1.632} = \frac{7.4 \pm 7}{1.632}$$

$$= \frac{1.40}{1.63} \quad \text{or} \quad 9$$

$$\therefore \quad u = .0245 \quad , \quad u = 9 \text{ is rejected}$$

$$\therefore \quad x = x_0 = uh = 65 + (+0.245) \times 10$$

$$= 65.245$$

From (1), Maximum value of $y = 0.272\,(.0245)^3 - 3.7\,(0.245)^2 + 2.03\,(0.245) + 42.2.=$ etc. Thus $f(x)$ is maximum at $x = 65.245$.

5·3. Approximate expressions for the derivative of a function.

We know that $\delta y_i = y_{i+\frac{1}{2}} - y_{i-\frac{1}{2}}$

and $\quad \mu y_i = \frac{1}{2}(y_{i+\frac{1}{2}} + y_{i-\frac{1}{2}})$

Hence

$$\mu\delta y_i = \mu\,(y_{i+\frac{1}{2}} - y_{i-\frac{1}{2}}) = \mu\, y_{i+\frac{1}{2}} - \mu\, y_{i-\frac{1}{2}} = \frac{1}{2}[(y_{i+1} + y_i) - (y_i - y_{i-1})]$$

$$= \frac{1}{2}(y_{i+1} - y_{i-1})$$

$$= \frac{1}{2}(e^{hD} - e^{-hD})\, y_i$$

since $\quad y_{i+1} = Ey_i = e^{hD}\, y_i$

$$\Rightarrow \quad \delta\mu = \sinh(hD) = hD + \frac{1}{3!}h^3D^3 + \cdots$$

$= hD$ provided that D^3 and higher derivatives can be neglected ...(1)

Thus $\frac{1}{2}(y_{i+1} - y_{i-1}) = hDy_i$

$$\Rightarrow \quad hDy_i = \frac{1}{2}(y_{i+1} - y_{i-1}) \text{ approx.} \qquad ...(2)$$

Next, $\quad \delta^2 y_i = \delta\,(y_{i+\frac{1}{2}} - y_{i-\frac{1}{2}}) = y_{i+1} - 2y_i + y_{i-1}$

$$= (e^{hD} + e^{-hD} - 2)\, y_i$$

$$= 2(\cosh hD - 1)\, y_i$$

$$= 2[\tfrac{1}{2}\, h^2D^2 + \cdots]\, y_i = h^2D^2 y_i \text{ approx.}$$

$$\Rightarrow \quad h^2D^2 y_i = \delta^2 y_i + y_{i+1} - 2y_i + y_{i-1} \qquad ...(3)$$

Again, $h^3D^3 y_i = (hD)(h^2D^2 y_i)$

$$= hD\,\delta^2 y_i$$

$$= \mu\delta(y_{i+1} - 2y_i + y_{i-1})$$

$$= \frac{1}{2}(y_{i+2} - 2y_{i+1} + y_i - y_i + 2y_{i-1} - y_{i-2}), \; \mu\delta = \frac{1}{2}(\Delta + \nabla)$$

$$= \frac{1}{2}(y_{i+2} - 2y_{i+1} + 2y_{i-1} - y_{i-2}) \qquad \text{...(4)}$$

Again $h^4D^4y_i = (h^2D^2)(h^2D^2y_i)$

$$= h^2D^2(y_{i+1} - 2y_i + y_{i-1})$$

$$= \delta^2(y_{i+1} - 2y_i + y_{i-1})$$

$$= \delta[\delta(y_{i+1} - 2y_i + y_{i-1})]$$

$$= y_{i+2} - 4y_{i+1} + 6y_i - 4y_{i-1} + y_{i-2} \qquad \text{...(5)}$$

Summarising, we have

$$y_i' = \frac{1}{2h}(y_{i+1} - y_{i-1}),$$

$$y_i'' = \frac{1}{h^2}(y_{i+1} - 2y_i + y_{i-1})$$

$$y_i''' = \frac{1}{2h^3}(y_{i+2} - 2y_{i+1} + 2y_{i-1} - y_{i-2})$$

$$y_i^{iv} = \frac{1}{h^4}(y_{i+2} - 4y_{i+1} + 6y_i - 4y_{i-1} + y_{i-2})$$

Solved Examples

Example 1. *Assuming Bessel's interpolation formula obtain the following result :*

$$\frac{d}{dx}y_x = \Delta y_{x-\frac{1}{2}} - \frac{1}{24}\Delta^3 y_{x-3/2} + \cdots$$

Sol. Bessel's formula is

$$y_x = \frac{y_0 + y_1}{2} + (x - \tfrac{1}{2})\,\Delta y_0 + \frac{x(x-1)}{2\,!}\cdot\frac{\Delta^2 y_{-1} + \Delta^2 y_0}{2} + \frac{\left(x - \frac{1}{2}\right)x\,(x-1)}{3\,!}\Delta^3 y_{-1} + \cdots$$

Changing x to $x + \frac{1}{2}$, we have

$$y_{x+1/2} = \frac{y_0 + y_1}{2} + x\Delta y_0 + \frac{\left(x + \frac{1}{2}\right)\left(x - \frac{1}{2}\right)}{2\,!}\cdot\frac{\Delta^2 y_{-1} + \Delta^2 y_0}{2} + \frac{x\left(x + \frac{1}{2}\right)\left(x - \frac{1}{2}\right)}{3\,!}\Delta^3 y_{-1} + \cdots$$

$$\therefore \quad \frac{dy_{x+1/2}}{dx} = \Delta y_0 + x.\frac{\Delta^2 y_{-1} - \Delta^2 y_0}{2} + \left(\frac{x^2}{2} - \frac{1}{24}\right)\Delta^3 y_{-1} + \cdots \qquad \text{...(6)}$$

Putting $x = 0, \dfrac{dy_{1/2}}{dx} = \Delta y_0 - \dfrac{\Delta^3 y_{-1}}{24} + \cdots$

Shifting the origin to $x - \frac{1}{2}$ we get

$$\frac{dy_x}{dx} = \Delta y_{x-1/2} - \frac{1}{24}\Delta^3 y_{x-3/2} + \cdots$$

$$= \delta y_x - \frac{1}{24}\delta^3 y_x + \cdots \qquad \ldots(6)$$

where $\qquad \delta y_x = y_{x+1/2} - y_{x-1/2}.$

Ex. Differentiating with respect to x and then putting $x = 0$ in the Stirling's formula

$$y_x = y_0 + x\frac{\Delta y_0 + \Delta y_{-1}}{2} + \frac{x^2}{2!}\Delta^2 y_{-1} + \frac{x(x^2-1)}{3!}\left[\frac{\Delta^3 y_{-1} + \Delta^3 y_{-2}}{2}\right] + \cdots$$

and changing the origin to x, show that, as far as third differences,

$$\frac{dy_x}{dx} = \frac{2}{3}(y_{x+1} - y_{x-1}) - \frac{1}{12}(y_{x+2} - y_{x-2})$$

Hence, if the unit of differencing be h, a first approximation, is

$$\frac{dy_x}{dx} = \frac{y_{x+h} - y_{x-h}}{2h}.$$

Example 2. *Find the first and second derivatives of the function tabulated below at the point $x = 3{\cdot}0$.*

x	y	Δy	$\Delta^2 y$	$\Delta^3 y$	$\Delta^4 y$
3·0	– 14·000				
		3·968			
3·2	– 10·032		·768		
		4·736		·048	
3·4	– 5·296		·816		0
		5·552		·048	
3·6	– 0·256		·864		0
		6·416		·048	
3·8	6·672		·912		
		7·328			
4·0	14·000				

Sol. Newton-Gregory formula is

$$f(a + hx) = f(a) + x\Delta f(a) + \frac{x(x-1)}{2!}\Delta^2 f(a)$$

$$+ \frac{x(x-1)(x-2)}{3!}\Delta^3 f(a) + \cdots$$

Differentiating w.r.t. x twice and then putting $x = 0$ in the equations obtained, we get

$$hf'(a) = \Delta f(a) - \tfrac{1}{2}\Delta^2 f(a) + \tfrac{1}{3}\Delta^3 f(a) - \tfrac{1}{4}\Delta^4 f(a) + \tfrac{1}{5}\Delta^5 f(a),$$

$$h^2 f''(a) = \Delta^2 f(a) - \Delta^3 f(a) + \tfrac{11}{12}\Delta^4 f(a) - \tfrac{5}{6}\Delta^5 f(a) + \cdots$$

Here $a = 3,\ h = 0{\cdot}2$

$\therefore\quad 0{\cdot}2 f'(3) = 3{\cdot}968 - \tfrac{1}{2} \times {\cdot}768 + \tfrac{1}{2} \times {\cdot}48 = 3{\cdot}6$

$\therefore\qquad f'(3) = 18$

Also $(0{\cdot}2)^2 f''(3) = {\cdot}768 - {\cdot}048 = {\cdot}72$

$\therefore\qquad f''(3) = 18$

$\therefore\quad \dfrac{dy}{dx}$ and $\dfrac{d^2y}{dx^2}$ at $x = 3$ are both equal to 18.

Note. The function tabulated above is

$$y = x^3 - 9x - 14$$

Hence $\dfrac{dy}{dx} = 3x^2 - 9,\ \dfrac{d^2y}{dx^2} = 6x$

Putting $x = 3$, we get

$$\frac{dy}{dx} = 18,\ \frac{d^2y}{dx^2} = 18.$$

Thus the values obtained by numerical differentiation are correct.

Example 3. *Find the value of $f'(\cdot 04)$ from the following table.*

x	u	$y = f(u)$	Δy	$\Delta^2 y$	$\Delta^3 y$	$\Delta^4 y$
– 3	·01	·1023				
			24			
– 2	·02	·1047		0		
			24		1	
– 1	·03	·1071		1		– 1
			25		0	
0	·04	·1096		1		– 1
			26		– 1	
1	·05	·1122		0		
			26			
2	·06	·1148				

Sol. Since we want to find the derivative at a point near the middle of the table, we may use any central difference formula, the interval being equal. Gauss's formula is

$$y_x = y_0 + x\Delta y_0 + \frac{x^2 - x}{2}\Delta^2 y_{-1} + \frac{x(x^2-1)}{6}\Delta^3 y_{-1}$$
$$+ \frac{x^4 - 2x^3 - x^2 + 2x}{24}\Delta^4 y_{-2} + \cdots$$

Differentiating w.r.t. x and putting $x = 0$, we get

$$hy'_u = \Delta y_0 - \frac{1}{2}\Delta^2 y_{-1} - \frac{1}{6}\Delta^3 y_{-1} + \frac{1}{12}\Delta^4 y_{-2} + \cdots$$

$$\therefore \quad 01 f'(\cdot 04) = \cdot 0026 - \frac{1}{2} \times (\cdot 0001) - \frac{1}{6} \times (-\cdot 0001) + \frac{1}{12} \times (-\cdot 0001)$$

$$= \cdot 00254$$

$$\therefore \quad f'(\cdot 04) = 0{\cdot}254.$$

Example 4. *Using the following data, find f (5) :*

x	$f(x)$	$\Delta' f(x)$	$\Delta'^2 f(x)$	$\Delta'^3 f(x)$
0	4			
		11		
2	26		7	
		32		1
3	58		11	
		54		1
4	112		16	
		118		1
7	466		22	
		228		
9	922			

Sol. Newton's divided difference when third differences are constant, is

$$f(x) = f(x_0) + (x - x_0) f(x_0, x_1) + (x - x_0)(x - x_1) f(x_0, x_1, x_2)$$
$$+ (x - x_0)(x - x_1)(x - x_2) f(x_0, x_1, x_2, x_3)$$

Differentiating w.r.t. x, we have

$$f'(x) = f(x_0, x_1) + (x - x_0 + x - x_1) f(x_0, x_1, x_2) + [(x - x_1)(x - x_2)$$
$$+ (x - x_2)(x - x_0) + (x - x_0)(x - x_1)] f(x_0, x_1, x_2, x_3)$$

Put $x = 5$, $x_0 = 0$, $x_1 = 2$, $x_2 = 3$, $x_3 = 4$. We have

$$f'(5) = 11 + 8 \times 7 + [6 + 15 + 10] \times 1$$
$$= 11 + 56 + 31$$
$$= 98.$$

Example 5. *Find dy/dx at $x = 1$ from the following table by constructing a central difference table.*

X	0·7	0·8	0·9	1·0	1·1
y	0·644218	0·717356	0·783327	0·841471	0·891207
X	1·2	1·3			
y	0·932039	0·963558			

Sol. We form a difference table

X	x	y	$\Delta\begin{pmatrix}=\delta\\ x+\frac{1}{2}h\end{pmatrix}$	Δ^2	Δ^3	Δ^4
0·7	– 3	0·644218				
			73138			
0·8	– 2	0·717356		– 7167		
			65971		– 660	
0·9	– 1	0·783327		– 7827		79
			58144		– 581	
1·0	0	0·841471		– 8408		85
			48736		– 496	
1·1	1	0·891207		– 8904		87
			40832		– 409	
1·2	2	0·932039		– 9313		
			31519			
1·3	3	0·963558				

Use the formula I (c), page 110,

$$\frac{dy}{dx}=\frac{1}{2h}\left[y_1-y_{-1}-\frac{1}{6}(\Delta^2 y_1-\Delta^2 y_{-1})\right.$$
$$\left.+\frac{1}{30}(\Delta^4 y_1-\Delta^4 y_{-1})-\cdots\right]$$

$$\therefore\quad\left(\frac{dy}{dx}\right)=\frac{1}{0\cdot 2}\left[0\cdot 891207-0\cdot 783327+\frac{1}{6}(0\cdot 008904\right.$$
$$\left.-0\cdot 007827)+\frac{1}{30}(0\cdot 000087-0\cdot 000079+\cdots)\right]$$
$$=0\cdot 54030.$$

Note. The function tabulated above is $y=\sin x$ so that dy/dx at $x=1$ is cos 1. The tabulated value $=0\cdot 540302$.

Example 6. *Determine the constants a, b, c and d such that the interpolating polynomial* $y_x=y(x_0+sh)=ay_0(x_0-h)+by_1(x_0+h)+\cdots$

Sol. We have

$$y(x_0+sh)=y(x_0)+shy'(x_0)+\frac{s^2h^2}{2\,!}y''(x_0)+\frac{s^3h^3}{3\,!}y'''(x_0)+\cdots \qquad ...(1)$$

Putting $s=-1$ and $s=+1$ successively, we have

$$y(x_0-h)=y(x_0)-hy'(x_0)+\frac{h^2}{2\,!}y''(x_0)-\frac{h^3}{3\,!}y'''(x_0)+\cdots$$

and $$y(x_0+h)=y(x_0)+hy'(x_0)+\frac{h^2}{2\,!}y''(x_0)+\frac{h^3}{3\,!}y'''(x_0)+\cdots$$

Also $$y'(x_0-h)=y'(x_0)-hy''(x_0)+\frac{h^2}{2\,!}y'''(x_0)-\frac{h^3}{3\,!}y''''(x_0)+\cdots$$

and $$y'(x_0+h)=y'(x_0)+h(y'')(x_0)+\frac{h^2}{2\,!}y''(x_0)+\frac{h^3}{3\,!}y'''(x_0)+\cdots$$

Substituting these values, in equation (1) and then equating the coefficients of constant terms, h, h^2 and h^3, we have $1 = a + b$, $s = -a + b + c + d$, etc.

Solving these equations, we find

$$a = \tfrac{1}{4}(s-1)(s^2+s-2),$$
$$b = \tfrac{1}{4}(s+1)(2+s-s^2),$$
$$c = \tfrac{1}{4}(s-1)^2(s+1) \text{ and}$$
$$d = \tfrac{1}{4}(s-1)(s+1)^2.$$

EXERCISE 5

1. Prove the following formulae which give derivatives in terms of differences.

(*a*) $y' = \dfrac{dy}{dx} = \dfrac{1}{h}(\Delta y - \tfrac{1}{2}\Delta^2 y + \tfrac{1}{3}\Delta^3 y - \tfrac{1}{4}\Delta^4 y + \cdots)$

and $y'' = \dfrac{d^2y}{dx^2} = \dfrac{1}{h^2}(\nabla^2 y + \nabla^3 y + \tfrac{11}{12}\nabla^4 y + \cdots)$

(*b*) $y' = \dfrac{1}{h}\left(\delta y - \dfrac{1}{24}\delta^3 y + \dfrac{3}{640}\delta^5 y - \cdots\right)$

and $y'' = \dfrac{1}{h^2}\left(\delta^2 y - \dfrac{1}{12}\delta^4 y + \dfrac{1}{90}\delta^6 y - \cdots\right)$

(*c*) $y' = \dfrac{\mu}{h}\left(\delta y - \dfrac{1}{6}\delta^3 y + \dfrac{1}{30}\delta^5 y - \cdots\right)$

$= \dfrac{1}{2h}\left[(y_1 - y_{-1}) - \dfrac{1}{6}(\delta^2 y_1 - \delta^2 y_{-1}) + \dfrac{1}{30}(\delta^4 y_1 - \delta^4 y_{-1})\right]$

and $y'' = \dfrac{1}{h^2}\left[\delta^2 y - \dfrac{1}{12}\delta^4 y + \dfrac{1}{90}\delta^6 y - \cdots\right]$

[**Hint.** (*a*) $\dfrac{dy}{dx} = Dy = \dfrac{1}{h}[\log(1+\Delta)]\,y = $ etc.

$$\frac{d^2y}{dx^2} = D^2 y = \frac{1}{h}\cdot hDy' = -\frac{1}{h}\log(1-\nabla)\,y'$$
$$= \frac{1}{h}\log(1-\nabla) - \frac{1}{h}\log(1-\nabla)\,y = \frac{1}{h^2}[\log(1-\nabla)]^2 y$$
$$= \frac{1}{h^2}(\nabla + \tfrac{1}{2}\nabla^2 + \tfrac{1}{3}\nabla^3 + \cdots)^2\, y = \text{etc.}$$

(*b*) and (*c*) See Ex. 2 (*b*) and (*c*) page 88.]

2. Deduce from Bessel's formula the following approximations :

$$\frac{d}{dx}y_x = \Delta y_{x-1/2} - \frac{1}{24}\Delta^3 y_{x-3/2}.$$
$$\frac{d^2}{dx^2}y_x = \tfrac{1}{2}(\Delta^2 y_{x-3/2} + \Delta^2 y_{x-1/2}).$$

$$\frac{d^3}{dx^3} y_x = \Delta^2 y_{x-3/2}.$$

[**Hint.** See Example 1, page 105. Differentiate (6) page 105 again]

3. Prove :

(*a*) $\frac{d}{dx} y_x = \frac{2}{3} (y_{x+1} - y_{x-1}) - \frac{1}{12} (y_{x+2} - y_{x-2})$ approx.

(*b*) $\frac{d}{dx} y_x = \frac{1}{h} (y_{x+h} - y_{x-h}) - \frac{1}{2h} (y_{x+2h} - y_{x-2h})$

$$+ \frac{1}{3h} (y_{x+3h} - y_{x-3h}) - \cdots \text{ approx.}$$

$\left[\textbf{Hint.}\ (b)\ \text{R.H.S.} = \frac{1}{h} \log \left(\frac{1+E}{1+E^{-1}} \right) y_x = \frac{1}{h} \log \cdot E y_2 = D y_x \right]$

4. Show that the expression given below are approximations to the third derivative of y_x :

(*a*) $\Delta^3 y_0 + (x - \frac{2}{3}) \Delta^4 y_0$;

(*b*) $\Delta^3 y_{-1} + (x - \frac{1}{2}) \cdot \frac{1}{2} (\Delta^4 y_{-2} + \Delta^4 y_{-1})$

[**Hint.** For (*a*), differentiate three times w.r.t. *x*, Newton-Gregory formula and for (*b*), differentiate Bessel's formula.]

5. (*a*) Find $\frac{dy}{dx}$ and $\frac{d^2 y}{dx^2}$ at $x = 1$ from the following table by constructing a central difference table

(*i*)

x :	1	2	3	4	5	6
y :	198669	295520	389418	479425	564642	644217

(*ii*)

x :	1	2	4	8	10
y :	0	1	5	21	27

(*iii*)

x :	1	1·2	1·4	1·6	1·8	2
y :	− 1	0·128	1·544	3·296	5·432	8

(*b*) Obtain $f'(0{\cdot}160)$ from the following table :

x :	0·145	0·150	0·155	0·165	0·175	0·180
y :	0·144492	0·149438	0·154380	·164252	0·174108	0·179030

6. (*a*) Find the first and second derivatives of $\log_e x$ at $x = 500$.

x	$\log x$
50C	6·214608
510	6·234411
520	6·253829
530	6·272877
540	6·291569
550	6·309918

(*b*) Take 10 figure logarithm to base 10 from $x = 300$ to $x = 310$ by unit increment. Calculate the first three derivatives of log x when $x = 300, x = 305, x = 310$.

7. (*a*) Find $\frac{dy}{dx}$ and $\frac{d^2y}{dx^2}$ of $y = x^{1/3}$ at $x = 50$ from the following table :

x	$y = x^{1/3}$
50	3·6840
51	3·7084
52	3·7325
53	3·7563
54	3·7798
55	3·8030
56	3·8259

(*b*) Given the following pairs of values of x and $y = f(x)$

x :	1	2	4	8	10
y :	0	1	5	21	25

Determine numerically the first and second derivatives of $f(x)$ as $x = 4$.

8. From the following table of values compute the third derivative of $f(\theta)$ when the argument θ has the value 5 :

θ	$f(\theta)$
2	57
4	1345
9	66340
13	402052
16	1118209
21	4287844
29	21242820

9. From the following data, find $f'(10)$:

x :	3	5	11	27	34
$f(x)$:	– 13	23	899	17315	35606

10. Use Stirling's Interpolation formula, to find the first derivative of the function $y = 2e^x - x - 1$ tabulated below, at the point $x = 0{\cdot}6$.

x	y
0·4	1·5836494
0·5	1·7974426
0·6	2·0442376
0·7	2·3275054
0·8	2·6510818

11. From the following table find the value of x for which $f(x)$ is a maximum and the maximum value of $f(x)$ in the range of x :

(*a*)

x :	60	75	90	105	120
$f(x)$:	28·2	38·2	43·2	40·9	37·7

[Put $X = \frac{x - 90}{15}$; $h = 15$. Use Stirling's formula to find $f(X)$]

(*b*)

x :	1·2	1·3	1·4	1·5	1·6
$f(x)$:	0·9320	0·9636	0·9855	0·9975	0·9996

12. Calculate $f'(7{\cdot}50)$ from the values using Bessel's formula :

x :	7·47	7·48	7·49	7·50	7·51	7·52	7·53
$f(x)$:	0·193	0·195	0·198	0·201	0·203	0·206	0·208

[Find $f(x)$ by using Bessel's formula

$$f'(x) = \Delta f(0) - \tfrac{1}{4}\Delta^2 f(-1) + \Delta^2 f(0) + \frac{1}{12}\Delta^3 f(-1) + \cdots \text{ and } X = \frac{x - 7{\cdot}50}{{\cdot}01}$$]

13. Find the first and second derivatives of the function tabulated below at the point $x = 0{\cdot}6$:

x :	0·4	0·5	0·6	0·7	0·8
y :	1·5836	1·7974	2·0442	2·3275	2·6511

14. Calculate $f'(3)$ and $f''(3)$ from

x :	3·0	3·2	3·4	3·6	3·8	4·0
$f(x)$:	– 14·000	– 10·032	– 5·296	0·256	6·672	14·000

[The function tabulated is $y = x^3 - 9x - 143$.]

15. (*a*) From the following table of values, find $f'''(5)$:

x :	2	4	9	13	16	21	29
$f(x)$:	57	1345	66340	402052	1118209	4217844	21242820

[**Hint.** 1st diff. : 644, 12999, 83928, 238719, 633927, 2119372

2nd diff. 1765, 7881, 22113, 49401, 114265

3rd diff. 556, 1186, 2274, 4054 , 4th diff. 45, 64, 89, 5th diff. 1, 1

$f'''(x) = 3x^2 f(x_0, x_1, x_2, x_3) + 4x^3 f(x_1, x_1, x_2, x_3, x_4) + \cdots$]

(*b*) Find $f'(1.5), f''(1.5)$ and $f'''(1.5)$ with the help of the following table

x :	1.5	2.0	2.5	3.0	3.5	4.0
$f(x)$:	3.375	7.00	13.675	24.00	38.875	59.00

16. A rocket is launched from the ground; its velocities were recorded as :

t	0	10	20	30	40	50
v	0	200	600	1200	2000	3000

Find the acceleration at $t = 50$ *(A.M.I.E. 1990)*

Hint. Acc. $= \left(\dfrac{dv}{dt}\right)_{t=5.0}$ .So we need differentiation $\dfrac{dy}{dt} = \dfrac{1}{h}\left[\Delta v_0 - \dfrac{1}{2}\Delta^2 v_o + \dfrac{1}{3}\Delta^3 v_o\right]_{v=3000}$

17. In a certain machine a slider moves along a fixed straight rod. Its distance x cms along the rod is given below for various values of the time t seconds. Find (*i*) the velocity of the slider and (*ii*) its acceleration when $t = 0{\cdot}3$

t :	0.0	0·1	0·2	0·3	0·4	0·5	0·6
x :	3·013	3·162	3·287	3·364	3·395	3·381	3·324

18. Find the value of x for which y is minimum. *(A.M.I.E. 1995)*

x :	0.60	0.65	0.70	0.75
y :	0.6221	0.6166	0.6138	0.6170

19. Find the maximum and minimum values of $f(x)$ for the following table :

x :	0	1	2	3	4	5
$f(x)$:	0	0.25	0	0.25	16.00	56.25

(B.Sc. B.R. 1989)

6

NUMERICAL QUADRATURE OR NUMERICAL INTEGRATION

6·1. Introduction. Sometimes we come across integrals which cannot be easily or exactly evaluated. Further if the values of the integrand for a few values of the variables are given, we cannot integrate. In such circumstances the numerical methods for integration are very useful. Numerical quadrature is the process of finding or evaluating the value of a definite integral from a set of numerical values of the integrand. When applied to the integration of a function of a single variable, the process is known as *quadrature*. The problem of numerical integration is solved by first approximating the integrand by a polynomial with the help of an interpolation formula and then integrating this approximation between the smaller limits between which the entire interval is divided. The areas are then added up to obtain the integral for the entire range.

6·2. A General Quadrature Formula for Equidistant Ordinates. Let $I = \int_a^b y\, dx$ where $y = f(x)$. Suppose $f(x)$ is given for certain equidistant values of x say $x_0,\ x_0 + h,\ x_0 + 2h, \ldots$ Let the range (a, b) be divided into n equal parts, each of width h so that $b - a = nh$.

Let $x_0 = a,\ x_1 = x_0 + h = a + h,\ x_2 = a + 2h, \ldots, x_n = a + nh = b$. We have assumed that the $n + 1$ ordinates $y_0, y_1, \ldots, y_n$ are equidistant.

$$\therefore \quad I = \int_a^b y\, dx = \int_{x_0}^{x_0+nh} y_x\, dx = \int_0^n y_{x_0 + hu}\, h\, du$$

where $\quad u = \dfrac{x - x_0}{h},\ dx = h\, du$

or $$I = h \int_0^n \left[y_0 + u\Delta y_0 + \frac{u(u-1)}{2\,!} \Delta^2 y_0 + \frac{u(u-1)(u-2)}{3\,!} \Delta^3 y_0 + \cdots \right] du$$

$$= h\left[ny_0 + \frac{n^2}{2}\Delta y_0 + \left(\frac{n^3}{3} - \frac{n^2}{2}\right)\frac{\Delta^2 y_0}{2!}\right.$$

$$\left. + \left(\frac{n^4}{4} - n^3 + n^2\right)\frac{\Delta^3 y_0}{3!} + \cdots \text{ up to } (n+1) \text{ terms}\right] \qquad ...(6\cdot1)$$

This is the *general quadrature formula*, A number of formulae can be deduced from this by putting $n = 1, 2, 3, \ldots$

6·3. Some important Approximate Quadrature Formulae.

6·3·1. The Trapezoidal Rule.

Put $n = 1$ in (6·1) and neglect second and higher differences.

We get

$$\int_{x_0}^{x_0+h} y\,dx = h\left[y_0 + \tfrac{1}{2}\Delta y_0\right]$$

$$= h\left[y_0 + \frac{y_1 - y_0}{2}\right] = h\left(\frac{y_0 + y_1}{2}\right)$$

Similarly $$\int_{x_0+h}^{x_0+2h} y\,dx = h\left(\frac{y_1 + y_2}{2}\right),$$

....

$$\int_{x_0+(n-1h)}^{x_0+nh} y\,dx = h\left(\frac{y_{n-1} + y_n}{2}\right)$$

Adding these n integrals, we obtain

$$\int_{x_0}^{x_0+nh} y\,dx = h\left[\tfrac{1}{2}(y_0 + y_n) + (y_1 + y_2 + \cdots + y_{n-1})\right]$$

= Distance between two consecutive ordinates

× [mean of the first and the last ordinate

+ sum of all the intermediate ordinates]

This rule is known as the Trapezoidal rule.

Note. Here we have a curve through two points (x_0, y_0) and $(x_0 + h, y_0 + k)$ only, so we suppose that y is a linear function in x, that is the equation of the curve is of the form

$y = a + bx$

6.3.2. Romberg Integration

We know that if h be the interval of differencing, then trapzoidal rule gives $\int_{x_0}^{x_0+h} dx = \frac{h}{2}(y_0 + y_1)$

Since this method gives only rough value of the integral, a modification is needed which is done with the help of truncation error. Suppose the error given by trapezoidal method is Ah^2 where A is a constant. Now let us evaluate I by Trapezoidal method

$$I = \int_{x_0}^{x_n} y dx$$

Assume that we are taking only two values of h, say h_1 and h_2 so that exact value

$$I = I_1 + \text{error} = I_1 + Ah_1^2 \quad \text{...(1)}$$

and $I = I_2 + \text{error} = I_2 + Ah_2^2$ ... (2)

As I_1 and I_2 are different approximate values obtained by Traperiodal rule so will be the two errors where A is a constant to be determined

From (1) and (2)

$$\Rightarrow \quad A = (I_1 - I_2)/(h_2^2 - h_1^2) \quad \text{...(3)}$$

Clearly the value of I_1 which is better than either of the two values (1) and (2) is obtained by finding the arithemetic mean or mean values of the two values (1) and (2)

$$\therefore \quad I = \frac{1}{2}\left(I_1 + I_2\right) + \frac{1}{2}A\left(h_1^2 + h_2^2\right)$$

$$= \frac{I_1 + I_2}{2} + \frac{(I_1 - I_2)}{2} \cdot \frac{h_1^2 + h_2^2}{(h_2^2 - h_1^2)}, \text{ using (3)}$$

$$= \frac{(I_1 + I_2)(h_2^2 - h_1^2) + (I_1 - I_2)(h_1^2 + h_2^2)}{2(h_2^2 - h_1^2)} = \frac{I_1 R_2^2 - I_2 h_1^2}{h_2^2 - h_1^2}$$

For particular values of $h_1 = h$ and $h_2 = \frac{1}{2}h$,

$$I = \frac{I_1\left(\frac{h^2}{4}\right) - I_2(h^2)}{\frac{1}{4}h^2 - h^2} = \frac{4I_2 - I_1}{3} = I_2 + \frac{1}{3}I_2 - \frac{1}{3}I_1$$

$$\therefore \quad I = I_2 + \frac{1}{3}(T_2 - I_1)$$

This formula gives us much improved value of *I*. The above process can be improved further by giving values of h_1 and h_2 till we arrive at two values of I which are very near to each other. This process of improvement is known as Romber g method.

Example. Find the value of the integral $I = \int_0^1 f(x)\,dx$ where $f(x) = \dfrac{1}{1+x}$, using Romberg integration.

Solution. We take $h = \frac{1}{2}, \frac{1}{4}$ and $\frac{1}{8}$ and calculate the values of I

$$h = \frac{1}{2}\begin{cases} x: & 0 \quad ½ \quad 1 \\ y & 1 \quad ⅔ \quad ½ \end{cases},\; h = \frac{1}{4}\begin{cases} x_0 & ¼ \quad ½ \quad ¾ \quad 1 \\ y_1 & ⅘ \quad ⅔ \quad 4/7 \quad ½ \end{cases}$$

Table

N	h	I
2	$\frac{1}{2}$	0.70833
4	$\frac{1}{4}$	0.69702
8	$\frac{1}{8}$	0.69412

$$h = \frac{1}{8},\quad \begin{matrix} x & 0 & 1/8 & 2/8 & 3/8 & 4/8 & 5/8 & 6/8 & 7/8 & 1 \\ \frac{1}{1+x}: & 1 & 8/9 & 8/10 & 8/11 & 8/12 & 8/13 & 8/14 & 8/15 & 8/16 \\ & f_0 & f_1 & f_2 & f_3 & f_4 & f_5 & f_6 & f_7 & f_8 \end{matrix}$$

$I_1 = \dfrac{h}{2}$ [Sum of the first and the last ordinate + 2 (Sum of the remaining ordinates]

$$= \frac{1}{4}\left[\left(\frac{1}{2}+1\right) + 2(0.6666)\right] = \frac{1}{4}\left[\frac{3}{2} + 1.3332\right] = 0.70833$$

$$I_2 = \frac{\frac{1}{4}}{2}\left[\left(\frac{1}{2}+1\right) + 2(0.8 + 0.6666 + 0.5714)\right] = 0.69702$$

$$I_3 = \frac{1/8}{2}\left[\left(\frac{1}{2}+1\right) + 2\,(0.8889 + 0.8 + 0.7273 + 0.6667 + 0.6154 + 0.5333 + 0.5714]\right.$$

$= 0.69412$

$$\therefore \quad I = I_2 + \frac{1}{3}(I_2 - I_1) \qquad ...(2)$$

$$= 0.69702 + \frac{1}{3}(0.6702 - 0.70833) = 0.69325 \qquad ..(3)$$

and $I = 0.69412 + \frac{1}{3}(0.69412 - 0.69702) = 0.69315$ ...(4)

Taking values given by (3) and (4) and using formula (2), we get

$$I = 0.69315 + \frac{1}{3}(0.69315 - 0.69325)$$

$$= 0.69315$$

If we further improve, then the approximation

$$I = 0.69315 + \frac{1}{3}(0.69315 - 0.69315)$$

$$= 0.69315 = 0.6931 \text{ up to four decemal places}$$

This is best approximation to $I = \int_0^1 \frac{dx}{1+x}$

Exact value of $I = \log(1+x)\Big|_0^1 = \log_2 = 0.6931$ up to four decimal places.

6·3·3. Simpson's One-third Rule. Put $n = 2$ in (6·1) and neglect third and higher differences, we get

$$\int_{x_0}^{x_0+2h} y\,dx = h\left[2y_0 + 2(y_1 - y_0) + \frac{\left(\frac{8}{3} - 2\right)}{2}(y_2 - 2y_1 + y_0)\right]$$

$$= \frac{h}{3}(y_0 + 4y_1 + y_2)$$

Similarly, $\int_{x_0+2h}^{x_0+4h} y\,dx = \frac{h}{3}(y_2 + 4y_3 + y_4)$

....

$$\int_{x_0+(n-2)n}^{x_0+nh} y\,dx = \frac{h}{3}(y_{n-2} + 4y_{n-1} + y_n),$$

when n is even.

Adding all these integrals, we have when n is even

$$\int_{x_0}^{x_0+nh} y\,dx = \frac{h}{3}[(y_0 + y_n) + 4(y_1 + y_3 + \cdots + y_{n-1}) + 2(y_2 + y_4 + \cdots + y_{n-2})]$$

This formula is known as *Simpson's one-third rule.*

Note. Since we have neglected all differences above the second, y must be a polynomial of second degree only, that is

$$y = ax^2 + bx + c.$$

Here we are given three points whose abscissae are x_0, $(x_0 + h)$ and $(x_0 + 2h)$.

Example. *Show that Simpson's one-third rule applied thrice gives*

$$\int_0^6 u_x\,dx = \frac{1}{3}[(u_0 + u_6) + 4(u_1 + u_3 + u_5) + 2(u_2 + u_4)]$$

6·3·4. Simpson's Three-eighth Rule. Put $n = 3$ in (6·1) and neglect all differences above the third, then

$$\int_{x_0}^{x_0+3h} y\,dx = h\left[3y_0 + \tfrac{9}{2}(y_1 - y_0) + \tfrac{5}{4}(y_2 - 2y_1 + y_0) + \tfrac{3}{8}(y_3 - 3y_2 + 3y_1 - y_0)\right]$$

$$= \frac{3h}{8}[y_0 + 3y_1 + 3y_2 + y_3]$$

Similarly $$\int_{x_0+3h}^{x_0+6h} y\,dx = \frac{3h}{8}[y_3 + 3y_4 + 3y_5 + y_6]$$

Adding such expressions as these from x_0 to x_n where n is a multiple of 3, we have

$$\int_{x_0}^{x_0+nh} y\,dx = \frac{3h}{8}[(y_0 + y_n) + 3(y_1 + y_2 + y_4 + y_5 + \cdots + y_{n-1})$$

$$+ 2(y_3 + y_6 + \cdots + y_{n-3})]$$

and is known as *Simpson's three-eighth* rule

On applying twice this rule we obtain

$$\int_0^6 u_x\,du = \frac{3}{8}[(u_0 + u_6) + 3(u_1 + u_2 + u_4 + u_5) + 2u_3]$$

Here we have neglected all differences above the third, so that we have assumed that y is a polynomial of the third degree, that is of the form

$$y = ax^3 + bx^2 + cx + d.$$

Change of scale. Let $y = hx$ so that the values of y are at equidistant interval of h and

$$dy = hdx$$

$$\therefore \qquad \int_0^{2h} u_y \cdot \frac{dy}{h} = \frac{1}{3}(u_0 + 4u_h + u_{2h})$$

or $$\int_0^{2h} u_x d_x = \frac{h}{3}(u_0 + 4u_h + u_{2h})$$

Thus in order to make the scale h times the original scale we have to multiply the right side by h and also each suffix of u by h.

6·3·5. Weddle's Rule. Put $n = 6$ in (6·1) and neglect all differences above the sixth, we obtain

$$\int_{x_0}^{x_0+6h} y\,dx = h\left[6y_0 + 18\Delta y_0 + 27\Delta^2 y_0 + 24\Delta^3 y_0 + \frac{123}{10}\Delta^4 y_0 + \frac{33}{10}\Delta^5 y_0 + \frac{41}{140}\Delta^6 y_0\right]$$

If we replace the last term by $\frac{3}{10}\Delta^6 y_0$, the error made is $\frac{h}{140}\Delta^6 y_0$ which will be negligible if the value of h is such that the sixth differences are small. With this assumptions, we have on replacing the last term by $\frac{3}{10}\Delta^6 y_0$

$$\int_{x_0}^{x_0+6h} y\,dx = \frac{3h}{10}[y_0 + 5y_1 + y_2 + 6y_3 + y_4 + 5y_5 + y_6]$$

Similarly,

$$\int_{x_0+6h}^{x_0+12h} y\,dx = \frac{3h}{10}[y_6 + 5y_7 + y_8 + 6y_9 + y_{10} + 5y_{11} + y_{12}]$$

....

$$\int_{x_0+(n-6)h}^{x_0+nh} y\,dx = \frac{3h}{10}[y_{n-6} + 5y_{n-5} + y_{n-4} + 6y_{n-3} + y_{n-2} + 5y_{n-1} + y_n],$$ if n is a multiple of 6.

Adding all these integrals, we have if n is a multiple of 6,

$$\int_{x_0}^{x_0+nh} y\,dx = \frac{3h}{10}[y_0 + 5y_1 + y_2 + 6y_3 + y_4 + 5y_5 + 2y_6 + 5y_7 + y_8 + \cdots]$$

This formula is known as *Weddle's Rule* and is more accurate, in general, than Simpson's Rule, but it requires at least seven consecutive values of the function. Here we have assumed that y is of the form

$$y = ax^6 + bx^5 + cx^4 + dx^3 + ex^2 + fx + g$$

Thus if $f(x)$ is a polynomial of the 5th degree or lower, Weddle's rule gives an exact result. Weddle's rule is generally more accurate than any of the others. Of the two Simpson rules, the one-third rule is better.

Example 1. *Calculate by Simpson's rule an approximate value of* $\int_{-3}^{+3} x^4\,dx$ *by taking seven equidistant ordinates. Compare it with exact value and the value obtained by using the Trapezoidal rule.*

Sol. Here $h = 1$, $y_{-3} = (-3)^4$, $y_{-2} = (-2)^4$, $y_{-1} = (-1)^4$, $y_0 = 0$, $y_1 = 1^4$, $y_2 = 2^4$ and $y_3 = 3^4$.

Hence by Simpson's rule with $h = 1$,

$$\int_{-3}^{+3} x^4\,dx = \tfrac{1}{3}\,[(y_{-3} + y_3) + 2(y_{-1} + y_1) + 4(y_{-2} + y_0 + y_2)]$$

$$= \tfrac{1}{3}\,[(-3)^4 + 3^4 + 2\{(-1)^4 + 1^4\} + 4\{(-2)^4 + 0 + 2^4\}]$$

$$= \tfrac{1}{3}\,(162 + 4 + 128) = 98.$$

Again the exact value

$$\int_{-3}^{+3} x^4\,dx = \left[\frac{1}{5}x^5\right]_{-3}^{+3} = \frac{1}{5}\{3^5 - (-3)^5\} = 97{\cdot}2$$

By the trapezoidal rule, the approximate value

$$= \tfrac{1}{2}\,h\{y_1 + y_7 + 2(y_2 + y_3 + y_4 + y_5 + y_6)\}$$

$$= \tfrac{1}{2}\,[y_{-3} + y_3 + 2(y_{-2} + y_{-1} + y_0 + y_1 + y_2)]$$

$$= \tfrac{1}{2}\,[(-3)^4 + 3^4 + 2\{(-2)^4 + (-1)^4 + 0 + 1^4 + 2^4\}]$$

$$= \tfrac{1}{2}\,(162 + 68) = 115.$$

This shows that the trapezoidal rule does not give an accurate result in this case. In general Simpson's rule gives a better result than the trapezoidal rule.

Example 2. *Calculate the approximate value of*

$$\int_0^{\frac{1}{2}\pi} \sin x\,dx$$

(i) by Trapezoidal rule,

(ii) by Simpson's rule, using 11 ordinates

Sol. Divide the range $\left(0, \frac{1}{2}\pi\right)$ into ten equal parts, so that $h = \pi/20$, and from trigonometric tables :

$y_1 = \sin 0 \quad = 0{\cdot}0000, \qquad y_2 = \sin \pi/20 \quad = 0{\cdot}1564,$

$y_3 = \sin \pi/10 = 0{\cdot}3090,$ $y_4 = \sin 3\pi/20 = 0{\cdot}4540$

$y_5 = \sin \pi/5 = 0{\cdot}5878,$ $y_6 = \sin \pi/4 = 0{\cdot}7071,$

$y_7 = \sin 3\pi/10 = 0{\cdot}8090,$ $y_8 = \sin 7\pi/20 = 0{\cdot}8910,$

$y_9 = \sin 2\pi/5 = 0{\cdot}9511,$ $y_{10} = \sin 9\pi/20 = 0{\cdot}9877.$

$y_{11} = \sin \pi/2 = 1{\cdot}000$

Using the Trapezoidal rule, approximate value of the integral

$$= h\left\{\left(\tfrac{1}{2}y_1 + y_{11}\right) + (y_2 + y_3 + \cdots + y_{10}\right)\right\}$$

$$= (1/20)\pi(0{\cdot}5 + 5{\cdot}8531) = 0{\cdot}9981.$$

Using Simpson's rule, the approximate value of the integral

$$= (h/3)[y_1 + y_{11} + 2(y_3 + y_5 + y_7 + y_9) + 4(y_2 + y_4 + y_6 + y_8 + y_{10})$$

$$= (\pi/60)(1 + 5{\cdot}3138 + 12{\cdot}7848) = 19{\cdot}0986 \times 0{\cdot}0524 = 1{\cdot}0006$$

Actual value of $\int_0^{\pi/2} \sin x\, dx = [-\cos x]_0^{\pi/2} = 1.$

Hence error in calculation by Simpson's rule = 0·0006 in excess

" " " " " Trapezoidal " = 0·0019 by defection

There there is a greater accuracy in the calculation by Simpson's rule.

Example 3. *Find by Weddle's rule the value of the integral*

$$I = \int_4^{5\cdot 2} \log_e x\, dx$$

Sol. Since, for Weddle's rule, the number of sub-divisions must be a multiple of six, we divide the interval of integration, *viz.*, = 1·2 into six equal parts, each of width 0·2. Hence $h = 0{\cdot}2$. The value of log x for each point of sub-divisions is given below :

x	$\log_e x$	x	$\log_e x$
4·0	1·38629463	4·8	1·56861592
4·2	1·43508453	5·0	1·60943791
4·4	1·48160454	5·2	1·64865863
4·6	1·52605630		

$$\therefore\ I = (0{\cdot}3)(0{\cdot}2)[3{\cdot}03495299 + 5(30{\cdot}445244) + 3{\cdot}0522046$$
$$+ 6(1{\cdot}52605630)] + (1{\cdot}56861592 + 5(1{\cdot}60943791) + 1{\cdot}64865863$$
$$= 1{\cdot}9273674.$$

6·4. Euler-Maclaurin Summation Formula

The definite integral

$$\int_a^{a+nh} f(x)\,dx$$

can be regarded as the measure of the area included between the curve $y = f(x)$, the ordinates $x = a$ and $x = a + nh$, and the axis of x. Let D and C be the two points on the curve whose abscissae are a and $a + nh$ respectively so that the area required is the area of the figure $ABCD$. Divide the base AB into n equal parts, each part of length h, at the points L, M, ...and with the corresponding points, $P, Q, \ldots$ on the curve.

Area of the quadrilateral $ALPD = \frac{1}{2} h[y + y_{a+h}]$

Area of the quadrilateral $LMQP = \frac{1}{2} h[y_{a+h} + y_{a+2h}]$, etc.

Area $ABCD$ = sum of the areas of the quadrilaterals $ALPD$, $LMQP$, etc., as a rough approximation.

$$\therefore \quad \frac{1}{h}\int_a^{a+nh} y\,dx = \tfrac{1}{2} y_a + y_{a+h} + y_{a+2h} + \cdots + y_{a+(n-1)h}$$

$$+ \tfrac{1}{2} y_{a+nh} + R_n \qquad \text{...(6·2)}$$

where R_n is the correction term.

Consider the case when $y = e^{cx}$, c being a constant.

$$\therefore \quad \frac{1}{h}\int_a^{a+nh} e^{cx}\,dx = \tfrac{1}{2} e^{ca} + e^{c(a+h)} + e^{c(a+2h)} \ldots$$

$$+ e^{c(a+n-1h)} + \tfrac{1}{2} e^{c(a+nh)} + R_n$$

$$= [e^{ca} + e^{c(a+h)} + \cdots + e^{c(a+\overline{n-1}h)}]$$

$$+ \tfrac{1}{2}[e^{c(a+nh)} - e^{ca}] + R_n$$

$$\Rightarrow \frac{1}{ch}(e^{c(a+nh)} - e^{ca}) = e^{ca}\left(\frac{e^{nch} - 1}{e^{ch} - 1}\right) + \tfrac{1}{2} e^{ca}(e^{cnh} - 1) + R_n$$

$$\Rightarrow \qquad R_n = e^{ca}(e^{cnh} - 1)\left[\frac{1}{ch} - \frac{1}{2} - \frac{1}{e^{ch} - 1}\right]$$

$$\text{Now} \quad \frac{1}{e^{\theta} - 1} = \frac{1}{\theta + \dfrac{\theta^2}{2!} + \dfrac{\theta^3}{3!} + \cdots} = \frac{1}{\theta}\left(1 + \frac{\theta}{2} + \frac{\theta^2}{6} + \cdots\right)^{-1}$$

$$= \frac{1}{\theta} - \frac{1}{2} + \frac{B_1}{2!}\theta - \frac{B_2}{4!}\theta^3 + \frac{B_3}{6!}\theta^5 - \cdots$$

where the numbers B, which are called Bernoullian numbers, have the numerical values

$$B_1 = \frac{1}{6},\ B_2 = \frac{1}{30},\ B_3 = \frac{1}{42},\ B_4 = \frac{1}{30},\ B_5 = \frac{5}{66}, \cdots$$

$$\therefore \quad R_n = e^{ca}(e^{cnh} - 1)\left[-\frac{B_1}{2!}ch + \frac{B_2}{4!}c^3h^3 - \frac{B_3}{6!}c^5h^5 + \cdots\right]$$

Now $\quad y_{a+nh} - y_1 = e^{c(a+nh)} - e^{ca} = e^{ca}(e^{chn} - 1)$

$$y_x = e^{cx}$$

$$\therefore \quad y'_x = ce^{ex}$$

$$\therefore \quad y'_{a+nh} - y'_a = ce^{ca}(e^{cnh} - 1)$$

$$y''_x = c^2e^{cx}$$

$$\therefore \quad y''_{a+nh} - y''_a = c^2e^{ca}(e^{cnh} - 1), \text{ etc.}$$

$$\therefore \quad R_n = -\frac{B_1}{2!}, h[y'_b - y'_a] + \frac{B_2}{4!}h^3[y'''_b - y'''_a] - \frac{B_3}{6!}h^5[y^v_b - y^v_a] + \cdots$$

where $b = a + nh$.

From (6·2), we have

$$\frac{1}{h}\int_a^{a+nh} y\,dx = \left[\tfrac{1}{2}y_a + y_{a+h} + y_{a+2h} + \cdots + y_{a+(n-1)h} + \tfrac{1}{2}y_{a+nh}\right]$$

$$-\frac{h}{12}(y'_{a+nh} - y'_a) + \frac{h^3}{720}(y'''_{a+nh} - y'''_a) - \cdots$$

$$\Rightarrow \quad \frac{1}{h}\int_a^b y\,dx = \sum_{r=0}^{n} y_{a+rh} - \tfrac{1}{2}(y_a + y_b) - R_n \qquad \ldots(6\cdot3)$$

where $\quad R_n = \frac{h}{12}[y'_b - y'_a] - \frac{h^3}{720}(y'''_b - y'''_a] + \frac{h^5}{30240}(y^v_b - y^v_a) - \cdots$

$$\Rightarrow \quad \sum_0^n y_{a+rh} = \frac{1}{h}\int_a^b y\,dx + \tfrac{1}{2}(y_a + y_b) + R_n,\ nh = b - a \qquad \ldots(6\cdot4)$$

This is *Euler-Maclaurin summation formula.* It is useful for finding the approximate sum of any number of consecutive values of a function when these values are given for equidistant values of x, provided the integral $\int_a^b f(x)\,dx$ can be easily evaluated.

Alternative Proof. We shall prove the result with the help of expansion of the operators :

If $\quad f(x) = \Delta F(x)$, then

$$\sum_{x=0}^{n-1} f(x) = f(0) + f(1) + \cdots + f(n-1)$$

$$= \Delta F(0) + \Delta F(1) + \cdots + \Delta F(n-1)$$

$$= [F(1) - F(0)] + [F(2) - F(1)] + \cdots + [F(n) - F(n-1)]$$

$$= F(n) - F(0)$$

Now $f(x) = \Delta F(x) \Rightarrow F(x) = \Delta^{-1} f(x) = (E-1)^{-1} f(x)$

$$F(x) = \left[D + \frac{D^2}{2!} + \frac{D^3}{3!} + \cdots \right]^{-1} f(x) \qquad [\because E = e^D]$$

$$= D^{-1} \left[1 - \frac{D}{2} + \frac{D^2}{12} - \frac{D^4}{720} + \cdots \right] f(x)$$

$$= \left[D^{-1} - \frac{1}{2} + \frac{D}{12} - \frac{D^3}{720} + \cdots \right] f(x)$$

$$= D^{-1} f(x) - \frac{1}{2} f(x) + \frac{1}{12} f'(x) - \frac{1}{720} f'''(x) + \cdots$$

$$= \int f(x) dx - \frac{1}{2} f(x) + \frac{1}{12} f'(x) - \frac{1}{720} f'''(x) + \cdots$$

$$\therefore \quad F(n) - F(0) = \int_0^n f(x) dx - \frac{1}{2} \left[f(n) - f(0) \right]$$

$$+ \frac{1}{12} \left[f'(n) - f'(0) \right] - \frac{1}{720} \left[f'''(n) - f'''(0) \right] + \cdots$$

But left hand side is equal to $f(0) + f(1) + \cdots + f(n-1)$

$$\therefore \quad \int_0^n f(x) dx = f(0) + f(1) + \cdots f(n-1) + \frac{1}{2} \left[f(n) - f(0) \right]$$

$$- \frac{1}{12} \left[f'(n) - f'(0) \right] + \frac{1}{720} \left[f'''(n) - f'''(0) \right]$$

$$= \frac{1}{2} f(0) + f(1) + \cdots + f(n-1) + \tfrac{1}{2} f(n) - \frac{1}{12} \left[f'(n) - f'(0) \right]$$

$$+ \frac{1}{720} \left[f'''(n) - f'''(0) \right] - \cdots$$

$$\Rightarrow \sum_0^n f(n) = \int_0^n f(x) dx + \frac{1}{2} \left[f(0) + f(n) \right] + \frac{1}{12} \left[f'(n) - f'(0) \right]$$

$$+ \frac{1}{720} \left[f'''(n) - f'''(0) \right] + \cdots$$

Example 4. (*a*) *Evaluate* :

$1^p + 2^p + \dots + n^p$ *where p is a positive integer.*

(b) *Show that* $\sum_1^n r^7 + \sum_1^n r^5 = 2\left(\sum_1^n r^3\right)^2$

(*B.A. Statistics Hons. Delhi, 1995*)

Sol. (a) Putting $y = f(x) = x^p$, $f'(x) = px^{p-1}$,

$f'''(x) = p(p-1)(p-2)x^{p-1}$, etc.; $a = 0, h = 1$ in (6·2), we get

$$\sum_0^n r^p = \int_0^n x^p\,dx + \frac{1}{2}(0 + n^p) + R_n$$

where $$R_n = \frac{1}{12}(pn^{p-1}) - \frac{1}{720}p(p-1)(p-2)\,n^{p-3}$$

$$+ \frac{1}{30240}p(p-1)(p-2)(p-3)(p-4)\,n^{p-5} - \dots$$

$$\Rightarrow \quad \sum_0^n r^p = \frac{n^{n+1}}{p+1} + \frac{n^p}{2} + \frac{p}{12}n^{p-1} - \frac{p(p-1)(p-2)}{720}n^{p-3} + \dots$$

(b) From the previous example, we have on putting $p = 7$ and 5 respectively,

$$\sum_1^n r^7 = \frac{1}{8}n^8 + \frac{1}{2}n^7 + \frac{7}{12}n^6 - \frac{7}{24}n^4 + \frac{1}{12}n^2 - \frac{1}{24}$$

and $$\sum_1^n r^5 = \frac{1}{6}n^6 + \frac{1}{2}n^5 + \frac{5}{12}n^4 - \frac{1}{12}n^2 + \frac{1}{252}$$

$$\therefore \quad \sum_1^n r^7 + \sum_1^n r^5 = \frac{1}{8}n^4(n+1)^4, \text{ on simplification}$$

$$= 2\left(\sum_1^n r^3\right)^2$$

Example 5. (i) *Find the sum of*

$$\frac{1}{51^2} + \frac{1}{53^2} + \frac{1}{55^2} + \dots + \frac{1}{99^2}$$

(ii) *Evaluate* : $\sum_{m=0}^{\infty} \frac{1}{(10+m)^2}$ (*B.Sc. Ranchi, 1996*)

Sol. (i) Put $f(x) = \frac{1}{x^2}$, $a = 51, b = 99 = 51 + 24h \quad \therefore h = 2$

Now $f'(x) = -\frac{2}{x^3}, f'''(x) = -\frac{24}{x^5}, f^{v}(x) = -\frac{72}{x^7}, \cdots$ in (6·3)

$$\sum_{x=51}^{99} \frac{1}{x^2} = \frac{1}{2}\int_{51}^{99} \frac{1}{x^2}\,dx + \frac{1}{2}\left[\frac{1}{51^5} + \frac{1}{99^2}\right]$$

$$+\frac{1}{3}\left[\frac{1}{51^3} - \frac{1}{99^3}\right] - \frac{4}{15}\left[\frac{1}{51^5} - \frac{1}{99^5}\right] + \frac{16}{21}\left[\frac{1}{51^7} - \frac{1}{99^7}\right]$$

$= 0{\cdot}0049988$, on simplification.

(*ii*) Put $f(x) = \frac{1}{(a+x)^2}$

$$\therefore \quad \int_0^\infty f(x)dx = \frac{1}{a},\ f'(0) = -\frac{2}{a^3},\ f'''(0) = -\frac{24}{a^5},\ f^{v}(0) = -\frac{720}{a^7}$$

Put $\quad a = 10,$

$$\therefore \quad \sum_0^\infty \frac{1}{(10+m)^2} = \frac{1}{10} + \frac{1}{2}\cdot\frac{1}{10^2} + \frac{1}{12}\cdot\frac{2}{10^3} - \frac{1}{720}\cdot\frac{24}{10^5}$$

$$= {\cdot}1 + {\cdot}005 + {\cdot}0006667 - {\cdot}0000003$$

$$= {\cdot}1051664.$$

Example 6. *Show that if $f(x)$ is a polynomial function,*

$$f(1) - f(2) + f(3) - f(4) + \cdots + (-1)^{x-1} f(x)$$

$$= (-1)^{x-1}\left[\frac{1}{2}f(x) + \frac{2^2-1}{2!}B_1 f'(x) - \frac{2^4-1}{4!}B_2 f'''(x) + \cdots\right] + \textit{a constant.}$$

Sol. We have

$$\frac{1}{h}\int_a^{a+nh} f(x)dx = \tfrac{1}{2}f(a) + f(a+h) + f(a+2h) + \cdots$$

$$+ f(a+\overline{n-1}h) + \tfrac{1}{2}f(a+nh) - \frac{B_1}{2!}h[f'(a+nh) - f'(a)]$$

$$+ \frac{B_2}{4!}h^3\,[f'''(a+nh) - f'''(a)] \qquad \text{...(1)}$$

Put $\quad n = 2m, a = 0, h = 1$ in (1),

$$\therefore \quad \int_0^{2m} f(x)dx = \tfrac{1}{2}f(0) + f(1) + f(2) + \cdots + f(2m-1) + \tfrac{1}{2}f(2m)$$

$$- \frac{B_1}{2!}[f'(2m) - f'(0)] + \frac{B_2}{4!}[f'''(2m) - f'''(0) - \cdots \qquad \text{...(2)}$$

Also, put $n = m, a = 0, h = 2$ in (1) we get

$$\int_0^{2m} f(x)dx = f(0) + 2[f(2) + f(4) + \cdots + f(2m-2)] + f(2m)$$

$$-\frac{2^2 B_1}{2\,!}[f'(2m) - f'(0)] + \frac{2^4 B_2}{4\,!}[f'''(2m) - f'''(0)] \qquad ...(3)$$

Subtracting (2) from (3), we get the required result.

Example 7. *Evaluate*

$$\int_0^1 \frac{dx}{1+x},$$

correct upto five decimal places by Euler-Maclaurin's formula.

Sol. In Euler-Maclaurin's formula, we put

$a = 0, a + nh = 1$ and $h = 0{\cdot}1 \therefore n = 10$,

$$f(x) = \frac{1}{1+x},\ f'(x) = -\frac{1}{(1+x)^2},\ f''(x) = -\frac{2}{(1+x)^3},\ f'''(x) = -\frac{6}{(1+x)^4},$$

$$\therefore I = \int_0^1 \frac{1}{1+x}\,dx = \frac{1}{2}\cdot\frac{1}{1} + \frac{1}{1{\cdot}1} + \frac{1}{1{\cdot}2} + \cdots + \frac{1}{1{\cdot}9}$$

$$+\frac{1}{2}\cdot\frac{1}{2} - \frac{{\cdot}1}{12}\left[-\frac{1}{2^3} + \frac{1}{1^3}\right] + \frac{({\cdot}1)^3}{720}\left[-\frac{6}{2^4} + \frac{6}{1^4}\right] - \cdots$$

$$= {\cdot}5 + {\cdot}90909 + {\cdot}8333 + \cdots + {\cdot}25000 - \frac{1}{120}\times\frac{3}{4} + \frac{1}{120000}\times\frac{15}{16}$$

$$= 6{\cdot}93773 - {\cdot}00625 + {\cdot}00001$$

$$= 6{\cdot}93149$$

$$\therefore \quad \int_0^1 \frac{1}{1+x}\,dx = 0{\cdot}69315 \text{ approx.}$$

Thus we get the value of $\log_e 2$ as $0{\cdot}69315$ correct to five places.

6·5. Newton-Cotes formula.

Consider Lagrange's formula

$$f(x) = \frac{(x-x_1)(x-x_2)...(x-x_n)}{(x_0-x_1)(x_0-x_2)...(x_0-x_n)} f(x_0)$$

$$+\frac{(x-x_0)(x-x_2)...(x-x_n)}{(x_1-x_0)(x_1-x_2)...(x_1-x_n)} f(x_1)$$

$$+\cdots+\frac{(x-x_0)(x-x_1)...(x-x_{n-1})}{(x_n-x_0)(x_n-x_1)...(x_n-x_{n-1})} f(x_n)$$

Integrating between the limits x_0 and $x_0 + nh$, we have

$$I = \int_{x_0}^{x_0+nh} f(x)dx$$

$$= H_0 f(x_0) + H_1 f(x_1) + \cdots + H_i f(x_i) + \cdots + H_n f(x_n) \quad ...(1)$$

where $$H_i = \frac{1}{(x_i - x_0)(x_i - x_1)...(x_i - x_n)} \times \int_{x_0}^{x_0+nh} \frac{(x - x_0)(x - x_1)(x - x_2)...(x - x_n)}{x - x_i} dx$$

Let $x_{i=1} x_i + h$ for all i so that $x_i = x_0 + ih$. Further $x = x_0 + th$ so that $dx = hdt$, t being the new real variable.

$$\therefore \quad H_i = \frac{1}{(ih)(i-1)h....1.h(-1)h(-2)h...[-(n-i)]h} \times \int_0^n \frac{th(t-1)h...(t-n)h.hdt}{(t-i)h}$$

$$= \frac{1}{i\,!\,(-1)^{n-i}(n-i)\,!\,h^n} \int_0^n \frac{t(t-1)...(t-n)}{h(t-i)} h^{n+2} dt$$

$$\Rightarrow \quad H_i = \frac{(-1)^{n-i}\,h}{i\,!\,(n-i)\,!} \int_0^n \frac{t(t-1)...(t-n)}{t-i} dt \quad ...(2)$$

(1) is known as Cote's formula with H_i given by (2).

Replacing H_i by $nh.\, C_i^n$, we have

$$I = nh. \sum_{i=0}^{n} f(x_i) C_i^n = (x_n - x_0) \sum_0^n f(x_i) C_i^n = (b - a) \sum_{i=0}^{n} f(x_i)\, C_i^n$$

The numbers C_i^n , $0 \le i \le n$, are called Cote's numbers.

Properties of Cote's numbers.

(*i*) $\overset{n}{\underset{k}{C}} = \overset{n}{\underset{n-k}{C}}$ or $H_i = H_{n-i}$ (*ii*) $\sum_{k=0}^{n} \overset{n}{\underset{k}{C}} = 1$

We shall prove (*i*) and leave (*ii*) for reader to prove.

(*i*) In (2), put $n - i$ for i and $t = n - s$, then we get

$$H_{n-i} = \frac{(-1)^i\, h}{i\,!\,(n-i)\,!} \int_n^0 \frac{(-1)^{n-2}(s-n)(s-n+1)...(s)}{(n-s)-(n-i)} (-ds)$$

$$= \frac{(-1)^i(-1)^n h}{i!\,(n-i)!} \int_0^n \frac{s(s-1)...(s-n)}{(s-i)} ds$$

$$= \frac{(-1)^{n-i} h}{i!\,(n-i)!} \int_0^n \frac{s(s-1)...(s-n)}{s-i} ds, = H_i \qquad [\because (-1)^i = (-1)^{-i}]$$

Note. Thus the coefficients of Newton-Cotes formula are symmetric from both the ends. Also it is sufficient to calculate only half the coefficients and the remaining half are obtained by property of symmetry.

Cor. *Derivation of Trapezoidal and Simpson's one-third rule.*

Put $n = 1$ in (1) on page 134, we get

$$I = \int_{x_0}^{x_0+h} f(x)dx = H_0 f(x_0) + H_1 f(x_1)$$

where $$H_0 = h \int_0^1 \frac{(t-1)}{(0-1)} dt = \frac{h}{2}$$

and $$H_0 = \int_0^1 \frac{(t-0)}{(1-0)} dt = \left| \frac{t^2}{2} \right|_0^1 = \frac{h}{2}$$

Thus $$\int_{x_0}^{x_0+h} f(x)\,dx = \frac{h}{2}[f(x_0) + f(x_1)]$$

which is Trapezoidal rule.

Put $n = 2$ in (1) page 126, so that we have

$$\int_{x_0}^{x_0+2h} f(x)dx = H_0 f(x_0) + H_1 f(x_1) + H_2 f(x_2)$$

where $$H_0 = \frac{(-1)^2 h}{0!\,2!} \int_0^2 (t-1)(t-2)\; dt = \tfrac{1}{3} \cdot h,$$

$$H_1 = \frac{(-1)^1}{1!\,1!} - h \int_0^2 (t-2)\,dt = \tfrac{4}{3}h, \text{ and}$$

$$H_2 = H_0 = \tfrac{1}{3}\,h.$$

$$\therefore \quad \int_{x_0}^{x_0+2h} f(x)dx = \frac{h}{3}[f(x_0) + 4f(x_1) + f(x_2)]$$

Similarly, Simpson's $\frac{3}{8}$-th rule and Weddle's rule can be deduced from (1) by putting $n = 3$ and $n = 6$ respectively.

Cote's numbers for some values of n

	C_0^n	C_1^n	C_2^n	C_3^n	C_4^n	C_5^n	C_6^n	Total $\sum_{k=0}^{n} {}^nC_k$	Remark ${}^nC_k = C_{n-k}^{n}$
1	$\frac{1}{2}$	$\frac{1}{2}$						1	
2	$\frac{1}{6}$	$\frac{4}{6}$	$\frac{1}{6}$					1	
3	$\frac{1}{8}$	$\frac{3}{8}$	$\frac{3}{8}$	$\frac{1}{8}$				1	
4	$\frac{7}{90}$	$\frac{32}{90}$	$\frac{12}{90}$	$\frac{32}{90}$	$\frac{7}{90}$			1	
5	$\frac{19}{288}$	$\frac{75}{288}$	$\frac{50}{288}$	$\frac{50}{288}$	$\frac{75}{288}$	$\frac{19}{288}$		1	
6	$\frac{41}{840}$	$\frac{216}{840}$	$\frac{27}{840}$	$\frac{272}{840}$	$\frac{27}{840}$	$\frac{216}{840}$	$\frac{41}{840}$	1	

Example 8. *If y_x is a polynomial in x of the third degree, find an expression for $\int_0^t y_x\,dx$ in terms of y_0, y_1, y_2 and y_3. Use this result to show that*

$$\int_1^2 y_x\,dx = \frac{1}{24}(-y_0 + 13y_1 + 13y_2 - y_3).$$

Sol. We have $y_x = \frac{(x-1)(x-2)(x-3)}{(-1)(-2)(-3)} y_0 + \frac{(x(x-2)(x-3)}{1(-1)(-2)} y_1$

$$+ \frac{x(x-1)(x-3)}{(2)(1)(-1)} y_2 + \frac{x(x-1)(x-2)}{(3)(2)(1)} y_3$$

$$= -\tfrac{1}{6}(x^3 - 6x^2 + 11x - 6)\, t_0 + \tfrac{1}{2}(x^3 - 5x^2 + 6x)\, y_1$$

$$- \tfrac{1}{2}(x^3 - 4x^2 + 3x)\, y_2 + \tfrac{1}{6}(x^2 - 3x^2 + 2x)\, y_3$$

$$\therefore \quad \int_0^t y_x\,dx = \left[-\frac{1}{6}\left(\frac{x^4}{4} - 2x^3 + \frac{11x^3}{2} - 6x\right) y_0 \right.$$

$$+ \frac{1}{2}\left(\frac{x^4}{4} - \frac{5}{3}x^3 + 3x^2\right) y_1 - \frac{1}{2}\left(\frac{x^4}{4} - \frac{4}{3}x^3 + \frac{3x^2}{2}\right) y$$

$$\left. + \frac{1}{6}\left(\frac{x^4}{4} - x^3 + x^2\right) y_3 \right]_0^t$$

$$\therefore \quad \int_0^1 y_x\,dx = -\frac{1}{6}\left(\frac{-9}{4}\right)y_0 + \frac{1}{2}\left(\frac{19}{12}\right)y_1 - \frac{1}{2}\left(\frac{5}{12}\right)y_2 + \frac{1}{6}\left(\frac{1}{4}\right)y_3$$

and
$$\int_0^2 y_x\,dfx = -\frac{1}{6}(-2)\,y_0 + \frac{1}{2}\left(\frac{8}{3}\right)y_1 - \frac{1}{2}\left(-\frac{2}{3}\right)y_2 + \frac{1}{6}(0)\,y_3$$

Subtracting, we get

$$\int_1^2 y_x\,dx = \frac{1}{24}(-y_0 + 13y_1 + 13y_2 - y_3).$$

6·6. Stirling's Formula for Approximation to Factorials. In Euler-Maclaurin formula 6·3, put $a = 1, h = 1$, and write

$$n-1 \quad \text{for} \quad n, f(x) = \log_e x,\ f'(x) = \frac{1}{x},\ f'''(x) = \frac{2}{x^3}, \cdots$$

We get,

$$\int_1^n \log x\,dx = \frac{1}{2}\log 1 + \log 2 + \log 3 + \cdots + \log(n-1)$$

$$+\frac{1}{2}\log n - \frac{1}{12}\left(\frac{1}{n} - \frac{1}{1}\right) + \frac{1}{720}\left(\frac{2}{n^3} - \frac{2}{1^3}\right) - \cdots$$

$$\Rightarrow \quad \Big[x\log x - x\Big]_1^n = \log(n\,!) - \frac{1}{2}\log n - \frac{1}{12n} + \frac{1}{360n^n} - \cdots$$

$$\Rightarrow \quad \log(n\,!) = \left(n + \frac{1}{2}\right)\log n - n + \frac{1}{12n} + \frac{1}{360n^3} + \cdots + c$$

where c is a constant independent of n. ...(6·5)

To find c, we makes use of Walli's formula in Trigonometry,

viz., $$\frac{\pi}{2} = \operatorname*{Lt}_{n\to\infty}\left[\frac{2\cdot2}{1\cdot3}\cdot\frac{4\cdot4 6\cdot6}{3\cdot5 5\cdot7}\cdots\frac{(2n)(2n)}{(2n-1)(2n+1)}\right]$$

$$= \operatorname*{Lt}_{n\to\infty}\frac{2^{4n}(n\,!)^4}{(2n\,!)(2n+1)}$$

$$\therefore \quad \log\frac{\pi}{2} = \operatorname*{Lt}_{n\to\infty}\left[4n\log 2 + 4\log(n\,!) - 2\log(2n\,!) - \log(2n+1)\right)$$

Substituting the values of $\log(n\,!)$ and $\log(2n\,!)$ from (6·5), we get

$$\log\frac{\pi}{2} = \operatorname*{Lt}_{n\to\infty}\left[4n\log 2 + 4\{(n+\tfrac{1}{2})\log n - n + c\}\right.$$

$$\left. - 2\{(2n+\tfrac{1}{2})\log 2n - 2n + e)\} - \log(2n+1)\right]$$

$$= \underset{n \to \infty}{\text{Lt}} [2c + 2\log n - \log 2n - \log(2n+1)]$$

$$= \underset{n \to \infty}{\text{Lt}} \left[2c + \log \frac{n^2}{4n^2 + 2n} \right] = 2c + \log\left(\tfrac{1}{4}\right)$$

$$= 2c - \log 4$$

$$c = \tfrac{1}{2}\log 2\pi = \log \sqrt{2\pi}$$

Substtituting the value of c in (6·5), we gett

$$\log(n!) = \left(n + \tfrac{1}{2}\right)\log n + \log\sqrt{2\pi} - n + \frac{1}{12n} - \frac{1}{360n^3} + \cdots$$

$$= \log\left(\sqrt{2\pi}\,.\,n^{n+1}\right) + \log e^{-n}\left(1 - \frac{1}{12n^2} + \frac{1}{360n^4} - \cdots\right)$$

$$\therefore \qquad n! = \sqrt{2\pi}\, n^{n+\frac{1}{2}}\,.\,e^{-n}\left(1 - \frac{1}{12n^2} + \frac{1}{360n^4} - \cdots\right)$$

$$\Rightarrow \qquad n! \simeq \sqrt{2\pi}\, e^{-n}\, n^{n+\frac{1}{2}}.$$

This is Stirling's formula for factorial n, when n is large.

Aliter. $\int_{x=1}^{k} \log x\, dx =$ area of the curve *PQRS*

$$= \frac{1}{2} QR(PQ + RS) + \varepsilon_k \ldots$$

$$= \frac{1}{2}\left[\log(k-1) + \log k\right] + \varepsilon_k \quad \ldots(1)$$

where ε_k is small.

Y

y = log x

P

O k–1 Q' Rk X

Put $k = 2, 3, 4, \ldots$ and add

$$\int_1^n \log x\, dx = \log n! - \frac{1}{2}\log n + \sum_2^n \varepsilon_i$$

$$(n\log n - n + 1) = \log n! - \tfrac{1}{2}\log n + \sum_2^n \varepsilon_i$$

$$\Rightarrow \qquad \log n! < \left(n + \tfrac{1}{2}\right)\log n - n + 1 + \Sigma_2^n\, \varepsilon_i$$

$$\Rightarrow \qquad n! < n^n\,.\,\sqrt{n}\,e^{-n} - \varepsilon_k$$

Also $\int_{k-1}^{k} \left(\frac{1}{x} - \frac{1}{k}\right)^2 dx > 0$ since integrand is positive,

$$\Rightarrow \qquad \frac{1}{k(k-1)} + \frac{1}{k^2} - \frac{2}{k}\log\frac{k}{k-1} > 0 \qquad \ldots(2)$$

$$\Rightarrow \quad \frac{2}{k}\log\frac{k}{k-1} < \frac{k}{k(k-1)} + \frac{1}{k^2} \Rightarrow \log\frac{k}{k-1} < \frac{2k-1}{2k(k-1)}$$

From (1), $\Big[n \log n - n \Big]_{k}^{k-1} = \frac{1}{2} = [\log k + \log (k-1)] + \varepsilon_k$

$$\therefore \quad \left(k - \frac{1}{2}\right)\log\frac{k}{k-1} - 1 = \varepsilon_k \Rightarrow (k - \tfrac{1}{2})\frac{2k-1}{2k\,(k-1)} - 1 > \varepsilon_k$$

$$\Rightarrow \quad \frac{(2k-1)\,(2k-1)}{4k\,(k-1)} - 1 > \varepsilon_k \Rightarrow \frac{1}{4k\,(k-1)} > \varepsilon_k$$

$$\Rightarrow \quad \sum_{k=2}^{n} \varepsilon_k < \frac{1}{4}\left(1 - \frac{1}{n}\right) < \frac{1}{4}$$

$$\log n\,! > \left(n + \frac{1}{2}\right)\log n - n + 1 - \frac{1}{4}$$

i.e., $$\log n\,! > \left(n + \frac{1}{2}\right)\log n - n + \frac{3}{4}$$

$$n\,! > n^n n\,!\, e^{-n} e^{3/4} \qquad \text{...(3)}$$

From (2) and (3) $n^n\, n\,!\, e^{-n} e^{3/4} < n\,! < n^n \sqrt{n}\, e^{-n}$

$$n\,! \simeq \sqrt{2\pi n}\, e^{-n} n^n.$$

This result is usefully applied in Statisties.

6·7. Gauss-Legendre formula for numerical integration. Sometimes it is not desired or not possible to express the integral $\int_a^b F(x)\,dx$ in terms of ordinates at equally spaced values of the abscissa, but it is possible to express in the following form at some number of points, but with differentt spacing and with better accuracy, Gauss's formula is expressed in the form

$$\int_a^b F(x)\,dx = \frac{b-a}{2}\Big[\, w_0 F(x_0) + w_1\, F(x_1) + \cdots + w_n\, F(x_n) \Big]$$

$$= \frac{b-a}{2} \sum_{i=0}^{n} w_i F(x_i) \qquad \text{...(1)}$$

where the w_i's are to be determined and x_i's are not equally spaced and w_i's and x_i's are called the weights and abscissae respectively.

Now we have $2n+2$ constants w_i and x_i to determine. Consequently we assume $F(x)$ to be a polynomial of degree $2n+2$.

Let $t = [2x - (a+b)]/(b-a)\,]$ with the result that the limits of integration become -1 to $+1$.

Hence $$\int_{-1}^{+1} F(t)\, dt = \sum_{i=0}^{n} w_i F(t_i) \qquad ...(2)$$

where the w_i's are to be determined, and x_i's are symmetrical with respect to the middle point of the interval.

Now $$\int_{-1}^{+1} F(t)\, dt = w_o F(t_0) + w_1 F(t_1) + \cdots + w_n F(t_n)$$

where $t_0, t_1, ..., t_n$ are the points of sub-division of the interval $(-1, 1)$. In other words, once the t_i's are fixed, then w_i's are uniquely determined,

where $$f(t) = f\left\{\frac{(b-a)t + (b+a)}{2}\right\}$$

Let $G_n(t)$ be the polynomial of degree n passing through the $(n+1)$ ordinates $F(t_i)$. The polynomial is unique and is of the form

$$G_n(t) = \sum_{i=0}^{n} a_i(t)\, F(t_i)$$

so that $$\int_{-1}^{+1} G_n(t) dt = \sum_{i=0}^{n} \int_{-1}^{+1} a_i(t)\, dt\, F(t_i)$$

$$F(t) = C_0 + C_1 t + C_2 t^2 + \cdots + C_{2n-1} t^{2n-1}$$

$$\therefore \quad \int_{-1}^{+1} F(t)\, dt = \int_{-1}^{+1} \left\{ C_0 + C_1 t + C_2 t^2 + \cdots + C_{2n-1} t^{2n-1} \right\} dt$$

$$= 2C_0 + \frac{2}{3} C_2 + \frac{2}{5} C_4 + \cdots \qquad ...(1)$$

Now $$F(t_i) = C_0 + C_1 t_i + C_2 t_i^2 + \cdots + C_{2n-1} t_i^{2n-1}$$

$$\therefore \quad \int_{-1}^{+1} F(x)\, dt = w_1 \left[C_0 + C_1 t_1 + C_2 t_1^2 + C_3 t_1^3 + \cdots + C_{2n-1} t_1^{2n-1} \right]$$

$$+ w_2 \left[C_0 + C_1 t_2 + C_2 t_2^2 + \cdots + C_{2n-1} t_2^{2n-1} \right]$$

$$+ w_3 \left[C_0 + C_1 t_3 + C_2 t_3^2 + C_{2n-1} t_3^{2n-1} \right]$$

$$+ \cdots\cdots\cdots\cdots\cdots\cdots\cdots\cdots$$

$$+ w_n \left[C_0 + C_1 t_n + C_2 t_n^2 + \cdots + C_{2n-1} t_n^{2n-1} \right]$$

$$= C_0 \left[w_1 + w_2 + w_3 + \cdots + w_n \right] + C_1 \left[w_1 t_1 + w_2 t_2 + \cdots + w_n t_n \right]$$

$$+ C_2\left[w_1t_1^2 + w_2t_2^2 + \cdots + w_n t_n^2 \right] + \cdots$$
$$+ \underset{2n-1}{C}\left[w_1t_1^{2n-1} + w_2t_2^{2n-1} + \cdots + w_n t_n^{2n-1} \right] \quad ...(2)$$

Since (1) and (2) are identical for all values of C_i. Hence comparing the coefficientts of C_i, we obtain $2n$ equations in $2n$ unknowns w_i and t_i :

$$\left.\begin{array}{r} w_1 + w_2 + \cdots + w_n = 2 \\ w_1t_1 + w_2t_2 + \cdots + w_nt_n = 0 \\ w_1t_1^2 + w_nt_2^2 + \cdots + w_nt_2^2 = \frac{2}{3} \\ \\ w_1t_1^{2n-1} + w_1t_2^{2n-1} + \cdots + w_nt_n^{2n-1} = 0 \end{array}\right\} \quad ...(3)$$

From (3) we can determine $2n$ unknowns w_i's and t_i's.

Thus if (2) is to be exact for polynomials of degree n, we should have

$$w_i = \int_{-1}^{+1} a_i(t)dt = \frac{1}{\underset{j \neq i}{\Pi}(t_j - t_i)} \int_{-1}^{+1} \prod_{j \neq i} (t - t_i)\, dt$$

$$\Rightarrow \qquad \int_{-1}^{+1} f(x)\, dx = \sum_{i=0}^{n} w_i f(t_i)$$

In other words, once the t_i^{2s} are fixed, then w_i are uniquely determined by the requirement that the formula holds for all polynomials of degree n.

Note 1. It is found that the x's are the zero's or the roots of the Legendre's polynomial $P_{2n+1}(x) = 0$ where

$$P_{n+1}(x) = \frac{1}{2^{n+1}(n+1)!} \frac{d^{n+1}}{dx^{n+1}} [(x^2 - 1)^{n+1}],$$

2. The following Table gives the values of the abscissas and weights for $n = 2$ to 6. for Gaussian integration

n	$\pm a_i$	w_i
2	0·5773	1·00000
3	0·00000	0·88888
	0·77459	0·55555
4	0·33998	0·652145
	0·86113	0·347854
5	0·000000	0·568888
	0·5384693	0·4786286
	0·9061798	0·2369268

	0·2386191861	0·4679139346
6	0·6612093865	0·3607615730
	0·9324695742	0·1713244934

Example 9. *Compute approximately the value of the integral* $I = \int_0^3 x^3 \cos x\, dx$ *using three point quadrature formula for the given numerical values of t's and w's :*

$$t_{-1} = -\sqrt{\frac{3}{5}}, \quad t_0 = 0 \quad t_1 = +\sqrt{\frac{3}{5}}$$

$$w_{-1} = \frac{5}{9} \quad w_0 = \frac{8}{9}, \quad w_1 = \frac{5}{9}$$

Sol. Here $a = 0$, $b = 3$. $\therefore\ x = \dfrac{(b-a)t + (b+a)}{2} = \dfrac{3}{2}(t+1)$

$$\therefore \quad dx = \frac{3}{2}\, dt.$$

$$\therefore \quad I = \int_0^3 x^2 \cos x\, dx = \frac{3}{2}\int_{-1}^{+1}\left[\frac{3}{2}(t+1)\right]^2 \cos\left[\frac{3(t+1)}{2}\right] dt$$

Using three point Gaussian quadrature formula, we find

$$I = \frac{3}{2}\left[\frac{5}{9}\left\{\frac{3}{2}(t_{-1}+1)\right\}^2 \cos\frac{3(t_{-1}+1)}{2} + \frac{8}{9}\left\{\frac{3}{2}(t_0+1)\right\}^2 \cos\frac{3(t_0+1)}{2} + \frac{5}{9}\left\{\frac{3}{2}(t_1+1)\right\}^4 \cos\frac{3(t_1+1)}{2}\right]$$

$$\Rightarrow \int_0^3 x^3 \cos x\, dx = -4{\cdot}936$$

Exact Value $= -4{\cdot}952$.

$\therefore$ Error in value up to three decinal places. $= -0{\cdot}016$.

Example 10. *Compute* $I = \int_0^1 x\, dx$ *by Gauss's formula with* $n = 4$ *up to 5 decimal places.*

Sol. We have $I = \int_0^1 x\, dx$. Here $a = 0, b = 1$

we change the limits of integration by the substitution

$$t = [2x - (a+b)]/(b-a).$$

$\therefore \quad x = \dfrac{1}{2}(t+1)$ so that when $x = 0, t = -1$ and when $x = 1, t = 1$

$$\therefore \qquad I = \frac{1}{4}\int_{-1}^{+1}(t+1)\,dt = \frac{1}{4}\sum_{i=j}^{n} w_i\, F(t_i)$$

Now choose $n = 4$ and use the table giving the values of the abscissac and weights.

Thus $$I \approx \tfrac{1}{4}[(-0{\cdot}86114+1)(0{\cdot}34785)+(-0{\cdot}33998+1)(0{\cdot}65214)$$
$$+(0{\cdot}33998+1)(0{\cdot}65214)+(0{\cdot}86114+1)(0{\cdot}34785)]$$
$$= \tfrac{1}{4}[0{\cdot}483024 + {\cdot}4304254 + {\cdot}8738545 + {\cdot}6473975]$$
$$= 0{\cdot}49999.$$

Exact value $= \int_0^1 x\,dx = 0{\cdot}5$. Error $= 0{\cdot}00001$ up to 5 decimal places.

EXERCISE 6

1. (*a*) Show that the area of a curve, divided into n parts by $n+1$ equidistant ordinates $y_0, y_1, \ldots, y_n$, is given aproximately by the series

$$ny_0 + \frac{n^2}{2}\Delta y_0 + \left(\frac{n^3}{3} - \frac{n^2}{2}\right)\frac{\Delta^2 y_0}{2\,!}$$
$$+ \left(\frac{n^4}{4} - n^3 + n^2\right)\frac{\Delta^3 y_0}{3\,!} + \ldots \text{ to } n+1 \text{ terms.}$$

$$\left[\textbf{Hint. } \int_0^n y_x dx = \int_0^n E^x y_0\, dx = \int_0^n (1+\Delta)^x y_0 dx = \text{etc.}\right]$$

(*b*) Show that by putting

$$x = \frac{x_0 + x_n}{2} + (x_n - x_0)\,t,\ f(x) = \phi(t),$$

$$I = \int_{x_0}^{x_n} f(x)\,dx = (x_n - x_0)\int_{-1/2}^{1/2}\phi(t)dt.$$

2. (*a*) If $y = A + Bx + Cx^2$, and y_0, y_1, y_2 are the values of y corresponding to $x = a, a+h, a+2h$ respectively, prove that

$$\int_a^{a+2h} ydx = \tfrac{1}{3}h\,(y_0 + 4y_1 + y_2)$$

For four values $a, a+h; a+2h, a+3h$ derive the formula

$$\int_a^{a+3h} ydx = \frac{3h}{8}[f(a) + 3f(a+h) + 3f(a+2h) + f(a+3h)]$$

(*b*) Derive Simpson's one third rule and find the effect of (*i*) change of origin and (*ii*) change of scale on this formula.

(*c*) A river is 80 feet wide. The depth d (in feet) of the river at a distance

x from one bank is given by the following table :

x	0	10	20	30	40	50	60	70	80
d	0	4	7	9	12	15	14	8	3

Find approximately the area of the cross section of the river using (*i*) five suitable points, (*ii*) all the points of the data. (*B.Sc. Ranchi, 1996*)

(*d*) A rocket is launched from the ground. Its acceleration is registered during the first 80 seconds and is given in the table below :

t (secs) :	0	10	20	30	40	50	60	70	80
a(m/sec^2) :	30·00	31·63	33·34	35·47	37·75	40·33	43·25	46·69	50·67

Find the velocity and height of the rocket at time $t = 80$. (*A.M.I.E. 1994*)

(*e*) The velocity of a train which starts from rest is given by the following table, the time being responed in minutes from start and speed in Km per hour.

Time (min)	2	4	6	8	10	12	14	16	18	20
Speed (Km/h)	10	18	25	29	32	20	11	5	2	8

Estimate approx. by Simpson's rule, the total distance in 20 minutes.

[**Hint.** (1) when $t = 0$, $x = 0$ (2) Minutes are to be converted into hour]

[**Hint.** For finding distance, apply Simpson's one-third rule, embering

$$S = \int_0^{20} v \text{ dt}.$$

3. (*a*) Show that if $f(x)$ is a cubic, then

$$\int_0^{2h} f(x)\,dx = \frac{1}{3}h\{f(0) + 4f(0) + 4f(h) + f(2h)\}.$$

Hence, deduce Simpson's rule for an approximate value of the area between two ordinates, the x-axis and the curve.

(*b*) Show that 1·62 is an approximate value of

$$\int_1^5 \frac{dx}{x}$$

if $h = 1$ in Simpson's rule.

(*c*) Apply Simpson's rule to estimate the value of the integral

$$\int_1^2 \frac{dx}{x},$$

by dividing the interval (1, 2) into 4 equal parts.

(*d*) Find using 7 ordinates the value of the integral

$$\int_2^3 \frac{dx}{x}$$

by Simpson's one-third rule correct to four places of decimals.

[**Hint.** Divide the interval (2, 3) into 6 equal parts.]

4. (*a*) State and prove Simpson's one-third rule for approximate integration or mechanical quadrature.

If $f(x)$ is a polynomial in x of third degree and

$$u_{-1} = \int_{-3}^{-1} f(x)\,dx, \quad u_0 = \int_{-1}^{1} f(x)\,dx, \quad u_1 = \int_{1}^{3} f(x)\,dx, \text{ then}$$

$$f(0) = \frac{1}{2}\left(u_0 - \frac{\Delta^2 u_{-1}}{12}\right)$$

(*b*) Calculate (upto 3 places of decimals)

$$\int_2^{10} \frac{dx}{1+x}$$

by dividing the range into eight equal parts.

(*c*) If u_x is a polynomial in x of second degree; then show that

(*i*) $\displaystyle\int_0^n u_x\,dx = n\left[\frac{3}{8}u_0 + \frac{1}{24}\left(19u_1 + \frac{5}{24}u_2 + \frac{1}{24}u_3\right)\right]$

(*ii*) $\displaystyle\int_0^1 u_x\,dx = \frac{1}{12}(5u_1 + 8u_0 - u_{-1})$ approx. (*B.Sc. Ranchi 1990*)

(*d*) Given that

$$e^0 = 1,\ e^1 = 2{\cdot}72,\ e^2 = 7{\cdot}39,\ e^3 = 20{\cdot}09,\ e^4 = 54{\cdot}60,$$

verify Simpson's rule by finding an approximate value of

$$\int_0^4 e^x\,dx$$

and compare it with the exact value. (*B.Sc. Orissa 1991*)

5. (*a*) Use Simpson's three-eights rule to obtain the value of

(*i*) $\displaystyle\int_0^{0\cdot3} (1 - 8x^3)^{1/2}\,dx.$ (*B.Sc. Guwahati, 1996*)

(*ii*) $\displaystyle\int_{1\cdot6}^{2\cdot2} f(x)\,dx$ given

x:	1·6	1·8	2·0	2·2
$f(x)$:	4·953	6·050	7·389	9·025

(*b*) Evaluate

$$\int_0^1 \frac{dx}{1+x^2}$$

(*i*) taking four equal parts and using Simpson's one-third rule and (*ii*) by using three-eights rule. Hence obtain the approximate value of π in each case. (*A.M.I.E. 1991, 1997*)

(*c*) Evaluate

$$\int_0^1 \sqrt{\sin x + \cos x}\,dx$$

correct to two decimal places using seven ordinates.

(*d*) Calculate the value of $\log_e 2$ by finding

$$\int_0^1 \frac{2x}{1+x^2}$$

using Simpson's rule by dividing the interval into four equal parts.

(*A.M.I.E. 1986* Antilog $z = 0{\cdot}69$)

6. (*a*) Compute the value of the definite integral

$$\int_{0\cdot2}^{1\cdot4} (\sin x - \log_e x + e^x)\, dx$$

by (*a*) the trapezoidal rule, (*b*) Simpson's one-third rule, (*c*) Simpson's three-eights rule and (*d*) Weddle's rule. After finding the true value of the integral, compare the error in the four cases.

(*b*) Calculate the approximate value of

$$\int_0^6 \frac{dx}{1+x^2}$$

using the Newton-Cote seven-point formula and compare the results with the true value $\tan^{-1} 6 = 1{\cdot}406$.

(*c*) Find the value of

$$\int_0^6 f(x)\, dx, \text{ using}$$

(*i*) Simpson's Rule (*ii*) Weddle's Rule from the following data :

x :	0	1	2	3	4	5	6
$f(x)$:	6·9897	7·4036	7·7815	8·1291	8·4510	8·7506	9·0309

7. (*i*) Evaluate

$$\int_0^1 e^x\, dx$$

approximately, in steps of 0·05 using (*a*) trapezoidal rule and (*b*) Simpson's one-third rule.

(*ii*) Evaluate

$$\int_{0\cdot5}^{0\cdot7} x^{1/2} e^{-x}\, dx$$

approximately by using a suitable formula and at least five points.

(*iii*) Evaluate $\int_0^1 \frac{dx}{1+x}$ approximately in step of 0·025 by Simpson's $\frac{1}{3}$ rule.

8. Obtain approximation to the value of or evaluate

(*i*) $\int_0^6 \frac{dx}{1+x^2}$

by applying (*a*) Weddle's rule (*b*) Simpson's one-third rule (*c*) Simpson's $\frac{3}{8}$ rule and compare the results with the true value $\tan^{-1} 6 = 1{\cdot}406$.

(*ii*) $\int_0^{\pi/2} (\sin x)\, dx$

(*iii*) $\int_{-1}^{+1} \sin (\log x)\, dx.$

[**Hint.** (*i*) (*b*) $\int_0^6 y_x\, dx = \int_0^2 y_x\, dx + \int_2^4 y_x\, dy + \int_4^6 y_x\, dx$

$$= \tfrac{1}{3}\,(y_0 + 4y_1 + y_2) + \cdots + \cdots$$

$$= \tfrac{1}{3}\,[(y_0 + y_6) + 2(y_2 + y_4) + 4(y_1 + y_3 + y_5)].$$

9. (*a*) Use Simpson's one-third rule to prove that $\log_e 7$ is approximately 1·95.

(*b*) If $u_x = a + bx + cx^2$ prove that

$$\int_1^3 u_x\, dx = 2u_2 + \frac{1}{12}\,(u_0 - 2u_2 + u_4)$$

and hence show that ·9921 is an approximation for

$$\int_{-\frac{1}{2}}^{+\frac{1}{2}} e^{-x^2/10}\, dx$$

[**Hint.**: (*a*) Change the origin to 1 in Simpson's formula and we get

$$\int_1^7 y_x\, dx = \tfrac{1}{3}\,[y_1 + 4y_2 + 2y_3 + 4y_4 + 2y_5 + 4y_6 + y_7]$$

and take $y_x = \frac{1}{x}$.]

(*b*) Shifting the origin to -2, we have

$$\int_{-1}^{+1} u_x\, dx = 2u_0 + \frac{1}{12}\,(u_{-2} - 2u_0 + u_2) \qquad \text{...(1)}$$

Change the scale to $\frac{1}{2}$ in (1) and take $u_x = e^{-x2/10}$

10. *Hardy's formula.* Show that if

$$y_x = a + bx + cx^2 + dx^3 + ex^4 + fx^5,$$

then $\int_{-3}^{+3} y_x\,dx = 2{\cdot}2y_0 + 1{\cdot}62(y_{-2}+y_2) + 0{\cdot}28(y_{-3}+y_3)$

or $$\int_0^6 y_x\,dx = 0{\cdot}28(y_0+y_6) + 1{\cdot}62(y_1+y_5) + 2{\cdot}2y_3$$

Using this, calculate the value of

$$\int_0^{\frac{1}{2}} (1-x^2)^{-1/2}\,dx$$

correct to four places of decimals.

11. Prove that

(*i*) $$\int_{-\frac{1}{2}}^{+\frac{1}{2}} f(x)\,dx = \frac{1}{2}\left\{f\left(-\frac{1}{2}\right)+f\left(\frac{1}{2}\right)\right\} + \frac{1}{24}\left\{\Delta f\left(-\frac{3}{2}\right) - \Delta f\left(\frac{1}{2}\right)\right\}$$

$$= \frac{1}{24}\left[13f\left(-\frac{1}{2}\right)+13f\left(\frac{1}{2}\right)-f\left(\frac{3}{2}\right)-f\left(-\frac{3}{2}\right)\right] \text{ approximately.}$$

(*ii*) $$\int_{-1}^{+1} f(x)\,dx = \frac{13}{12}[f(1)+f(-1)-f(3)-f(-3)] \text{ approximately.}$$

[**Hint.** (*i*) Use Lagrange's formula, given

$$f\left(-\frac{3}{2}\right), f\left(-\frac{1}{2}\right), f\left(\frac{1}{2}\right), f\left(\frac{3}{2}\right)$$

Aliter. $f(x) = f\left(-\frac{3}{2}+x+\frac{3}{2}\right) = E^{x+3/2} f\left(-\frac{3}{2}\right) = (1+\Delta)^{x+3/2} f\left(-\frac{3}{2}\right) =$ etc.]

12. Which of the two following formulae would you expect to give the better approximation for $\int_0^4 y_x\,dx$?

(*a*) $\frac{1}{9}\{5(y_0+y_4)+4(y_1+y_3)+18y_2\}$.

(*b*) $\frac{2}{15}\{2(y_0+y_2+y_4)+12(y_1+y_3)\}$.

[**Hint.** (*a*) Assume y_x to be of second degree.

(*b*) Assume y_x to be of first degree.]

13. Prove that, if $f(x)$ is a continuous function whose fifth differences are constant, then

$$\int_{-1}^{+1} f(x)\,dx = \frac{8}{9}f(0) + \frac{5}{9}\{f(\sqrt{0{\cdot}6}) + f(-\sqrt{0{\cdot}6})\}$$

[**Hint.** $f(x) = a + bx + cx^2 + dx^3 + ex^4 + fx^5$] (*B.A. Math. Hons. Delhi, 1984*)

14. (*a*) If the fourth differences of $f(x)$ can be neglected and

$$\int_{-1}^{+1} f(x)\,dx = \frac{2}{3}[f(x_1)+f(x_2)+f(x_3)]$$

find the values of x_1, x_2 and x_3.

(*b*) Find x_1 and x_2 such that the following approximate formula becomes exact to the highest possible order :

$$\int_{-1}^{+1} f(x)\,dx = \frac{1}{3}[f(-1) + 2f(x_1) + 3f(x_2)]$$

(*c*) If $f(x)$ is a function whose fifth differences are constant, then

$$\int_{-1}^{+1} f(x)\,dx$$

can be expressed in the form

$$pf(-\alpha) + qf(0) + pf(\alpha).$$

Find the values of p, q, and α. Use this formula to obtain the value of $\log_e 2$ to four places of decimals.

[**Hint.** (*a*) Let $f(x) = A + Bx + Cx^2 + Dx^3$, then

$$\int_{-1}^{+1} f(x)\,dx = 2A + \frac{2}{3}C = \frac{2}{3}\left[3A + B\Sigma x_1 + C\Sigma x_1^2 + D\Sigma x_1^3\right]$$

Compare the coefficients of B, C, and D and solve the equations for x_1, x_2 and x_3. In order to obtain log 2, change the origin to unity and make the scale $\frac{1}{2}$ and $f(x) = \frac{1}{1+x}\cdot$

(*c*) $f(x) = a + bx + cx^2 + dx^3 + ex^4 + fx^5$

15. (*a*) If third differences are constant, prove that

$$\int_0^2 f(x)\,dx = \frac{1}{24}\left[f\left(-\frac{1}{2}\right) + 23f\left(\frac{1}{2}\right) + 23f\left(\frac{3}{2}\right) + f\left(\frac{5}{2}\right)\right]$$

(*b*) Adapt the formula to find an approximate value of $\log_e 2$ from

$$\int_1^2 \frac{1}{x}\,dx \quad \text{or} \quad \int_0^1 \frac{1}{x+1}\,dx.$$

[**Hint.** (*a*) Use Newton's advancing difference formula with origin at $-\frac{1}{2}\cdot$

Aliter. $f(x) = f\left(-\frac{1}{2} + x + \frac{1}{2}\right) = E^{x+1/2} f\left(-\frac{1}{2}\right) = (1+\Delta)^{x+1/2} f\left(-\frac{1}{2}\right) =$ etc.]

or take $f(x) = a + bx + cx^2 + dx^3$

and show that $\int_0^4 f(x)\,dx = \frac{1}{12}[f(-1) + 23f(1) + 23f(3) + f(5)]$

so that $\int_{-2}^{+2} f(x)\,dx = \frac{1}{12}[f(-3) + 23f(-1) + 23f(1) + f(3)]$.

16. Calculate the sum of :

(*i*) $\frac{1}{100} + \frac{1}{101} + \frac{1}{102} + \frac{1}{103} + \frac{1}{104}$

(*ii*) $\dfrac{1}{201^2}+\dfrac{1}{203^2}+\cdots+\dfrac{1}{299^2}=\sum_{x=1}^{99}\dfrac{1}{(200+x)^2}$

(*iii*) $\dfrac{1}{400}+\dfrac{1}{402}+\cdots+\dfrac{1}{498}+\dfrac{1}{500}$.

by using Euler-Maclaurin summation formula. (*B.Sc. Ranchi 1996*)

17. (*a*) If $f(x)$ and its derivative $f'(x)$ are continuously increasing in the range $0<x<n$, show that

$$\int_0^n f(x)\,dx<\tfrac{1}{2}[f(0)+f(n)]+\sum_{x=0}^{n-1}f(x)$$

by putting $a=0$, $h=1$ in Euler-Maclaurin formula.

(*b*) If $f(x)$ either continuously increases or decreases as x increases, show that

$$\int_1^n f(x)\,dx<f(1)+f(n)+\sum_{x=1}^{n-1}f(x).$$

Taking $f(x)=\dfrac{1}{x}$ and using this result show that

$$\log n<1+\left\{1+\frac{1}{2}+\frac{1}{3}+\cdots+\frac{1}{n-1}\right\},$$

however great n may be.

(*c*) If $f'(x)$ is continuous for $x>0$, then

$$\sum_0^n f(x)=\int_0^n f(x)\,dx+\tfrac{1}{2}f(0)+\tfrac{1}{2}f'(x)-\int_0^n p(x)f'(x)\,dx$$

where $p(x)=x-[x]-\frac{1}{2}$ where $[x]$ means the greatest integer.

$$\left[\textbf{Hint. } \int_0^n p(x)f'(x)\,dx=\sum_{k=1}^{n}\int_{k-1}^{k}p(x)f'(x)\,dx=\sum_{k=1}^{n}\int_{k-1}^{k}\left(x-k+\tfrac{1}{2}\right)f'(x)\,dx\right]$$

18. (*a*) Obtain the approximate quadrature formula :

$$\int_{-1/2}^{3/2}y_x\,dx=\frac{1}{24}(27y_0+17y_1+5y_2-y_3).$$

(*b*) Use Euler-Maclaurin summation formula to find the value of π from

(*i*) $\dfrac{\pi}{4}=\displaystyle\int_0^1\frac{dx}{1+x^2}$

(*ii*) $\log 2$ from $\displaystyle\int_0^1\frac{dx}{1+x}$ correct up to three decimal places.

(*c*) Establish the approximate quadrature formula :

$$\int_0^1 \frac{dx}{1+x^2} = a\sum_{x=1}^{a} \frac{1}{x^2+a^2} + \frac{1}{4a}\left(\frac{1}{6a}+1\right)$$

[**Hint.** (*c*) Use Euler-Maclaurin formula with $u_x = \dfrac{a}{a^2+x^2}$

$$\int_0^a \frac{a}{a^2+x^2}\,dx = \sum_0^a \frac{a}{a^2+x^2} - \frac{1}{2}\left[\frac{1}{2}+\frac{1}{2a}\right] + \frac{1}{12}\cdot\frac{1}{2a^2}$$

and range from 0 to a. Put $x = ay$ in the integral].

19. Derive the following quadrature formula :

$$\int_{-a}^{+b} f(x)\,dx = \frac{a+b}{6ab}\,[b(2a-b)f(-a) + (a+b)^2 f(0) + a(2b-a)f(b)]$$

applicable when the intervals in a table are unequal.

[**Hint.** Let $f(x) = Ax^2 + Bx + C$.

Obtain the values of A, B and C in terms of $f(-a), f(0)$ and $f(b)$].

Aliter. $f(x) = \dfrac{(x^2-bx)}{a(a+b)}f(-a) + \dfrac{x^2+ax-bx-ab}{ab}f(0) + \dfrac{x(x+a)}{b(b+a)}f(a).$

20. Assuming Lagrange's interpolation formula or otherwise obtain a formula for method of integration in the form

$$\int_a^b f(x)\,dx = (b-a)\sum_{r=0}^{n} k_r f(x_r) + R_{n+1}$$

Hence deduce Simpson's one-third rule with the help of this result.

21. The two radii which form a diameter of a circle are bisected and perpendicular ordinates are raised at the points of the intersection. Find the area of the portion of the circle included between the two ordinates, the diameter and the curve, the radius being unity. Compare the result found by Weddle's rule with the exact value.

$$\left[\textbf{Hint. } S = S_{-\frac{1}{2}}^{\frac{1}{2}} \sqrt{1-d_n} = 2\int_0^{1/2}\sqrt{1-x^2}\,dx,\ n=6,\ h=\frac{1}{2},\ y_0 = 1\right]$$

22. Calculate the total arc length of the ellipse $x^2 + y^2/4 = 1$ correct up to six places of decimals.

$$\left[\textbf{Hint. } S = 4\int_0^1 \sqrt{\frac{1+3x^2}{1-x^2}}\,dx = 4\int_0^{\pi/2} \sqrt{1+3\sin^2\theta}\,d\theta\right]$$

23. Apply Simpson's one-third and three-eights rule thrice and twice respectively to obtain Weddle's rule.

[**Hint.** Divide in the ratio 9 : 4]

24. (*a*) Show that the Cotes Numbers H_j's, occurring in the Newton-Cotes formula

$$\int_{-1}^{+1} f(x)\,dx = \sum_{j=1}^{n} H_j f(a_j)$$

for equally spaced values $a_1, a_2, ..., a_n$, are given by

$$H_j = h \int_0^{n-1} \binom{t}{j-1}\binom{n-t-1}{n-j}\,dt, \quad j = 1, 2, ..., n$$

(*b*) Prove :

(*i*) $H_j = H_{n-j+1}$ for $j = 1, 2, ..., n$, $H_i = n_{0_i}$

(*ii*) $\sum_{j=1}^{n} H_j = 1,$ (*B.Sc. Agra, 1990*)

where

n	h_0	h_1	h_2	H_0	H_1	H_2
1	$-\frac{1}{\sqrt{3}}$	0	$\frac{1}{\sqrt{3}}$	1	1	1
2	$-\frac{\sqrt{3}}{5}$	0	$\frac{\sqrt{3}}{5}$	$\frac{5}{9}$	$\frac{8}{9}$	$\frac{5}{9}$

25. If $\int_0^m f(x)\,dx = a_1 f(x_1) + a_2 f(x_2) + \cdots + a_m f(x_m)$, ...(1) and a's are constants where $f(x)$ is a polynomial in x of the $(n-1)$th degree, $n > 2$, prove that

$3\sum_{r=1}^{m} a_r x_r^2 = m^3$. (*B.A. Math. Hons., Delhi 1991 ; B.Sc. Bangalore 1993*)

[**Hint.** Put $f(x) = x^2$ in (1)]

26. Compute the integral $\int_5^{12} \frac{dx}{x}$

using Gauss's quadrature formula for the given numerical values of t's and the w's for $n = -1$ to 1 :

w :	$\frac{4}{9}$	$\frac{5}{18}$	$\frac{5}{18}$
t :	0·4372	0·3873	0·3873.

27. Evaluate the following integrals by using the Gaussian quadrature formula for three points :

(*i*) $\int_0^4 \frac{e^{-x2/2}}{\sqrt{2\pi}}\,dx$ (*ii*) $\int_0^1 \frac{dx}{1+x^2}$ (*iii*) $\int_0^{\pi} \frac{\sin x}{x}\,dx$

(*iv*) $\int_1^2 \frac{dx}{1+x^3}$, you may take $n = 4$.

28. Evaluate the integrals (*i*) $I = \int_0^1 \frac{dx}{1+x}$ (*ii*) $\int_0^1 x\,dx$ using Gauss formula, upto five decimal places, with $n = 4$.

[**Hint.** (*i*) First we transfer the interval [0, 1] to the interval [– 1, 1]. with the help of the new variable $t = px + q$.

Thus $-1 = q,\ 1 = p + q \therefore\ p = 2,\ q = -1$ and $t = 2x - 1$

$$\therefore\quad I = \int_0^1 \frac{dx}{1+x} = \int_{-1}^{+1} \frac{dt}{t+3} = \frac{8}{9} f\left(-\sqrt{\frac{3}{8}}\right) + \frac{8}{9} f(0) + \frac{5}{9} f\left(\sqrt{\frac{3}{8}}\right)$$

$$= \frac{1}{9}\left[\frac{5}{\frac{-\sqrt{3}}{8}+3} + \frac{8}{0+3} + \frac{5}{\frac{+\sqrt{3}}{8}+3}\right] = \frac{121}{189} = 0{\cdot}693122$$

Exact solution $= \log(1+x)]_0^1 = \log_e 2 = 0{\cdot}693147$

Error $= 0{\cdot}000025 = 25 \times 10^{-6}$]

(*ii*) Put $x = \frac{1}{2}(u+1)$ to change the limits of integration (0, 1) to (– 1, 1). Thus

$I = \frac{1}{4}\int_{-1}^{+1} (u+1)du = \frac{1}{4}\sum_{i=1}^{n} w_i F(u_i)$ where $F(u_i) = u_i + 1$. Choose $n = 4$, see table given on page 142

29. Using the following table of values, approximate by Simpson's rule the arc length of the graph $y = 1/x$ between the points (1, 1) and $(5, \frac{1}{5})$

x :	1	2	3	4	5
$y = \sqrt{\dfrac{1+x^4}{x^4}}$:	1·414	1·031	1·007	1·002	1·001

$$\left[\textbf{Hint. } \text{Reqd. arc length} = \int_1^5 \sqrt{1 + \left(\frac{dy}{dx}\right)^2}\, dx = \int_1^5 \sqrt{\frac{1+x^4}{x^4}}\, da\right]$$

30. Use the method of Gauss to verify the following computation

$$\int_0^{\pi/2} \frac{d\theta}{\sqrt{1+\sin^2\theta}} = 1{\cdot}311028.$$ (*I.A.S. 1990*)

$$\left[\textbf{Hint. } \text{Put } \sin\theta = x, \text{ the given integral reduces to } \int_0^1 \frac{dx}{\sqrt{1-x^4}}\right]$$

31. The integral $I = \int_a^b f(x)\,dx$ is to be estimated by Trapezoidal type rule using two base points x_1 and x_2 that do not necessarily coincide with the integration limits a and b. Show that the required approximation is

$$I = \frac{b-a}{x_2 - x_1}\left[x_2 f(x_1) - x_1 f(x_2) + \frac{b+a}{2}\{f(x_2) - f(x_1)\}\right]$$

(*M.Sc., Meerut, 1972*)

32. What is the advantage of Romberg's method over Trapezoidal rule ? Using Romberg's method, evaluate the integral

$$I = \int_0^1 \frac{dx}{1+x^2}$$

and hence find the approximate value of π correct to four decimal places.

33. A curve passes through the points (1,2); 1.5, 2.4); (2.0, 2.7), (2.5, 2.8), (3,3) 3.5, 2.6) and (4.0, 2.1). Obtain the area bounded by the curve, the x – axis and $x = 1$ and $x = 4$ (M.S., 1983)

[**Hint.** Area = $\int_1^4 y d_x$. Take $h = 0.5$ and apply Simpson" one-third rule.]

34. A curve passes through the points (0, 0), (1,2), (2,2.5) (3,2.3), (4, 2), (5, 1.7), (6, 1.5)

Calculate $\bar{x} = \left(\int_0^6 xydx \div \int_0^6 yd_x \right)$

35. The velocity v of a particle at distance s from a point on its path is given by

S (meters):	0	10	20	30	40	50	60
v (m/sec) :	47	58	64	65	61	52	38

Estimate the time taken to travel 60 meters by using Simpson" one third rule.

[**Hint :** $v = \frac{ds}{dt} : t = \int_0^{60} \frac{1}{v} ds$, find the values of $\frac{1}{v}$ for different]

Ans. 1.0635 sects.

7

SUMMATION OF SERIES

7·1. Introduction. A very important application of the calculus of finite differences is the problem of finding a formula for the sum to n terms of the given series. We shall see in this chapter that the summation problem is the problem of finite integration.

7·2. To find the sum to n terms of the series whose general term is the first difference of a function.

Let the series whose sum to n terms is required be

$$u_1, u_2, u_3, \ldots$$

where u_a is the first difference of the function $f(x)$ that is,

$$u_x = \Delta f(x)$$

$$\Rightarrow \qquad f(x) = \Delta^{-1} u_x,$$

We are to find the sum

$$u_1 + u_2 + \cdots + u_n \quad \Rightarrow \quad \sum_{x=1}^{n} \bar{u}_x$$

We have if h be the interval of differencing,

$$u_a = \Delta f(a) = f(a+h) - f(a),$$

$$u_{a+h} = \Delta f(a+h) = f(a+2h) - f(a+h),$$

$$\ldots\ldots \quad \ldots\ldots \quad \ldots\ldots \quad \ldots\ldots \quad \ldots\ldots$$

$$u_{a+(n-1)h} = \Delta f(a+n-1h) = f(a+nh) - f(a+\overline{n-1}\,h),$$

Adding,

$$u_a + u_{a+h} + \cdots + u_{a+(n-1)h} = f(a+nh) - f(a) \qquad \ldots(7\cdot1)$$

By a suitable change of origin and scale, we can make $h = 1, a = 1$ and therefore (7·1) reduces to

$$\sum_{1}^{n} u_x = f(n+1) - f(1) = [f(x)]_1^{n+1} \qquad \ldots(7\cdot2)$$

where $\Big[f(x)\Big]_1^{n+1}$ means the difference between the values of $f(x)$ at the upper and the lower limits.

Thus a sufficient condition for finding the sum Σu_x is to find a function $f(x)$ such that $\Delta f(x) = u_x$. We shall call $f(x)\ [= \Delta^{-1} u_x]$ the indefinite integral of u_x and when the definite limits are applied we have the definite integral of u_x.

Thus

$$\Sigma u_x = \Delta^{-1} u_x = f(x) + c \quad \text{is an indefinite integral.}$$

But $$\sum_a^b u_x = \Big[f(x)\Big]_a^{b+1} \quad \text{is a definite integral.}$$

Note 1. It can be easily proved that if m be a positive integer, and $\overset{x}{\underset{m}{C}} = \binom{x}{m}$, then

1. $\Delta\left[\dfrac{a^x}{a^h - 1}\right] = a^x$, the interval of differencing being h.

2. $\Delta\left[\dfrac{1}{amh}(ax+b)^{(m)}\right] = (ax+b)^{(m-1)}$

3. $\Delta\left[\dfrac{-1}{amh}(ax+b)^{(-m)} = (ax+b)^{-m-1)}\right]$

4. $\Delta\left[\binom{x}{m}\right] = \binom{x}{m-1}$, $h = 1$

5. $\Delta\left[\dfrac{\cos(ax+b-\frac{1}{2}a)}{2\sin\frac{1}{2}a}\right] = \sin(ax+b),\ a \neq 0$

6. $\Delta\left[\dfrac{\sin(ax+b-\frac{1}{2}a)}{2\sin\frac{1}{2}a}\right] = \cos(ax+b),\ a \neq 0$

7. $\Delta\left[\dfrac{a^{x+1}\sin b(x+1) - a^x \sin bx}{a^2 - 2a\cos b + 1}\right] = a^x \sin bx$

8. $\Delta\left[\dfrac{\cosh(ax+b-\frac{1}{2}a)}{2\sinh\frac{1}{2}a}\right] = \sin h\,(ax+b),\ a \neq 0$

2. The operator Δ converts exponential functions into exponential functions, polynomials into polynomials, trigonometrical functions into trigonometrical functions.

3. The formula (7·2) can be used to find the sum to n terms of the series whose nth term can be expressed as ka^x, $kx^{(-m-1)}$ $k\begin{pmatrix} x \\ m-1 \end{pmatrix}$ or a polynomial in x.

Example 1. *Use the method of finite differences to find the sum to n terms of the series whose xth term is*

(*a*) $(x+3)(x+4)(x+5)$,

(*b*) $x(x+2)(x+4)$.

(*c*) $(3x-2)(3x+1)(3x+4)$.

Sol. (*a*) $u_x = (x+5)(x+4)(x+3) = (x+5)^{(3)}$

$$\therefore \quad \sum_{x=1}^{n} u_r = \sum_{1}^{n} (x+5)^{(3)} = \left[\frac{1}{4}(x+5)^{(4)}\right]_1^{n+1}, h = 1$$

$$= \frac{1}{4}[n+6]^{(4)} - 6^{(4)}]$$

$$= \frac{1}{4}[(n+6)(n+5)(n+4)(n+3) - 300].$$

(*b*) $$u_x = x(x+2)(x+4) = x^3 + 6x^2 + 8x$$

$$= x^{(3)} + 9x^{(2)} + 15x^{(1)}$$

$$\therefore \quad \sum_{1}^{n} u_x = \sum_{1}^{n} x^{(3)} + 9\sum_{1}^{n} x^{(2)} + 15\sum_{1}^{n} x^{(1)}$$

$$= \left[\frac{1}{4}x^{(4)}\right]_1^{n+1} + 9\left[\frac{1}{3}x^{(3)}\right]_1^{n+1} + 15\left[\frac{1}{2}x^{(2)}\right]_1^{n+1}$$

$$= \left[\frac{1}{4}(n+1)^{(4)} + \frac{9}{3}(n+1)^{(3)} + \frac{15}{2}(n+1)^{(2)}\right] - \left[\frac{1}{4}\times 1^{(4)} + \frac{9}{3}\times 1^{(3)} + \frac{15}{2}\times 1^{(2)}\right]$$

$$= (n+1)n(n-1)(n-2) + \frac{2}{8}(n+1)n(n-1) + \frac{15}{2}(n+1)n - 0$$

$$= \frac{n(n+1)}{12}[3(n-1)(n-2) + 36(n-1) + 90)$$

$$= \frac{n(n+1)}{4}[n^2 + 9n + 20]$$

$$\Rightarrow \qquad \sum_{1}^{n} u_x = \frac{1}{4} n (n+1)(n+4)(n+5)$$

(c) $$u_x = (3x+4)(3x+1)(3x-2)$$

$$= (3x+4)^{(3)}, \text{ interval of differencing} = 1$$

$$\therefore \qquad \sum_{x=1}^{n} u_x = \sum_{1}^{n} (3x+4)^{(3)} = \left[\frac{(3x+4)^{(4)}}{3\times 4}\right]_1^{n+1}, h = 1$$

$$= \frac{1}{12} [(3x+7)^{(4)} - 7^{(4)}]$$

$$= \frac{1}{12} [(3n+7)(3n+4)(3n+1)(3n-2) - 840]$$

Example 2. *Find* $\sum_{1}^{n} u_x$ *where*

(i) $u_x = \dfrac{1}{(x+3)(x+4)}$, (ii) $u_x = \dfrac{x+3}{x(x+1)(x+2)}$

Sol. (i) $u_x = \dfrac{1}{(x+3)(x+4)} = (x+2)^{(-2)}$

$$\therefore \qquad \sum_{1}^{n} u_x = \sum_{1}^{n} (x+2)^{(-2)} = \left[\frac{(x+2)^{(-1)}}{-1}\right]_1^{n+1}$$

$$\sum_{1}^{n} u_x = -\left[(n+3)^{(-1)} - 3^{(-1)}\right] = -\left[\frac{1}{(n+4)} - \frac{1}{4}\right] = \frac{n}{4(n+4)}.$$

(ii) $$u_2 = \frac{1}{(x+1)(x+2)} + \frac{3}{x(x+1)(x+2)}$$

$$= x^{(-2)} + 3(x-1)^{(-3)}$$

$$\therefore \qquad \sum_{1}^{n} u_x = \left[\frac{x^{(-1)}}{-1} + 3\frac{(x-1)^{(-2)}}{-2}\right]_1^{n+1}$$

or $$\sum_{1}^{n} u_x = \left[\frac{-(n+1)^{-1}}{1} - \frac{3}{2} n^{(-2)}\right] - \left[\frac{1^{(-1)}}{-1} - \frac{3}{2} 0^{(-2)}\right]$$

$$= -\left[\frac{1}{n+1} + \frac{3}{2(n+2)(n+1)}\right] + \left[\frac{1}{2} + \frac{3}{2\times 1\times 1}\right]$$

$$= \frac{2n+5}{2(n+1)(n+2)} + \frac{5}{4}$$

$$= \frac{5n^2 + 11\,n}{4\,(n+1)\,(n+2)}.$$

Example 3. *Sum the following series to n terms :*

2, 12, 36, 80, 150, 252,....

Sol. The difference table is

u_x	Δu_x	$\Delta^2 u_x$	$\Delta^3 u_x$
2			
	10		
12		14	
	24		6
36		20	
	44		6
80		26	
	70		6
150		32	
	102		
252			

Third difference are constant. Hence

$$u_x = u_0 + x_{(1)}\Delta u_0 + x_{(2)}\Delta^2 u_0 + x_{(3)}\Delta^3 u_0,\ x_{(r)} \equiv x_{c_r}$$

$$= 2 + 10x_{(1)} + 14x_{(2)} + 6x_{(3)}$$

If the series is $u_0, u_1, u_2, \ldots, u_x, \ldots, u_{n-1}$,

then we require $\sum_{x=0}^{n-1} u_x$.

We have $\sum_0^{n-1} u_x = \sum_0^{n-1} [2 + 10x_{(1)} + 14x_{(2)} + 6x_{(3)}]$

$$\Rightarrow \quad \sum_0^{n-1} u_x = \Big[\, 2x_{(1)} + 10x_{(2)} + 14x_{(3)} + 6x_{(4)} \,\Big]_0^n$$

$$= 2n + 10n_{(2)} + 14n_{(3)} + 6n_{(4)}$$

$$= 2n + 10\frac{n\,(n-1)}{2} + 14\frac{n\,(n-1)\,(n-2}{6} + 6\frac{n\,(n-1)\,(n-2)\,(n-3)}{24}$$

$$= \frac{1}{6}\,n(n+1)\,(n+2)\,(3n+1), \quad \text{on reduction.}$$

7·3. Let us consider the series in which the general term is of the form

$$u_x = \phi\,(x) + ar^x$$

where $\phi(x)$ is a rational integral function of x of degree n, and from this series let us form the successive differences. Each term in any of these differences in the sum of two parts, one arising from terms of the form $\phi(x)$ in the original series and the other from terms of the form ar^x. Now since $\phi(x)$ is of degree n, the part arising from $\phi(x)$ will be zero in the $(n+1)$ the and succeeding differences and therefore these differences will be in geometrical progression (G.P.) with common ration r. For example, if

$$u_x = c + bx + ar^x$$

then

$$\Delta u_x = b + a(r-1)r^x$$

$$\Delta^2 u_x = a(r-1)^2 r^x$$

$$\Delta^3 u_x = a(r-1)^3 r^x$$

$$\ldots\ldots\ldots\ldots\ldots\ldots$$

$$\Delta^n u_x = a(r-1)^n r^x$$

so that the second and all the succeeding differences are in G.P. with common ratio r.

Hence if the first few terms of a series are given, and if the nth differences of these terms form a G.P. whose common ratio is r, then we may assume that the general term of the given series is $ar^x + f(x)$, where $f(x)$ is a rational integral function of x of degree $(n-1)$. Now in order to find the sum to n term of such a series, it will be found convenient to express $f(x)$ in factorial notation and use the following results :

$$\sum_1^n x^{(r)} = \left[\frac{x^{(r+1)}}{r+1}\right]_1^{n+1} \quad \text{and} \quad \sum_1^n a^x = \left[\frac{a^x}{a-1}\right]_1^{n+1}$$

Example 4. *Find the general term and the sum to n terms of the series :*

1, 2, 4, 8, 17, 40, 104, ...

Sol. The difference table is

u_x	Δu_x	$\Delta^2 u_x$	$\Delta^3 u_x$
1			
	1		
2		1	
	2		1
4		2	
	4		3
8		5	
	9		9
17		14	
	23		27
40		41	
	64		
104			

Since third differences are in G.P. with common ratio 3, we assume that

$$u_x = a.3^x + bx^{(2)} + cx^{(1)} + d,$$

$$\therefore \quad \Delta u_x = a\,(3-1)\,.\,3^x + 2bx^{(1)} + c = 2a3^x + 2bx^{(1)} + c,$$

$$\Delta^2 u_x = a(3-1)^2\,.\,3^x + 2b = 4a\,.\,3^x + 2b,$$

$$\Delta^3 u_x = 8a\,.\,3^x.$$

Putting $x = 1$ in these results, we have

$$3a + c + d = 1,$$

$$6a + 2b + c = 1,$$

$$12a + 2b = 1,$$

$$24a = 1.$$

These equations give

$$a = \frac{1}{24}\,;\, b = \frac{1}{4}\,;\, c = \frac{1}{4}\,;\, d = \frac{5}{8}$$

Hence $$u_x = \frac{1}{24}\,.\,3^x + \frac{1}{4}x^{(2)} + \frac{1}{4}x^{(1)} + \frac{5}{8}$$

$$\therefore \quad \sum_1^n u_x = \left[\frac{1}{24}\frac{3^x}{3-1} + \frac{1}{4\times 3}x^{(3)} + \frac{1}{4\times 2}x^{(2)} + \frac{5}{8}x^{(1)}\right]_1^{n+1}$$

$$\sum_1^n u_x = \left[\frac{1}{48}\,.\,3^{n+1} + \frac{(n+1)^{(3)}}{12} + \frac{(n+1)^{(2)}}{8} + \frac{5}{8}(n+1)^{(1)}\right] - \left[\frac{1}{24}\times\frac{3}{2} + 0 + 0 + \frac{5}{8}\right]$$

$$= \frac{1}{8}\left[\frac{3^n}{2} + \frac{2}{3}(n+1)\,n\,(n-1) + (n+1)\,n + 5\,(n+1) - \frac{1}{2} - 5\right]$$

$$= \frac{1}{8}\left[\frac{1}{2}(3^n - 1) + 5n + \frac{n\,(n+1)}{3}\,(2n - 2 + 3)\right]$$

$$= \frac{1}{8}\left[\frac{1}{2}(3^n - 1) + 5n + \frac{1}{3}n\,(n+1)\,(2n + \quad\right]$$

7·4. Summation by Parts. Let us consider the series in which the general term is the product of two functions, that is, the general term is of the form

$$u_x = f_1\,(x)\,f_2(x)$$

Let $$\Delta g_1 = f_1(x) \text{ and } \Delta\, g_2(x) = f_2\,(x)$$

We have $\Delta[f_1(x)g_2(x)] = f_1(x+1)g_2(x+1) - f_1(x)g_2(x)$

$$= f_1(x+1)g_2(x+1) - f_1(x)g_2(x+1) + f_1(x)g_2(x+1) - f_1(x)g_2(x)$$

$$= g_2(x+1)\Delta f_1(x) + f_1(x)\Delta g_2(x)$$

$$\therefore \quad \sum_1^n f_1(x)\,\Delta g_2(x) = [f_1(x)\,g_2(x)]_1^{n+1} - \sum_1^n \Big[g_2(x+1)\Delta f_1(x) \Big]$$

This formula is analogous to the rule of Integration by Parts and is known as Summation by Parts. This formula is applicable if we can determine $g_2(x)$ from $\Delta g_2(x)$ and if we can find

$$\sum_1^n [g_2(x+1)\,\Delta f_1(x)]$$

Example 5. *Sum to n terms the series whose xth term is*

$$2^x(x^3+x).$$

Sol. Let $\qquad f_1(x) = x^3 + x$

and $\qquad \Delta g_2(x) = 2^x.$

Then $\Delta f_1(x) = 3x^2 + 3x + 2,\ g_2(x) = \dfrac{2^x}{2-1} = 2^x.$

$$\therefore \quad g_2(x+1) = 2^{x+1}$$

$$\sum_1^n (x^3+x)2^x = \Big[2^x(x^3+x) \Big]_1^{n+1} - \sum_1^n 2^{x+1}(3x^2+3x+2)$$

$$= \Big[2^x(x^3+x) \Big]_1^{n+1} - \Big[2^{x+1}(3x^2+3x+2) \Big]_1^{n+1} + \sum_1^n 2^{x+2}(6x+6)$$

by applying again the formula to the second summation.

$$\therefore \quad \sum_1^n 2^x(x^3+x) = \Big[2^x(x^3+x) \Big]_1^{n+1} - \Big[2^{x+1}(3x^2+3x+2) \Big]_1^{n+1}$$

$$+ \Big[2^{x+2}(6x+6) \Big]_1^{n+1} - \sum_1^n 2^{x+3}\,.\,6$$

$$= \Big[2^x(x^3+x) \Big]_1^{n+1} - \Big[2^{x+1}(3x^2+3x+x) \Big]_1^{n+1}$$

$$+ \Big[2^{x+2}(6x+6) \Big]_1^{n+1} - 6\Big[2^{x+3} \Big]_1^{n+1}$$

$$= 2^{n+1}(n^3 - 3n^2 + 10n - 14) + 28, \text{ on reduction.}$$

Example 6. *Evaluate*

$$\sum_{1}^{n} 4^x \left\{ \frac{x^2}{(x+1)(x+2)} \right\}$$

Sol. We have $\Sigma u_x = \Sigma \left[4^x \left\{ 1 + \frac{1}{x+1} - \frac{4}{x+2} \right\} \right]$

$$= \Sigma 4^x + \Sigma 4^x \left(\frac{1}{x+1} - \frac{4}{x+2} \right)$$

Now $\quad \Delta\left(\frac{4^x}{x+1} \right) = \frac{4^{x+1}}{x+1} - \frac{4^x}{x+1} = -4^x \left[\frac{1}{x+1} - \frac{4}{x+2} \right]$

$\therefore \quad \sum_{1}^{n} u_x = \sum_{1}^{n} 4^x - \Sigma\Delta\left(\frac{4^x}{x+1} \right)$

$$= \left[\frac{4^x}{4-1} \right] - \left[\frac{4^x}{x+1} \right]_1^{n+1} \left[\because \Sigma\Delta\left(\frac{4^x}{x+1} \right) = \frac{4^x}{x+1} \right]$$

$$= \frac{1}{3}\left(4^{n+1} - 4 \right) - \left(\frac{4^{n+2}}{n+1} - \frac{4}{2} \right)$$

$$= \frac{2}{3} + \frac{4^{n+1}}{2}\left(1 - \frac{3}{n+2} \right).$$

Example 7. *Sum to n terms and to infinity the series ;*

$$\frac{3}{2 \cdot 5 \cdot 8} + \frac{5}{5 \cdot 8 \cdot 11} + \frac{7}{8 \cdot 11 \cdot 14} + \cdots$$

Sol. Here $\quad u_x = \frac{2x+1}{(3x-1)(3x+2)(3x+5)} = (2x+1)(3x-4)^{(-3)}$

Let $\quad f_1(x) = 2x + 1, \Delta g_2(x) = (3x-4)^{(-3)}$

$\therefore \quad \Delta f_1(x) = 2, \; g_2(x) = \frac{(3x-4)^{(-2)}}{-2 \times 3}$

Therefore by the formula of Summation by Parts, the sum to n terms

$$= \left[\frac{(3x-4)^{(-2)}}{(-2)} \times (2x+1) \right]_1^{n+1} - \sum_{1}^{n} \left\{ \frac{(3x-1)^{(-2)}}{-6} \times 2 \right\}$$

$$= \left[\frac{(n-1)^{(-2)}}{-6} \times (2n+3) + \frac{(-1)^{(-2)}}{2} \right]$$

$$- \left[\frac{(3x-1)^{(-1)}}{-6 \times -2 \times -1} \times 2 \right]_1^{n+1}$$

$$= \left[\frac{-2n-3}{6\,(3n+2)\,(3n-5)} + \frac{1}{2 \times 2 \times 5} \right] - \left[\frac{(3n+2)^{(-1)}}{9} - \frac{2^{(-1)}}{9} \right]$$

$$= \frac{-2n-3}{6\,(3n+2)\,(3n+5)} - \frac{1}{9\,(3n+5)} + \frac{1}{20} + \frac{1}{45}$$

$$= \frac{13}{180} - \frac{12n+3}{18\,(3n+2)\,(3n+5)}.$$

7·5. To prove that $\sum_{1}^{n} u_x = \left[\Delta^{-1} u_x \right]_1^{n+1}$

where $\Delta^{-1} \equiv \frac{1}{\Delta}.$

We have $\sum_{1}^{n} u_x = u_1 + u_2 + \cdots + u_n$

$$= (1 + E + E^2 + \cdots + E^{n-1})\, u_1$$

$$= \frac{E^n - 1}{E - 1} u_1 = \frac{1}{\Delta} (u_{n+1} - u_1)$$

$$= \Delta^{-1} (u_{n+1} - u_1) = \Delta^{-1} u_{n+1} - \Delta^{-1} u_1$$

$$= \left[\Delta^{-1} u_x \right]_1^{n+1}$$

Note. The symbol Δ^{-1} is often denoted by the symbol Σ. It should be noted that $\Delta\Sigma \equiv 1$.

For, if $\Delta f(x) = u_x$,

then $f(x) = \Sigma u_x$

and $u_x = \Delta f(x) = \Delta\Sigma u_x$

so that $\Delta\Sigma \equiv 1.$

It can be proved that it does not follow from $\Delta\Sigma = 1$ that $\Sigma\Delta \equiv 1$. In fact $\Sigma\Delta \neq 1$.

7·6. To prove that

$$\Delta^{-1} [(x-b)^{-n}] = \frac{(-1)^{n-1}}{(n-1)\,!} \phi^{n-1} (x-b)$$

where $\phi(x) = \frac{\Gamma' x}{\Gamma x} \equiv \frac{d}{dx} \log \Gamma_x.$

We have $\Gamma(x+1) = x\Gamma x$

By logarithmic differentiation, we have

$$\frac{\Gamma'(x+1)}{\Gamma(x+1)} = \frac{1}{x} + \frac{\Gamma'(x)}{\Gamma(x)}$$

$$\Rightarrow \qquad \phi(x+1) = \frac{1}{x} + \phi(x)$$

$$\Rightarrow \qquad \Delta\phi(x) = \frac{1}{x}$$

$$\therefore \qquad \Delta^{-1}\left(\frac{1}{x}\right) = \phi(x) \qquad \ldots(1)$$

Differentiating (1), we have

$$\phi'(x+1) = -\frac{1}{x^2} + \phi'(x)$$

$$\Rightarrow \qquad \Delta\phi'(x) = -\frac{1}{x^2}$$

$$\therefore \qquad \Delta^{-1}\left(\frac{1}{x^2}\right) = -\phi'(x)$$

Proceeding in this way, we obtain

$$\Delta^{-1}\left[\frac{1}{(x-b)^n}\right] = \frac{(-1)^{1-n}\phi^{n-1}(x-b)}{(n-1)!}$$

Cor. We have

$$\Delta \log \Gamma x = \log \Gamma(x+1) - \log \Gamma x = \log x$$

$$\therefore \qquad \Delta^{-1} \log x = \log \Gamma x.$$

Hence $\displaystyle\sum_{1}^{n} \log x = \left[\Delta^{-1} \log x\right]_1^{n+1} = \log \Gamma(n+1).$

Example 8. *Sum the series* :

$$a^{n+1} - n(a+b)^{n+1} + \frac{n(n-1)}{2!}(a+2b)^{n+1}$$
$$- \frac{n(n-1)(n-2)}{3!}(a+3b)^{n+1} + \cdots$$

Sol. Taking the interval of differencing to be b, the series may be written as

$$(1 - {}^nc_1 E + {}^nc_2 E^2 - \ldots)\, a^{n+1}$$
$$= (1-E)^n a^{n+1} = (-1)^n \Delta^n a^{n+1}$$

Now $\quad a^{n+2} = a^{(n+1)} + \frac{1}{2} n(n+1) a^{(n)} b + \text{terms in } a^{(r)} b^{n+1-r}, r < n$

[See solution to Q. 36, exercise 1]

$$\therefore \quad \Delta^n a^{n+1} = (n+1)!\, b^n a^{(1)} + \frac{1}{2} n(n+1) b^{n+1} n!$$

$$= \frac{1}{2} b^n (n+1)!\,(2a+nb)$$

$\therefore$ Sum of the series $\frac{1}{2} b^n (n+1)!\,(2a+nb)$

Example 9. *Sum to n terms the series whose xth term is*

$$\frac{x \,.\, 2^x}{(x+2)!}$$

Sol. We wish to find V_x such that

$$\Delta V_x = u_x = \frac{x \,.\, 2^x}{(x+2)!} \quad ...(1)$$

Assume that $\quad V_x = \dfrac{2^x \,.\, f(x)}{(x+1)!}$

where $f(x)$ is an undetermined rational integral function of x.

$$\therefore \quad \frac{2^{x+1} f(x+1)}{(x+2)!} - \frac{2^x f(x)}{(x+1)!} = \frac{2^x \,.\, x}{(x+2)!}$$

or $$2f(x+1) - (x+2) f(x) = x \quad ...(2)$$

Since the right-hand side of this equation is of the first degree; the left-hand side should also be of the first degree and therefore $f(x)$ must be equal to constant, say k,

i.e., $\quad f(x) = k$

$\therefore \quad f(x+1) = k$

$\therefore$ from (2),

$$2k - (x+2)k = x$$

This gives $\quad k = -1, \Rightarrow f(x) = -1$

$$\therefore \quad V_x = -\frac{2^x}{(x+1)!}.$$

Hence the sum to n terms $= \displaystyle\sum_1^n \frac{x \,.\, 2^x}{(x+1)!}$

$$= \left[\frac{-2^x}{(x+1)!}\right]_1^{n+1} = 1 - \frac{2^{n+1}}{(n+2)!}$$

Alternative Method :

$$u_x = \frac{x.2^x}{(x+2)!} = \frac{[(x+2)-2]2^x}{(x+2)!}$$

$$= \left[\frac{-2^{x+1}}{(x+2)!} + \frac{2^x}{(x+1)!}\right] = \Delta\left(\frac{2^x}{(x+1)!}\right)$$

$$\therefore \quad \sum_1^n u_x = \left[\frac{-2^x}{(x+1)!}\right]_1^{n+1} = 1 - \frac{2^{n+1}}{(n+2)!}$$

Example 10. *If fourth and higher differences are ignored, prove that the sum of successive terms of a function, of which u_0 is the central term, is*

$$nu_0 + \frac{1}{24}(n^3 - n)\,\Delta^2 u_{-1},$$

where n is odd.

Sol. Since u_0 is the central term, a central difference formula will be convenient so that Gauss's forward formula gives

$$u_x = u_0 + {}^xC_1\Delta u_0 + {}^xC_2\Delta^2 u_{-1} + {}^{x+1}C_3\Delta^3 u_{-1} + \cdots$$

$$= u_0 + x^{(1)}\,\Delta u_0 + \frac{1}{2}x^{(2)}\,\Delta^2 u_{-1} + \frac{1}{6}(x+1)^{(3)}\,\Delta^3 u_{-1}$$

$$\therefore \quad \Sigma u_x = c + xu_0 + \frac{1}{2}x^{(2)}\Delta u_0 + \frac{1}{6}x^{(3)}\,\Delta^2 u_{-1} + \frac{1}{24}(x+1)^{(4)}\Delta^3 u_{-1}$$

Summing between the limits $-\frac{1}{2}(n-1)$ and $\frac{1}{2}(n-1)$ and remembering that the coefficients of Δu_0 and Δu_{-1} cancel, we get

$$\sum_{-\frac{1}{2}(n-1)}^{\frac{1}{2}(n-1)} u_x = \left(c + xu_0 + \frac{1}{2}x^{(2)}\Delta u_0 + \frac{1}{6}x^{(3)}\Delta^2 u_{-1} + \frac{1}{24}(x+1)^{(4)}\,\Delta^3 u_{-1} \right]_{-\frac{1}{2}(n-1)}^{\frac{1}{2}(n-1)}$$

$$= \left[\frac{n+1}{2} - \left(\frac{-n-1}{2}\right)\right] u_0 + \frac{1}{6}\left(\frac{n+1}{2}\cdot\frac{n-1}{2}\cdot\frac{n-3}{2} - \left(\frac{-n-1}{2}\cdot\frac{n+1}{2}\cdot\frac{n+3}{2}\right)\right]\Delta^2 u_{-1}$$

$$= nu_0 + \frac{1}{24}(n^3 - n)\,\Delta^2 u_{-1}.$$

7·7. A Useful Result

Theorem. *If $\phi(E)$ is a polynomial in E and $f(x)$ is a function of x, then*

$$\phi(E)\,[a^x f(x)] = a^x \phi(aE)[f(x)].$$

Proof. Let $\phi(E) \equiv A_0 + A_1 E + A_2 E^2 + \cdots + A_r E^r + \cdots$

A's being constants.

Now $\quad A_r E^r\,[a^x f(x)]$

$$= A_r\, a^x a^r f(x+r)$$

$$= a^x A_r\, a^r E^r f(x) = a^x A_r\,(aE)^r f(x)$$

$$\therefore \quad \phi(E)[(a^x f(x)] = \sum_{r=0} [A_r a^r E^r \{a^x f(x)\}]$$

$$\therefore \quad \phi(E) a^x f(x) = \sum_{r=0} [a^x A_r (aE)^r f(x)]$$

$$= a^x . \sum_{r=0} A_r (aE)^r . f(x)$$

$$= a^x \phi(aE) f(x).$$

Cor. $(E-a)^n a^x f(x) = a^x (aE-a)^n f(x) = a^x a^n \Delta^n f(x)$

Example 11. *Use the above result to evaluate.*

(*i*) $\sum_0^\infty a^x f(x)$.

(*ii*) $2{\cdot}2 + 7{\cdot}4 + 14{\cdot}8 + 23{\cdot}16 + 34{\cdot}32 + \cdots$ *to n terms.*

Sol. (*i*) $\sum_0^\infty [a^x f(x)] = \Delta^{-1} [a^x f(x)] = (E-1)^{-1} a^x f(x)$

so that $\phi(E) \equiv (E-1)^{-1}$.

$$\therefore \quad \sum_0^\infty a^x f(x) = a^x (aE-1)^{-1} f(x)$$

$$= a^x (a + a\Delta - 1)^{-1} f(x)$$

$$= a^x (a-1)^{-1} \left[1 + \frac{a\Delta}{a-1}\right]^{-1} f(x)$$

$$= \frac{a^x}{a-1}\left[1 - \frac{a\Delta}{a-1} + \frac{a^2\Delta^2}{(a-1)^2} - \cdots\right] f(x) + \text{constant}$$

(*ii*) Each term is the product two factors. Let us find the general term of the series 2, 7, 14, 23, 34.

The difference table is

$f(x)$	$\Delta f(x)$	$\Delta^2 f(x)$
2		
	5	
7		2
	7	
14		2
	9	
23		2
	11	
34		

$$\therefore \quad f(x) = f(1) + (x-1)\Delta f(1) + \frac{(x-1)(x-2)}{2!} \Delta^2 f(1)$$

$$= x^2 + 2x - 1.$$

Hence the general term of the given series is

$$2^x (x^2 + 2x - 1)$$

$$\therefore \quad \sum 2^x (x^2 + 2x - 1) = \frac{2^x}{2-1} [1 - 2\Delta + 4\Delta^2] (x^2 + 2x - 1)$$

by example 11 (*i*)

Now $\Delta(x^2 + 2x - 1) = 2x + 3, \Delta^2 (x^2 + 2x - 1) = 2$

$\therefore$ Sum to n terms

$$= \Big[2^x (x^2 + 2x - 1) - 2^{x+1}(2x + 3) + 2^{x+2} \times 2 \Big]_1^{n+1}$$

$$= 2^{n+1}[(n + 1)^2 + 2 (n + 1) - 1] - 2^{n+2} [2 (n + 1) + 3]$$

$$- 2^2 (2 + 3) + 2 \times 2^{n+3} - 2^3 \times 2$$

$$= n^2 . 2^{n+1}, \text{ on simplification.}$$

EXERCISE 7

1. (*a*) Use the method of finite differences to find the sum of the first. (*i*) n squares, (*ii*) n cubes.

(*b*) Evaluate :

$$1^4 + 2^4 + \cdots + n^4.$$

(*c*) Prove : $\Delta^{-1} x^4 = \frac{x^{(5)}}{5} + \frac{3}{2} x^{(4)} + \frac{7h}{3} x^{(2)} + \frac{h^3 x^{(2)}}{2} + b(x)$ where $b(x)$ is any function with period h.

2. Sum to n terms the series whose xth term is :

(*i*) $x (x - 1) (x - 2)$,

(*ii*) $x (x + 1) (x + 3)$,

(*iii*) $x (x + 2) (x + 5)$,

(*iv*) $(3x - 1) (3x + 2) (3x + 5) (3x + 8)$.

3. Sum to n terms the series whose xth terms are :

(*i*) $\frac{1}{(x + 1) (x + 2) (x + 3)}$, (*ii*) $\frac{1}{(3x - 2) (3x + 1) (3x + 4)}$, (*iii*) $\frac{1}{4x^2 - 1}$.

Find also the sum to infinity of the series (*iii*).

4. (*a*) Derive the formula :

$$\sum_{n=1}^{\infty} x^{(-m)} = \frac{1}{(m - 1) [(m - 1)] !}$$

where $(n - 1)^{(-m)} = 1 \div n (n + 1) (n + 2)\ldots(n + m - 1)$ and m is positive integer greater than 1.

(*b*) Evaluate : $\Delta^{-1} \left[\frac{x + 4}{(x + 2) (x + 3) (x + 5) (x + 6)} \right]$

Sum the following series to n terms :

5. $2{\cdot}5 + 5{\cdot}8 + 8{\cdot}1\ 11{\cdot}14 + \ldots$

6. $1{\cdot}3{\cdot}5 + 3{\cdot}5{\cdot}7 + 5{\cdot}7{\cdot}9 + \cdots$

7. $1\cdot3\cdot5+2\cdot4\cdot6+3\cdot5\cdot7+\cdots$

8. $\frac{1}{1\cdot4}+\frac{1}{4\cdot7}+\frac{1}{7\cdot10}+\cdots$

9. $\frac{4}{1\cdot2\cdot3}+\frac{5}{2\cdot3\cdot4}+\frac{6}{3\cdot4\cdot5}+\ldots$

10. (*i*) $\frac{1}{1\cdot2\cdot4}+\frac{1}{2\cdot3\cdot5}+\frac{1}{3\cdot4\cdot6}+\cdots$

(*ii*) $\frac{3}{1\cdot2\cdot4}+\frac{4}{2\cdot3\cdot5}+\frac{5}{3\cdot4\cdot6}+\cdots$

Sum the following series to n terms :

11. 0, 10, 33, 77, 150,......

12. 12, 40, 90, 168, 280, 432,......

13. 5, 10, 17, 28, 47, 82,......

14. 1, 6, 11, 18, 31, 58, 115,......

15. 1, 0, 1, 8, 29, 80, 193,......

16. (*i*) 4, 13, 35, 94, 262, 755,......

(*ii*) 5, 10, 23, 60, 169, 494, 1467,......

[**Hint** : (*ii*) $y_x=a\,.\,3^x+bx^{(3)}+cx^{(2)}+dx^{(1)}+e$]

Find the general term and the sum to n terms of the series :

17. $2\cdot2+6\cdot4+12\cdot8+12\cdot16+30\cdot32+\cdots$

18. $\frac{5}{1\cdot2}\cdot\frac{1}{3}+\frac{7}{2\cdot3}\cdot\frac{1}{3^2}+\frac{9}{3\cdot4}\cdot\frac{1}{3^3}+\frac{11}{4\cdot5}\cdot\frac{1}{3^4}+\ldots$

Find the sum to n terms the series :

19. $a+2a^2+3a^3+4a^2+\ldots$

20. $1^2+\frac{3^2}{2}+\frac{5^2}{2^2}+\frac{7^2}{2^3}+\ldots$

21. Show that $\Delta\,(x\,!)=x\,(x\,!)$ and hence sum to n terms the series :

(*i*) $1+3\,(2\,!)+7\,(3\,!)+13\,(4\,!)+21\,(5\,!)+\cdots$

(*ii*) $(x^2-x+1)\,(x\,!)$

[**Hint.** (*ii*) $u_x=(x^2-x+1)\,(x\,!)=[(x+1)^2-3x]\,(x\,!)$

$=[(x+1)\,(x+1)\,!]-3x\,(x\,!)=\text{etc.}$]

22. Prove that

$$\sum_{1}^{n}(x^2+1)\,(x\,!)=n\,[(n+1)\,!]$$

[**Hint** : $(x^2+1)\,(x\,!)=(x+1)^2\,(x\,!)-2x\,(x\,!)$'

$=(x+1)\,[(x+1)\,!]-2x\,(x\,!)$

$=\Delta\,[(x+1)\,!]-2\Delta\,(x\,!)$.

23. Sum the following series :

$$\frac{1}{x+1}-\frac{n}{(x+1)\,(x+2)}+\frac{n\,(n-1)}{(x+1)\,(x+2)\,(x+3)}-\ldots+\frac{(-1)^n\,n\,!}{(x+1)\,(x+2)\cdots(x+n+1)}$$

[**Hint.** The given series

$$= x^{(-1)} + {}^nc_1 \Delta x^{(-1)} + {}^nc_2 \Delta^2 x^{(-1)} + \cdots + {}^nc_n \Delta^n x^{(-1)}$$

$$= (1 + \Delta)^n x^{(-1)} = \text{etc.}]$$

24. Sum the series :

$$x^3 + (x+1)^3 + (x+2)^3 + \cdots + (x+n-1)^3$$

25. Show that

(i) $\Delta^{-1}\left[\dfrac{1}{x(x+1)}\right] = C - \dfrac{1}{x},$

(ii) $\Delta^{-1}\left[\dfrac{x \, . \, 2^x}{(x+1)(x+2)}\right] = \dfrac{1}{x+1} \, . \, 2^x + C,$

(iii) $\Delta^{-1}\left[\dfrac{2x-1}{2^{x-1}}\right] = C - \dfrac{2x+1}{2^{x-2}},$

(iv) $\Delta^{-1}\left[2^x \, . \, x \, . \, \dfrac{x\,!}{(2x+1)\,!}\right] = C - 2^{x-2}\,\dfrac{(x-1)\,!}{(2x-1)\,!}$

C being a constant.

[**Hint.** (iv) $2^x \, x \, . \, \dfrac{x\,!}{(2x+1)\,!} = \dfrac{2^{x-1}\,(x\,!)\,(2x+1-1)}{(2x+1)\,!}$

$$= \frac{2^{x-1}\,(x\,!)}{2x\,!} - \frac{(2^{x-1}\,x\,!)}{(2x+1)\,!}$$

$$= \frac{2^{x-1}\,(x-1)\,!}{(2x-1)\,!} - \frac{2^{x-1}\,(x\,!)}{(2x+1)\,!}$$

$$= \Delta\left(-\frac{2^{x-2}\,(x-1)\,!}{(2x-1)\,!}\right) = \text{etc.} \,]$$

26. Prove that

$$\sum_{0}^{n-1} y_r x^r = \frac{y_0 - x^n y_n}{1-x} + \frac{x}{(1-x)^2}(\Delta y_0 - x^n \Delta y_n)$$

$$+ \frac{x^2}{(1-x)^3}(\Delta^2 y_0 - x^n \Delta^2 y_n) + \cdots$$

Use it to obtain $\displaystyle\sum_{n}^{r=1} r(r+1)\,x^{r-1}$.

27. Establish the following formula for the difference of a product of two functions :

$$\Delta(u_x v_x) = u_x \Delta v_x + v_{x+1} \Delta u_x,$$

and deduce the formula, for the sum of a product,

$$\sum_{x=0}^{n-1} (u_x \Delta v_x) = (u_n v_n - u_0 v_0) - \sum_{x=0}^{n-1} (v_{x+1} \Delta u_x)$$

By means of this formula, find the value of

$$\sum_{x=0}^{n-1} x^2.$$

28. Show that

(*i*) $\Delta^{-1}(u_x v_x) = u_x \Delta^{-1} v_x - \Delta^{-1}(\Delta u_x \Delta^{-1} v_{x+1})$.

(*ii*) $\Delta^{-1}[v(x)\,\Delta u(x)] = u(x)\,v(x) - \Delta^{-1}[u(x)\,\Delta v(x)]$

Use the above result to evaluate $\Delta^{-1}(xa^x)$.

29. If u_n is the nth term of the series 1, 2, 3, 5, 8, 13,... in which each term after the second is the sum of the two preceding terms, prove that

$$u_n^2 - u_{n-1}\,u_{n+1} = (-1)^n.$$

[**Hint.** Prove the result by Mathematical Induction.

$$u_{n+1}^2 - u_n \cdot u_{n+2} = u_{n+1}(u_{n-1} + u_n) - u_n(u_n + u_{n+1})$$
$$= u_{n-1}u_{n+1} - u_n^2$$
$$= (-1)^{n+1} \text{ using the given result.}$$

Show that the result is true for $n = 2, n = 3$, etc.]

30. Evaluate the sum of the double series

(*i*) $\displaystyle\sum_{s=1}^{n} \sum_{r<1} s(s+2)\,r(r+1)$ (*ii*) $\displaystyle\sum_{s=1}^{n} \sum_{r<s} s(s+2)\,r(r+1)$

[**Hint.** (*i*) Given expression

$$= \sum_{s=1}^{n} s(s+2) \sum_{r=0}^{s-1} r(r+1)$$

$$= \sum_{s=1}^{n} s(s+2) \sum_{r=1}^{s-1} r(r+1)$$

$$= \sum_{s=1}^{n} s(s+2)\,\frac{1}{3}(s-1)\,s\,(s+1)$$

$$= \frac{1}{3}\sum_{1}^{n} (s+2)^{(4)}\,[(s+3) - 3]$$

$$= \frac{1}{3}\sum_{1}^{n} (s+3)^{(5)} - \sum_{1}^{n} (s+2)^{(4)} = \text{etc.}]$$

8

DIFFERENCE EQUATIONS

8·1. Definitions. Difference equation, its order and solution. A *difference equation* is an equation which involves an independent variable, dependent variable and the successive differences of the dependent variable. Since the successive differences of dependent variable can be expressed in terms of the successive values of the dependent variable. A difference equation may also be defined as an equation which expresses a relation between an independent variable and the successive values of the dependent variable. The study of difference equations is useful in dealing with Mathematical and Economic models, etc.

The relation or equation

$$F[x, f(x), \Delta f(x), ..., \Delta^n f(x)] = 0 \qquad ...(i)$$

where the function F is given and the function f is unknown, is said to be a difference equation of order n, if the relation (i) after interchanging the increment by its expression in terms of f explicitly contains both $f(x+n)$ as well as $f(x)$. We shall prefer to write u_x for $f(x)$.

The following are some examples of difference equations :

(1) $\Delta^3 u_x - 3\Delta^2 u_x + 2\Delta u_x + u_x = \cos \pi x$,

(2) $u_{x+2} - 3u_{x+1} + 5u_x = x^3$,

(3) $u_{x+3} - 5u_{x+2} = 2^x$,

where Δ is the operator such that $\Delta u_x = u_{x+1} - u_x$.

The first difference equation involves successive differences of the dependent variable u_x while the second equation involves the successive values of the dependent variable. In practice it will be found more convenient to deal with difference equations involving successive values of the dependent variables and not successive differences.

Note. A difference equation can be transformed in terms of operator by the relation $D = E - 1$. Thus the difference equation (1) can be transformed into the following :

$$(E-1)^3 u_x - 3(E-1)^2 u_x + 2(E-1)\, u_x + u_x = \cos \pi x$$

or $$E^3 u_x - 6E^2 u_x + 11Eu_x - 5u_x = \cos \pi x.$$

A difference equation not involving successive values of u_x greater than u_{x+n} is said to be of *order n*. It is the difference between the largest and smallest arguments n appearing in an equation. Then (2) is a difference equation of order 2. By putting $u_{x+2} = v_x$, the difference equation (3) becomes

$$v_{x+1} - 5v_x = 2^x$$

and, there, (3) is of order 1.

A *solution* of a difference equation over a set S is a relation between the independent variable and the dependent variable which *satisfies* the equation or is an identity over S. Such a relation on substitution in the equation makes the left-hand and right-hand members identically equal. Thus c_1 and c_2 are arbitrary constants of periodic functions of period 1, $u_x = c_1 3^x + c_2(-1)^x$ is a solution of equation (1), page 165 where the *initial condi- tions* are $c_1 = 2$ and $c_2 = -1$.

Conversely, if we have $u_x = c_1\, 3^x + c_2(-1)^x$ ...(1)

and we want to form a difference equation, then putting $x+1$ and $x+2$ for x we obtain

$$u_{x+1} = c_1\, 3^{x+1} + c_2(-1)^{x+1} = 3c_1 \cdot 3^x - c_2(-1)^x \qquad ...(2)$$

and $$u_{x+2} = c_1\, 3^{x+2} + c_2(-1)^{x+2} = 9c_1 \cdot 3^x + c_2(-1)^x \qquad ...(3)$$

Eliminating c_1 and c_2 from the three equations, we have

$$\begin{vmatrix} u_x & 1 & 1 \\ u_{x+1} & 3 & -1 \\ u_{x+2} & 9 & 1 \end{vmatrix} = 0$$

$$\Rightarrow \qquad u_{x+2} - 2u_{x+1} - 3u_x = 0$$

which is difference equation of order 2 of which (1) is the solution.

An equation over a set S of the form where p_i's and $R(x)$ are functions of x or constants defined for all values of x in the set S is called a *linear difference* equation over S of order n. If $R(x) = 0$, the equation is called linear *homogeneous*, otherwise it is called linear *non-homogeneous equation.*

Example. *Show that* $y_x = 1 - 2x^{-1}$, $x = 1, 2, 3, \ldots$ *is a solution of the equation* $(x+1)y_{x+1} + xy_x = 2x - 3.$

Solution of an equation of order n, containing n arbitrary periodic constants $c_1, c_2, \ldots, c_n$. A solution containing arbitrary constants equal in number to the order of the equation is called the *general solution* or *complete primitive.* Any solution obtained from the general solution by assigning particular values to one or more of the arbitrary constants or periodic functions of period 1 is called *particular solution.* Thus $2 \cdot 3^x - (-1)^x$ is a particular solution of equation (1).

Linear Difference Equations

An equation of the form

$$u_{x+n} + p_1 u_{x+n-1} + \cdots + p_n u_x = R(x)$$

where p_i's and $R(x)$ are functions of x or constants, is called a *linear difference equation* of order n. If $R(x) = 0$, the equation is called *linear homogeneous*; otherwise it is called linear *non-homogeneous equation.* Some difference equations have infinitely many solutions while others have no solution at all. The linear difference equation has at last one solution and only one solution under certain conditions.

Ex. 1. *Show that*

$$u_x = 2^x \left(c_1 \cos \frac{2\pi x}{3} + c_2 \sin \frac{2\pi x}{3} \right)$$

is a solution of $u_{x+2} + 2u_{x+1} + 4u_x = 0$, *where* c_1 *and* c_2 *are arbitrary periodic functions of period unity.*

2. *Show that* $(u_{2+1} - u_x)^2 + u_x^2 = -1$ *has no real-valued solution.*

8·2. Solution of Simple (Easy) Difference Equations. Many difference equations can be solved either by a mere change of form or by a simple substitution or on multiplication by some suitable function known as *integrating factor* and thus reducing the equation to an integral form.

Example 1. *Solve* :

$$u_{x+1} - au_x = 0, \quad a \neq 1.$$

Sol. Multiplying by the factor a^{-x-1}, we obtain

$$a^{-(x+1)} u_{x+1} - a^{-x} u_x = 0$$

$$\Rightarrow \quad \Delta(a^{-x} u_x) = 0$$

$$\therefore \quad a^{-x} u_x = c$$

$$\Rightarrow \quad u_x = ca^x,$$

where c is a periodic function of x of period 1.

Example 2. *Solve* :

$$u_{x+1} - u_x = ca^x$$

where c *is a periodic function of period 1 and* $a \neq 1$.

Sol. We have

$$\Delta u_x = ca^x$$

$$u_x = \Delta^{-1}(ca^x) + c'$$

$$= \frac{ca^x}{a-1} + c' ;$$

where c is a periodic function.

Example 3. *Solve* :

$$u_{x+1} - bu_x = ca^x$$

where c is a periodic function of period 1, when

(*i*) $b \neq a$. (*ii*) $b = a$.

Sol. (*i*) $b \neq a$.

Let $u_x = kca^x$ be a solution of the given equation.

then $$kca^{x+1} - bkca^x = ca^x$$

$\therefore$ $$k = \frac{1}{a-b} \cdot$$

Hence $\dfrac{ca^x}{a-b}$ is a particular solution of the equation.

Aliter. (*i*) We have

$$b^{-x-1}u_{x+1} - b^{-x}u_x = ca^x b^{-x-1}$$

$\Rightarrow$ $$\Delta(b^{-x}u_x) = cb^{-1}(ab^{-1})^x$$

$\therefore$ $$u_x = cb^{x-1}\Delta^{-1}(ab^{-1})^x$$

$$= \frac{cb^{x-1}a^x b^{-x}}{ab^{-1} - 1} = \frac{ca^x}{a-b} \cdot$$

(*ii*) $b = a$

The equation is

$$u_{x+1} - au_x = ca^x$$

Let $u_x = kca^{x-1}x$ be a solution of the given equation.

$\therefore$ $$kca^x(x+1) - akca^x x = ca^x$$

$\Rightarrow$ $$k = 1$$

Hence the equation is

$$u_x = cxa^{x-1}$$

Aliter. (*ii*) Multiply both sides by $a^{-(x+1)}$

$$a^{-(x+1)}u_{x+1} - a^{-x}u_x = ca^{-1}$$

$\Rightarrow$ $$\Delta(a^{-x}u_x) = ca^{-1}$$

$\Rightarrow$ $$a^{-x}u_x = ca^{-1}\Delta^{-1}(1) = cxa^{-1}$$

$\Rightarrow$ $$u_x = cxa^{x-1}.$$

Example 4. *Solve* :

$$u_{x+1} - au_x = bx$$

Sol. Multiplying by the factor $a^{-(x+1)}$ we get

$$a^{-(x+1)}u_{x+1} - a^{-x}u_x = bxa^{-x-1}$$

Now the left member is exact and the right member can be made exact by undeterminate coefficients as follows :

$$a^{-(x+1)}u_{x+1} - a^{-x}u_x = -b\left(\frac{x+1}{a-1} + \frac{1}{(a-1)^2}\right)a^{-x-1}$$
$$+ b\left[\frac{x}{a-1} + \frac{1}{(a-1^2)}\right)a^{-x}$$

$$\therefore \qquad a^{-x}u_x = -b\left(\frac{x}{a-1} + \frac{1}{(a-1)^2}\right)a^{-x} + c$$

$$\Rightarrow \qquad u_x = a^x c - \frac{bx}{a-1} - \frac{b}{(a-1)^2}.$$

Aliter. We have

$$\Delta(a^{-x}u_x) = bxa^{-x-1}$$

$$\therefore \qquad a^{-x}u_x = b\Delta^{-1}[xa^{-x-1}]$$

$$= ba^{-1}\Delta^{-1}(xa^{-x}) = ba^{-1}(E-1)^{-1}(xa^{-x})$$

$$= -ba^{-1}[1 + E + E^2 + \cdots]\, xa^{-x}$$

$$= -ba^{-1}[xa^{-x} + (x+1)\, a^{-(x+1)} + \cdots]$$

$$= -ba^{-1}xa^{-x}[1 + a^{-1} + a^{-2} + \cdots] - ba^{-1} - a^{x+1}[1 + 2a^{-1} + 3a^{-2}]$$

$$= -\frac{bx^{-1}xa^{-x}}{1-a^{-1}} - ba^{-1}\, a^{-(x+1)}(1-a^{-1})^{-2}$$

$$\Rightarrow \qquad a^{-\alpha}u_x = \frac{bxa^{-x}}{a-1} - \frac{ba^{-x}}{(a-1)^2} + c$$

$$\therefore \qquad u_x = \frac{bx}{a-1} - \frac{b}{(a-1)^2} + ca^x$$

Example 5. [*Method of undetermined coefficients*] *Solve* :

$$u_{x+1} - u_x = (x^2 - 2x)\, 2^x.$$

Sol. Assume $u_x = (Ax^2 + Bx + C)\, 2^x$

$$\therefore \qquad u_{x+1} - u_x = [Ax^2 + (4A + B)\, x + (2A + 2B + C)]\, 2^x$$

Comparing coefficients, we have

$$A = 1,\ B = -6,\ C = 10$$

whence $\qquad u_x = (x^2 - 6x + 10)\, 2^x + C'(x).$

Aliter. $\qquad \Delta u_x = (x^2 - 2x)\, 2^x$

$$\therefore \qquad u_x = \Delta^{-1}[(x^2 - 2x)\, 2^x]$$

Now see Example 5, Chapter 7.

8·3. Homogeneous Linear Equations.

8·3·1. First Order Homogeneous Linear Equations.

Consider the homogeneous linear equation of order one *viz.*

$$\Delta u_x + B(x)\, u_x = 0$$

$$\Rightarrow \qquad u_{x+1} - A(x)\, u_x = 0 \qquad ...(1)$$

where $A(x) = 1 - B(x)$.

$\therefore \qquad u_1 = A(0)\, u_0$, on putting $x = 0$

Hence $\qquad u_2 = A(1)\, u_1 = A(1)\, A(0)\, u_0$

$$u_3 = A(2)\, u_2 = A(2)\, A(1)\, A(0)\, u_0$$

$$\dots \quad \dots \quad \dots \quad \dots \quad \dots \quad \dots$$

$$u_x = A(x-1) \dots A(2)\, A(1)\, A(0)\, u_0$$

$$\Rightarrow \qquad u_x = C \prod_{x=0}^{x-1} A(x)$$

where C is the arbitrary value assigned to u_0.

Aliter. From (1),

$$\log u_{x+1} = \log A(x) + \log u_x$$

$$\Rightarrow \qquad \Delta \log u_x = \log A(x)$$

$$\therefore \qquad \log u_x = \Delta^{-1}[\log A(x)]$$

$$\Rightarrow \qquad u_x = C \exp\,[\Delta^{-1} \log A(x)]$$

where C is an arbitrary periodic function of x of period 1.

Example 6. *Solve* :

$$u_{x+1} - \frac{1}{x}\, u_x = 0, \quad x > 0.$$

Sol. Here, $\qquad A(x) = \dfrac{1}{x}\cdot$

$$\therefore \qquad \prod_{x=1}^{x-1} A(x) = \prod_{x=1}^{x-1} \frac{1}{x} = \frac{1}{(x-1)\,!} = \frac{1}{\Gamma(x)}\cdot$$

$\therefore$ Solution is $u_x = \dfrac{C}{(x-1)\,!} = \dfrac{C}{\lceil x}$

where C is an arbitrary periodic function of x of period 1.

Aliter. We have $u_{x+1} = \dfrac{1}{x}\, u_x \Rightarrow \dfrac{u_{x+1}}{u_x} = \dfrac{1}{x} = x^{-1}$

$$\therefore \qquad \log u_{x+1} - \log u_x = -\log x$$

$$\Rightarrow \quad \Delta \log u_x = -\log x$$

$$\Rightarrow \quad \log u_x = \Delta^{-1}(-\log x) = \log\left\{\frac{C}{\Gamma x}\right\}$$

$$\therefore \quad u_c = \frac{C}{\Gamma x}$$

Example 7. *Solve :*

$$u_{x+1} = 3^x u_x$$

First Method.

Sol. Here $A(x) = 3^x$

and $$\prod_{x=0}^{x-1} A(x) = 3^0 \cdot 3^1 \ldots 3^{x-1} = 3^{\frac{1}{2}x(x-1)}$$

Hence solution is $u_x = C\ 3^{\frac{1}{2}x(x-1)}$, C being a periodic function of x of period 1.

Second Method

$$u_{x+1} = 3^x u_x.$$

Taking logarithms to the base 3, we have

$$\log_3 u_{x+1} - \log_3 u_x = x \quad [\because \log_3 3 = 1]$$

$$\Rightarrow \quad \Delta \log_3 u_x = x$$

$$\therefore \quad \log_3 u_x = \Delta^{-1} x + C = \frac{x^{(2)}}{2} + C$$

$$\therefore \quad u_x = C \cdot 3^{\frac{x^{(2)}}{2}} = C \cdot 3^{\frac{1}{2}x(x-1)}$$

Property of linear homogeneous equation

Theorem. *If γ is a solution of the linear homogeneous equation*

$$y_{x+n} + p_1 y_{x+n-1} + \cdots + p_{n-1} y_{x+1} + p_n y_x = 0 \qquad ...(1)$$

and y^* is a solution of the complete equation

$$y_{x+n} + p_1 y_{x+n-1} + \cdots + p_{n-1} y_{x+1} + p_n y_x = R(x) \qquad ...(2)$$

then $\gamma + y^*$ is a solution (2).

Proof. Since γ is a solution of (1), we have

$$\gamma_{x+n} + p_1 \gamma_{x+n-1} + \cdots + p_{n-1} \gamma_{x+1} + p_n \gamma_x = 0 \qquad ...(3)$$

Also as y^* is a solution of (2). we have

$$y^*_{x+n} + p_1 y^*_{x+n-1} + \cdots + p_{n-1} y^*_{x+1} + p_n y^*_n = R(x) \qquad ...(4)$$

Adding (3) and (4), we get

$$(\gamma_{x+n} + y^*_{x+n}) + p_1(\gamma_{x+n-1} + y^*_{x+n-1}) + \cdots + p_n(\gamma_x + y^*_n) = R(x)$$

which implies that $\gamma + y^*$ is also a solution of (2). This solution is called the complete solution *i.e.*, the solution of (2) consists of the sum of the general solution of (1) together with the particular solution of the equation (2).

Hence General Solution = *C.F.* + *P.I.*

Definition. The solutions $y^{(1)}, y^{(2)}, ..., y^{(n)}$ are said to form a fundamental set of solution of *n*th order homogeneous difference equation

$$y_{x+n} + p_1 y_{x+n-1} + p_2 y_{x+n-2} + \cdots + p_n y_x = 0 \qquad ...(1)$$

if $\Delta \neq 0$ where

$$\Delta = \begin{vmatrix} y_0^{(1)} & y_0^{(2)} \cdots y_0^{(n)} \\ y_1^{(1)} & y_1^{(2)} \cdots y_1^{(n)} \\ y_{n-1}^{(1)} & y_{n-1}^{(2)} \cdots y_{n-1}^{(n)} \end{vmatrix}$$

Theorem. *Let* $y^{(1)}$ *and* $y^{(2)}$ *be two solutions of the second order homogeneous difference equation.*

$$y_{x+2} + p_1 y_{x+1} + p_2 y_x = 0$$

and let $\gamma = c_1 y^{(1)} + c_2 y^{(2)}$ *where* c_1, c_2 *are arbitrary constants, then* γ *is the general solution of (1) provided that*

$$\begin{vmatrix} y_0^{(1)} & y_0^{(2)} \\ y_1^{(1)} & y_1^{(2)} \end{vmatrix} \neq 0 \quad i.e. \ y_0^{(1)} y_1^{(2)} - y_0^{(2)} y_1^{(1)} \neq 0$$

Proof. We know that if $y_{(1)}$ and $y_{(2)}$ are two solutions of (1), then $\gamma = c_1 y_1^{(1)} + c_2 y_2^{2}$, c_1 and c_2 are constants, is also a solution of (1).

Now we shall prove that if y is any solution of (1), then the two constants c_1 and c_2 can be determined such that γ and y are identical. By uniqueness theorem, it follows that the two solutions of γ_0 and γ_1 the linear difference equation of order 2 are identical if they have the same value for two consecutive values of x, say $x = 0$ and $x = 1$. Thus we have

$$\gamma_0 = c_1 y_0^{(1)} + c_2 y_0^{(2)} = y_0 \qquad ...(2)$$

and

$$\gamma_1 = c_1 y_1^{(1)} + c_2 y_1(2) = y_1 \qquad ...(3)$$

Now (2) and (3) will have unique solution when solved if and only if

$$\begin{vmatrix} y_0^{(2)} & y_0^{(2)} \\ y_1^{(1)} & y_1^{(2)} \end{vmatrix} \neq 0$$

Thus for each choice of y_0 and y_1. we could find a unique pair of values of c_1 and c_2. Hence γ is a general solution of (1).

8·3·2. General Solution of Higher Order Homogeneous Linear Equations with Variable Coefficients. The homogeneous equation of order n is of the form

$$u_{x+n} + p_1 u_{x+n-1} + \cdots + p_n u_x = 0, 1, 2, \ldots \quad \ldots(8\cdot2)$$

where p's are variables.

Analogue to the theorems in ordinary differential equations, we have the following theorems which can be easily proved :

(*a*) If $u_1(x)$ is a solution of (8·2), so is $c_1\ u_1\ (x)$

(*b*) $u_1(x), u_2(x), \ldots, u_n(x)$ are n solutions of (8·2), then

$$c_1 u_1(x) + c_2 u_2(x) + \cdots + c_n u_n(x),$$

is also a solution, where $c_1, c_2, \ldots, c_n$ are arbitrary periodic constants.

Since linear equations with variable coefficients are solvable under severe restrictions upon the coefficients, we consider linear equations with constant coefficients, which are of the form :

$$A_0 u_{x+n} + A_1 u_{x+n-1} + \cdots + A_n u_x = 0 \quad \ldots(8\cdot3)$$

$$\Rightarrow \quad (A_0 E^n + A_1 E^{n-1} + \cdots + A_n)\ u_x = 0 \quad \ldots(8\cdot4)$$

where E is the operator such that $E^n u_x = u_{x+n}$; $A_0, A_1, A_2, \ldots, A_n$ are constants and $A_0 \neq 0,\ A_n \neq 0$. Without loss of generality, we take $A_0 = 1$, and then (8·4) can be written as

$$\left.\begin{aligned} & f(E)u_x = 0 \\ \text{where} \quad & f(E) = E^n + A_1 E^{n-1} + \cdots + A_n \end{aligned}\right\} \quad \ldots(8\cdot5)$$

and $\quad f(E) = 0$ is called the *auxiliary equation* of the equation (8·3).

Putting $u_x = E^x v_x$ in (8·5) we have

$$[E^n + A_1 E^{n-1} + A_2 E^{n-2} + \cdots + A_{n-1} E + A_n]\ v_x = 0$$

Now $f(E) = E^n + A_1 E^{n-1} + \cdots + A_{n-1} E + A_n$, is known as the characteristic function of the given equation.

Case I. *Auxiliary equation has n distinct real roots.*

Let $\quad \alpha_1, \alpha_2, \ldots, \alpha_n$ be the n distinct real roots.

$$\therefore \quad (E - \alpha_1)(E - \alpha_2)\ldots(E - \alpha_n)\ u_x = 0$$

Hence $(E - \alpha_1)\ u_x = 0,\ (E - \alpha_2)\ u_x = 0, \ldots, (E - \alpha_n)\ u_x = 0$

Consider $(E - \alpha_r)\ u_x = 0, \quad r = 1, 2, \ldots, n$

or $\quad u_{x+1} - \alpha_r u_x = 0$

Solution is $u_x = c_r\, \alpha_r^x$ [See Ex. 1, page 167]

where c_r is a periodic function of period 1.

Hence the solution of the given equation is

$$u_x = c_1\, \alpha_1^x + \cdots + c_r\, \alpha_r^x + \cdots + c_n\, \alpha_n^x.$$

Case II. *Auxiliary equation has repeated real roots.*

Consider $u_{x+2} - 2\alpha u_{x+1} + \alpha^2 u_x = 0.$

Here $\alpha_2 = \alpha_1 = \alpha$ and we have

$$(E - \alpha)^2 u_x = 0 \qquad ...(1)$$

To solve this equation, we make the substitution

$$u_x = \alpha^x v_x,$$

where v_x is a new dependent variable, (1) may be written as

$$\alpha^{x+2} v_{x+2} - 2\alpha^{x+2} v_{x+1} + \alpha^{x+2} v_x = 0$$

or $\Delta^2 v_x = 0, \quad \alpha \neq 0$

Hence $v_x = c_1 + c_2 x$

$\therefore$ $u_x = (c_1 + c_2 x)\, \alpha^x$

where c_1 and c_2 are periodic functions of x with period 1.

Thus the solution corresponding to this pair of equal roots is

$$(c_1 + c_2 x)\, \alpha^x.$$

Aliter. $(E - \alpha)^2 u_x = 0 \quad \Rightarrow \quad (E - \alpha)(E - \alpha)\, u_x = 0$

Let $(E - \alpha)\, u_x = v_x$

Then $(E - \alpha)\, v_x = 0, \quad \Rightarrow \quad v_x = c_1\, \alpha^x$

Hence $(E - \alpha)\, u_x = c_1\, \alpha^x.$

$\Rightarrow$ $u_{x+1} - \alpha\, u_x = c_1\, \alpha^x.$

Multiply both sides by $\alpha^{-(x+1)}$

$$\alpha^{-(x+1)} u_{x+1} - \alpha^{-x} u_x = c_1\, \alpha^{-1}$$

$$\Delta(\alpha^{-x} u_x) = c_1\, \alpha^{-1}$$

$\therefore$ $\alpha^{-x} u_u = c_1\, \alpha^{-1} x + c_2$

$\Rightarrow$ $u_x = (c_1 x + c_2)\, \alpha^x$

where c_1 and c_2 are periodic functions of x with period unity.

Similarly, it can be proved that if the root α is repeated k times, then the solution of the equation

$$(E - \alpha)^k\, u_x = 0$$

corresponding to this repeated root α is

$$u_x = (c_0 + c_1 x + \cdots + c_{k-1} x^{k-1})\, \alpha^x = \alpha^x \sum_{r=0}^{k-1} c_r\, x^r.$$

Case III. *Auxiliary equation has non-repeated complex roots.*

Let $\alpha_1 = \alpha + i\,\beta, \ \alpha_2 = \alpha - i\,\beta, \ i = \sqrt{-1}$

then the corresponding two terms of the solution will be

$$c_1(\alpha + i\,\beta)^x + c_2(\alpha - i\,\beta)^x.$$

This can be written as

$$r^x(C_1 \cos x\,\theta + C_2 \sin x\,\theta)$$

where $r = +\sqrt{\alpha^2 + \beta^2}$, $\theta = \tan^{-1}\left(\dfrac{\beta}{\alpha}\right)$, $C_1 = c_1 + c_2$, $C_2 = i(c_1 - c_2)$

Note. If $\alpha = 0$, $\beta = 1$ then $u_x = c_1(i)^x + c_2(-i)^x$

$$= c_1\, e^{i\pi x/2} + c_2\, e^{-i\pi x/2} = C_1 \cos\frac{\pi x}{2} + C_2 \sin\frac{\pi x}{2}$$

Case IV. *Auxiliary equation has repeated complex roots.*

It can be easily proved that in such a case, the solution is

$$r^x[(c_1 + c_2 x)\cos x\,\theta + (c_3 + c_4 x)\sin x\,\theta]$$

where $r = +\sqrt{\alpha^2 + \beta^2}$, $\theta = \tan^{-1}\left(\dfrac{\beta}{\alpha}\right)$.

Example 8. *Solve the equations* :

(*i*) $u_{x+2} - 4u_x = 0$

(*ii*) $16u_{x+2} - 8u_{x+1} + u_x = 0$ (*B.A. Math. Hons., Delhi, 1983*)

(*iii*) $u_{x+2} - 4u_{x+1} + 13u_x = 0.$

Sol. (*i*) $u_{x+2} - 4u_x = 0.$ or $(E^2 - 4)u = 0$

$$E^2 - 4 = 0$$

$$\therefore \quad E = 2, -2.$$

$\therefore$ Solution is

$u_x = c_1(-2)^x + c_2 2^x$, c_1 and c_2 are arbitrary periodic functions.

(*ii*) We have $16E^2 - 8E + 1 = 0$ or $(4E - 1)^2 = 0$, $\therefore\ E = \frac{1}{4}, \frac{1}{4}$

$\therefore$ Solution is $u_x = (c_1 + c_2 x)\left(\dfrac{1}{4}\right)^x$

(*iii*) We have $E^2 - 4E + 13 = 0$

$$E = 2 \pm 3i$$

$$\therefore \quad \alpha = 2, \beta = 3.$$

$$\therefore \quad r = +\sqrt{\alpha^2 + \beta^2} = \sqrt{13},\ \theta = \tan^{-1}\left(\frac{3}{2}\right)$$

$\therefore$ Solution is $u_x = 13^{\frac{1}{2}x}(c_1 \cos\theta x + c_2 \sin\theta x)$, $\theta = \tan^{-1}\left(\dfrac{3}{2}\right)$

Example. *Show that the function y given by*

$$y_x = 1 - \frac{2}{x},\ x = 1, 2, 3, \ldots$$

is a solution of the first order difference equation

$$(x+1)\,y_{x+1} + xy_x = 2x - 3,\ x = 1, 2, 3, \ldots$$

Sol. We have $y_x = 1 - \dfrac{2}{x}$ and $y_{x+1} = 1 - \dfrac{2}{x+1}$

Substituting the values of y_x and y_{x+1} in L.H.S. *i.e.*,

$$(x+1)\,y_{x+1} + xy_x = 2x - 3, \text{ we obtain}$$

$$= (x+1)\left[1 - \frac{2}{x+1}\right] + x\left[1 - \frac{2}{x}\right]$$

$$= x + 1 - 2 + x - 2 = 2x - 3$$

$$= \text{R.H.S.}$$

Hence $y_x = 1 - \dfrac{2}{x}$ is a solution of the given equation.

8·4. Non-homogeneous equation reducible to homogeneous equation. Suppose the difference equation is

$$u_{x+2} + 2pu_{x+1} + qu_x = c$$

where p, q and c are constants.

To reduce this equation to a homogeneous form, we make the transformation

$$u_x = v_x + D$$

where D is a constant, such that

$$(v_{x+2} + D) + 2p(v_{x+1} + D) + q\,(v_x + D) = c.$$

It reduces to

$$v_{x+2} + 2pv_{x+1} + qv_x = 0.$$

where $\qquad D = c/(1 + 2p + q)$

and $\qquad 1 + 2p + q \neq 0$

Now $\qquad v_x = A\alpha^x + B\beta^x$

is the general solution of the homogeneous equation so that

$$u_x = \frac{c}{1 + 2p + q} + A\alpha^x + B\beta^x$$

where A and B are arbitrary constants.

Note. If $1 + 2p + q = 0$,

then $\qquad q = -(1 + 2p)$.

In this case make the transformation $u_x = v_x + \delta x$ where δ is a constant so chosen that

$$v_{x+2} + \delta(x+2) + 2p\,[v_{x+1} + \delta(x+1)] - (1+2p)[v_x + \delta x] = 0.$$

It is reduced to

$$v_{x+2} + 2pv_{x+1} - (1+2p)v_x = 0$$

and $$\delta = c/2\,(1+p)$$

8·5. Non-homogeneous Linear Equations. Just as in the case of linear differential equations the general solution of the non-homogeneous difference equation is found by adding a particular solution of the non-homogeneous equation to the general solution of the associated homogeneous equation. That is

General Solution = Complementary Function + Particular Integral.

We shall employ two methods to find particular integrals or solutions.

(*a*) the method of undetermined coefficients, and

(*b*) the method of operators.

The non-homogeneous (general linear) difference equation with constant coefficients is of the form

$$f(E)u_x = \phi(x),\ E = 1+\Delta \qquad \text{...(1)}$$

The particular integral of this equation is

$$u_x = \frac{1}{f(E)}\,\phi(x)$$

where the right-hand member is the expression such that when operated upon by $f(E)$ produces $\phi(x)$.

8·5·1. Method of evaluating $\frac{1}{f(E)}\phi(x)$. These methods are of a general character and of universal application in so far as the sums exist. The labour of applying the general methods even in simple cases may be very considerable. However, Boob has given the methods in which the right side of the complete equation is one of the following forms :

(*i*) V_x, (*ii*) a^x, (*iii*) $a^x \cdot V_x$,

where V_x is a polynomial in x of degree m, say.

(*i*) $\phi(x) = x^m$, m zero or a positive integer.

Hence $u_x = \frac{1}{f(1+\Delta)}\,x^m$.

Now suppose that the characteristic equation does not admit the root unity.

Then $f(1+\lambda) = \lambda^n + A_1\lambda^{n-1} + \cdots + A_{n-1}\lambda + A_n$.

If we expand $\frac{1}{f(1+\lambda)}$ in ascending powers of λ as far as λ^m, we get

$$\frac{1}{f(1+\lambda)} \equiv B_0 + B_1\lambda + \cdots + B_m\lambda^m + \frac{g(\lambda)\lambda^{m+1}}{f(1+\lambda)}$$

where $g(\lambda)$ is a polynomial.

Thus $$1 \equiv f(1+\lambda)(B_0 + B_1\lambda + \cdots + B_m\lambda^m) + g(\lambda)\lambda^{m+1}$$

Writing Δ for λ, we get

$$x^m = f(1+\Delta)(B_0 + B_1\Delta + B_2\Delta^2 + \cdots + B_m\Delta^m)\, x^m + g(\Delta)\Delta^{m+1}\, x^m$$

Now $\Delta^{m+1}x^m = 0$

Hence $(B_0 + B_1\Delta + B_2\Delta^2 + \cdots + B_m\Delta^m)\, x^m$

satisfies the equation

$$f(1+\Delta)u_x = x^m$$

$$\Rightarrow \qquad f(E)u_x = x^m$$

and is therefore a particular solution of the equation. This actual expansion of $\dfrac{1}{f(1+\Delta)}$ can be performed by ordinary long division or resolving into partial fractions.

If the characteristic equation admits the root unity of **multiplicity** r, then equation (1) becomes

$$f_1(E)(E-1)^r\, u_x = x^m$$

where $f_1(\rho)$ is a polynomial of degree $n - r$ in ρ.

Putting

$$(E-1)^r\, u_x = \Delta^r u_x = v_x$$

the equation for v_x is

$$f_1(E)v_x = x^m$$

Now $\qquad f_1(1+\lambda) = C_0 + C_1\lambda + C_2\lambda^2 + \cdots + C_{n-r}\,\lambda^{n-r},$

where $C_0 \neq 0$.

Thus $\qquad V_x = (b_0 + b_1\Delta + b_2\Delta^2 + \cdots + b_m\,\Delta^m)\, x^m,$

which is a polynomial of degree m in x.

Writing this as

$$V_x = d_0\,{}^xC_m + d_1\,{}^xC_{m-1} + \cdots + d_{m-1}\,{}^xC_0,$$

a particular solution of the given equation is obtained from

$$\Delta^r u_x = d_0\,{}^xC_m + d_1\,{}^xC_{m-1} + \cdots + d_{m-1}\,{}^xC_1 + d_m\,{}^xC_0.$$

Hence particular solution is

$$\therefore \qquad u_x = d_0\,{}^xC_m + d_1\,{}^xC_{m+r-1} + \cdots + d_{m-1}\,{}^xC_{r+1} + d_m\,{}^xC$$

(*ii*) $\qquad \phi(x) = a^x$

we have $\qquad f(E)a^x = f(a)\, a^x,$

$$\therefore \qquad u_x = \frac{1}{f(E)}\, a^x = \frac{1}{f(a)}\, a^x.$$

provided that $f(a) \neq 0$.

Let a be a zero of order r of $f(\rho)$, we have

$$f(\rho) = (\rho - a)^r f_1(\rho)$$

where $$f_1(a) = \frac{f^{(r)}(a)}{r\,!}$$

Put $$u_x = a^x$$

Then the equation is

$$f_1(E)(E-a)^r . a^x v_x = a^x$$

$$\Rightarrow \quad a^x f_1(aE)(aE-a)^r v_x = a^x$$

or $$a^r \Delta^r v_x = \frac{1}{f_1(aE)} \cdot 1 = \frac{1}{f_1(a + a\Delta)} \cdot 1 = \frac{1}{f_1(a)} \cdot$$

Therefore,

$$\Delta^r v_x = \frac{a^{-r}\, r\,!}{f^{(r)}(a)} = \frac{a^{-r}\, r\,!}{f(r)(a)} C_0$$

so that $$v_x = {}^xC_r \frac{a^{-r}\, r\,!}{f^{(r)}(a)} \cdot$$

Hence $$u_x = a^x v_x = {}^xC_r \cdot \frac{a^{-r}.a^x.r\,!}{f^{(r)}(a)}$$

is the particular solution.

8·5·2. To prove

(*i*) $\dfrac{1}{f(E)} a^x v_x = a^x \cdot \dfrac{1}{f(aE)} v_x$

(*ii*) $\dfrac{1}{(E-a)^k f(E)} \cdot a^x = \dfrac{x^{(k)}\, a^{x-k}}{k\,! f(a)}, \; f(a) \neq 0$

(*iii*) $\dfrac{1}{(E-a)^n} a^{n+x} \Delta^n \left[\dfrac{f(x)}{a^n} \right] = f(x)$

Proof. (*i*) Now $E(a^x w_x) = a^{x+1} E(w_x) = a^x.aEw_x$,

$$E^2(a^x w_x) = E(Ea^x w_x) = E(a^x a^1 Ew_x) = a^{x+2}E^2 w_x = a^x (aE)^2 w_x$$

$\therefore$ Let $f(aE)w_x = v_x \quad \therefore \quad w_x = \dfrac{1}{f(aE)} v_x$

$$a^x w_x = \frac{1}{f(aE)} a^x f(aE) w_x = \frac{1}{f(aE)} a^x v_x$$

$$\Rightarrow \quad \frac{1}{f(E)} a^x v_x = a^x \cdot \frac{1}{f(aE)} \cdot v_x.$$

Aliter. Let $f(E)u_x = a^x v_x$

Put $u_r = a^x w_x$

$$\therefore \quad f(E)a^x w_x = a^x v_x$$

$$a^x f(aE) w_x = a^x v_x$$

But $\left\{\begin{array}{l} E(a^x v_x) = a^{x+1} E(v_x) = a^x.aEv_x. \\ f(E)a^x v_x = a^x f(aE) v_x. \end{array}\right\}$

Hence $f(aE)\, w_x = v_x \quad \Rightarrow \quad w_x = \dfrac{1}{f(aE)} v_x$

Hence $\dfrac{1}{f(E)} a^x.v_x = w_x = a^x \cdot \dfrac{1}{f(aE)} \cdot v_x$

Cor. If $v_x = 1$, then $\dfrac{1}{f(E)} a^x = \dfrac{1}{f(a)} a^x$

(*ii*) In (*i*) put $v_x = 1$ and instead of $f(E)$ take $(E-a)^k f(E)$

$$\therefore \quad \frac{1}{(E-a)^k f(E)} a^x.1$$

$$= a^x \frac{1}{(aE-a)^k f(aE)} \cdot 1$$

Now $f(aE).1 = f(a)$

since $f(aE)\, 1 = (A_0 + A_1 aE + A_2 a^2 E^2 + \cdots).\, 1 = A_0 .1 + aA_1 + a^2 A_2 + \cdots$
$= f(a)$

$$\therefore \quad \frac{1}{(E-a)^k f(E)} a^x = a^x \frac{1}{a^k \Delta^k} \cdot \frac{1}{f(a)} \cdot 1. = a^{x-k} \cdot \frac{1}{f(a)} \cdot \frac{1}{\Delta^k} \cdot 1$$

$$= a^{x-k} \frac{1}{f(a)} \cdot \frac{x^{(k)}}{k\,!}$$

(*iii*) We have $(E-a)^n f(x) = (E-a)^n a^x \dfrac{f(x)}{a^x}\} = a^x (aE-a)^n \dfrac{f(x)}{a^x}$

$$= a^{x+n} \Delta^n \left[\frac{f(x)}{a^x}\right]$$

Hence the result follows.

Example 9. *Solve the equation* :

$$u_{+2} - 4u_x = 9x^2$$

Sol. The complementary solution of the given equation is

$$u_x = c_1(2)^x + c_2(-2)^x$$

For the particular solution we assume

$$u_x = Ax^2 + Bx + C.$$

$$\therefore \quad [A(x+2)^2 + B(x+2) + C] - 4\,[Ax^2 + Bx + C] = 9x^2$$

$$\Rightarrow \quad (-3A-9)x^2 + x(4A-3B) + (4A+2B-3C) = 0$$

$$\therefore \quad -3A-9=0,\ 4A-3B=0,\ 4A+2B-3C=0$$

$$\therefore \quad A=-3, B=-4, C=-\frac{20}{3}.$$

$\therefore$ General solution is

$$u_x = c_1 2^x + c_2(-2)^x - 3x^2 - 4x - \frac{20}{3}.$$

Alternatively. Particular Integral

$$= \frac{1}{E^2-4} \cdot 9x^2$$

$$= \frac{1}{\Delta^2+2\Delta-3} \cdot 9x^3 \qquad (\because E = 1+\Delta)$$

$$= -3 \frac{1}{1-\left(\frac{2\Delta+\Delta^2}{3}\right)} \cdot x^2$$

$$= -3\left[1-\left(\frac{2\Delta+\Delta^2}{3}\right)\right]^{-1} x^2$$

$$\Rightarrow \quad \text{P.I.} = -3\left[1+\frac{2\Delta+\Delta^2}{3}+\frac{4\Delta^2}{9}\right](x^{(2)}+x), \quad \text{neglecting } \Delta^3 \text{ and higher operators.}$$

$$= -3\left[1+\frac{2\Delta}{3}+7\frac{\Delta^2}{9}\right][x^{(2)}+x]$$

$$= -3\left[x^{(2)}+\frac{2}{3}\cdot 2x^{(1)}+\frac{7}{9}\cdot 2+x+\frac{2}{3}\right]$$

$$= -3\left[x(x-1)+\frac{7}{3}x+\frac{20}{9}\right]$$

$$= -3\left[x^2+\frac{4}{3}x+\frac{20}{9}\right]$$

$$= -3x^2-4x-\frac{1}{3}\cdot 20$$

$\therefore$ General solution is

$$u_x = c_1 2^x + c_2(-2)^x - 3x^2 - 4x - \frac{20}{3}$$

Example 10. *Solve the equation* :

$$u_{x+4} - 5u_{x+3} + 9u_{x+2} - 7u_{x+1} + 2u_x = x^3 + 1.$$

Sol. Here $f(E) = E^4 - 5E^3 + 9E^2 - 7E + 2$

$$= (E-1)^3(E-2)$$

$\therefore \quad (\Delta - 1)\Delta^3 u_x = x^3 + 1$

Put $\quad \Delta^3 u_x = v_x$, we have

$$v_x = \frac{1}{\Delta - 1}(x^3 + 1) = -(1 - \Delta)^{-1}(x^3 + 1)$$

$$= (-1 - \Delta - \Delta^2 - \Delta^3 \ldots)(x^3 + 1)$$

$$= -(x^3 + 3x^2 + 9x + 14)$$

Thus $\quad \Delta^3 u_x = -6\,{}^x c_3 - 12\,{}^x c_2 - 13\,{}^x c_1 - 14\,{}^x c_0$

The particular solution is

$$-6\,{}^x c_6 - 12\,{}^x c_5 - 13\,{}^x c_4 - 14\,{}^x c_3.$$

Hence the general solution is

$$u_x = c_1 2^x + c_2 + c_3 x + c_4 x^2 - \frac{17}{24}x^3 - \frac{1}{4}x^4 + \frac{1}{40}x^5 - \frac{1}{120}x^6.$$

Example 11. *Solve the equation :*

$$u_{x+2} - 7u_{x+1} + 10u_x = 12 \cdot 4^x.$$

Sol. We first find the general solution of

$$u_{x+2} - 7u_{x+1} + 10u_x = 0.$$

The auxiliary equation is

$$a^2 - 7a + 10 = 0$$

$$\therefore \quad a = 2 \text{ or } a = 5$$

Hence the complementary function is

$$u_x = c_1 2^x + c_2 5^x.$$

Now we look for a *particular solution of the original non-homogeneous equation.*

Method I. (Method of Undetermined Coefficients). Let

$$u_x = A.4^x$$

where A is a constant to be determined.

Substituting in the difference equation, we get

$$A \cdot 4^x[4^2 - 7 \cdot 4 + 10] = 12 \cdot 4^x$$

whence $\quad A = -6.$

The general solution is

$$u_x = c_1 2^x + c_2 5^x - 6 \cdot 4^x$$

This method is useful in finding the particular solution of the complete equation when $R(x)$ is of special type. We set up a trial solution which consists of the number of unknown constant coefficients, corresponding to each term present in $R(x)$. The constant coefficients are to be determined by substitution in the difference equation.

S.No.	*Special type of R(x)*	*Trial solution*
1.	$a^x f(x)$	$a^x(A_0 + A_1 x + \cdots + A_n x^n)$
2.	$a^x \sin bx$ or $a^x \cos bx$	$a^x(A \sin bx + B \cos bx)$
3.	a^x	$A \cdot a^x$
4.	$\sin bx$ or $\cos bx$	$A \sin bx + B \cos bx.$

Method II. The Method of Operators.

We have $(E^2 - 7E + 10)u_x = 12 \cdot 4x$

$$\therefore \qquad \text{P.I.} = \frac{1}{E^2 - 7E + 10} 12 \cdot 4^x = 12 \cdot \frac{1}{4E^2 - 7E + 10} \cdot 4^x$$

$$= 12 \cdot 4^x \cdot \frac{1}{4 \cdot 4^2 - 7 \cdot 4 + 10}$$

$$= \frac{1}{-2} \times 12 \cdot 4^x = -6 \cdot 4^x$$

Example 12. *Solve the equation* :

$$u_{x+2} - 7u_{x+1} + 10u_x = 12 \cdot 5^x$$

(*B.A. Hons. Delhi, 1982*)

Sol. Here if we take $x_x = A \cdot 5^x$ then

$$A \cdot 5^x(5^2 - 7 \times 5 + 10) = 12 \cdot 5^x$$

which does not determine A.

In such a case we **assume a particular solution** of the form

$$u_x = A(x \cdot 5^x).$$

$$\therefore \qquad A[(x+2)5^{x+2}] - 7A[(x+1)\, 5^{x+1}] + 10A(x5^x) = 12 \cdot 5^x$$

$$\Rightarrow \qquad A[25(x+2) - 35(x+1) + 10x] = 12$$

$$\Rightarrow \qquad 15A = 12$$

$$\Rightarrow \qquad A = \frac{4}{5}$$

$\therefore$ Solution is $\qquad u_x = c_1 2^x + c_2 5^x + \frac{4}{5}(x \cdot 5^x).$

Alternatively. By the **method of operators;**

$$\text{P.I.} = \frac{1}{E^2 - 7E + 10} \cdot 12 \cdot 5^x$$

$$= \frac{1}{(E-5)(E-2)} \cdot 12 \cdot 5^x$$

$$= 12 \cdot \frac{1}{(E-6)} \cdot \tfrac{1}{3} \cdot 5^x$$

$$= 4 \frac{1}{E-5} \cdot 5^x$$

$$= 4 \cdot 5^x \frac{1}{\cdot 5E - 5} \cdot 1, \text{ (by 8·5·2.}(i)\text{, page 179)}$$

$$= 4 \cdot 5^{x-1} \cdot \frac{1}{E-1} \cdot 1 = 4 \cdot 5^{x-1} . \Delta^{-1} .1$$

$$= 4 \cdot 5^{x-1} x.$$

8·6. First order Non-homogeneous Linear Equations

$$u_{h+1} = Au_h + B, \; B \neq 0$$

Consider the following linear equation of first order

$$f_1(x)u(x+1) + f_2(x)u(x) = \phi(x), \; x = 0, 1, ... \qquad ...(8·6)$$

where $f_1(x)$, $f_2(x)$ and $\phi(x)$ are given function of $x. f_0(x) \neq 0, f_1(x) \neq 0$.

Let $u_1(x)$ be the particular function which satisfies the equation (8·6). Then we have

$$f_1(x)u_1(x+1) + f_2(x)u_1(x) = \phi(x) \qquad ...(8·7)$$

Set $u(x) = u_1(x) + v(x)$ in (8·6) so that

$$f_1(x)\,[u_1(x+1) + v(x+1)] + f_2(x)[u_1(x) + v(x)] = \phi(x)$$

or using (8·7). $f_1(x)v(x+1) + f_2(x)v(x) = 0$ ...(8·8)

which is a homogeneous equation. Thus (8·6) can be solved.

Aliter. Consider the equation $y_{h+1} = Ay_h + B, \; B \neq 0$.

or $f_1(h)y_{h+1} + f_2(h)y_h = \phi(h)$

We re-write as

$$y_{h+1} = \frac{-f_2(h)}{f_1(h)} y_h + \frac{\phi(h)}{f_1(h)} = Ay_h + B, A \neq 0$$

Putting $h = 0$ and $h = 1$ in (1) respectively, we have

$$y_1 = Ay_0 + B, y_2 = Ay_1 + B = A\,(Ay_0 + B) = A^2 y_0 + B(1 + A)B,$$

Also $y_3 = Ay_2 + B = A^3 y_0 + B(1 + A + A^2)$

In general, $y_h = A^h y_0 + B(1 + A + \cdots + A^{h-1})$

$$= A^h y_0 + B\frac{(1 - A^h)}{1 - A}, \; A \neq 1$$

$$= y_0 + Bh, A = 1$$

Example 13. *Solve* : $u(x+1) - au(x) = \cos nx$ ...(1)

Sol. Let $u_1(x)$ be the particular function.

$\therefore \; u_1(x+1) - au_1(x) = \cos nx.$

Set $u(x) = u_1(x) + v(x)$ in (1)

$\therefore \; u(x+1) = u_1(x+1) + v(x+1)$

Hence $v(x+1) = av(x)$,

the solution of which is $v(x) = c.a^x$ where c is an arbitrary function.

The equation (1) can be written as

$$Eu(x) - au(x) = \cos nx$$

$$\therefore \quad u_1(x) = \frac{1}{E-a} \cos nx$$

$$= \text{Real part of} \frac{1}{E-a} (e^{in})^x.1$$

$$= \text{Real part of } e^{inx} \frac{1}{e^{in} - a} \cdot 1$$

$$\left[\because \frac{1}{f(E)} a^x = a^x \frac{1}{f(a)} \cdot 1, f(a) \neq 0 \right]$$

$$= \text{Real part of} \frac{e^{inx}(e^{-in} - a)}{1 - a(e^{in} + e^{-in}) + a^2}$$

$$= \frac{\cos (x-1)n - a \cos nx}{1 - 2a \cos n + a^2}$$

Hence the complete solution is

$$u(x) = \frac{\cos (x-1)\, n - a \cos nx}{1 - 2a \cos n + a^2} + c(x)a^x.$$

8.7 First order Linear Equation with Rational Coefficients

Consider the equation $u(x+1) = xu(x)$ ...(8·9)

Here $\quad u(x) = (x-1)u(x-1)$

$$= (x-1)(x-2)u(x-2)$$

....

$$= \Gamma(x),\ u_0 = 1, \text{ say}.$$

Thus $u(x) = \Gamma(x)$ is the particular solution of the equation (8·9).

Now consider the equation with rational coefficient

$$u(x+1) = f(x)u(x)$$

where $f(x)$ is a rational function.

We suppose that $f(x) = d \dfrac{(x-a_1)(x-a_2)...(x-a_k)}{(x-b_1)(x-b_2)...(x-b_l)}$

Since $\quad \Gamma(x+1-a_i) = (x-a_i)\Gamma(x-a_i)$

and $\quad d^{x+1} = d.\, d^x,$

we find that (8·9) has the particular solution

$$u(x) = d^x \frac{\Gamma(x-a_1)\Gamma(x-a_2)...\Gamma(x-a_k)}{\Gamma(x-b_1)\Gamma(x-b_2)...\Gamma(x-b_l)}$$

The general solution of (8·9) is

$$u(x) = c(x)d^x \prod_{i=1}^{k} \Gamma(x-a_i) \Big/ \prod_{i=1}^{l} \Gamma(x-b_i).$$

Example 14. *Solve* :

$$(x^2 - 2x)u(x+1) = (2x+1)u(x)$$

Sol. Here $$u(x+1) = \frac{2x+1}{x^2-2x}\, u(x)$$

so that $$f(x) = \frac{2x+1}{x^2-2x} = \frac{2\left(x+\frac{1}{2}\right)}{x(x-2)} \quad \text{[Make Coeff. of } x = 1]$$

$$= 2 \cdot \frac{\Gamma\left(x+\frac{1}{2}\right)}{\Gamma x \Gamma(x-2)}.$$

Hence the general solution of the given equation is

$$u(x) = c(x).\, 2^x \cdot \frac{\Gamma\left(x+\frac{1}{2}\right)}{\Gamma x \Gamma(x-2)}.$$

8·8. Higher Order Homogeneous Linear Equations with Constant Coefficients

We give below the solution of a linear homogeneous equation of order n due to Seliwanoffi.

Consider the homegenous linear equation

$$U_{x+n} + A_1 U_{x+n-1} + A_2 U_{x+n-2} + \cdots + A_n U_x = 0 \qquad \ldots(8\cdot10)$$

Let $U_x = a^x V_x$, where V_x is a polynomial of mth degree.

Generalising this result

$$U_{x+1} = a^{x+1}\, V_{x+1} = a^{x+1}[V_x + \Delta V_x]$$

$$U_{x+2} = a^{x+2}\, V_{x+2} = a^{x+2}[V_x + 2\Delta V_x + \Delta^2 V_x]$$

$$U_{x+3} = a^{x+3}\, V_{x+3} = a^{x+3}[V_x + 3\Delta V_x + 3\Delta^2 V_x + \Delta^3 V_x]$$

$$\cdots\cdots\cdots\cdots\cdots\cdots\cdots\cdots\cdots\cdots$$

$$\cdots\cdots\cdots\cdots\cdots\cdots\cdots\cdots\cdots\cdots$$

$$U_{x+n} = a^{x+u} V_{x+n}[V_x + x\Delta V_x + \frac{x^{(2)}}{2\,!}\Delta^2 V_x + \cdots + \Delta^n V_x]$$

Therefore equation (8·10) becomes

$$a^x[a^n + A_1 a^{n-1} + \cdots + A_n]V_x$$

$$+ a^{x+1}[na^{n-1} + (n-1)A_1 a^{n-2} + (n-2)A_2 a^{n-3} + \cdots + A_{n-1}]\,\Delta V_x$$

$$+ a^{x+2}[n(n-1)a^{n-2} + (n-1)(n-2)A_1 a^{n-3} + \cdots + 2A_{n-2}]\,\frac{\Delta^2 V_x}{2\,!}$$

$$+ \cdots + a^{x+n} n\,!\, \frac{\Delta^n V_x}{n\,!} = 0$$

$$\Rightarrow \quad a^x f(a) V_x + a^{x+1} f'(a) \Delta V_x + a^{x+2} f''(a) \frac{\Delta^2 V_x}{!} + \cdots$$

$$+a^{x+n} f^{(n)}(a) \frac{\Delta^n V_x}{n!} = 0 \quad (8\cdot11)$$

where $f(a)$ is the auxiliary function

$$f(a) = a^n + A_1 a^{n-1} + A_2 a^{n-2} + \cdots + A_n.$$

If $a = a_1$ is simple root of $f(a) = 0$ and $V_x = 1$, the equation (8·11) is satisfied and a_1^x is a particular solution of (8·10). If $a_1, a_2 \ldots a_n$ are n simple roots of $f(a) = 0$, particular solution of (8·10) are $(a_1)^x, (a_2)^x, \ldots, (a_n)^x$ and the general solution is

$$U_x = c_1(a_1)^x + c_2(a_2)^x + \cdots + c_n(a_n)^x$$

If a_1 is k times repeated root then particular solution of (8·10) is

$$U_x = (c_1 + c_2 x + c_3 x^3 + \cdots + c_k x^{k-1}) a_k^x$$

Hence general solution is

$$U_x = (c_1 + c_2 x + c_3 x^2 + \cdots + c_k x^{k-1}) a_k^x + c_{k+1} a_{k+1}^x + \cdots + c_n a_n^x.$$

8·9. First order linear equation with variable coefficients.

Consider the equation

$$u(x+1) - \phi(x) u(x) = f(x) \quad \ldots(8\cdot12)$$

Set

$$u(x) = u_1(x) v(x) \quad \ldots(8\cdot13)$$

so that from (8·12), we have

$$u_1(x+1) v(x+1) - \phi(x) u_1(x) v(x) = f(x) \quad \ldots(8\cdot14)$$

The complementary function of (8·12) is given by

$$u(x+1) - \phi(x) u(x) = 0$$

so that if $u_1(x)$ be the *C.F.*, then from (8·14)

$$u_1(x+1) v(x+1) - u_1(x+1) v(x) = f(x)$$

or

$$u_1(x+1) \Delta v(x) = f(x)$$

$$\Delta f(x) = f(x) / u_1(x+1)$$

$$\therefore \quad v(x) = \Delta^{-1} \left[\frac{f(x)}{u_1(x+1)} \right]$$

By §8·3, we know that

$$u_1(x) = C(x) \exp [\Delta^{-1} \log \phi(x)].$$

Hence on substituting the values of $u_1(x)$ and $v(x)$ in (8·13) we get the complete solution of (8·12) as

$$u(x) = C(x) \Delta^{-1} \left[\frac{f(x)}{u_1(x+1)} \right] \exp [\Delta^{-1} \log \phi(x)].$$

Example 15. *Solve* :

$$u(x+1) - e^{2x-1} u(x) = 5x e^{x^2}$$

Sol. The complementary function is given by

$$u(x+1) = e^{2x-1}u(x)$$

$\therefore$ $$\log u(x+1) - \log u(x) = 2x - 1$$

$\therefore$ $$\Delta \log u(x) = 2x - 1$$

$\therefore$ $$u_1(x) = C(x) \exp [\Delta^{-1}(2x-1)]$$

$$= C(x) \exp (x-1)^2$$

Putting $u(x) = \exp (x-1)^2\, v(x)$ in the given equation we have

$$e^{x^2} v(x+1) - e^{2x-1}\, e^{(x-1)^2}\, v(x) = 5x\, e^{x2}$$

$\Rightarrow$ $$v(x+1) - v(x) = 5x$$

$\Rightarrow$ $$\Delta v(x) = 5x$$

$\therefore$ $$v(x) = \Delta^{-1}\, 5x = \tfrac{5}{2}\, x(x-1).$$

Hence the complete solution of the given equation is

$$u(x) = \left[\; C(x) \cdot \frac{5}{2}\, x(x-1) \;\right] e^{(x-1)^2}$$

8·10. Equation Homogeneous in $u(x)$.

The general type of an equation homogeneous in $u(x)$ is

$$\phi\left(\frac{u(x+1)}{u(x)},\; x \right) = 0$$

Solving for $\dfrac{u(x+1)}{u(x)}$, we get a linear equation. Now *see* §8·7, page 185.

Example 16. *Solve* :

$$2[u(x+1)]^2 + u(x+1)u(x) - [u(x)]^2 = 0$$

Sol. The given equation reduces to

$$u(x+1) = \tfrac{1}{2}\, u(x) \quad \text{or} \quad u(x+1) = -\, u(x)$$

the solutions of which are

$$u(x) = C(x) \left(\tfrac{1}{2}\right)^x \quad \text{or} \quad u(x) = C(x)\, (-1)^x$$

Example 17. *Solve* :

$$\{u(x+1)\}^2 - xu(x+1)u(x) - 6x^2\{u(x)\}^2 = 0$$

Sol. This given equation reduces to

$$u(x+1) = 3x.u(x) \quad \text{or} \quad u(x+1) = -\, 2x.u(x)$$

the solutions of which are

$$u(x) = C(x).3x.\Gamma x \quad \text{or} \quad u(x) = C(x)(-2)^x \Gamma x.$$

8·11. Equations reducible to linear equations with constant coefficients.

The equation

$$u(x+n) + A_1\phi(x)u(x+n-1) + A_2\phi(x)\phi(x-1)u(x+n-2) + \cdots$$
$$+ A_x\phi(x)\phi(x-1)\ldots\phi(x+n-1)u(x) = r(x),\; \phi(x) \neq 0$$

with A's constants, can be reduced to an equation with constant coefficients by the substitution

$$v(x+j) = \phi(x+j-n-1)\ldots\phi(x)u(x+j),\ 0 \le j \le n$$

The transformed equation is

$$u(x+n) + A_1\, v(x+n-1) + \cdots + A_u\, v(x)$$
$$= r(x)/[\phi(x)\phi(x-1)\ldots\phi(x_{n-1})]$$

The equation

$$u(x+1)u(x) + \alpha(x)u(x+1) + \beta(x)u(x) = \gamma(x)$$

can, with the help of the substitution

$$u(x) = \frac{v(x+1)}{v(x)} - \alpha(x),$$

be reduced to

$$v(x+2) + \{\beta(x) - \alpha(x+1)\}v(x+1) - \{\alpha(x) + \beta(x) + \gamma(x)\}v(x) = 0$$

This equation may be an equation with constant coefficients.

Example 18. *Solve* :

$$u_{x+1}\, u_x + (x+2)\, u_{x+1} + xu_x = -2 - 2x - x^2$$

Sol. We have

$$[u_x + (x+2)][u_{x+1}] + xu_x = -2\ 2x - x^2$$

Put $\quad u_x = \dfrac{v(x+1)}{v(x)} - (x+2)$

The given equation becomes

$$\frac{v(x+1)}{v(x)}\left[\frac{v(x+2)}{v(x+1)} - (x+3)\right]$$
$$+ x\,\frac{v(x+1)}{v(x)} - x^2 - 2x = -2 - 2x - x^2$$

$$\Rightarrow \quad \frac{v(x+2)}{v(x)} - (x+3)\,\frac{v(x+1)}{v(x)} + x\,\frac{v(x+1)}{v(x)} + 2 = 0$$

$$\Rightarrow \quad v(x+2) - 3v(x+1) + 2v(x) = 0$$

$$\Rightarrow \quad (E^2 - 3E + 2)v(x) = 0$$

$$\therefore \quad E = 1 \quad \text{or} \quad 2.$$

Hence $\quad v(x) = C_1 + C_2\, 2^x$

The complete solution of the given equation is

$$u(x) = \frac{C_1 + C_2\, 2^{x+1}}{C_1 + C_2\, 2^n} - (x+2).$$

Note. $\quad \Delta^{-1}(a^{kx}) = \dfrac{a^{kx}}{a^k - 1}$

MISCELANEOUS SOLVED EXAMPLES

Example 19. *Find the sequence of numbers starting with zero and unity where each term after the second is the sum of the preceding two terms i.e., 0, 1, 2, 3, 5, 8, 13, ... (Fibonacci numbers).*

Sol. Here $u_1 = 0, u_2 = 1$, Let us find the solution of the difference equation

$$u_{x+2} = u_{x+1} + u_x, \qquad x \geq 1$$

Auxiliary equation is $E^2 - E - 1 = 0$ so that $E = \frac{1 \pm \sqrt{5}}{2}$

Hence solution is $u_x = c_1\left(\frac{1+\sqrt{5}}{2}\right)^x + c_2\left(\frac{1-\sqrt{5}}{2}\right)^x$

wi... $c_1 + c_2 = 0$ and $c_1 + c_2 + \sqrt{5}\,(c_1 - c_2) = 2$

Solving, we get

$$c_1 = \frac{1}{\sqrt{5}}, \quad c_2 = \frac{-1}{\sqrt{5}}$$

Thus $u_x = \frac{1}{\sqrt{5}}\left(\frac{1+\sqrt{5}}{2}\right)^x - \frac{1}{\sqrt{5}}\left(\frac{1-\sqrt{5}}{2}\right)^x$

Example 20. *The seeds of a certain plant when one year old produce ten-fold and when two years old or more produce eighteen-fold. A seed is planted and every seed subsequently produced is planted as soon as it is produced. Prove that the number of grains at the end of the nth year is*

$$\frac{1}{a}\left\{\left(\frac{11+a}{2}\right)^n - \left(\frac{11-a}{2}\right)^n\right\}, \quad a = 3\sqrt{17}$$

Sol. Let u_n be the number of grains at the end of the nth year, then the difference equation is

$$u_n = u_{n-1} + 18u_{n-2} + 10\,(u_{n-1} - u_{n-2})$$

$$\Rightarrow \qquad u_n - 11u_{n-1} - 8u_{n-2} = 0$$

The auxiliary equation is

$$E^2 - 11E - 8 = 0 \text{ which gives}$$

$$E = \frac{11 \pm a}{2} \text{ where } a = 3\sqrt{17}$$

Hence the solution is

$$u_x = c_1\left(\frac{11+a}{2}\right)^n + c_2\left(\frac{11-a}{2}\right)^n$$

Now we determine the constants c_1 and c_2 given that

$$u_0 = 1, u_1 + 1 + 10 = 11.$$

Hence $1 = c_1 + c_2$ and $11 = c_1\left(\frac{11+a}{2}\right) + c_2\left(\frac{11-a}{2}\right)$, we find $c_1 = \frac{1}{2a}, \; c_2 = -\frac{1}{2a}$.

Example 21. *Find and solve a linear equation which is obtained if the coefficient of the expansion of the function* $\frac{1}{t^2+t+1}$ *are obtained from a recurrence relation.*

Sol. Setting

$$\frac{1}{t^2+t+1} = f(0) + f(1)t + f(2)t^2 + \cdots + f(x)t^x + f(x+1)t^{x+1} + f(x+2)t^{x+2} + \cdots$$

so that $\quad 1 = (t^2+t+1)[f(0) + f(1)t + \cdots]$

whence $\left.\begin{aligned} f(0) &= 1 \\ f(0) + f(1) &= 0 \;\; i.e.,\; f(1) = -1 \end{aligned}\right\}$ initial conditions

Hence $f(x+2) + f(x+1) + f(x) = 0, \qquad \ldots(1)$

the linear equation we sought.

Auxiliary equation of (1) is $E^2 + E + 1 = 0$

$$\Rightarrow \quad E = \frac{-1 \pm \sqrt{1-4}}{2} = e^{\pm \frac{i2\pi}{3}}$$

The solution of this equation can be easily shown to be

$$f(x) = C_1\left(\cos\frac{2\pi}{3} + i\sin\frac{2\pi}{3}\right)^x + C_2\left(\cos\frac{2\pi}{3} - i\sin\frac{2\pi}{3}\right)^x$$

with $\quad 1 = C_1 + C_2$ and $-1 = (C_1 - C_2)\sqrt{3}$.

i.e., $\quad C_1 = \frac{1}{2}\left(1 - \frac{1}{\sqrt{3}}\right), \; C_2 = \frac{1}{2}\left(1 + \frac{1}{\sqrt{3}}\right)$.

EXERCISE 8

1. Write down the order of the following difference equations :

(*i*) $u(x+2) + 3u(x+1) + 2u(x) = x^2$,

(*ii*) $u(x+4) - 5u(x+2) + 6u(x) = 0$,

(*iii*) $\Delta^2 u(x) + 8\Delta u(x) + 3u(x) = \cos \pi x$,

(*iv*) $\Delta^2 u(x) + 2\Delta u(x) + u(x) = x + 3$.

(*v*) $\Delta^2 u_{x+1} + \frac{1}{2}\Delta^2 u_x = 0$.

2. Eliminate the constants A and B from

(*i*) $y_n = A.2^n + B.3^n \quad$ (*ii*) $u_n = A.3^n + B.4^n$

and derive the corresponding difference equation of the lowest possible order. (*Cal. Univ. 1995*)

(*iii*) α, β, u_0 and u_1 are given numbers. For $k = 0, 1, 2, \ldots$ we have

$u_{k+2} = \alpha u_{k+1} + \beta u_k. \qquad \ldots(1)$

Putting $U_k = u_k\, u_{k+2} - U_{k+1}^2$, show that (1) can be reduced to

$$U_k U_{k+2} - U_{k+1}^2 = 0$$

i.e., $$\frac{U_{k+2}}{U_{k+1}} = \frac{U_{k+1}}{U_k}.$$

(*iv*) Form the difference equation corresponding to two-parameter family $y = ax^2 - bx$, a and b being constants.

3. (*a*) Show that :

(*i*) $u(x) = c_1 \cos\left(\frac{2\pi x}{3}\right) + c_2 \sin\left(\frac{2\pi x}{3}\right)$ is a solution of $u(x+2) + u(x+1) + u(x) = 0$

(*ii*) $u(x) = 3^{x/2}\,[c_1 \cos(\theta x) + c_2 \sin(\theta x)]$ is a solution of $u(x+2) + 2u(x+1) + 3u(x) = 0$, where $\theta = \tan^{-1}(-\sqrt{2})$

(*iii*) $u_x = c_1 e^{\alpha x} + c_2^{-\alpha x}$ or $c_3 \cosh \alpha x + c_4 \sinh \alpha x$ is a solution of $u_{x+1} - 2u_x \cosh \alpha + u_{x-1} = 0$

(*iv*) $u_x = c_1 \cos \alpha x + c_2 \sin \alpha x$ where $\sin \alpha \neq 0$ is the general soution of $u_{x+1} - 2u_x \cos \alpha + u_{x-1} = 0$

(*v*) $u_x = a^x\left[c_1 \sin\frac{\pi x}{4} + c_2 \cos\frac{\pi x}{4} + c_2 \sin\frac{3\pi x}{4} + c_4 \cos\frac{3\pi x}{4}\right]$ is a solution of $u_{x+4} + a^4 u_x = 0$

(*iv*) $u_1 = \frac{1}{2} x(x-1)$ is a solution of the equation $y_{x+1} - y_x = x$

(*b*) Obtain the solution of the equation $u_{x+1} - e^{mx} u_x = 0$, in the form $u_x = c\, e^{x(x-1)m/2}$

(*c*) Prove that the solution of the equation $(E - a)^m f(m) = 0$

is $$f(x) = \sum_{r=0}^{m-1} c_r a^x x^r$$

where a's are constants and c_r are periodic functions with period 1.

Solve the following equations, by the method of differences :

4. $u_{x+1} - 2u_x = 2x$.

5. $u_{x+1} - 2u_x = 3^x$.

6. (*i*) $u_{x+1} - u_x = x^3,\ u_0 = 1$

(*ii*) $u_{x+1} - 3u_x = x,\ u_0 = 1$.

7. $3u_{x+1} - u_x = x$.

8. (*i*) $\Delta^2 u_x + \Delta u_x = x + \sin x$

(*ii*) $(x+3)\,\Delta u_x + 2u_x = x + 1$

(*iii*) $\Delta u_x = \left(\frac{1}{x} - 1\right) u_x$ with initial condition $u_1 = 1$.

(*iv*) $u_{x+1} - 2 \sin x\, u_x + u_{x-1} = 0$, where $u_0 = x$ and $u_1 = \cos x$

9. $u_{x-1} - 3u_x = 2^x$. **10.** $u_{x+1} - xu_x = 0$.

11. $u_{x+1} - x^2 u_x = 0$. **12.** $(x+1)u_{x+1} - xu_x = 0$.

13. $u_{x+2} - 8u_{x+1} + 15u_x = 0$. **14.** $u_{x+3} - 3u_{x+1} - 2u_x = f0$.

15. (*i*) $u_{x+2} + 2u_{x+1} + 4u_x = 0$. (*ii*) $y_{n+2} - 2y_{n+1} + (1+n^2)y_n = 0$

16. (*i*) $u_{x+2} + a^2 u_x = \cos ax, a = 0, a = 1$.

$$\left[\text{P.I.} = \text{R.P. of}\left(\frac{e^{iax}}{e^{isa} + a^2}\right) = \text{etc.}\right]$$

(*ii*) $u_{x+2} + u_x = \sin \pi x$

(*iii*) $u_{x+2} - 7u_{x+1} + 12u_x = \cos x$ with $u_0 = 0 = u_1$.

(*iv*) $u_{x+1} - au_x = \sin bx$.

17. (*i*) $u_{x+4} + u_x = 0$

(*ii*) $u_{x+4} + u_x = 5{\cdot}2^x$ with $u_0 = 1, u_1 = 0$

18. (*i*) $u_{x+2} - 5u_{x+1} + 6u_x = 5^x$.

(*ii*) $u_{x+1} = x(u_x + u_{x-1}), u_0 = 1$.

[**Hint.** (*ii*) $u_{x+1} - (x+1)u_x = -[u_x - xu_{x-1}]$

$\therefore\ u_{x+1} - (x+1)u_x = u_0(-1)^x$ Now apply, 8·9, page 187]

19. (*i*) $u_{x+2} - 3u_{x+1} - 4u_x = 3^x$.

(*ii*) $u_{x+1} - 3u_x + 2u_{x-1} = 1 + a^x, a \neq 1$ or 2.

[**Hint.** (*ii*) Assuming a particular solution in the form $u_x = Ax + Ba^x$ the substitution of which is the given equation leads to

$$-A + \left(a - 3 + \frac{2}{a}\right)Ba^x = 1 + a^x$$

and, therefore, $A = -1, B = a/[(a-1)(a-2)]$.

20. $u_{x+2} - 2u_{x+1} + u_x = x^{(3)} + x(-1)^x$

21. $u_{x+2} - 7u_{x+1} - 8u_x = x^{(2)}.2^x$

22. $u_{x+2} - 4u_x = 5{\cdot}3^x$.

23. $u_{x+2} - 4u_x = 2^x$.

24. $u_{x+2} - 4u_{x+1} + 4u_x = 2^x$.

25. $u_{x+2} - 2u_x = x^2.2^x$.

26. (*a*) $u_{x+3} - 5u_{x+2} + 8u_{x+1} - 4u_x = x.2^x$.

$$\left[\textbf{Hint. } \text{P.I.} = \frac{1}{E^3 - 5E^2 + 8E - 4}x2^x = \frac{1}{(E-1)(E-2)^2}x2^x\right.$$

$$= 2^x \frac{1}{(2E-1)(2E-2)^2}x$$

$$\left. = \frac{2^x}{4} \cdot \frac{1}{1 + 2\Delta} \cdot \frac{1}{\Delta^2}x = \text{etc.}\right]$$

(*b*) Obtain particular solution of the following by the method of undetermined coefficients :

$$u_{x+2} - 2u_{x+1} + u_x = 2^x (x-1) + \sin 2x\pi + x^2 + 3x + 5 \qquad ...(1)$$

[**Hint.** Let the trial solution corresponding to the term $2^x(x-1)$ on the right side of (*i*) be $u_x = 2^x(A_0 + A_1 x)$; show that $A_0 = -5, A_1 = 1$. For $\sin 2\pi x$, trial solution is $y_x = x^2 [B \sin 2\pi x + B_1 \cos 2\pi x]$, find B_0 and B_1. Similarly for the remaining part.]

27. (*a*) Show that $u_x = a^{\frac{1}{2}x^2}$ is a P.I. of

$$u_{x+2} - a(a^x + 1) u_{x+1} + a^{x+1} u_x = 0.$$

(*b*) Show that $(x-2)!$ is a P.I. of the equation

$$u_{x+2} - 3x\, u_{x+1} + 2x(x-1)\, u_x = 0.$$

28. $u_x - u_{x-1} + 2u_{x-2} = x + 2^x.$

29. (*i*) $u_{x+1} - (x+1)\, u_x = 2^x(x-1).$ (*B.Sc. Statistics Hons. Delhi, 1995*)

[**Hint.** Assume particular integral of the form $(Ax + B)2^x$]

(*ii*) $u_{x+1} - (x+1)\, u_x = (x+1)!$ with $u_0 = 2.$

30. $u_{x+1} - pa^{2x} u_x = q\, a^{x2}$

31. (*i*) $u_x u_{x+1} - au_x + b = 0.$

[Put $u_x = v_{x+1}/v_x$]

(*ii*) Show that the non-linear equation

$$u_{x+1} = \frac{u_x}{1 + u_x} \text{ has the solution } u_x = \frac{c}{1 + cx}$$

where *c* is a constant.

[**Hint.** We have $u_{x+1}\, u_x = -(u_{x+1} - u_x)$; try the solution

$$u_x = \frac{A}{1 + B_x} \cdot \quad \text{Show that } B = A$$

32. $u_{x+1} - e^{2x} u_x = x^{(2)} c^{x^2 + x + \frac{1}{4}}$

[**Hint.** Let $A_x = e^{2x}$; then $Z_x = \prod_{i=0}^{x-1} A_i$

$$= e^{x(x-1)};\ Z_{x+1} = e^{(x+1)}$$

Now $V_x = B' \dfrac{B_x}{Z_{x+1} + c} w$, where $B_x = x^{(i)}\, e^{x^2 + x + 1/4}$]

33. $u_{x+1} - a\, u_x = (2x+1)\, a^x.$

34. $u_{x+2} - 2m\, u_{x+1} + (m^2 + n^2) u_x = m^x.$ (*B.A. Statistics Hons. Delhi, 1996*)

35. (*i*) $u_{x+1} - 2u_x^2 + 1 = 0.$

(*ii*) $u_{x+1} = 2u_x \sqrt{1 - u_x^2}$

[**Hint.** (*i*) Put $u_x = \cos v_x$ so that $\cos v_{x+1} = \cos 2v_x$

$\therefore\ v_{x+1} = 2n\pi + 2v_x$, n being an integer, positive or negative

$\therefore\ 2^{-(x+1)} v_{x+1} - 2^{-x} v_x = n\pi . 2^{-x}$

$$\Rightarrow \qquad \Delta(2^{-x} v_x) = n\pi.2^{-x} \Rightarrow 2^{-x} v^x = \left(\frac{n\pi.2^{-x}}{2-1}\right)\Bigg]$$

(*ii*) Put $u_x = \sin v_x$ then $\sin v_{x+1} = \sin 2v_x \Rightarrow v_{x+1} = 2v_x$

36. $u_{x+1} - au_x = a^x (x+1)^{(-2)}$

37. $u_{x+1} - e^{-2ax} u_x = e^{-ax2}, a \neq 0.$

38. (*i*) $u_{x+1}\, u_x - a^x (u_{x+1} - u_x) - 1 = 0.$

(*ii*) $u_{x+1}\, u_x + (x+2) u_{x+1} + xu_2 + x^2 + 2x + 2 = 0.$

(*iii*) $x(x+1)\, u_x - 2x\, u_{x+1} + u_x = 0.$

(*iv*) $u_{k+2} - 2u_{k+1} + u_k = 2^x (x-1) + \sin 2x\pi + x^2 + 3x - 0$

(B.Sc. Meerut, 1979)

$$\left[\textbf{Hint.}\ (i)\ \frac{u_{x+1} - u_x}{1 + u_{x+1}\, u_x} = \frac{1}{a^x} = a^{-x}.\ \text{Put } u_x = \tan v_x\right]$$

39. Solve $u_{x+1}\, u_x + \sqrt{(1 - u_{x+1}^2)(1 - u_x^2)} = a_x$

[**Hint.** Put and we have

$$a^x = \cos u_{x+1} \cos v_x + \sin v_{x+1} \sin v_x$$

$$= \cos (v_{x+1} - v_x) = \cos \Delta v_x$$

whence finally $\quad u_x = \cos [c + \Delta^{-1} (\cos^{-1} a^x)].$

40. $(ax + b) u(x+1) + (cx + d) u(x) = 0.$

$$\left[\textbf{Hint.}\ u(x+1) = \left(-\frac{c}{a}\right) \frac{x + \frac{d}{c}}{x + \frac{b}{a}} \cdot u(x)\right.$$

Hence solution is $u(x) = C(x)(-c/a)^x \dfrac{\Gamma(x + d/c)}{\Gamma(x + b/a)}\Bigg]$

41. (*i*) $(x+1)\, u_{x+1}^2 - (x-1) u_x^2 = 0.$

(*ii*) $u_{x+1}^2 - a^x u_x^2 = a^{\frac{1}{2}x(x-1)}$

(*iii*) $u_{x+1}^2 - 3u_{x+1}\, u_x + 2u_x^2 = 0.$

$$\left[\textbf{Hint.}\ (i)\ \text{Put } v_x = u_x^2 \ \therefore\ v_{x+1} = \frac{x-1}{x+1} v_x \quad (ii)\ v_x = u_x^2\right]$$

42. (*i*) $u_{x+2} - 2(x+1) u_{x+1} + x(x+1)\, u_x = 0.$

(*ii*) $u_x\, u_{x+1}\, u_{x+2} = a(u_x + u_{x+1} + u_{x+2})$

(*iii*) $u_{x+1}\, u_{x+2} + u_{x+2}\, u_x + u_x u_{x+1} = x^3$

[**Hint.** (*i*) Put $u_x = \Gamma(x) v_x$.

(*ii*) Put $u_x = \sqrt{a} \tan v_x$

(*iii*) Divide both sides by $u_x u_{x+1} u_{x+2}$]

43. $u_x = \sqrt{[u_{x-1} + \sqrt{\{u_{x-2} + \sqrt{(u_{x-3} + \cdots)}\}}]}$

[**Hint.** Here $u_x^2 = u_{x-1} + \sqrt{\{u_{x-2} + \sqrt{(u_{x-3} + \cdots)}\}}$

$$= u_{x-1} + u_{x-1} = 2u_{x-1}$$

Hence $2 \log u_x = \log 2 + \log u_{x-1}$

Let $\log u_x = u_x$, then

$$2v_x - v_{x-1} = \log 2$$

Let $w_x + \log 2 = v_x$

so that $2w_x - w_{x-1} = 0$

$\therefore$ $w_x = A.2^{-x}$ so that $v_x = A.2^{-x} + \log 2$

Hence $\log u_x = A2^{-x} + \log 2$

$\therefore$ $\frac{1}{2} u_x = \exp(A2^{-x})$

Thus $u_x = B.2^{-x}$.

where B is an arbitrary constant.

44. $u_x . u_{x+2} = u_{x+1}^2$

[**Hint.** Put $u_{x+1} u_x = v_x$ $\therefore$ Given equation reduces to $v_{x+1} = v_x \Rightarrow \Delta v_x = 0$

Hence $u_{x+1} = C u_x$ $\therefore$ $u_x = AC^x$.

Aliter. The given equation can be written as

$$\frac{u_{x+2}}{u_{x+1}} - \frac{u_{x+1}}{u_x} = 0 \text{ or } \Delta\left(\frac{u_{x+1}}{u_x}\right) = 0$$

45. $u_{x+3}^2 - 3u_{x+2}^2 - 3u_{x+1}^2 - u_x^2 = 0$

[**Hint.** Let $v_x = u_x^2$]

46. $\Delta^2 u_x = (\Delta u_x)(\Delta u_{x+1})$

[**Hint.** Put $\Delta u_x = v_x$, $v_x \neq 0$

$\therefore$ Given equation reduces to $\Delta v_x = v_x v_{x+1}$

$\Rightarrow$ $v_{x+1} - v_x = v_x v_{x+1}$. Divide both sides by $v_x v_{x+1}$

$$\Rightarrow \frac{1}{v_{x+1}} - \frac{1}{v_x} = -1$$

$$\therefore \quad \Delta\left(\frac{1}{v_x}\right) = -1 \quad \therefore \quad v_x^{-1} = -(x + A)$$

$$\therefore \quad v_x^{-1} = B - \Delta^{-1}\left(\frac{1}{x+A}\right) = B' - \sum_{1}^{x-1} \frac{1}{x+A} \cdot]$$

47. $u_{x+4} u_{x+2}^2 u_x = u_{x+1}^4 u_{x+3}^4$

[**Hint.** $\log u_{x+4} + 6 \log u_{x+3} + \log u_x = 4 \log u_{x+1} + 4 \log u_{x+3}$

Write $v_x = \log u_x$,

$v_{x+4} + 6v_{x+2} + v_x = 4v_{x+1} + 4v_{x+3}$].

48. $u_x u_{x+2}^3 = u_{x+1}^3 u_{x+3}$

where $u_0 = 1/\sqrt{2\pi}, u_1 = 1/\sqrt{2\pi e}, u_2 = 1/\sqrt{2\pi e^4}$.

[**Hint.** $v_{x+3} - 3v_{x+2} + 3v_{x+1} - v_x = 0 \Rightarrow (E-1)^3 v_x = 0$

when $v_x = \log u_x \therefore v_x = A + Bx + Cx^2$

$\Rightarrow \quad u_x = e^{xv} = e^{Bx + Cx^2}$

Thus $\quad u_0 = 1/\sqrt{2\pi} = \alpha,$

$u_1 = 1/\sqrt{2\pi e} = \alpha e^{B+C} \Rightarrow e^{B+C} = 1/\sqrt{e}$

Again $\quad u_2 = 1/\sqrt{2\pi e^4} = \alpha e^{2B+4C}$

$\Rightarrow \quad e^{2B+4C} = e^{-2}$

Hence $\quad B + C = -\frac{1}{2}, \ B + 2C = -1.$

$\therefore \quad B = 0, \ C = -\frac{1}{2}.$

Thus $\quad v_x = (1/\sqrt{2\pi})\, e^{-\frac{1}{2}x^2}$]

49. Find the general solution of the equation

$$u_{x+1} - 2u_x + 4u_{x-1} = f(x)$$

in each of the following cases :

(*i*) $f(x) = a^x, a \neq 1$ (*ii*) $f(x) = e^{bx}, b \neq 0$ (*iii*) $f(x) = \sin cx$

(*iv*) $f(x) = 1$ (*v*) $f(x) = x$ (*vi*) $f(x) = xe^{bx}, b \neq 0.$

50. Find the general solution of the equation

$$u_{x+1}(t) - (1+t)u_x(t) + tu_{x-1}(t) = f_x(t)$$

in each of the following cases :

(*i*) $f_x(t) = 1$ (*ii*) $f_x(t) = t$ (*iii*) $f_x(t) = t^x$ (*iv*) $f_x(t) = x^t$.

51. Show that the function y given by

$$y_h = 1 - \frac{2}{h}, \ h = 1, 2, 3, \ldots$$

is a solution of the first order equation

$$(h+1)y_{h+1} + hy_h = 2h - 3.$$

52. Solve the difference equation

$$y_{h+1} = 2y_h - 1, \ h = 0, 1, 2, \ldots$$

with initial condition $y_0 = 5$

[**Hint.** This equation is of the form $y_{h+1} = Ay_h + B$. The solution of which is $y_h = A^h y_0 + B\dfrac{1-A^h}{1-A}$, $y_0 = 5, A = 2, B = -1$]

53. Solve the system of equations :

$$x_{h+2} - x_h + 2y_{h+1} = 0$$

$$y_{k+1} - y_k - 2x_k = 2^k$$

[**Hint.** The given equations can be written as

$$(E-1)x_k + 2Ey_k = 0 \quad \text{and} \quad -2x_k + (E-1)y_k = 2^k$$

operate these equations with $(E-1)$ and 2E respectively and then subtract so that we have on simplification

$$(E+1)^2 x_k = -4.2^k$$

which is a linear difference equation with constant coefficients.

Now put $x_k = m^4$ in (1) and find that auxiliary equations is $(m+1)^2 k = 0$

$\therefore \quad m = -1, -1$

Hence C.F. $= (c_1 + c_2 k)(-1)^k$

$$\text{P.I.} = \frac{1}{(E+1)^2} - 4.2^k = -4.2^k \cdot \frac{1}{(2+1)^2} = -\frac{4}{9} \cdot 2^k$$

$\therefore \quad x_k =$ general solution of (1) = C.F. + P.I. = etc. New get y_k]

54. Determine λ for which a non-trivial solution exists when y satisfies the difference equation

$y_{k+1} - \lambda y_h + y_{k-1} = 0$, $k = 1, 2, 3$ with the conditions $y_0 = y_4 = 0$

[Hint. $y_2 - \lambda y_1 + y_0 = 0, y_3 - \lambda y_2 + y_1 = 0, y_4 - \lambda y_3 + y_2 = 0.$

Also $y_0 = 0 = y_4$

$$\therefore \quad \begin{vmatrix} 0 & 1 & -\lambda \\ 1 & -\lambda & 1 \\ -\lambda & 1 & 0 \end{vmatrix} = 0 \text{ which gives } \lambda = 0_1 \pm \sqrt{2},$$

9

GENERATING FUNCTIONS OF SEQUENCES EXISTENCE AND UNIQUE THEOREM REGARDING SOLUTION

9.1. Introduction. We know that the solution of a difference equation defined over the set of k-values 0, 1, 2, ... is a sequence. In order to study the sequence, the method of generating functions is very useful and was introduced by Laplace.

Def. Generating Function. If $y_0, y_1, y_2, \dots$ is a sequence of real numbers, the function Y (defined over an interval containing zero) with value at s given by the series

$$Y(s) = y_0 + y_1 s + y_2 s^2 + \cdots + y_k s^k + \cdots = \sum_{k=0}^{\infty} y_k s^k$$

is called the generating function of the sequence $\{y_k\}$.

Example 1. *The function given by*

$$Y(s) = \frac{1}{1 - As}, \quad A \neq 0$$

is the generating function of the sequence $\{A_k\}$.

Sol. For, $\dfrac{1}{1 - As} = (1 - As)^{-1} = 1 + As + A^2 s^2 + \cdots + A^k s^k + \cdots$

and so $\quad y_0 = 1, \ y_1 = A, \ y_2 = A^2, \dots; \ y_k = A^k, \dots$

Note 1. If $A = 1$, then the *g.f.* of the constant sequence 1, 1, 1, ... is $\dfrac{1}{1 - s}$ or $(1 - s)^{-1}$.

2. The function given by

$Y(s) = (A + Bs)^n$, n being a positive integer, is the generating function of the sequence $\{y_k\}$

For, $(A + Bs)^n = \sum_{k=0}^{n} \binom{n}{k} A^{n-k} B^k s^k, \quad k = 0, 1, 2, \dots, n$

and hence $$y_k = \binom{n}{k} A^{n-k} B^k, \quad k = 0, 1, 2, ..., n$$
$$= 0, \quad k > n.$$

Ex. Show that

(*i*) $Y(s) = s/(1-s)^2$ (*ii*) $Y(s) = 2s/(1-s)^2$

is the generating function of the sequence $\{y_k\}$ where

(*i*) $y_k = k$ (*ii*) $y_k = k(k+1)$

[**Hint.** (*i*) $y(s) = s(1-s)^{-1} = 0 + 1 \cdot s + 2s^2 + 3s^3 + \cdots$ so that $y_0 = 1, y_1 = 1, y_2 = 2, \ldots$]

Theorem 1. *If* $\{Y_k^{(1)}\}$ *has the generating function* Y_1 *and* $\{Y_k^{(2)}\}$ *has the generating function* Y_2, *then* $\{c_1 Y_k^{(1)} + c_2 Y_k^{(2)}\}$ *has the generating function* $C_1 Y_1 + C_2 Y_2$ *where* C_1, C_2 *are any constants.*

Proof. $$Y_1(s) = y_0^{(1)} + y_1^{(1)} s + y_2^{(1)} s^2 + \cdots + y_k^{(1)} s^k + \cdots \quad ...(1)$$
$$Y_2(s) = y_0^{(2)} + y_1^{(2)} s + y_2^{(2)} s^2 + \cdots + y_k^{(2)} s^k + \cdots \quad ...(2)$$

Multiplying (1) by C_1 and (2) by C_2 and adding, we have

$$C_1 Y_1(s) + C_2 Y_2(s) = (C_1 y_0^{(1)} + C_2 y_0^{(2)}) + (C_1 y_1^{(1)} + C_2 y_1^{(2)}) s + \cdots$$
$$+ (C_1 y_k^{(1)} + C_2 y_k^{(2)}) s^k + \cdots$$

which shows that

$C_1 Y_1 + C_2 Y_2$ is the *g.f.* of the sequence with general term $C_1 y_k^{(1)} + C_2 y_k^{(2)}$

Theorem 2. *If the generating function of the sequence* $\{y_k\}$ *is known, to find the g.f. of the sequence* $\{y_{k+1}\}$, $\{y_{k+2}\}$s, ...

Proof. Let $Y(s) = y_0 + y_1 s + y_2 s^2 + \cdots + y_k s^k + \cdots$ be the *g.f.* of the sequence $\{y_k\}$.

Transposing y_0 to the L.H.S. and dividing both sides by $s \neq 0$, we get

$$\frac{Y(s) - y_0}{s} = y_1 + y_2 s + y_3 s^2 + \cdots + y_{k+1} s^k + \cdots$$

Thus $\dfrac{Y(s) - y_0)}{s}$ is the g.f. of the sequence $\{y_{k+1}\}$.

Also the g.f. of $\{\Delta y_k\}$ or $\{\Delta y_k - y_k\}$ $$= \frac{Y(s) - y_0}{s} - Y(s)$$
$$= \frac{1}{s} [(1-s) Y(s) - y_0]$$

Similarly, we obtain

$$\frac{Y(s)-y_0-y_1 s}{s^2}=y_2+y_3 s+\cdots+y_{k+2}s^k+\cdots$$

Thus L.H.S. is the g.f. of the sequence $\{y_{k+2}\}$.

9·3. The solution of the first order difference equation with the help of the generating function.

Consider the first order equation

$$y_{k+1}=Ay_k+B \qquad k=0,1,2,\ldots \qquad \ldots(1)$$

where A and B are constants.

We can show by successively substituting the values of $y_k, y_{k-1}, \ldots, y_k$, that the unique solution of (1) is given by

$$y_k=\begin{cases} A^k y_0+B\left(\dfrac{1-A^k}{1-A}\right), & A\neq 1\\ y_0+Bk \quad, & A=1. \end{cases}$$

where y_0 is the prescribed initial value.

The above solution can also be written as

$$y_k=\begin{cases} A^k(y_0-y_x)+y^x, & A\neq 1\\ y_0+Bk\,, & A=1 \end{cases}$$

where $\qquad y_x=\dfrac{B}{1-A}\cdot$

We now obtain this solution by the method of generating functions. We see, from theorem 2 that

$y_{k+1}=Ay_k+B$ has the *g.f.* $y(s)$ given by

$$\frac{Y(s)-y_0}{s}=AY(s)+B\cdot\frac{1}{1-s} \qquad \ldots(9\cdot 2)$$

$$\Rightarrow \qquad Y(s)=\frac{y_0}{1-As}+\frac{Bs}{(1-s)(1-As)}, \text{ on simplification}$$

$$=\frac{y_0}{1-As}+\frac{B}{1-A}\left(\frac{1}{1-s}-\frac{1}{1-As}\right), \quad A\neq 1$$

$$=(y_0-y_x)\frac{1}{1-As}+y_x\frac{1}{1-s}$$

Thus the solution of (1) or $Y(s)$ is the generating function of the sequence $\{y_k\}$ where

$$y_k=(y_0-y_x)A^k+y_x\,, \quad A\neq 1$$

If $A=1$, then from (9·2)

$$\frac{Y(s)-y_0}{s}=Y(s)+\frac{B}{1-A}$$

$$\Rightarrow \qquad Y(s) = \frac{y_0}{1-s} + \frac{Bs}{(1-s)^2}$$ [see Ex. page 200]

Thus the solution is $y_k = y_0 + Bk$.

Example 2. *Use the method of generating function to find the solution of the difference equation*

$$y_{k+2} - 5y_{k+1} + 6y_k = 2.$$

with initial condition $y_0 = 1,\ y_1 = 2.$

Sol. By Theorem 2, we have for the generating function,

$$\frac{Y(s) - 1 - 2s}{s^2} - 5\,\frac{Y(s) - 1}{s} + 6Y(s) = \frac{2}{1-s}$$

$$\Rightarrow \qquad Y(s) - 1 - 2s - 5s(Y(s) - 1) + 6s^2Y(s) = \frac{2s^2}{1-s}$$

$$\Rightarrow \qquad (1 - 5s + 6s^2)Y(s) = 1 - 3s + \frac{2s^2}{1-s}$$

$$Y(s) = \frac{1-3s}{(1-2s)(1-3s)} + \frac{2s^2}{(1-s)(1-2s)(1-3s)}$$

$$= \frac{1}{1-2s} + \frac{1}{1-s} - \frac{2}{1-2s} + \frac{1}{1-3s}$$

$$= \frac{1}{1-s} + \frac{-1}{1-2s} + \frac{1}{1-3s}$$

Hence the solution is $y_k = 1 - 2^k + 3^k$.

Example 3. *Solve* : $2y_{k+1} - y_k = 4,\ \ k = 0, 1, 2, \ldots$ *with initial condition* $y_0 = 3.$

Sol. Writing the given equation in the form $y_{k+1} = Ay_k + C$, we get

$$A = \tfrac{1}{2},\ C = 2.$$

∴ Solution is

$$y_k = A^k\, y_0 + C\,\frac{1-A^k}{1-A}$$

$$= \left(\tfrac{1}{2}\right)^k 3 + 4\left[1 - \left(\tfrac{1}{2}\right)^k\right] = 4 - \left(\tfrac{1}{2}\right)^k, \qquad k = 0, 1, 2, \ldots$$

Hence we get the following sequence of values

$$3,\ 7/2,\ 15/4,\ 31/8, \ldots$$

Example 4. *Solve* :

(*i*) $y_{x+1} - 2y_x = 5$

(*ii*) $y_{x+1} - 2y_x = x + 1$

Sol. (*i*) The corresponding homogeneous equation is

$$y_{x+1} - 2y_x = 0 \Rightarrow y_{x+1} = 2y_x$$

which when compared with $y_{x+1} = A\ y_x + C$, given $A = 2,\ C = 0$

Hence the solution is $y_x = \lambda A^x + C \dfrac{(1 - A^x)}{1 - A} = \lambda\, 2^x$

Let us now obtain particular solution by hit and trial method.

Putting $y_x = A$ in the given equation, we get

$$A - 2A = 5 \Rightarrow A = -5$$

Hence $y_x = -5$ is a particular solution.

Thus the general solution of (1) is given by

$$y_x = \lambda \cdot 2^x - 5$$

(*ii*) Let us find the particular solution of (*ii*) by hit and trial method, so we try to find whether the constant function is a solution or not. Clearly we can show that $y_x = A$ cannot be a solution.

Hence we take $y_x = Ax + B$

$\therefore$ From (1), we have $A(x + 1) + B - 2(Ax + B) = x + 1$

$\Rightarrow \qquad (-A - 1)x + (A - B - 1) = 0$

This is an identity which is true for all x.

i.e., $\qquad -A - 1 = 0$ and $A - B - 1 = 0$

We find $A = -1,\ B = -2$

$$y_x = -x - 2$$

Hence the General Solution of the given equation is given by

$$y_x = \lambda\, 2^x - x - 2.$$

EXERCISE 9·1

1. (*a*) Find the generating function of the sequence $\{y_k\}$ where

(*i*) $y_k = k^2 + 6k + 4, \quad k = 1, 2, 3, \ldots$

(*ii*) $y_k = qp^{k-1}, \quad k = 1, 2, 3, \ldots$

(*iii*) $y_0 = y_1 = 0,\ \ y_k = 1$ if $k > 1$

(*iv*) $y_3 = 0,\ y_1 = 1,\ y_k = 0$ if $k > 1$

(*v*) $y_{k+1} = y_k$.

(*vi*) $y_k = \frac{1}{2} k(k - 1)$

(*b*) Apply the *g.f.* technique to solve the initial value problem.

$$y_{k+2} - y_{k+1} + y_k = 0,\ y_0 = 0,\ y_1 = 1.$$

Ans. (*a*) (*i*) $y_k = (k + 1)(k + 2) + 3k + 2,\ Y(s) = \dfrac{2}{(1 - s)^2} + \dfrac{3s}{(1 - s)^2} + \dfrac{2}{1 - s}$

(*ii*) $Y(s) = \dfrac{qs}{(1 - ps)}$ $\qquad$ (*iii*) $Y(s) = \dfrac{s^2}{(1 - s)}$

(*iv*) $Y(s) = s$ (*v*) $\frac{1}{s}[(1-s)Y(s) - y_0]$

(*vi*) $y_k = \frac{1}{2}k(k-1) - k\ y(s)\ \frac{k}{(1-s)^2} - \frac{s}{(1-s)^2}$

(*b*) $y_k = -\frac{\sqrt{5}}{5}\left(\frac{1-\sqrt{5}}{2}\right)^k + \frac{\sqrt{5}}{5}\left(\frac{1+\sqrt{5}}{2}\right)^k$

2. Find the sequence $\{y_k\}$ of which the *g.f.* $Y(s)$ is given by

(*i*) $Y(s) = \frac{s}{1-2s} + \frac{s}{(1-s)^2}$ (*ii*) $Y(s) = \frac{s^2}{1-s}$

(*iii*) $Y(s) = (3-s)^n$

Ans. (*i*) $y_0 = 0,\ y_k = 2^{k-1} + k + 1$ if $k > 0$

(*ii*) $y_0 = y_{-1} = 0,\ y_k = 1$ if $k > 1$

(*iii*) $y_k = \binom{n}{k}(-1)^k 3^{n-k}$ if $k = 0, 1, \ldots, n,\ y_k = 0$ if $k > n$

3. Solve : (by the method of generating function)

(*i*) $y_{k+1} = y_k + k$ with $y_0 = 1$.

(*ii*) $y_{k+2} - 2y_{k+1} + y_k = 2^k$ with $y_0 = 2$ and $y_1 = 1$

Ans. (*i*) $Y(s) = \frac{s^2}{(1-s)^2} + \frac{1}{1-s},\ y_k = \frac{1}{2}k(k-1) + 1.$

(*ii*) $Y(s) = \frac{1}{(1-s)^2} - \frac{3s}{(2-s)^2} + \frac{1}{1-2s},$

$y_k = (k+1) - 3k + 2^k = 1 - 2k + 2^k.$

4. If $\{y_k\}$ has a *g.f.* $y(s)$, show that the *g.f.* of $\{Dy_k\}$ is given by

$\left(\frac{1}{s}\right)[(1-s)y(s) - y_0]$

[**Hint.** $G.F. = \frac{y(s) - y_0}{s} - y(s) = \frac{(1-s)y(s) - y_0}{s}$]

9·4. An Existence and Uniqueness Theorem regarding solution of linear difference sequence.

We know that some linear difference equations of order n have infinitely many solutions whereas others have no solution at all. We know that for the class of linear difference equations we can always find at least one solution and under certain conditions, only one solution. Theorems establishing such results are referred to as existence and uniqueness theorem. They are of fundamental importance in many branches of mathematics.

Theorem. *The linear difference equation of order n*

$$f_0(x)y_{x+n-1} + \cdots + f_{n-1}(x)y_{x+1} + f_n(x)y_x = g(x) \qquad \ldots(9{\cdot}3)$$

over a set S of consecutive integral values of x has one, and only one, solution y for which values at n consecutive x-values are arbitrarily prescribed.

Proof. By hypothesis, S is a set of one of the types $a \le x$ or $a \le x \le b$, a, b, non-negative integers. Suppose first that $y_a, y_{a+1}, \ldots, y_{a+n-1}$ are the n prescribed values of y. We shall now prove with the help of Mathematical Induction that the value of y at each point of S is uniquely determined. The prescribed values determine the values y_{a+n} uniquely as from (9·3) when $x = a$, we have

$$y_{a+n} = \frac{1}{f_0(a)} [g(a) - f_1(a) y_{a+n-1} \cdots - f_{n-1}(a) y_{a+1} - f_n(a) y_0],\ f_0(a) \neq 0$$

which determines y_{a+n} uniquely.

We now set up the hypothesis (to be proved by Induction) that y is known for all x-values in S upto and including y_{a+j} where $j \ge n$.

Putting $x = x_1 = a + j + 1 - n$ in (9·3), we obtain

$$f_0(x_1) y_{a+j-1} = g(x_1) - f_1(x) y_{a+j} - f_2(x_1) y_{a+j-1} - \cdots - f_n(x_1) y_{a+j-n+1}$$

Since $y_{a+j}, y_{a+j-1}, \ldots, y_{a+j-n+1}$ are known and $f_0(x_1) \neq 0$ since f_0 is never zero is S, we conclude that y_{a+j+1} is uniquely determined. *i.e.*, y is uniquely determined for $x = a + j + 1$. Thus we have proved by induction that y is uniquely determined for all x is S provided that the values $y_a, y_{a+1}, \ldots, y_{a+n-1}$ are known or prescribed.

Now if y_m $(m > a)$ is the first of the n consecutive prescribed values of y instead of y, then we can successively determine unique values for $y_{m+1}, y_{m+2}, \ldots, y_{a+1}, y_a$ and then show as in the preceding part that all other values of y are uniquely determined. Putting $x = m - 1$ in (9·3) we obtain

$$f_n(m-1) y_{m-1} = g(m-1) - f_0(m-1) y_{m-1+n} - \cdots - f_{n-1}(m-1) y_m.$$

Since $y_m, y_{m+1}, \ldots, y_{m+n-1}$ are supposed to be the prescribed values of y, the right-hand side and, ∴, y_{m-1} is determined as f_n is never zero. In any case whether S is the finite set $a \le x \le b$ or the infinite set $x \ge a$ and no matter n consecutive values of y are prescribed, we have a unique value of y determined for every x-value in S and, ∴, have a unique function which satisfies the difference equation and assumes the prescribed value.

Ex. 1. *Show that the equation*

$$x y_{x+1} - y_x = 0. \quad x = 0, 1, 2, \ldots$$

is linear but not of first order over the indicated set of x-values. If the initial condition $y_0 = 0$ is prescribed show that y_1 is not uniquely determined and that there are infinitely many solutions of the difference *equation with* $y_0 = 0$.

Ex. 2. *Show that*

$$y_{x+2} - xy_{x+1} - y_x = 0, \quad x = 0, 1, 2, \ldots$$

has no solution when $y_0 = 1$ *and* $y_2 = 1$ *but there are infinitely many solutions if* $y_0 = y_2 = 1$, *say.*

Theorem *Let* $y^{(1)}$ *and* $y^{(2)}$ *be two solutions of the homogeneous difference equation*

$$y_{k+2} + a_1 y_{k+1} + a_2 y_k = 0 \qquad \ldots(1)$$

and let $Y = C_1 y^{(1)} + C_2 y^{(2)}$ *where* C_1 *and* C_2 *are arbitrary constants.*

If
$$\begin{vmatrix} y_0^{(1)} & y_0^{(2)} \\ y_1^{(1)} & y_1^{(2)} \end{vmatrix} \neq 0, \qquad \ldots(2)$$

then Y is the general solution of (1).

Proof. Since $y^{(1)}$ and $y^{(2)}$ are the solutions of (1), it can be proved that $C_1 y^{(1)} + C_2 y^{(2)}$ is also a solution. Hence Y is a solution of (1). Next, we shall prove that if y is any solution of (1), then C_1 and C_2 may be determined and hence Y and y are identical. By the uniqueness theorem, it is clear that we should prove that Y and y are equal at $k = 0$ and $k = 1$ Thus we should determine C_1 and C_2 so that $Y_0 = y_0$ and $Y_1 = y_1$ for any choice of y_0 and y_1. But

$$Y_0 = C_1 y_0^{(1)} + C_2 y_0^{(2)} \quad \text{and} \quad Y_1 = C_1 y_1^{(1)} + C_2 y_1^{(2)}$$

so C_1 and C_2 must satisfy the equations

$$y_0^{(1)} C_1 + y_0^{(2)} C_2 = y_0; \; y_1^{(1)} C_1 + y_1^{(2)} C_2 = y_1$$

Since
$$\begin{vmatrix} y_0^{(1)} & y_0^{(2)} \\ y_1^{(1)} & y_1^{(2)} \end{vmatrix} \neq 0,$$

we find a first finite unique pair of values of C_1 and C_2 for each choice of y_0 and y_1. This completes the proof.

Def. Two solutions $y^{(1)}$ and $y^{(2)}$ of the above theorem satisfying condition (2) are said to form a fundamental set or system of solution of (1).

Ex. 3. *Show that* $y^{(1)} = 2^k$ *and* $y^{(2)} = k2^k$ *are the solutions of the following homogeneous equation. Also show that they form a fundamental set of solutions.*

$$y_{k+2} - 4y_{k+1} + 4y_k = 0 \qquad \ldots(1)$$

$y^{(1)} = 2^k$ is solution of (1) since we have on putting $y_k^{(1)} = 2^k$,

$$2^{k+2} - 4 \cdot 2^{k+1} + 4 \cdot 2^k = 2^k(4 - 4 \cdot 2 + 4) = 0$$

Also $y^{(2)} = k \cdot 2^k$ is a solution (1).

For then L.H.S. of (1) $= (k+2)\, 2^{k+2} - 4(k+1)\, 2^{k+1} + 4k\, 2^k$

$= 2^k\, [4k + 8 - 8k - 8 + 4k] = 0$

Thus $y^{(1)} = 2^k$ and $y^{(2)} = k\, 2^k$ are the solutions of (1)

Further, $\begin{vmatrix} y_0^{(1)} & y_0^{(2)} \\ y_1^{(1)} & y_1^{(2)} \end{vmatrix} = \begin{vmatrix} 1 & 0 \\ 2 & 2 \end{vmatrix} = 2 \neq 0,$

Hence $y^{(1)}$ and $y^{(2)}$ form a fundamental set of (1).

EXERCISE 9·2

1. Show that $y^{(1)} = \left(\frac{1}{2}\right)^k$, $y^{(2)} = (-2)^k$ are the solutions of the following homogeneous equation and that they form fundamental set of solutions.

$$2y_{k+2} + 3y_{k+1} - 2y_k = 0$$

2. Obtain the fundamental set of solutions of the following equations.

$$9y_{k+2} - 6y_{k+1} + y_k = 0$$

$$y_{k+2} + 6y_{k+1} + 25y_k = 0.$$

3. Solve the system of equations :

$$x_{h+1} - x_h + 2y_{h+1} = 0 \qquad ...(i)$$

$$y_{h+1} - y_h - 2x_h = 2^h \qquad ...(ii)$$

[**Hint.** The given equations can be written as

$$(E-1)\, x_h + 2Ey_h = 0 \qquad ...(1)$$

and $$2x_h + (E-1)\, y_h = 2^h \qquad ...(2)$$

Operate both sides of the first equation by $(E-1)$ and second by $2E$ and then subtract, we find, on simplification,

$$(E+1)^2\, x_h = -4 \cdot 2^h$$

C.F. of this equation is $(c_1 + c_2 h)\, (-1)^h$

The P.I. $= \dfrac{1}{(E+1)^2}(-4 \cdot 2^x) = -4 \cdot \dfrac{2^h}{(2+1)^2} = -\dfrac{4}{9}\, 2^x$

Hence the general solution $x_h = C.F. + P.I. =$ etc, ...(3)

From (i), we have $2y_{h+1} = x_x - x_{h+1}$

$$= \left[(c_1 + c_2 h)(-1)^h - \frac{4}{9}\, 2^h \right] - \left[c_1 + c_2(h+1)(-1)^{h+1} - \frac{4}{9}\, 2^{h+1} \right]$$

$$= (2c_1 + c_2 + 2c_3 h)(-1)^h + \frac{4}{9} \cdot 2^h$$

$$\therefore \quad y_{h+1} = \left(c_1 + \frac{1}{2} c_2 + c_3 h \right)(-1)^h + \frac{1}{9} \cdot 2^{h+1}]$$

(3) and (4) jointly are the solutions of (1) and (2)

4. Prove that $f_0(h) y_{h+2} + f_1(h) y_{h+1} + f_2(h) y_h = g(h)$, $h = 0, 1, 2, ...,$ $y_0 = \alpha$, $y_1 = \beta$ has one and only one solution where $f_0, f_1, f_2 \neq 0$.

Hence write down all possible solutions of

$$y_{h+2} + \sin h\; y_{h+1} + (h^2 + 1) y_h = 0,\; h = 2, 3, 4, ...,\; y_2 = 0,\; y_3 = 0.$$

10

DIFFERENCE EQUATIONS BY MATRIX METHOD

10·1. Solution of second order difference equations by Matrix Method

Consider a sequence of trials, each with only two possible outcomes, which takes place at times $t = 0, 1, 2, \ldots$ Suppose the system is in state 1 or state 2 at any time t according as the trial taking place at that time results in one or the other of the possible outcomes. The initial probabilities p_0 and $q_0 (= 1 - p_0)$ for the system to be in state 1 and state 2 respectively at $t = 0$ are prescribed as are the constant transition probabilities for—the system to pass from any state to any state from one trial to the next *i.e.*, in the time interval between consecutive t-values.

$$P(1 \to 1) = 1 - \alpha \qquad P(1 \to 2) = \alpha \qquad 0 \le \alpha, \beta \le 1$$

$$P(2 \to 1) = \beta \qquad P(2 \to 2) = 1 - \beta,$$

With p_i and q_i denoting the probabilities of finding the system in state 1 and state 2 respectively at time t, we are led to the system of difference equations

$$p_{t+1} = (1 - \alpha)\, p_i + \beta q_i \qquad t = 0, 1, 2, \ldots \qquad (i)$$

$$q_{i+1} = \alpha p_i + (1 - \beta) q_t$$

We want to find expressions for p_i and q_i in terms of the given initial probabilities p_0 and q_0, the probabilities α and β, and the time t. That is, we want to find the probability distribution of the state of the system at any time t from the prescribed data : the probability distribution of the initial state of the system at $t = 0$; the transition probabilities governing the random fluctuations in the state of the system from trial to trial, and the elapsed time from $t = 0$ (or the number of trials which have taken place).

The system of two simultaneous difference equations (1) can be written as a single matrix difference equation.

Let V_i denote the column vector with two components p_i and q_t *i.e.*,

$$V_t = \begin{pmatrix} p_t \\ q_t \end{pmatrix} \qquad t = 0, 1, 2, \ldots$$

Let A denote the matrix of transition probabilities

$$A = \begin{pmatrix} 1-\alpha & \beta \\ \alpha & 1-\beta \end{pmatrix}$$

Then the system (1) may be written

$$V_{t+1} = AV_t \qquad t = 0, 1, 2, - \qquad \ldots(2)$$

are $\begin{pmatrix} p_{i+1} \\ q_{t+1} \end{pmatrix} = \begin{pmatrix} 1-\alpha & \beta \\ \alpha & 1-\beta \end{pmatrix} \begin{pmatrix} p_i \\ q_i \end{pmatrix}$

$$= \begin{pmatrix} (1-\alpha)\, p_i + \beta q_t \\ \alpha p_t + (1-\beta)\, q_t \end{pmatrix}$$

Thus $\quad p_{t+1} = (1-\alpha)\, p_t + \beta q_t$

$$q_{t+1} = \alpha p_t + (1-\beta)\, q_t$$

Putting $t = 0$ in (2), $V_1 = AV_0$

Putting $t = 1$ in (2), $V_2 = AV_1 = A(AV_0) = (AA)\, V_0 = A^2 V_0$

Similarly $\quad V_3 = AV_2 = A(A^2 V_0) = A^3 V_0$

Hence by Mathematical induction, we have

$$V_t = A^t V_0 \qquad t = 0, 1, 2, \ldots$$

Example. *Solve the second order homogeneous difference equation*

$$x_{t+1} - 3x_t + 2x_{t-1} = 0 \qquad t = 1, 2, 3, \ldots \qquad \ldots(1)$$

with initial values $x_0 = 0, x_1 = 1$.

Solution. We write the given difference equations as a system of two first order equations

$$x_{t+1} = 3x_t - 2y_t,\ y_{t+1} = x_t \qquad t = 1, 2, 3, \ldots \qquad \ldots(2)$$

We define the column vectors

$$V_t = \begin{pmatrix} x_t \\ y_t \end{pmatrix} \qquad t = 1, 2, 3, \ldots$$

the equations (2) can written on matrix form as

$$V_{t+1} = AV_t \qquad t = 1, 2, 3, \ldots$$

where $\quad A = \begin{pmatrix} 3 & -2 \\ 1 & 0 \end{pmatrix}$

$$V_{t+1} = AV_t = A^2 V_{t-1} = \cdots = A^t V_1 \qquad t = 0, 1, 2, \ldots$$

$$V_1 = \begin{pmatrix} x_i \\ y_1 \end{pmatrix} = \begin{pmatrix} x_1 \\ x_0 \end{pmatrix} = \begin{pmatrix} i \\ 0 \end{pmatrix}$$

To find A^t, the characteristic equation is

$$\begin{vmatrix} 3-\lambda & -2 \\ 1 & -\lambda \end{vmatrix} = \lambda^2 - 3\lambda + 2 = 0$$

The characteristic roots of A are $\lambda_1 = 1, \lambda_2 = 2$. Hence

$$A^t = aI + bA$$

The constants a and b are determined by

$$1 = a + b,\ 2^t = a + 2b$$

$$0 = 2 - 2^t,\ \ b = 2^t - 1$$

$$A^t = (2 - 2^t)\begin{pmatrix} 1 & 0 \\ 0 & 1 \end{pmatrix} + (2^t - 1)\begin{pmatrix} 3 & -2 \\ 1 & 0 \end{pmatrix}$$

$$V_{t-1} = A^t V_1 = \begin{pmatrix} 2^{t+1} - 1 \\ 2^t - 1 \end{pmatrix},\ t = 0, 1, 2, \ldots$$

Since $x_i = y_{t+1}$ is the second component of this column vector we find the solution of (1) given as

$$x_t = 2t - 1 \qquad t = 0, 1, 2, \ldots$$

Ex. Solve using the matrix method, the following second order difference equation by first replacing each by a system of two first order equations.

(*i*) $x_{i+1} - 4x_t + 3x_{t-1} = 0 \qquad t = 1, 2, 3, \ldots$

$x_0 = 0, x_1 = 1$ (*B.Sc. Hons. Meerut, 1981*)

(*ii*) $x_{t+1} - x_t - 2x_{t-1} = 0 \qquad t = 1, 2, 3, \ldots$

$x_0 = 1, x_1 = 0$ (*B.Sc. Hons. Meerut, 1980*)

(*iii*) $x_{t+1} - x_{t-1} = 0 \qquad t = 1, 2, 3, \ldots$

$x_0 = 1, x_1 = 3.$ (*B.Sc. Hons. Meerut, 1979*)

Answer. (*i*) $x_t = \frac{1}{2}(3^t - 1)$ (*ii*) $x_t = \frac{1}{3}[2^t + 2(-1)^t]$

(*iii*) $x_t = 2 - (-1)^t$; $t = 0, 1, 2.$

10·2. Solution of System of Linear Difference Equations by Matrix Method

Consider a system of linear difference equations

$$\left.\begin{aligned} x_{k+1} &= ax_k + by_k \\ y_{k+1} &= cx_k + dy_k \end{aligned}\right\} \qquad \ldots(1)$$

with the initial conditions $x_0 = e$ and $y_0 = f$ where a, b, c, d are constants.

Let $V_k = \begin{bmatrix} x_k \\ y_k \end{bmatrix},\ A = \begin{bmatrix} a & b \\ c & d \end{bmatrix},\ V_0 = \begin{bmatrix} x_0 \\ y_0 \end{bmatrix} = \begin{bmatrix} e \\ f \end{bmatrix}$

The system of equations (1) can be written in the matrix from

$$\begin{bmatrix} x_{k+1} \\ x_{k+1} \end{bmatrix} = \begin{bmatrix} a & b \\ c & d \end{bmatrix} \begin{bmatrix} x_k \\ y_k \end{bmatrix}$$

$$\Rightarrow \quad V_{k+1} = A \,.\, V_k \quad ...(2)$$

By the Mathematical Induction, we shall now prove that

$$V_k = A^k V_0 \quad ...(3)$$

Putting $k = 0, V_0 = A^\circ V_0 = V_0$ which is true. Hence (3) is true for $k = 0$.

When $\quad k = 1, V_1 = A' V_0 = A V_0$

Let (3) be true for $k = m$ so that

$$V_m = A^m V_0$$

Also from (2), $V_{m+1} = A V_m = A.A^m V_0 = A^{m+1} V_0$

Thus (3) is true for $k = m + 1$.

Hence by mathematical induction, (3) is true $\forall\ k, k = 0, 1, 2, ..$

Next, we determine A^k.

The characteristic equation of the matrix A is given by

$$f(\lambda) = |A - \lambda I| = a_0 + a_1 \lambda + a_2 \lambda^2 = 0 \quad ...(4)$$

By Cayley-Hamilton theorem, the matrix A satisfies (4)

$$\therefore \quad f(A) = A - A - 0 \Rightarrow f(A) = 0$$

Next we have

$$\Rightarrow \quad \lambda^k = q(\lambda) f(\lambda) + r(\lambda)$$

where $q(\lambda)$ is the quotient and $r(\lambda)$ is the remainder.

Since $f(\lambda)$ is quadratic in λ, $r(\lambda)$ must be linear in λ *i.e.*, $a + b\lambda$ so that $\quad \lambda^k = q(\lambda) + a + b\lambda$

By Cayley-Hamilton Theorem,

$$A^k = q(A) f(A) + a + bA$$

$$= 0 + a + b \,.\, A$$

$$\Rightarrow \quad A^k = a I_2 + bA$$

Thus we want to determine a and b.

Let λ_1, λ_2 be the characteristics of A and let $f(\lambda_1) = 0 = f(\lambda_2)$

$$\therefore \quad \lambda_1^k = a + b\lambda_1, \lambda_2^k = a + b\lambda_2,$$

$$\begin{vmatrix} 1 & \lambda_1 \\ 1 & \lambda_2 \end{vmatrix} \neq 0, \lambda_1 \neq \lambda_2.$$

Solving these we get the values of arbitrary constants a and b and thus we determine A^k, I_2 and A_1 being known.

$\therefore$ From (3), we get V_k when V_0 is known.

Example. *Solve* :

$$\left.\begin{aligned} x_{n+1} &= 7x_n + 10y_n \\ y_{n+1} &= x_n + 4y_n \end{aligned}\right\} \qquad ...(1)$$

with $x_0 = 3,\ y_0 = 2.$

Sol. The system of equations (1) can be written in the matrix form

$$\begin{bmatrix} x_{n+1} \\ y_{n+1} \end{bmatrix} = \begin{bmatrix} 7 & 10 \\ 1 & 4 \end{bmatrix} \begin{bmatrix} x_n \\ y_n \end{bmatrix}$$

$\Rightarrow$ $$V_{k+1} = A^k V_0$$

The solution is $V_k = A^k \cdot V_0.$

The characteristic equation of the matrix A is given by

$$|A - \lambda I| = \begin{vmatrix} 7 - \lambda I & 10 \\ 1 & 4 - \lambda \end{vmatrix} = 0 \Rightarrow (7-\lambda)(4-\lambda) - 10 = 0$$

$\Rightarrow$ $$\lambda^2 - 11\lambda + 18 = 0$$

$\therefore$ $$\lambda_1 = 2,\ \lambda_2 = 9$$

Hence $$\begin{bmatrix} x_k \\ y_k \end{bmatrix} = \begin{bmatrix} \lambda_1^k \\ \lambda_2^k \end{bmatrix} \begin{bmatrix} x_0 \\ y_0 \end{bmatrix}$$

$$= \begin{bmatrix} 2^k \\ 9^k \end{bmatrix} \begin{bmatrix} 3 \\ 2 \end{bmatrix}$$

$$x_k = 3 \cdot 2^k + 2 \cdot 9^k$$

Ex 1. *Solve* :

$$\left.\begin{aligned} x_{n+1} &= 2x_n + 10y_n \\ y_{n+1} &= x_n + 5y_n \end{aligned}\right\} \text{ with } x_0 = 1,\ y_0 = 1$$

Ex 2. *Use matrix method to solve*

$$x_{t+1} - 3x_t + 2x_{t-1} = 0,\ t = 1, 2, 3, \ldots$$

$$x_0 = 0,\ x_1 = 0$$ *(M.Sc. Meerut, 1992, 88, 87, 86)*

[**Hint.** Introduce the new function y defined for $t = 1, 2, \ldots$ such that the second order diff. equation. (1) can be written as a system of two first order diff. eqn. $\left.\begin{aligned} x_{t+1} &= 3x_t - 2y_t \\ y_{t+1} &= x_t \end{aligned}\right\} t = 1, 2, \ldots$

This system can be written in matrix form as

$$\begin{bmatrix} x_{t+1} \\ y_{t+1} \end{bmatrix} = \begin{bmatrix} 3 & -2 \\ 1 & 0 \end{bmatrix} \begin{bmatrix} x_t \\ y_t \end{bmatrix} \quad \text{or} \quad V_{t+1} = AV_t$$

where $V_t = \begin{bmatrix} x_t \\ y_t \end{bmatrix}$ and $A = \begin{bmatrix} 3 & -2 \\ 1 & 0 \end{bmatrix}$

Now $V_1 = \begin{bmatrix} x_1 \\ y_1 \end{bmatrix} = \begin{bmatrix} x_1 \\ x_0 \end{bmatrix} = \begin{bmatrix} 1 \\ 0 \end{bmatrix}$ $\therefore$ $V_2 = AV_1,\ V_3 = AV_2 = A^2V_1$

In general $V_{t+1} = A^t V_1 = A^t \begin{bmatrix} 1 \\ 0 \end{bmatrix}$ $\therefore$ $A^t = aI + bA$ where a and b are obtained from,

$\lambda_1^t = a + b\lambda_1$ and $\lambda_2^t = a + b\lambda_2$ where λ_1, λ_2 are the roots of

$$|A - \lambda I| = \begin{vmatrix} 3-\lambda & -2 \\ 1 & 0-\lambda \end{vmatrix} = 0$$

$\therefore$ $\lambda_1 = 1, \lambda_2 = 2$ and then $b_1 = 2 - 2^t$ and $b' = 2^t - 1$.

11

BERNOULLI AND EULER POLYNOMIALS

11·1. The ϕ Polynomials. The ϕ polynomial $\phi_\nu^{(n)}(x)$ of degree ν and order n is defined by the relation

$$f_n(t)\, e^{xt+g(t)} = \sum_{\nu=0}^{\infty} \frac{t^\nu}{\nu\,!}\, \phi_\nu^{(n)}(x) \qquad \text{...(11·1)}$$

where $f_n(t)$ and $g(t)$ are functions of t such that the right side is a uniformly convergent series in t for certain range of x.

Putting $x=0$ and writing, $\phi_\nu^{(n)}$ for $\phi_\nu^{n}(0)$, we have

$$f_n(t)\, e^{g(t)} = \sum_{\nu=0}^{\infty} \frac{t^\nu}{\nu\,!}\, \phi_\nu^{(n)}$$

Writing $x+y$ for x in (11·1), we get

$$\sum_{\nu=0}^{\infty} \frac{t^\nu}{\nu\,!}\, \phi_\nu^{(n)}(x+y) = e^{xt} \sum_{\nu=0}^{\infty} \frac{t^\nu}{\nu\,!}\, \phi_\nu^{(n)}(y)$$

$$= \left(1 + tx + \frac{t^2x^2}{2\,!} + \cdots + \frac{t^\nu x^\nu}{\nu\,!} + \cdots\right)$$

$$\times \left(\phi_0^{(n)}(y) + \frac{t}{1\,!}\, \phi_1^{(n)}(y) + \cdots + \frac{t^\nu}{\nu\,!}\, \phi_\nu^{(n)}(y) + \cdots\right)$$

Equating the coefficients of t^ν, we have

$$\phi_\nu^{(n)}(x+y) = \phi_\nu^{(n)}(y) + x\, {}^\nu c_1\, \phi_{\nu-1}^{(n)}(y) + x^2\, {}^\nu c_2\, \phi_{\nu-2}^{(n)}(y) + \cdots + x^\nu\, \phi_0^{(n)}(y).$$

Putting $y=0$, we obtain

$$\psi_\nu^{(n)}(x) = \phi_\nu^{(n)} + x\, {}^\nu c_1\, \phi_{\nu-1}^{(n)} + x^2\, {}^\nu c_2\, \phi_{\nu-2}^{(n)} + \cdots + x^\nu\, \phi_0^{(n)}\,, \qquad \text{...(11·2)}$$

which is a polynomial in x of degree ν.

Symbolically

$$\phi_\nu^{(n)}(x) = (\phi^{(n)} + x)^\nu \qquad ...(11\cdot 2)$$

Thus $$\frac{d}{dx} \cdot \phi_{\nu+1}(x) = (\nu+1)(\phi^{(n)} + x)^\nu = (\nu+1)\,\phi_\nu^{(n)}(x) \qquad ...(11\cdot 3)$$

and $\therefore$ $$\int_a^x \phi_\nu^{(n)}(t)\,dt = \frac{1}{\nu+1}\left[\phi_{\nu+1}^{(n)}(x)\right]_a^x$$

$$= \frac{1}{\nu+1}\left[\phi_{\nu+1}^{(n)}(x) - \phi_{\nu+1}^{(n)}(a)\right]$$

Operating on (11·1) with Δ, we have

$$\sum_{\nu=0}^{\infty} \frac{t^\nu}{\nu\,!}\,\Delta\,\phi_0^{(n)}(x) = (e^t - 1)\,f_n(t)e^{xt + g(t)} \qquad ...(11\cdot 3')$$

11·2. The β Polynomials. The β Polynomial $\beta_\nu^{(n)}$ of degree ν and n is defined by the relation

$$t^n(e^t - 1)^{-n} e^{xt + g(t)} = \sum_{\nu=0}^{\infty} \frac{t^\nu}{\nu\,!}\,\beta_\nu^{(n)}(x) \qquad ...(11\cdot 4)$$

Using (11·3′).

$$\sum_{\nu=0}^{\infty} \frac{t^\nu}{\nu\,!}\,\Delta\beta_\nu^{(n)}(x) = (e^t - 1)t^n(e^t - 1)^{-n} e^{xt + g(t)}$$

$$= t \sum_{\nu=0}^{\infty} \frac{t^\nu}{\nu\,!}\,\beta_\nu^{(n-1)}(x)$$

$$= t \sum_{\nu=1}^{\infty} \frac{t^{\nu-1}}{(\nu-1)\,!}\,\beta_{\nu-1}^{(n-1)}(x).$$

Equating the coefficients of t^ν, we get

$$\Delta\beta_\nu^{(n)}(x) = \nu\beta_{\nu-1}^{(n-1)}(x)$$

or $$[\beta^{(n)} + x + 1]^\nu - [\beta^{(n)} + x]^\nu = \nu[\beta_{\nu-1}^{n-1} + x]^{\nu-1}$$

Putting $x = 0$

we have $$(\beta^{(n)} + 1)^\nu - \beta_\nu^n = \nu\beta_{(\nu-1)}n - 1 \qquad ...(11\cdot 5)$$

which is the recurrence relation between the β numbers of orders n and $n-1$.

11·3. Bernoulli's Polynomials. Bernoulli's polynomial $\beta_\nu^{(n)}(x)$ of order n and degree ν is defined by the relation

$$t^n(e^t - 1)^{-n} e^{xt} = \sum_{\nu=0}^{\infty} \frac{t^\nu}{\nu\,!}\,\mathbf{B}_\nu^{(n)}(x) \qquad ...(11\cdot 6)$$

Putting $x = 0$,

$$t^n(e^t - 1)^{-n} = \sum_{\nu=0}^{\infty} \frac{t^\nu}{\nu\,!} \mathbf{B}_\nu^{(n)} \qquad \text{...(11·7)}$$

where $\mathbf{B}_\nu^{(n)} = \mathbf{B}_\nu^{(n)}(0)$ and are called Bernoulli's numbers of order n.

From (11·7), equating the coefficients of various powers of t, we obtain

$$\mathbf{B}_0^{(n)} = 1,\ \mathbf{B}_1^{(n)} = -\tfrac{1}{2}n,\ \mathbf{B}_2^{(n)} = \frac{1}{12}n(3n-1), \text{ etc.} \qquad \text{...(11·8)}$$

Proceeding as in (11·1), we easily obtain

$$\mathbf{B}_\nu^{(n)}(x) = (\mathbf{B}^{(n)} + x)^\nu. \qquad \text{...(11·9)}$$

Hence $\Delta \mathbf{B}_\nu^{(n)}(x) = \nu \mathbf{B}_{\nu-1}^{(n-1)}(x)$ ...(11·10)

$$= \nu(\nu-1)\mathbf{B}_{\nu-1}^{(n-1)}(x) = \cdots$$

$$= \nu(\nu-1)...(\nu-n+1)\mathbf{B}_{\nu-n}^{0},\ \nu > n.$$

$$\Delta^n \mathbf{B}_\nu^{(n)}(x) = \nu(\nu-1)(\nu-2)\ldots(\nu-n+1)\, x^{\nu-n} \qquad \text{...(11·10)}'$$

since $\mathbf{B}_{\nu-n}^{0} = x^{\nu-n},\ \nu > n$

We can establish that

$$\mathbf{B}_\nu^{(n)}(x+1) = \mathbf{B}_\nu^{(n)}(x) + \nu \mathbf{B}_{\nu-1}^{(n-1)}(x) \qquad \text{...(11·11)}$$

$\therefore$ $\mathbf{B}_\nu^{(n)}(1) = \mathbf{B}_\nu^{(n)} + \nu \mathbf{B}_{\nu-1}^{(n-1)}$

11·3·1. To prove that

$$\mathbf{B}_\nu^{(n)}(n-x) = (-1)^\nu \mathbf{B}_\nu^{(n)}(x).$$

Writing $n - x$ for x in (11·6), we have

$$\sum_{\nu=0}^{\infty} \frac{t^\nu}{\nu\,!} \mathbf{B}_\nu^{(n)}(n-x) = t^n(e^t-1)^{-n} e^{(n-x)t}$$

or

$$\sum_{\nu=0}^{\infty} \frac{t^\nu}{\nu\,!} B_\nu^{(n)}(n-x) = (-t)^n e^{-xt}(e^{-t}-1)^{-n}$$

$$= \sum_{\nu=0}^{\infty} \frac{(-t)^\nu}{\nu\,!} \mathbf{B}_\nu^{(n)}(x)$$

Equating the coefficient of t^ν, we have the required result which is the complementary argument theorem.

11·3·2. To prove that

$$\mathbf{B}_\nu^{(n+1)}(x) = \left(1 - \frac{\nu}{n}\right)\mathbf{B}_\nu^{(n)}(x) + \nu\left(\frac{x}{n} - 1\right)\mathbf{B}_{\nu-1}^{(n)}(x).$$

Differentiating (11·6) with respect to t and then multiplying by t, we have

$$\sum_{\nu=0}^{\infty} \frac{t^\nu}{(\nu-1)!} \mathbf{B}_\nu^{(n)}(x) = \frac{nt^n e^{xt}}{(e^t-1)^n} + \frac{xt^{n+1} e^{xt}}{(e^t-1)^n} - \frac{nt^{n+1} e^{(x+1)t}}{(e^t-1)^{n+1}}$$

$$= n \sum_{\nu=0}^{\infty} \frac{t^\nu}{\nu!} \mathbf{B}_\nu^{(n)}(x) + xt \sum_{\nu=0}^{\infty} \frac{t^\nu}{\nu!} \mathbf{B}_\nu^{(n)}(x) - n \sum_{\nu=0}^{\infty} \frac{t^\nu}{\nu!} \mathbf{B}_\nu^{(n+1)}(x+1)$$

Equating the coefficient of t^ν, we have from (11·11)

$$\nu\mathbf{B}_\nu^{(n)}(x) = n\mathbf{B}_\nu^{(n)}(x) + \nu\mathbf{B}_{\nu-1}^{(n)}(x) - n\mathbf{B}_\nu^{(n+1)}(x+1) \qquad \text{...(11·12)}$$

$$\mathbf{B}_\nu^{(n+1)}(x+1) = \mathbf{B}_\nu^{(n+1)}(x) + \nu\mathbf{B}_{\nu-1}^{(n)}(x)$$

$$\mathbf{B}_\nu^{(n+1)}(x) = \left(1 - \frac{\nu}{n}\right)\mathbf{B}_\nu^{(n)}(x) + \nu\left(\frac{x}{n} - 1\right)\mathbf{B}_{\nu-1}^{(n)}(x) \qquad \text{...(11·13)}$$

Cor. 1. Putting $x = 0$,

$$\mathbf{B}_\nu^{(n+1)} = \left(1 - \frac{\nu}{n}\right)\mathbf{B}_\nu^{(n)} - \nu\mathbf{B}_{\nu-1}^{(n)},$$

Also on putting $x = 0$ in (11·12)

$$\mathbf{B}_\nu^{(n+1)}(1) = \left(1 - \frac{\nu}{n}\right)\mathbf{B}_\nu^{(n)}$$

or $$\mathbf{B}^{\nu(n+\nu+1)}(1) = \frac{n}{n+\nu}\mathbf{B}_\nu^{(n+\nu)} \quad \text{on writing } n+\nu \text{ for } n.$$

Cor. 2. Putting $\nu = n$ in (11·12) we obtain

$$\mathbf{B}_n^{(n+1)}(x) = (x-n)\mathbf{B}_{n-1}^{(n)}(x)$$
$$= (x-n)(x-n+1)\mathbf{B}_{n-2}^{(n-1)}(x) = \cdots$$
$$= (x-n)(x-n+1)\ldots(x-2)(x-1).\ \mathbf{B}_0^{(1)}(x)$$
$$= (x-1)^{(n)}.$$

11·4. Bernoulli's Polynomial and Number of the First Order.

Putting $n = 1$ in (11·6) and writing $\mathbf{B}_\nu(x)$ for $\mathbf{B}_\nu^{(1)}(x)$, we have

$$\frac{te^{xt}}{e^t - 1} = \sum_{\nu=0}^{\infty} \frac{t^\nu}{\nu!}\mathbf{B}_\nu(x) \qquad \text{...(11·14)}$$

and therefore $$\frac{t}{e^t - 1} = \sum_{0}^{\infty} \frac{t^\nu}{\nu}\mathbf{B}^\nu \qquad \text{...(11·15)}$$

(11·14) and (11·15) are referred to as the generating function of the Bernoulli's polynomials and Bernoulli's numbers, B_ν, of the first order. Properties and analogous to (11·2′), (11·3) and (11·3′) can be established as follows :

Differentiating (11·14) with respect to x and dividing by t, we obtain the relation

$$B'_\nu(x) = \nu B_{\nu-1}(x) \qquad ...(11\cdot16)$$

Also
$$\int_a^x B_\nu(x)dx = \frac{1}{\nu+1}[B_{\nu+1}(x) - B_{\nu+1}(a)]$$

Writing $1 - x$ for x in (11·14) we have

$$\frac{te^{(1-x)t}}{e^t - 1} = \sum_{\nu=0}^{\infty} \frac{t^\nu}{\nu\,!} \mathbf{B}_\nu(1-x) \qquad ...(11\cdot17)$$

But
$$\text{left side} = \frac{te^{-xt}}{1 - e^{-t}} = \frac{-te^{-t\infty}}{e^{-t} - 1}$$

$$= \sum_{\nu=0}^{\infty} \frac{(-t)^\nu}{\nu\,!} B_\nu(x) \qquad ...(11\cdot18)$$

Equating coefficients of t^ν in (11·7) and (11·18) we obtain

$$\mathbf{B}_\nu(1-x) = (-1)^\nu \mathbf{B}_\nu(x).$$

If $\nu = 2k$, then

$$B_{2k}(1-x) = B_{2k}(x).$$

We derive the first few Bernoulli's polynomials as follows :

Multiplying both sides of (11·14) by $e^t - 1$, we get

$$t\left(1 + tx + \frac{t^2x^2}{2\,!} + \cdots\right) = \left(t + \frac{t^2}{2\,!} + \cdots\right) \times \left(B_0(x) + tB_1(x) + \frac{t^2}{2\,!}B_2(x) + \cdots\right)$$

Equating coefficients of various powers of t, we obtain

$$\left.\begin{aligned} B_0(x) &= 1, \\ B_1(x) &= x - \tfrac{1}{2}, \\ B_2(x) &= x^2 - x + \tfrac{1}{6}, \\ B_3(x) &= x^3 - \tfrac{3}{2}x^2 + \tfrac{1}{2}x, \\ B_4(x) &= x^4 - 2x^3 + x^2 - \tfrac{1}{30}, \end{aligned}\right\} \qquad ...(11\cdot19)$$

and so on.

We have

$$\frac{t}{2} + \sum_{\nu=0}^{\infty} \frac{t^\nu}{\nu\,!} B_\nu = \frac{t}{2} \cdot \frac{e^t + 1}{e^t - 1}.$$

The function on the right is even since the change of $-t$ for t leaves the function unaltered. It follows that the expansion contains no odd powers of t, and hence

$$B_{2k+1} 0 =, k > 0,$$

$$B_1 = -\tfrac{1}{2} \cdot$$

Example 1. *Evaluate* $\sum_{s=1}^{n} s^{\nu}$

by the use of Bernoulli's numbers.

Sol. We have

$$\int_{s}^{s+1} B_\nu(x)dx = \frac{1}{\nu+1}\,[B_{\nu+1}(s+1) - B_{\nu+1}(s)] = s^{\nu}$$

Thus $$\sum_{s=1}^{n} s^{\nu} = \int_{0}^{n+1} B_\nu(x)dx = \frac{1}{\nu+1} \cdot [B_{\nu+1}(n+1) - B_{\nu+1}]$$

If $\nu = 3$, $$\sum_{s=1}^{n} s^{3} = \tfrac{1}{4}\,[B_4(n+1) - B_4]$$

$$= \tfrac{1}{2}\,[(n+1)^4 - 2(n+1)^3 + (n+1)^2]$$

$$= \tfrac{1}{2}\,[n(n+1)]^2.$$

Example 2. [Multiplication Theorem]. *Show that*

$$B_\nu(mx) = m^{\nu-1} \sum_{s=0}^{m-1} B_\nu\left(x + \frac{s}{m}\right).$$

Sol. Writing $x + \dfrac{s}{m}$ for x in (11·14) and summing for s and 0 to $m-1$, we have

$$\sum_{\nu=0}^{\infty} \frac{t_\nu}{\nu\,!} \sum_{s=0}^{m-1} B_\nu\left(x + \frac{s}{m}\right)$$

$$= \sum_{s=0}^{m-1} \frac{te^{\left(x+\frac{s}{m}\right)t}}{e^{t} - 1}$$

$$= \frac{te^{xt}(e^t - 1)}{(e^t - 1)(e^{t/m} - 1)} = \frac{m.t/m\, e^{mxt/m}}{e^{t/m} - 1}$$

$$= \sum_{v=0}^{\infty} \frac{mt^v}{m_v\, v\,!} B_v(mx)$$

Thus $$B_v(mx) = m^{v-1} \sum_{s=0}^{m-1} B_v\left(x + \frac{s}{m}\right).$$

Cor. Putting $x = 0$ and 2 respectively, we have

$$\sum_{s=1}^{m-1} B_v\left(\frac{s}{m}\right) = -\left(1 - \frac{1}{m^{v-1}}\right) B^v$$

$$B_v\left(\frac{1}{2}\right) = -\left(-\frac{1}{2^{v-1}}\right) B_v, \quad v = 1, 2, \ldots$$

Example 3. *Prove that*

(*i*) $B_v^{(m+n)}(x+y) = [B^{(m)}(x) + B^{(n)}(y)]^v$

(*ii*) $B_v^{(n)}(x+y) = \sum\limits_{p=0}^{v} {}^vC_p\, B_p^{(n)}(x) y^{v-p}$,

(*iii*) $B_v^{(n)}(x) = \sum\limits_{p=0}^{v} C_p^a \quad B_s'\; B_{v-p}^{(n-1)}(x)$

Sol. (*i*) by definition of $B_v^{(n)}(x)$ we have, replacing n by $m+n$ and x by $x+y$,

$$\frac{t^{m+n} e^{t(x+y)}}{(e^t - 1)^{m+n}} = \sum_{v=0}^{\infty} \frac{t^v}{v\,!} B_v^{(m)}(x+y) \qquad \ldots(1)$$

$$\Rightarrow \quad \frac{t^m e^{tx}}{(e^t-1)^m} \cdot \frac{t^n e^{ty}}{(e^t-1)^n} = \sum_{v=0}^{\infty} \frac{t^v}{v\,!} B_v^{(m+n)}(x+y)$$

$$\Rightarrow \quad \left[\sum_{v=r}^{\infty} \frac{t^{v-r}}{(v-r)\,!} B_{v-r}^{(m)}(x)\right]\left[\sum_{v=r}^{\infty} \frac{t^v}{v\,!} B_v^{(n)}(y)\right]$$

$$= \sum_{v=0}^{\infty} \frac{t^v}{v\,!} B_v^{(m+n)}(x+y) \qquad \ldots(2)$$

Equating the coefficient of t^v on both sides, we get

$$\sum_{r=0}^{v} \frac{v\,!}{(v-r)\,!\,r\,!} B_{v-r}^{(m)}(x)\, Br_r^{(n)}(y) = B_v^{(m+n)}(x+y) \qquad \ldots(3)$$

$$\Rightarrow \quad [B^{(m)}(x) + B^{(n)}(y)]^v = B_v^{(m+n)}(x+y).$$

Hence the result follows.

(*ii*) Replacing r by p, v by p, y by x, x by y and putting $m = 0$ in (3), we get

$$\sum_{p=0}^{v} C_p^v B_{v-p}^0 B_p(x) = B_v^{(n)}(x+y)$$

$$B_v^{(n)}(x+y) = \sum_{p=0}^{v} {}^vC_p B_p^{(n)}(x) y^{v-p}.$$

(*iii*) Writing $n = 1$, $m = n - 1$ in (2), page 221 we have

$$\left[\sum_{v=p}^{\infty} \frac{t^{v-p}}{(v-p)!} B_{v-p}^{(n-1)}(x)\right]\left[\sum_{p=0}^{\infty} \frac{t^p}{p!} B_p^{(1)}\right]$$

$$= \sum_{v=0}^{\infty} \frac{t^v}{v!} B_v^{(n)}(x)$$

Equating the coefficients of t^v on both sides, we have

$$\sum_{p=0}^{v} {}^vC_p B_p^{(1)} B_{v-p}^{(n-1)}(x) = B_v^{(n)}(x).$$

Example 4. *Obtain the formulae :*

(*i*) $(x+y)^v = \sum_{p=0}^{v} {}^vC_x B_p^{(n)}(x)\, B_{v-p}^{(-n)}(y)$

(*ii*) $\sum_{p=0}^{v} {}^vC_p B_p^{(n)} B_{v-p}^{(-n)} = 0, \quad v > 0.$

Sol. (*i*) Replace m by $-n$, r by p and x by y, y by x, in (3) of Ex. 3 above and we get

$$\sum_{p=0}^{v} {}^vC_p B_{v-p}^{(-n)} B_p^{(x)}(x) = B_v^0(x+y)$$

or

$$\sum_{p=0}^{v} {}^vC_p B_p^{(n)}(x) B_{v-p}^{(-n)}(y) = (x+y)^v.$$

(*ii*) Putting $x = 0$, $y = 0$ and $m = -n$ in (2), page 220, we get

$$\left(\sum_{v=0}^{\infty} \frac{t^{v-p}}{(v-p)!} B_{v-p}^{-n}\right)\left(\sum_{p=0}^{\infty} \frac{p}{p!} B_p^{(-n)}\right) = \sum \frac{t^v}{v} B_v^0 \cdot$$

Equating the coefficient of t^v on both sides, we have

$$\sum_{p=0}^{\nu} {}^{\nu}C_p\, B_p^{(n)}\, B_{\nu-p}^{(-n)} = B_\nu^0 = 0, \quad \nu > 0.$$

Example 5. *Prove :*

(*i*) $1 + \frac{1}{2} + \frac{1}{3} + \cdots + \frac{1}{n} = \frac{(-1)^{n-1}}{(n-1)\,!} B_{n-1}^{(n+1)}$

(*ii*) $\frac{1}{t} \log(1+t) = \sum_{n=0}^{\infty} \frac{t^\nu}{(\nu+1)\,!} B_n^{(n+1)}$

Sol. (*i*) We have, by the Binomial Theorem,

$$(1+t)^{x-1} = \sum_{n=0}^{\infty} \frac{(x-1)(x-2)\ldots(x-n)t^n}{n\,!}$$

$$= \sum_{n=0}^{\infty} \frac{t^n}{n\,!} B_n^{(n+1)} \quad \ldots(1)$$

Differentiate with respect to x and we obtain

$$(1+t)^{x-1} \log(1+t) = \sum_{n=0}^{\infty} \frac{t^n}{(n-1)\,!} B_n^{(n+1)}(x)$$

Putting $x = 0$, we get

$$[\log(1+t)](1+t)^{-1} = \sum_{n=0}^{\infty} \frac{t^n}{(n-1)\,!} B_{n-1}^{(n+1)}$$

or
$$\left(t - \frac{t^2}{2} + \cdots + \frac{-1^{n-2}t^{n-1}}{n-1} + \frac{(-1)^{n-1}t^n}{n} + \cdots\right)$$

$$\times (1 - t + t^2 - \cdots) = \sum_{n=1}^{\infty} \frac{t^n}{(n-1)\,!} B_{n-1}^{(n+1)}$$

Equating the coefficient of t^n on both sides, we get

$$(-1)^{n-1} \cdot \frac{1}{n} + (-1)^{n-1} \frac{1}{n-1} + \cdots = \frac{1}{(n-1)\,!} B_{n-1}^{(n+1)}$$

$$\Rightarrow \quad \frac{1}{n} + \frac{1}{n-1} + \cdots + \frac{1}{2} + 1 = \frac{(-1)^{n-1}}{(n-1)\,!} B_{n-1}^{(n+1)}$$

(*ii*) Differentiate (1) n times with respect to x and we get

$$(1+t)^{x-1}[\log(1+t)]^n = \sum_{\nu=n}^{\infty} \frac{t^\nu}{(\nu-n)\,!} B_{\nu-n}^{(\nu+1)}(x).$$

Putting $x = 1$ and dividing by t^n, we have

$$\left[\frac{\log(1+t)}{t}\right]^n = \sum_{\nu=0}^{\infty} \frac{t^\nu}{\nu!} B_\nu^{(n+\nu+1)}(1)$$

$$= \sum_{\nu=0}^{\infty} \frac{t^\nu}{\nu!} \cdot \frac{n}{n+\nu} B_\nu^{(n+\nu)}$$

Putting $x=0$ and writing $n+\nu$ for n, we get

$$B_\nu^{(n+\nu+1)}(1) = \frac{n}{n+\nu} B_\nu^{(n+\nu)}$$

Putting $n=1$, we get

$$\frac{\log(1+t)}{t} = \sum_{\nu=0}^{\infty} \frac{t^\nu}{(\nu+1)!} B_\nu^{(\nu+1)}$$

Example 6. *Prove that*

$$\frac{t}{e^t-1} = \frac{\log(1+\Delta)}{\Delta} e^{\theta t}$$

Hence deduce

$$B_{2n-1} = (-1)^{n+1}\left\{1 - \frac{\Delta}{2} + \frac{\Delta^2}{3} - \dots + \frac{\Delta^{2n}}{2n+1}\right\} 0^{2n}$$

Sol. We have

$$\frac{t}{e^t-1} = \frac{\log e^t}{e^t-1} = \frac{\log E}{E-1} e^{0.t}$$

$$= \frac{\log(1+\Delta)}{\Delta} e^{0.t}$$

Equating coefficients, we find

$$(-1)^{n+1} \frac{B_{2n+1}}{\lfloor 2n} = \frac{\log(1+\Delta)}{\Delta} \cdot \frac{0^{2n}}{\lfloor 2n}$$

$$\therefore \quad B_{2n-1} = (-1)^{n+1}\left(1 - \frac{\Delta}{2} + \frac{\Delta^2}{3} - \dots + \frac{\Delta^{2n}}{2n+1} \dots\right) 0^{2n}$$

Example 7. *Prove that*

$$B_{2n-1} = (-1)^{n+1}\left(\frac{\Delta}{1^2} - \frac{\Delta^2}{2^2} + \frac{\Delta^3}{3^2} - \dots\right) 0^{2n+1}$$

Sol. We will use the fact that

$$\Delta^{n-1} 0^{x-1} = \frac{1}{n} \Delta^n 0^x - \cdot \Delta^n 0^{x-1},$$

in the previous result.

We obtain

$$B_{2n-1} = (-1)^{n+1}\left\{\left(\Delta - \frac{\Delta^2}{2^2} + \frac{\Delta^3}{3^2} - \cdots\right)0^{2n+1} - \left(\Delta - \frac{\Delta^2}{2} + \frac{\Delta^3}{3} - \cdots\right)0^{2n}\right\}$$

$$= (-1)^{n+1}\left(\Delta - \frac{\Delta^2}{2^2} + \frac{\Delta^3}{3^2} - \cdots\right)0^{2n+1}$$

since the other term is

$$\log(1+\Delta)\,0^{2n} = D.0^{2n} = 0.$$

Example 8. *Show that*

$$\Delta^n O^v = \frac{v\,!}{(v-n)\,!}\,B_{(v-l}^{(-n)}$$

Sol. Since $B_v^{(n)}(x+h)$ is a polynomial of degree v, we have by Newton's interpolation formula

$$B_v^{(n)}(x+h) = B_x^{(n)}(h) + \sum_{s=1}^{v} x^{(s)}\Delta^s B_v^{(n)}(h)$$

$$= \sum_{s=0}^{v} {}^vC_s x^{(s)}\, B_{(v-s)}^{(n-s)}(h)$$

since $\qquad \Delta^n B_v^{(n)}(x) = v(v-1)\ldots(v-s+1)x^{v-n}.$

Putting $h = 0$, we get

$$B^{(n)}(x) = \sum_{s=0}^{v} {}^vC_s x^{(s)} B_{v-s}^{(n-s)}$$

Further putting $n = 0$ and remembering that $B_v^{(0)}(x) = x^v$, we obtain

$$x^v = \sum_{s=0}^{v} {}^vC_s x^{(s)} B_{v-s}^{(n-s)}.$$

Operating with Δ^n, $n \le v$ and then putting $x = 0$, we get

$$\Delta^n O^v = \frac{v\,!}{(v-n)\,!}\,B_{(v-n)}^{(-n)}.$$

11·4. The η Polynomials. The η polynomials are defined by the relation

$$2^n e^{xt+g(t)}(e^t+1)^{-n} = \sum_{v=0}^{\infty} \frac{t^v}{v\,!}\,\eta_v^{(n)}(x)$$

$$\therefore \qquad \sum \frac{t^{\nu}}{\nu\,!}\,\Delta\eta_{\nu}^{(n)} = \sum_{\nu=0}^{\infty} \frac{t^{\nu}}{\nu}\,,\ \eta_{\nu}^{(n-1)}(x)$$

Hence $\qquad \Delta\eta_{\nu}^{(n)}(x) = \eta_{\nu}^{(n-1)}(x).$

11·5. Euler's Polynomials. Euler's polynomial of order n and degree ν is defined by the relation

$$2^{n} e^{te}(e^{t}+1)^{-n} = \sum_{\nu=0}^{\infty} \frac{t^{\nu}}{\nu\,!} E_{\nu}^{(n)}(x) \qquad \text{...(11·21)}$$

The Euler's numbers, $E_{\nu}^{(n)}$ of order n are defined by the relation

$$E_{\nu}^{(n)} = 2^{\nu} E_{\nu}^{(n)}(x)\left(\frac{1}{2}n\right)$$

Set $\qquad (E_{\nu}^{(n)}(0) = 2^{-\nu} C_{\nu}^{(n)}.$

The generating function for C numbers is therefore

$$\frac{2^{n}}{(e^{t}+1)^{n}} = \sum_{\nu=0}^{\infty} \frac{t^{\nu}}{\nu\,!}\frac{1}{2^{\nu}} C_{\nu}^{(n)}$$

11·5·1. To prove that

$$E_{\nu}^{(n+1)}(x) = \frac{2}{n} E_{\nu+1}^{(n)}(x) + \frac{2}{n}(n-x)\, E_{\nu}^{(n)}(x)$$

Differentiating (11·21) with respect to t and then multiplying by t, we have

$$\sum_{\nu=1}^{\infty} \frac{t^{\nu}}{(\nu-1)} E_{\nu}^{(n)}(x) = \frac{2^{n}\cdot xt\, e^{xt}}{(e^{t}+1)^{n}} - \frac{2^{n}\cdot nt\, e^{(x+1)t}}{(e^{t}+1)^{n+1}}$$

$$= x\sum_{\nu=1}^{\infty} \frac{t}{(\nu \quad 1)} E_{\nu-1}^{(n)}(x) - \tfrac{1}{2}\, n \sum_{\nu=1}^{\infty} \frac{t^{\nu}}{(\nu \quad 1)\,!} E_{\nu-1}^{(n+1)}(x+1)$$

Equating the coefficients of $t^{\nu+1}$

$$E_{\nu+1}^{(n)}(x) = xE_{\nu}^{(n)}(x) - \tfrac{1}{2}\, nE_{\nu}^{(n+1)}(x+1)$$

But $\qquad \nabla E_{\nu}^{(n)}(x) = E_{\nu}^{(n-1)}(x)$

or $\qquad E_{\nu}^{(n)}(x+1) = 2E_{\nu}^{(n-1)}(x) - E_{\nu}^{(n)}(x)$

or $\qquad E_{\nu}^{(n+1)}(x+1) = 2E_{\nu}^{(n)} - E_{\nu}^{(n)}(x)$

Therefore we get the recurrence relation

$$E_{\nu}^{(n+1)}(x) = \frac{2}{n}\cdot E_{\nu+1}^{(n)}(x) + \frac{2}{n}(n-x)E_{\nu}^{(n)}(x)$$

Writing $\qquad x = 0$, we obtain

$$C_{\nu}^{(n+1)} = \frac{2}{n} \cdot C_{\nu+1}^{(n)} + 2C_{\nu}^{(n)}.$$

11·5·2. To prove that

$$\mathbf{E}_{\nu}^{(n)}(n-x) = (-1)^{\nu} E_{\nu}^{(n)}(n)$$

[Complementary Argument Theorem]

From (11·21), have

$$\sum_{\nu=0}^{\infty} \frac{t^{\nu}}{\nu\,!} E_{\nu}^{(n)}\,(n-x) = \frac{2^n e^{(n-x)t}}{(e^t+1)^n} = \frac{2^n e^{-xt}}{(e^{-t}+1)^n}$$

$$= \sum_{\nu=0}^{\infty} \frac{(-t)^{\nu}}{\nu\,!} \cdot E_{\nu}^{(n)}(x)$$

Hence $E_{\nu}^{(n)}(n-x) = (-1)^{\nu} E_{\nu}^{(n)}(x)$

Put $x = 0,\ \nu = 2\mu,\ E_{2\mu}^{(n)}(n) = E_{2\mu}^{(n)}(0) = 2^{-2\mu}\, C_{2\mu}^{(n)}$

Thus $E_{2\mu}^{(n)} - 2^{-2\mu} C_{2\mu}^{(n)}$ has zeros at $x = 0,\ x = n$.

Again, put $x = \frac{1}{2} n,\ \nu = 2\mu + 1$, we obtain

$$E_{2\mu+1}^{(n)} = -E_{2\mu+1}^{(n)}$$

or

$$E_{2\mu+1}^{(n)} = 0.$$

Thus *Euler's numbers with an odd suffix are all zero.*

EXERCISE 11

1. Prove that

(*i*) $E_{\nu}^{(m+n)}(x+y) = [E^{(n)}(y) + E^{m}(x)]^{\nu}$

(*ii*) $(x+y)^{\nu} = \sum_{p=0}^{\nu} {}^{\nu}C_p\, E_p^{(n)}(x)\, E_{\nu-p}^{-n}(y)$

(*iii*) $B_{\nu}^{(n+1)} = \dfrac{\nu\,!}{n\,!} \dfrac{d^{n-\nu}}{d^{x-p}} [(x-1)(x-2)...(x-n)]$

2. Prove that

(*i*) $x^{(n)} = \sum_{1=p}^{n} {}^{n-1}C_{p-1} x^p\, B_{n-p}^{(n)}$.

(*ii*) $E_{\nu}^{(n)}(x) = x - \dfrac{1}{2} n + \dfrac{1}{2} E_{\nu}^{(n)}$.

(*iii*) $E_{\nu}^{\prime}(n)\left(\dfrac{n+x}{2}\right) = \left(\dfrac{x+E^{(n)}}{2}\right)^{\nu}$

Putting $x = 1,\ x = -1$ and adding, deduce that

$$(E^{(n)}+1)^u+(E^{(n)}-1)^x = 2^x E_v^{(n)}\left(\frac{n+1}{2}\right)+2^v E_v^{(n)}\left(\frac{n-1}{2}\right)$$
$$= 2^{v+1}\nabla E_v^{(n)}\left(\frac{n-1}{2}\right)$$
$$= 2^{v+1}E_v^{(n-1)}\left(\frac{n-1}{2}\right)$$
$$= 2E_v^{(n1)}$$

3. Prove, for $E_v(x)$ Euler's polynomials of the first order

(i) $\dfrac{2e^{xt}}{e^t+1}\sum_{v=1}^{\infty}\dfrac{t^v}{v!}E_v(x), \dfrac{2}{e^t+1}=\sum_{x=0}^{\infty}\dfrac{t^v}{v!}\dfrac{1}{2^v}C^v$

(ii) $\dfrac{2e^{1/2t}}{e^t+1}=\sum_{v=0}^{\infty}\dfrac{t^v}{v!}\dfrac{E^v}{2^v}$

(iii) $\nabla E^{(n)}=x^v, DE_v(x)=vE_{-1}$

(iv) $E_v^{(n)}=(-1)^v E_v(x)$

4. Show that $\sum_{s=1}^{\infty}(-1)^v s^v=\frac{1}{2}\cdot(-1)^n E_v(n+1)-\frac{1}{2}E_v(1)$

5. Prove that $E_v^{(n)}=\sum_{p=0}^{v}\dfrac{E^{p(n)}}{2^p}\,{}^vC_p\left(x-\frac{1}{2}n\right)^{v-p}$

6. Prove that

(i) $B_n=\sum_{m=0}^{n}\dfrac{(-1)^m}{m+1}\Delta^m 0^n.$

(ii) $B_{2n}=\dfrac{\Delta O^{2n+1}}{1^2}-\dfrac{\Delta^2 O^{2n+2}}{2^2}+\cdots+\dfrac{\Delta^{2n+1}O^{2n+1}}{(2n+1)^2}.$

(iii) $B_{2n}=(-1)^{n-1}\left(\frac{1}{2}-\frac{1}{3}\Delta+\frac{1}{4}\Delta^2-\cdots\right)^{2n-1}$

(iv) $B_{2n-1}=(-1)^{n-1}\left\{\dfrac{\Delta O^{2n+1}}{1^2}-\dfrac{\Delta^2 O^{2n+1}}{2^2}+\cdots+\dfrac{\Delta^{2n+1}O^{2n+1}}{(2n+1)^2}\right\}$

(v) $B_n=\left(\dfrac{1}{2}-\dfrac{\Delta}{3}+\dfrac{\Delta^2}{4}-\cdots\right)1^n,$

(vi) $B2_{n-1}=\dfrac{2n}{2^{2n}-1}\cdot\dfrac{(-1)^{n-1}}{2+\Delta}O^{2n-1}$

7. Prove that if $f_n(x)$ is a polynomial of degree n having the property that $Df_n(x)=f_{n-1}(x)$, then $f_n(x)$ must have the form

$$f_n(x)=\frac{c_0}{n!}x^n+\frac{c_1}{(n-1)!}x^{n-1}+\cdots+c_{n-1}x+c_n$$

where $c_0, c_1, \ldots$ are contants independent of n.

12

REMAINDER TERMS IN (OR ERRORS ASSOCIATED WITH) INTERPOLATION FORMULAS

12·1. Introduction. In earlier chapters, we have represented the given function $y=f(x)$ over the interval $x_0 \le x \le x_n$ by a polynomial $P_n(x)$ of the nth degree which passes through $n+1$ consecutive pairs of values $(x_0, y_0), (x_1, y_1), ...(x_n, y_n)$. Clearly, the polynomial, called the interpolating polynomial, will approximate the given function more closely if we increase the number of coinciding points. But when such number increases indefinitely the interpolating formulas become infinite series, called interpolation series. An interpolation series like a power series converges and represents the given function over a certain interval but fails to represent it outside the interval. It is beyond the scope of this book to test the convergence of an interpolating series. We shall, therefore, derive expressions for the errors or remainder terms in polynomial approximation formulas. We shall need Rolle's theorem which states that :

If the function $f(x)$ be continuous and possesses continuous derivative in the interval $x_0 \le x \le x_n$ whose x_0, x_n are two real roots of the equation $f(x)=0$, then the equation $f'(x)=0$ has at least one real root lying in the interval (x_0, x_n).

12·2·1. Remainder Terms in Lagrange's and Newton's Forward Interpolation Formulas.

Let $P_n(x)$ be the polynomial of the nth degree in x representing the given function $f(x)$. Assume that $f(x)$ possesses continuous $(n+1)$th order derivative in the interval of interpolation $x_0 \le x \le x_n$. Since $P_n(x)$ has the same values as $f(x)$ at the points $x_0, x_1, ..., x_n$,

$$f(x)=P_n(x)+g(x)$$

where $g(x)$ has the roots $x_0, x_1, ..., x_n$.

$$\therefore \quad f(x) = P_n(x) + K(x)(x - x_0)(x - x_1)\ldots(x - x_n)$$

where $K(x)$ is to be determined in terms of the function $f(x)$ which can be done as follows :

Consider the function

$$\phi(t) = f(t) - P_n(t) - K(x)[t - x_0)(t - x_1)\ldots(t - x_n)] \qquad ...(12\cdot1)$$

Now $\phi(t) = f(t)$ vanishes for the $n+2$ real roots $x, x_0, x_1, ..., x_n$. Therefore by Rolle's theorem $\phi'(t)$ has at least $n+1$ roots lying between the smallest and greatest of the above roots. Likewise $\phi''(t)$ has at least n roots in this very interval, and finally the $(n+1)$th derivative $\phi^{n+1}(t)$ has at least one root $t = \xi$ in the interval x_0 to x_n.

Also from (12·1).

$$\phi^{n+1}(t) = f^{n+1}(t) - K(x)[(n+1)\,!]$$

since $$P_n^{n+1}(t) = \frac{d^{n+1}}{dt^{n+1}} P_n(t) = 0.$$

Hence $$K(x) = \frac{1}{(n+1)\,!} f^{n+1}(\xi), \quad x_0 < \xi < x_n \qquad ...(12\cdot2)$$

Substituting this value of $K(x)$ in (12·1), we have

$$f(x) = P_n(x) + f^{n+1}(\xi) \cdot \frac{(x - x_0)(x - x_1)\ldots(x - x_n)}{(n+1)\,!}$$

Hence Truncation error $= R_n = f(x) - P_n(x)$

$$= \frac{(x - x_0)(x - x_1)\ldots(x - x_n)}{(n+1)\,!} f^{n+1}(\xi)$$

$$= \frac{f^{n+1}(\xi)}{(n+1)\,!} \prod_{i=0}^{n} (x - x_i) \qquad ...(12\cdot3)$$

This is the remainder term in Lagrange's formula.

We have $x - x_0 = x - a = hu$, $x - x_1 = x - a + h = h(u - 1)$, etc.

Substituting the values of $x - x_0, x - x_1$, etc. in (12·3), we get

$$R_n = \frac{u(u-1)\ldots(u-n)}{(n+1)\,!} h^{n+1} f^{n+1}(\xi)$$

$$= \frac{u^{(n+1)}}{(n+1)\,!} h^{n+1} f_{n+1}(\xi), \; x_0 < \xi < x_n \qquad ...(12\cdot4)$$

which is the remainder term in Newton's formula for forward interpolation.

12·2·2. Remainder Term in Newton Backward Interpolation Formula.

Hence or otherwise show that if (error function or probability integral)

$$f(x) = P_n(x) + K(x)(x - x_n)(x - x_{n-1}) : (x - x_0)$$

where $K(x)$ is to be expressed in terms of $f(x)$.

Consider the function

$$\phi(t) = f(t) - P_n(t) - K(x)(t - x_n)(t - x_{n-1})\ldots(t - x_0)$$

Since $\quad \phi^{n+1}(t) = 0 \quad$ for $\quad t = \xi, \; x_0 < \xi < x_n,$

we have
$$0 = f^{n+1}(\xi) - 0 - K(x)\,[(n+1)\,!]$$

or
$$K(x) = f^{n+1}(\xi)/(n+1)\,!$$

Hence truncation error $= R_n = f^{n+1}(\xi), \dfrac{(x - x_n)(x - x_{n-1})\ldots(x - x_0)}{(n+1)\,!}$ ...(12·5)

We have $\quad \dfrac{x - x_n}{h} = u, \quad \dfrac{x - x_{n-1}}{h} = u + 1, \ldots, \dfrac{x - x_0}{h} = u + n$

Substituting the values for $x - x_n, x - x_{n-1}$ etc. in (12·5), we get

$$R_n = u(u+1)(u+2)\ldots(u+n)\frac{h^{n+1}}{(n+1)\,!}f^{n+1}(\xi) \qquad \text{...(12·6)}$$

This is the remainder term in Newton's formula for Backward Interpolation.

12·3. Remainder Terms in Central-difference Formulas.

12·3·1. Remainder Term in Newton's Divided Difference Formula. Equation (2·7) can be written in the form

$$f(x) = P_n(x) + R_n,$$

Assume that $f^{n+1}(x)$ exists and is continuous, the function $f(x, x_0, x_1, \ldots, x_n)$ is finite, and therefore by (2·7)′ R_n vanishes at the $(n+1)$ points $x_0, x_1, \ldots, x_n$. Thus by Rolle's theorem $R_n'(x)$ vanishes at n points lying between the smallest and largest of these arguments and so on. Finally $R_n(x)$ vanishes for some point $x = \xi$ *i.e.*, $R^n(\xi) = 0$ and therefore

$$f^n(\xi) = (n\,!)\,f(x_0, x_1, \ldots, x_n)$$

or
$$f(x_0, x_1, \ldots, x_n) = \frac{1}{n\,!}f^n(\xi), \; x_0 < \xi < x_n.$$

or
$$f(x, x_0, x_1, \ldots, x_n) = \frac{1}{(n+1)\,!}f^{n+1}(\xi), \; x_0 < \xi < x_n$$

$\therefore$
$$R_n(x) = \frac{1}{(n+1)\,!}f^{n+1}(\xi)\prod_{i=0}^{n}(x - x_i)$$

Notes 1. This formula gives the upper and lower limits for the value of a divided difference.

2. This is the same remainder as for Lagrange's formula. That they must agree can be seen from the fact that they represent the difference between $f(x)$ and a polynomial of the nth degree that has the same value as $f(x)$ at $x = x_0, x_1, ..., x_n$. Since the polynomial is unique, this difference is unique.

12·3·2. Remainder Term in Stirling's Formula

In Stirling formula there are $2n+1$ terms and the polynomial $P_{2n+1}(x)$ coincides with the given function $f(x)$ at the $2n+1$ points

$$u = -n, -(n-1), -(n-2), ..., -2, -1, 0, 1, 2, ..., n-2, n-1, n$$

or
$$x = x_0 - nh, x_0 - (n-1)h, ..., x_0 - h, x_0, x_0 + h..., x_0 + (n-1)h, x_0 + nh$$

$$\therefore \quad f(x) = P_{2n+1}(x) + K(x)(x - x_0)(x - x_{-1})(x - x_1)...(x - x_{-n})(x - x_n) \quad ...(12·7)$$

where $K(x)$ is to be expressed in terms of $f(x)$.

Consider the function

$$\phi(t) = f(t) - P_{2n+1}(t) - K(x)(t - x_0)(t - x_1)(t - x_{-1})...(t - x_n)(t - x_{-n})$$

Now $\phi(t)$ vanishes for the $2n+2$ values

$$t = x, x_0, x_1, x_{-1}, ..., x_n, x_{-n}.$$

Hence by Rolle's theorem $\phi_{2n+1}(\xi) = 0$ for some value $t = \xi$ in the interval x_{-n} to x_n.

Also $$\phi^{2n+1}(t) = f^{2n+1}(t) - K(x)[(2n+1)\,!]$$

$$\therefore \quad K(x) = f^{2n+1}(\xi)/(2n+1)\,!, \; x_{-n} < \xi < x_n$$

Substituting the value of $K(x)$ in (12·7), we have

$$f(x) = P_{2n+1}(x) + f^{2n+1}(\xi)\,\frac{(x - x_0)(x - x_1)(x - x_{-1})...(x - x_n)(x - x_{-n})}{(2n+1)\,!}$$

$$\therefore \text{ Error} = R_n = f(x) - P_{2n+1}(x)$$
$$= f^{2n+1}(\xi)\,\frac{(x - x_0)(x - x_1)(x - x_{-1})...(x - x_n)(x - x_{-n})}{(2n+1)\,!}$$

In order to obtain the value of R_n in terms of u, we substitute in the above formula the following :

$$x - x_0 = hu,$$
$$x - x_1 = h(u-1),$$
$$x - x_{-1} = x - (x_0 - h) = x - x_0 + h = hu + h = h(u+1),$$
$$\dots\dots\dots\dots\dots\dots\dots\dots\dots\dots$$
$$x - x_n = h(u-n), \; x - x_{-n} = h(u+n).$$

We get $$R_n = \frac{h^{2n+1} f^{2n+1}(\xi)}{(2n+1)\,!}\,.$$
$$\times\, u(u^2 - 1^2)(u^2 - 2^2)(u^3 - 3^2)...(u^2 - n^2), \; x_{-n} < \xi < x_n.$$

12·3·3. Remainder Terms in Bessel's Formula.

In Bessel's formula there are $2n+2$ terms and the polynomial coincides with the given function at the $2n+2$ points

$$u = -n, -(n-1), \ldots, -1, 0, 1, \ldots, n-1, n, n+1$$

or
$$x = x_0 - nh, x_0 - (n-1)h, \ldots, x_0 - h, x_0, x_0 + h, \ldots$$
$$x_0 + nh, x_0 + (n+1)h.$$

Now

$$f(x) = P_{2n+1}(x) + K(x)(x - x_1)(x - x_{-1}) \ldots (x - x_n)(x - x_{-n})(x - x_{n+1})$$

where $K(x)$ is to be determined in terms of $f(x)$.

Consider the function

$$\phi(t) = f(t) - P_{2n+2}(t)$$
$$- K(x)(t - x_0)(t - x_1)(t - x_{-1}) \ldots (t - x_n)(t - x_{-n})(t - x_{n+1}).$$

Now $\phi(t)$ vanishes at the $2n+3$ points

$$t = x, x_0, x_1, x_{-1}, \ldots, x_n, x_{-n}, x_{n+1}$$

$\therefore$ $\phi^{2n+2}(t) = 0$ for some value ξ in the interval (x_n, x_{n+1})

But $\phi^{2n+2}(t) = f^{2n+2}(t) - K(x)(2n+1)\,!$

$\therefore$ $K(x) = f^{2n+1}(\xi)/(2n+1)\,!$

Hence $f(x) = P_{2n+2}(x) + \dfrac{f^{2n+2}(\xi)}{(2n+2)\,!} \times$

$$(x - x_0)(x - x_1)(x - x_{-1}) \ldots (x - x_n)(x - x_{-n})(x - x_{n+1}) \qquad \ldots(12\cdot8)$$

$\therefore$ Error $= R_n$

$$= \frac{f^{2n+2}(\xi)}{(2n+2)\,!}(x - x_0)(x - x_1)(x - x_{-1}) \ldots (x - x_n)(x - x_{-n})(x - x_{n+1})$$

Putting $x - x_0 = hu$, $x - x_1 = h(u-1)$, $x - x_{-1} = h(u+1)$, etc., we have

$$R_n = \frac{h^{2n+2}}{(2n+2)\,!} f^{2n+2}(\xi)$$
$$\times u(u-1)(u+1) \ldots (u-n)(u+n)(u-n-1) \qquad \ldots(12\cdot9)$$

On putting $u = v + \frac{1}{2}$, in (12·9)

we have

$$R_n = \frac{h^{2n+2}}{(2n+2)\,!} f^{2n+2}(\xi) \times \left(v^2 - \frac{1}{4}\right)\left(v^2 - \frac{9}{4}\right) \cdots \left(v^2 - \frac{(2n+1)^2}{4}\right)$$

$$R_n = \frac{\Delta^{2n+2} y_{-n-1} + \Delta^{2n+2} y_{-n}}{(2n+2)\,!} \times \left(v^2 - \frac{1}{4}\right)\left(v^2 - \frac{9}{4}\right) \cdots \left(v^2 - \frac{(2n+1)^2}{4}\right)$$

12·4. Remainder terms for the Quadrature Formulas.

Newton Cote's formula is

$$f(x) = f(x_0 + nh) = nh \sum_{i=0}^{n} f C_i^m + E^s \quad \text{where } E^s = h \text{ [e rror term}$$

$$= \frac{h^{n+2}}{(n+2)!} f^{n+2}(\xi)\, s(s-1)\ldots(s-n) \text{ when } n \text{ is even. } x_0 < \xi < x_n$$

12·4·1. Formulas for the Inherent or Truncation or Error in Simpson's and Weddle's Rules. Assume that $f(x)$ is continuous function in the interval $(x_0 - h, x_0 + h)$ which possesses continuous first four derivatives.

Let $\quad F(x) = \int_a^x f(x)\, dx,$

Then $\quad I = \int_{x_0 - h}^{x_0 + h} f(x)\, dx = F(x_0 + h) - F(x_0 - h)$

By Simpson's one-third Rule, we have

$$I_s = \frac{h}{3} [f(x_0 - h) + 4f(x_0) + f(x_0 + h)]$$

$$\therefore \quad \text{Error} = I - I_s = \phi(h) = F(x_0 + h) - F(x_0 - h) - \frac{h}{3} [f(x_0 - h) + 4f(x_0) + f(x_0 + h)]$$

Differentiate w.r.t. h, we get

$$\therefore \quad \phi'(h) = f(x_0 + h) + f(x_0 - h) - \frac{h}{3} [f'(x_0 + h) - f'(x_0 - h)] - \frac{1}{3} [f(x_0 + h) + 4f(x_0) + f(x_0 - h)]$$

$$\therefore \quad \phi'''(h) = f'(x_0 + h) - f'(x_0 - h) - \frac{h}{3} [f''(x_0 + h) + f''(x_0 - h)] - \frac{1}{3} [f'(x_0 + h) - f'(x_0 + h)]$$

and $\quad \phi'''(h) = -\frac{h}{3} [f'''(x_0 + h) - f'''(x_0 - h)]$

$$= -\frac{2h^2}{3} f^{iv}(\xi) \quad \text{by the Mean value Theorem}$$

where $\quad x_0 - h < \xi_0 < x_0 + h. \quad$ ...(12·10)

Integrating (12·10) from 0 to h, nothing that though $f^{iv}(\xi)$ depends to some extent on the magnitude of the interval $(0, h)$, we replace it by its mean value in the interval and remove it from the integral sign, we have.

$$\phi''(h) = -\frac{2h^3}{9} f^{iv}(\xi),$$

$$\phi'(h) = -\frac{h^4}{18} f^{iv}(\xi),$$

$$\phi(h) = -\frac{h^5}{90} f^{iv}(\xi),$$

Thus the inherent error for a sub-interval of width, h, is

$$-\frac{h^5}{180} f^{iv}(\xi).$$

and that for an interval of width n is given by

$$E_s = \text{Error} = -\frac{nh^5}{180} f^{iv}(\xi)$$

$$= -\frac{(b-a)h^4}{180} f^{iv}(\xi), \quad a < \xi < b.$$

While deriving Weddle's Rule, we had neglected the quantity

$$\frac{h}{140}\Delta^6 y_0 \quad \text{or} \quad -\frac{h^7}{140} f^{iv}(\xi).$$

Here $E_s = \text{Error} = -\dfrac{h}{140}\Delta^6 y = -\dfrac{h^7}{140} f^{iv}(\xi).$

Note. The values found for E_w show that $E_w = 0$ when $f(x) = 0$. Hence when $f(x)$ is a polynomial of the 5th degree or lower, Weddle's Rule gives an exact value of

$$\int_a^b f(x)\, dx.$$

12·5. Error in the Linear Interpolation

Suppose the function $f(x)$ is tabulated for $x = a, a + h$. Then the error or the remainder term is

$$R_1 = \frac{(x-a)(x-a-h)}{2!} J''(\xi), \quad a < \xi < a + h$$

$$\therefore \qquad |R_1| \le \frac{|x-a|\,|x-a-h|\,M}{2!}$$

where $\quad M = \max\limits_{a \le x \le a+h} \left| f''(\xi) \right|$

Now $\pi(x) = (x-a)(x-a-h) = x^2 - x(2a+h) + a(a+h)$

The point of maximum is given by

$$2x-(2a+h)=0 \quad \therefore \quad i.e., \quad x=a+\tfrac{1}{2}h.$$

Thus the maximum occurs at $x=a+\frac{1}{2}h$ and the maximum value of $[\pi(x)]=\frac{1}{4}h^2$. Thus

$$|R| \le \frac{1}{8}h^2 M$$

where
$$M=\max_{a\le\xi\le a+h}\left|f''(\xi)\right|$$

Also
$$\begin{aligned}\Delta^2 f(x) &= f(x+2h)-2f(x+h)+f(x)\\ &= [f(x+2h)-f(x+h)]-[f(x+h)-f(x)]\\ &= hf'(\xi_1)-hf'(\xi_2),\end{aligned}$$

where $x+h<\xi_1<x+2h, \quad x<\xi_2<x+h$

$$\begin{aligned}\therefore \quad \Delta^2 f(x) &= h[f'(\xi_1)-f'(\xi_2)]\\ &= h^2 f''(\xi), \quad a\le\xi\le a+h\end{aligned}$$

so that
$$M=\frac{1}{h^2}\Delta^2 f(a)$$

Thus
$$|R_1| \le \tfrac{1}{8}\Delta^2 f(a)$$

Hence the error in the linear interpolation is not greater than $\frac{1}{8}$ of the second difference.

Example 1. *Interpolate sin x in a table of equally spaced points with step h using quadratic interpolation*

$$f(x)-R(x)=\frac{(x-x_0)(x-x_1)(x-x_2)}{3!}f'''(\xi)=\frac{[\pi(x)]}{3!}f'''(\xi)$$

Sol. Let $x_0=-h,\ x_1=0,\ x_2=h$

then
$$\pi(x)=x^3-h^2x \quad \text{or} \quad \pi(x)=3x^2-h^2=0$$

$\therefore$ When $x=\pm\frac{h}{\sqrt{3}}$, maximum occurs and the maximum value is $\frac{2h^3}{3\sqrt{3}}$.

$$\text{Error}=\max\left|f(x)-R(x)\right| \le \frac{\sqrt{3}}{27}h^3, -h\le x\le h$$

To obtain upto seven place decimals using quadratic interpolation h should be chosen such that

$$\frac{\sqrt{3}}{27}h^3<5\cdot10^{-3}. \quad \text{Hence } h \simeq 0\cdot07$$

Example 2. *Determine the maximum error in the interpolation of log 1·24* *by Newton's formula for forward interpolation, when the following table is given*

x	1·02	1·03	1·04	1·05	1·06
$f(x)$	0·0086002	0·0128372	0·0170333	0·0211893	0·0253059

Sol. Here $y_0 = 0{\cdot}0086002,\ \Delta y_0 = 0{\cdot}0042370,$

$$\frac{1}{2\,!}\Delta^2 y_0 = -\,0{\cdot}00002045,\ \frac{1}{3\,!}\Delta^3 y_0 = 0{\cdot}00000013.$$

Since $h = 0{\cdot}01 = 10^{-2},\ u = \dfrac{1{\cdot}024 - 1{\cdot}02}{8{\cdot}01} = 0{\cdot}4$

$$u - 1 = -\,0{\cdot}6,\ u - 2 = 1{\cdot}6$$

Newton formula for forward interpolation

$$y = y_0 + u^{(1)}\Delta y_0 + \frac{1}{2\,!}u^{(2)}\Delta^2 y_0 + \frac{1}{3\,!}u^{(3)}\Delta^3 y_0 + \cdots$$

$$= 0{\cdot}0086002 + 0{\cdot}4(\cdot 0043370) + \cdot 04(-\,0{\cdot}6)(-\,0{\cdot}00002045)$$
$$+ (\cdot 4)(-\,0{\cdot}6)(-\,1{\cdot}6) \times 0{\cdot}00000013$$

or $\log 1{\cdot}024 = 0{\cdot}01030000,$ on simplification.

$$\text{Error} = \frac{1}{4\,!}h^4 u^{(4)} f^4(\xi),\ \text{ since } u = 3$$

$$= \frac{10^{-8}}{4\,!}(0{\cdot}4)(-\,0{\cdot}6)(-\,1{\cdot}6)(-\,2{\cdot}6)\left(\frac{d^4 \log x}{dx^4}\right)_{x = \xi}$$

where $1{\cdot}02 < \xi < 1{\cdot}05.$

Since $\dfrac{d^4 \log x}{dx^4} = \dfrac{-\,3\,!}{x^4}\log_{10} e$

It is due to replacing the tabulated function $f(x)$ by an interpolating polynomial $P_n(x)$.

The truncation error ε in the interpolation must be in the range

$$0{\cdot}892 \times 10^8 < \varepsilon < 1{\cdot}002 \times 10^8$$

Thus we can find the rough upper limit for $|\,\varepsilon\,|$.

Example 3. *Determine the accuracy of the interpolation by Lagrange's formula for log 47 given.*

x	40	42	45	48	49	50
$y = \log x$	1·6020600	1·6232493	1·6532126	1·6812413	1·6901960	1·989700

Sol. Letting $x_0 = 40,\ x_1 = 42,\ x_2 = 45,\ x_3 = 48,\ x_4 = 49,\ x_5 = 50,$ and applying Lagrange's interpolation formula,

$$y = \frac{(5)(2)(-\,1)(-\,2)(-\,3)}{(-\,2)(-\,5)(-\,8)(-\,9)(-\,10)}y_0 + \frac{(7)(2)(-\,1)(-\,2)(-\,3)}{(2)(-\,3)(-\,6)(-\,7)(-\,8)}y_1$$
$$+ \cdots + \frac{(7)(5)(2)(-\,1)(-\,2)}{(10)(8)(5)(2)(1)}y_5$$

$$= \frac{1}{120} y_0 - \frac{1}{24} y_1 + \frac{7}{30} y_2 + \frac{35}{24} y_3 - \frac{5}{6} y_4 + \frac{7}{40} y_5$$

$= 1{\cdot}6720980$, on putting values of y's.

Here $n = 5$,

$$\therefore \text{ Remainder term } = \frac{(7)(5)(2)(-1)(-2)(-3)}{6\,!} f^6(\xi)$$

$$= \frac{-7}{12} f^6(\xi), \quad 40 < \xi < 50$$

Now $$f^6(x) = \frac{d^6}{dx^6} \log x = \frac{5\,!}{x^6} \log_{10} e$$

Hence $f^6(\xi) < \dfrac{70}{40^6} \log_{10} e = 0{\cdot}73 \times 10^{-8}$

and the interpolation error is less than one unit in the eighth place.

Example 4. *The following table gives the values of the function $y = 10x^5 + x^4$ for certain value of x. Find y when $x = 2{\cdot}27$ by using Bessel's formula. Obtain the error.*

x	y	Δy	$\Delta^2 y$	$\Delta^3 y$	$\Delta^4 y$
2·0	336·0000				
		91·8582			
2·1	427·8582		19·0724		
		110·9306		2·8266	
2·2	538·7888		21·8990		0·2664
		132·8216		3·0930	
2·3	671·6184		24·9920		0·2784
		157·8216		3·3714	
2·4	829·4400		28·3634		
		186·1850			
2·5	1015·6250				

Sol. Taking $x_0 = 2{\cdot}2$, $x = 2{\cdot}27$, $h = 0{\cdot}1$, we have

$$u = \frac{x - x_0}{h} = \frac{0{\cdot}7}{\cdot 1} = 0{\cdot}7$$

or $$v = u - \frac{1}{2} = 0{\cdot}2$$

$$y = \frac{538{\cdot}7888 + 671{\cdot}6184}{2} + 0{\cdot}2(132{\cdot}8296)$$

$$+ \left(\frac{0{\cdot}04 - 0{\cdot}25}{2}\right)\left(\frac{21{\cdot}8990 + 24{\cdot}9920}{2}\right) + 0{\cdot}2\left(\frac{0{\cdot}04 - 0{\cdot}25}{6}\right)(0{\cdot}0930)$$

$$= 605{\cdot}2036 + 26{\cdot}56592 - 2{\cdot}46178 - 0{\cdot}02165$$

$$= 629{\cdot}28609$$

We have $2n + 2 = 4 \quad \therefore \quad n = 1$

Here $\qquad y = f(x) = x^4 + 14x^5$.

Also $f^{iv}(x) = 24 + 1200x$

or $f^{iv}(\xi) = 24 + 1200\xi, \quad 2{\cdot}0 < \xi < 2{\cdot}1, \; -2{\cdot}25 < \eta < 2{\cdot}5$

$$f^{iv}(\xi) = f^{iv}(2{\cdot}25 + 0{\cdot}1\eta)$$

$$= 24 + 2700 + 120\eta = 2724 + 120\eta$$

$$\therefore \quad R_n = \frac{h^4 f^{iv}(\xi)}{4\,!}\left(v^2 - \frac{1}{4}\right)\left(v^2 - \frac{9}{4}\right)$$

$$= (0{\cdot}1)^4 \frac{(2724 + 120\eta)}{24}(0{\cdot}04 - 0{\cdot}25)(0{\cdot}04 - 2{\cdot}25)$$

$$= 0{\cdot}00527 + 0{\cdot}000232\eta = 0{\cdot}00527 \pm 0{\cdot}00058.$$

Example 5. *Find the value of integral*

$$\int_{30^\circ}^{90^\circ} \log_{10} \sin x \, dx$$

by Simpson's Rule taking ten sub-integrals.

Estimate the error in answer.

$x =$	30°,	36°,	42°,	48°,	54°	
$\log_{10} \sin x =$	− ·3020,	− ·2308,	− ·1745,	− ·1289,	− ·0920	
$x =$	60°,	66°,	72°,	78°,	84°,	90°
$\log_{10} \sin x =$	− ·0624,	− ·0333,	− ·01218,	− ·0096,	− ·0024,	0

Sol. Simpson's one-third rule is

$$\int_{x_0}^{x_0+nh} y\,dx = \frac{h}{3}\,[(y_0 + y_n) + 4(y_1 + y_3 + \cdots + y_{n-1})$$

$$+ 2(y_2 + y_4 + \cdots + y_{n-2})]$$

$$\therefore \quad \int_{30^\circ}^{90^\circ} \log_{10} \sin x \, dx = \int_{\pi/6}^{\pi/2} \log_{10} \sin x \, dx$$

$$= -\frac{\pi}{99}\,[(\cdot 3010 + 0) + 4(\cdot 2308 + \cdot 1289 + \cdot 0624 + \cdot 0218$$

$$+ \cdot 0024)] + 2[\cdot 1745 + \cdot 0920 + \cdot 0393 + \cdot 0096]$$

$$= -\cdot 09488, \text{ on simplification by putting } \pi = \frac{22}{7}$$

$$\text{Error term} = \frac{(b-a)h^4}{180} f^{iv}(\xi),$$

where $f(x) = \log_e \sin x \cdot \log_{10} e.$

EXERCISE 12

1. Find the maximum error in the interpolation of sin 52° by Newton's formula for forward interpolation.

2. Determine the accuracy of the interpolation by Lagrange's formula for $\log_{10} 656$.

3. Estimate the accuracy of your answer to Question 10, page 67

4. Complete by Simpson's rule the value of the integral

$$\int_{200}^{10000} \frac{dx}{\log_{10} x}$$

taking eight sub-intervals.

Estimate the inherent error in your answer.

5. Find by Weddle's rule the value of the integral

$$\int_{0\cdot4}^{1\cdot6} \frac{x}{\sinh x}\, dx$$

taking twelve sub-intervals.

Obtain the error in the evaluation of the integral.

6. Prove that

$$\int_a^b f(t)dt = \frac{b-a}{6}\left[f(a) + 4f\left(\frac{a+1}{2}\right) + f(b)\right] - \frac{(b-a)^5}{2880} f^{iv}(\xi)$$

where $a < \xi < b$.

If $|f^{iv}(x)| \le M$ for $0 \le x \le 1$, show that the error done in calculating

$$\int_0^1 f(t)\, dt \text{ by Simpson's rule } \not> \frac{M}{2880} \cdot \frac{1}{n^4}$$

the interval (0, 1) being dividing into $2n$ equal parts.

7. Determine the accuracy with which a linear interpolation based on the data of Ex. 3., page 236 would give the value of log 47.

[**Hint.** Lagrange's formula with remainder term is

$$y = \frac{x - x_1}{x_0 - x_1} y_0 + \frac{x - x_0}{x_1 - x_0} y_1 + \frac{1}{2}(x - x_0)(x - x_1)\left[\frac{d^2}{dx^2} \log x\right]_{x=\xi}$$

$\therefore$ Remainder term

$$= \frac{(47-45)(47-48)}{2} - \frac{1}{x^2}(\log_{10} e),\ 45 < \xi < 48$$

$$< \frac{0\cdot43}{45^2} = 2 \times 10^{-4}]$$

8. Obtain Euler-Maclaurin's formula of summation and show that

$$\sum_0^\infty \frac{1}{(x+s)^2} = \frac{1}{x+1} + \frac{1}{2(x+1)^2} + \frac{\theta}{6(x+1)^3},\quad 0 < \theta < 1.$$

9. Show that the error in the interpolation formula of $f(x)$ when the interpolation points are x_0, x_1, x_2 is given by

$$E_x = \frac{1}{6}(x - x_0)x(x - x_1)f'''(\xi),$$

where ξ lies in the interval limited by the largest and smallest of x_0, x_1 and x_2.

10. The following formula for approximate integration are correct to third differences,

$$\int_{-3}^{3} u_x\, dx = \frac{3}{4}(3u_{-2} + 2u_0 + 3u_2),$$

$$\int_{-3}^{3} u_x\, dx = (u_{-3} + 4u_0 + u_3).$$

Prove that if these formulas are applied to a function whose fifth differences are constant, the respective errors involved in the approximations are in the ratio 7 : 18 and are in opposite directions.

11. Prove Simpson's formula

$$\int_a^b f(x)\, dx = \frac{b-a}{3n}[f(x_0) + 4f(x_1) + 2f(x_2) + \cdots + f(x_{2n})] - E_s$$

where $x_0 = a,\ x_{2n} = b.$

[**Hint.** Prove general Simpson's $\frac{1}{3}$ rule *viz.*,

$$\int_{x_0}^{x_0+nh} f(x)\, dx = \frac{h}{3}[f(x_0) + 4f(x_0 + h) + \cdots + f(x_0 + nh)].$$

where $x_0 = a,\ x_0 + nh = b$ and then replace b by $\frac{1}{2}h$.]

12. The function $y = e^x$ is tabulated for $x = 0(0{\cdot}01)$. 1. Show that the maximal error on linear interpolation is 0·000034.

13. Show that the error associated with linear interpolation of $f(x)$, using ordinates at x_0 and x_1 with $x_0 \le x \le x_1$, is not greater than $\frac{1}{8}M(x_1 - x_0)^2$ where $M = Max.\ |f''(x)|$,

$$x_0 \le x \le x_1.$$

Hence or otherwise show that if (error function)

$$f(x) = \frac{2}{\sqrt{\pi}} \int_0^x e^{-\frac{t^2}{2}}\, dt,$$

the absolute error corresponding to the linear interpolation between x_0 and x_i cannot exceed

$$(x_1 - x_0)^2 / 2\sqrt{2\pi e}$$

and hence is smaller than $\frac{1}{8}(x_1 - x_0)^2$.

[**Hint.** If $f(x) = (x - x_0)(x - x_1) = x^2 - x(x_0 + x_1) + x_0x_1$ is given, when $x = \dfrac{x_0 + x_1}{2}$. The max. value of $f(x) = \frac{1}{8}(x_1 - x_0)^2$]

14. A table of values of the area of the standard normal curve namely $\dfrac{1}{\sqrt{2\pi}}\int_0^x e^{-\left(\frac{1}{2}\right)t^2}\,dt$ is to be constructed for values of x at an interval of h. Find the maximum value of h so that linear interpolation in the table gives an answer correct to at least four places of decimals.

15. Show that E_n = Error due to Newton-Cotes formula

$$f(x) = \int_0^{n-1} t(t-1)\ldots(t-n+1)\, f(t)dt$$

where $f^n(x)$ denotes the nth derivative of $f(x)$ and $0 < x < n - 1$.

16. Obtain the formula

$$f(x) = f(x_0) + (x - x_0) f'(x_0) + (x - x_0)^2 f(x_0, x_0, x_1)$$
$$+ (x - x_0)^2 (x - x_1) f(x_0, x_0, x_1) + R(x)$$

where $R(x) = \dfrac{1}{24}(x - x_0)^2 (x - x_1)^2 f'''(\xi)$, $(x_0 < x,\ \xi < x_1)$.

Hence show that if $x_1 - x_0 = h$, then

$$|R(x)| \le \frac{h^4}{384} \max_{x_0 \le x \le x_1} |f'''(x)|$$

(*b*) A table of the probability integral

$$y = \frac{2}{\sqrt{\pi}} \int_0^x e^{-(1/2)x^2}\, dx$$

from $x = 0$ to $x = 3$ has entries at intervals of 0·001. What is the maximum error of linear interpolation?

17. (*a*) The function $1/x$ is tabulated in Barlow's Tables at unit. Determine the maximum error of linear interpolation

(*b*) The function e^x is tabulated at intervals of 0·01 from 0 to 1. What is the maximum error of linear interpolation?

[**Hint.** (*a*) Here $f(x) = \dfrac{1}{x}$, $f''(x) = \dfrac{-2}{x^3}$, $h = 1$, $x = 650$

$$\therefore\quad |R_1| \le \frac{1}{8} h^3 M = \frac{1}{8} \cdot 1 \cdot \frac{-2}{(650)^3} = \frac{1}{1,098,500000}$$

$\therefore\quad |R_1| < 0{\cdot}00000001.$]

18. Show that corresponding to the interpolation of $f(x)$ when the interpolation points are $-h, 0, h$, the error is given by

$$R(x) = \frac{1}{6} x(x^2 - h^2) f'''(\xi) \qquad (-h < x,\ \xi < h)$$

and show also that

$\frac{h^3}{9\sqrt{3}} \max_{-h<x<h} |f'''(x)|$ is the largest possible value of $R(x)$.

19. (*a*) Determine m and hence h so that the absolute value of error in using Simpson's rule for $2m$ intervals to approximate

$$\int_0^2 \frac{dx}{1+x} \quad \text{is less than } 5 \times 10^{-5}.$$

(*b*) Determine the coefficients in the formula

$$\int_s^{2h} x^{-\frac{1}{2}} f(x)\,dx = (2h)^{1/2} [A_0 f(0) + A_1 f(h) + A_2 f(2h)] + B$$

and calculate the remainder R when $f'''(x)$, is constant.

[**Hint.** Take $f(x) = 1, x$ and x^2 and get $A_0 + A_1 + A_2 = 2$,

$$A_1 + 2A_2 = \frac{4}{3},\ A_1 + 4A_2 = \frac{8}{3} = \frac{C}{3\,!} f'''(\xi), 0 < \xi < 2h.]$$

20. Let $f(x) = \log(1 + x)$, $x_0 = 1$ and $x_1 = 1{\cdot}1$. Use linear interpolation to calculate an approximate value for $f(1{\cdot}04)$ and obtain a bound on truncation-error.

21. Determine the step-size that can be used in the tabulation of $f(x) = \sin x$ in the interval $[0, \pi/4]$ at equally spaced nodal points, so that the truncation-error of the quadratic interpolation is less the 5×10^{-8}.

[**Hint.** Let x_{i-1}, x_i, x_{i+1} denote three equi-spaced points with step-size $h \sin \theta\, |f''(x)| \le 1$, $x\, \varepsilon\, [0, \pi/4]$, the truncation error of the quadratic Lagrange interpolation is bounded by $|E_2(f, x)| \le 1/6\ |(x - x_{i-1})(x - x_i)(x - x_{i+1})|$, $x_{i-1} \le x \le x_{i+1}$. In order to find the minimum value of this expression we substitute $t = (x - x_i)/h$ and determine the minimum of $|(t-1)\,t\,(t+1)|$ $h^3/6$, $-1 \le t \le 1$. We obtain $\max_{-1 \le t \le 1} |(t-1)\,t\,(t+1)|\, h^3/6$, where

$$t(t^2 - 1) = 0 \Rightarrow t = \frac{1}{\sqrt{3}}$$

$$\therefore \text{Max}\left[t(t^2-1)\frac{h^3}{\sqrt{3}}\right]_{t=\frac{1}{\sqrt{3}}} = \frac{h^3}{9\sqrt{3}} \le 5 \times 10^{-8} \Rightarrow h \cong 0{\cdot}009.]$$

22. Find the maximum error in the interpolation of $\sin 52°$ by Newton formula for forward interpolation, gives $\sin 45° = 0{\cdot}7071$, $\sin 50° = 0{\cdot}7660$, $\sin 33° = 0{\cdot}8192$, $\sin 60° = 0{\cdot}8663$.

[**Hint.** We find $\sin 52° = 0{\cdot}788032$ by Newton's formula where

$$R_n = \frac{\Delta^{n+1} y_0}{(n+1)\,!} u(u-1)(u-2)\ldots(u-n),\ u = \frac{x - x_0}{n} = \frac{52 - 45}{5} = 1{\cdot}4,\ n = 2$$

$$R_2 = \frac{\Delta^3 y_0}{3\,!}(1{\cdot}4)({\cdot}4)(-{\cdot}06) = [-0{\cdot}0007 \times 1{\cdot}4 \times -{\cdot}4 \times -{\cdot}6]/6 = {\cdot}000392.]$$

23. Show that truncation error in Trapezoidal Rule is

$$\left(\frac{h^3}{12}\right) f'''(\eta),\ x_{n-1} < \eta < x_{n+1} \quad \text{or} \quad (b-a)\frac{h^L}{12} M$$

Hint. We have $\int_{xo}^{x_1} ydx = \int_{x_0}^{x_1}\left[y_0 + \frac{(x-x_0)}{1!}y_0' + \frac{(x-x_0)^2}{2!}y_0''\right.$

$$=\left[y_0 x + \frac{(x-x_0)^2}{2!}y_0' \mid \frac{(x-x_0)^3}{3!}y_0'' + -\right]_{x_o}^{x_1}$$

$$= y_0(x-x_1) + \frac{(x_1-x_0)^2}{2!}y_0' + \frac{(x_1-x_0)^3}{3!}y_0'''$$

$$= hy_0 + \frac{h^2}{2!}y_0' + \frac{h^3}{3!}y_0'' + - . \;, \quad h = \text{internal of differencing} \qquad (i)$$

But $A_0 = \int_{x_0}^{x_1} yd_x = \frac{h}{2}(y_0 + y_1)$ by Trapezoidal Rule

and $\quad y_1 == y_0 + \frac{h}{1!}y_0' + \frac{h^2}{2!}y_0'' + -$

$$\therefore \quad A_0 = \frac{h}{2}\left(2y_0 + \frac{h}{1!}y_0' + \frac{h^2}{2!}y_0'' + -\right) \qquad ...(2)$$

Now Error in the first interval (x_0, x_1) is given by ...(2)

$$\therefore \text{ Truncation Error } = \int_{y_0}^{y_1} y\,dx - A_0 = h^3 y_0''\left(\frac{1}{3} - \frac{1}{4}\right) = -\frac{1}{12}h^3 y_0'' + -$$

Similarly, Error in the with interval $= 1\frac{1}{12}h^3 y_0'' + ...$

$$\therefore \quad E = \text{total error} = -\frac{1}{12}h^3\left(y_0'' + y_1'' + ... y_{n\_1}''\right)$$

Hence $| E | < \frac{nh^3}{12} M$ where $M = \max \; (1y_0'' |, y_1'' |, ... \;] = \max\limits_{1=<} | y_n'')$

Cor. If the interval is (a, b), then $nh = b - a$ and,

$$\therefore \; | E | < (b-a)\frac{h^2}{12}M.$$

PART - II

NUMERICAL ANALYSIS

13

EIGEN VALUES AND EIGEN VECTORS OF A MATRIX

13·1. Introduction. Matrices play an important role in funding the solution of linear simultaneous equations, difference equations, etc.

13·2. Characteristic equation of a matrix. Let $A = (a_{ij})$ be a square matrix of order n and λ be a number. The determinantal equation

$$\phi(\lambda) = |A - \lambda I| = \begin{vmatrix} a_{11} - \lambda & a_{12} & \dots & a_{1n} \\ a_{21} & a_{22} - \lambda & \dots & a_{2n} \\ \vdots & & & \vdots \\ a_{n1} & a_{n2} & \dots & a_{nn} - \lambda \end{vmatrix} = 0 \qquad \text{...(1)}$$

is called the characteristic equation of A and is clearly of degree n in λ. The roots of this equation are called the characteristic roots or latent roots or the eigen values of A. A square matrix of order n has always n eigen values, some or all of them may coincide. From (1), we have

$(a_{11} - \lambda)(a_{22} - \lambda)\dots(a_{nn} - \lambda)$ + terms containing λ to at most $(n-2)$ power = 0.

$$\Rightarrow \quad \lambda^n - (a_{11} + a_{22} + \dots + a_{nn})\lambda^{n-1} + \dots + (-1)^n \det. A = 0$$

$$\text{Thus} \quad \sum_{i=1}^{n} \lambda_i = a_{11} + \dots + a_{nn} = \text{trace } A$$

$$\text{and} \quad \prod_{i=1}^{n} \lambda_i = \text{determinant } A = |A|.$$

Consider the following set of n homogeneous equations in n variables $x_1, x_2, \dots, x_n$:

$$\Rightarrow \quad \begin{matrix} (A - \lambda I)X = 0 \\ \text{or} \quad AX = \lambda X \end{matrix} = \begin{bmatrix} a_{11} - \lambda & \dots & a_{1n} \\ a_{21} & \dots & a_{2n} \\ \dots & \dots & \dots \\ a_{n1} & \dots & a_{nn} - \lambda \end{bmatrix} \begin{bmatrix} x_1 \\ x_2 \\ \dots \\ x_n \end{bmatrix} = \begin{bmatrix} 0 \\ 0 \\ \dots \\ 0 \end{bmatrix} \qquad \begin{matrix} \text{...(2)} \\ \\ \\ \text{...(3)} \end{matrix}$$

These equations will have a non-trivial solution iff

$$|A - \lambda I| = 0$$

i.e., iff λ is a root of the characteristic equation of A, the values of λ may be distinct or repeated.

The non-zero column vector X satisfying (2) is called the characteristic or eigen vector of A. Corresponding to each of the eigen value, there is a eigen vector of A. Thought the characteristic roots or eigen values of A and A' are same, their eigen vectors will be different unless A is symmetric. If X is a non-zero solution of (2), $kX, k \neq 0$ is also so. Hence (2) defines a eigen vector irrespective of its length. It is therefore customary to normalize the length of an eigen vector to unity. If λ is a distinct eigen value, the eigen vector X corresponding to it is independent of the others.

Premultiplying both sides of (2) by A,

$$A^2X = A(\lambda X) = \lambda(AX) = \lambda(\lambda X) = \lambda^2 X$$

and so on $A^m X = \lambda^m X$, m being positive integer.

Thus if A has eigen values $\lambda_1, ..., \lambda_n$, then A^m has eigen values $\lambda_1^m, ..., \lambda_n^m$ but A and A^m have the same eigen vectors.

Important properties of eigen values and eigen vectors of a matrix.

1. If $\lambda_1, \lambda_2, ..., \lambda_n$ are the eigen values of the square matrix A of order n, then the eigen values of $I - A$ are $1 - \lambda_1, 1 - \lambda_2, ..., 1 - \lambda_n$ and the eigen values of $A - kI$ are $\lambda_1 - k, \lambda_2 - k, ..., \lambda_n - k$, k being scalar.
2. The eigen values of a Hermitian matrix are real.
3. The eigen values of the polynomial $\phi(A)$ are $\phi(\lambda_1), \phi(\lambda_2), ..., \phi(\lambda_n)$.
4. A and $P^{-1}AP$ have the same eigen values where P is a non-singular matrix. A and P are called similar matrix.
5. All the eigen values and eigen vectors of a real symmetric matrix are real.
6. If A is a real symmetric matrix, then the eigen vectors corresponding to different eigen values are orthogonal.
7. If A is a real symmetric matrix of order n, then there exists an orthogonal matrix P of order n such that

$$P'AP = \text{diag}(\lambda_1, \lambda_2, ..., \lambda_n).$$

8. If an eigen value λ_i of a symmetric matrix A of order n has multiplicity $k\ (\geq 2)$, then there exists k orthonormal eigen vectors with eigen value λ_i.
9. Upper bounds for eigen values of a matrix

$$A = [a_{ij}],\ i,\ j = 1, 2, ..., n$$

$$|a_{ii} - \lambda| \leq |a_{i1}| + |a_{i2}| + \cdots |a_{i,i-1}| + |a_{i,i+1}| + \cdots + |a_{i,n}|$$

$$= \sum_{\substack{i=1 \\ j=i}}^{n} |a_{ij}|$$

Thus the eigen values of A lie in the union of the circles

$$|z - a_{ii}| \leq \sum_{i=1,\, j \neq i}^{n} |a_{ij}|, \quad 1 \leq i \leq n \quad \text{[Gersgorian circle]}$$

in the complex plane.

10. Every matrix satisfies its own characteristic equation.

Example 1. *Find the eigen values of the matrix*

$$A = \begin{bmatrix} 2 & -1 & -1 \\ 1 & 0 & -1 \\ -1 & 1 & 2 \end{bmatrix}$$

and verify that A satisfies its characteristic equation. Use this fact to find A^{-1}.

Sol. We have

$$|A - \lambda I| = \begin{bmatrix} 2-\lambda & -1 & -1 \\ 1 & -\lambda & -1 \\ -1 & 1 & 2-\lambda \end{bmatrix} = 2 - 5\lambda + 4\lambda^2 - \lambda^3$$

The eigen values are the roots of the equation $\lambda^3 - 4\lambda^2 + 5\lambda - 2 = 0$. Thus $\lambda = 2, 1, -1$.

$\Rightarrow$ the characteristic function is $2 - 5T + 4T^2 - T^3$

where T is the linear vector function corresponding to A.

Thus $2I - 5A + 4A^2 - A^3$

$$= \begin{bmatrix} 2 & 0 & 0 \\ 0 & 2 & 0 \\ 0 & 0 & 2 \end{bmatrix} - \begin{bmatrix} 10 & -5 & -5 \\ 5 & 0 & -5 \\ -5 & 5 & 10 \end{bmatrix}$$

$$+ \begin{bmatrix} 16 & -12 & -12 \\ 12 & -8 & -12 \\ -12 & 12 & 16 \end{bmatrix} - \begin{bmatrix} 8 & -7 & -7 \\ 7 & -6 & -7 \\ -7 & 7 & 8 \end{bmatrix}$$

$$= \begin{bmatrix} 0 & 0 & 0 \\ 0 & 0 & 0 \\ 0 & 0 & 0 \end{bmatrix}$$

Hence A satisfies its characteristic equation.

Also $2I = 5A - 4A^2 + A^3 = A(5I - 4A + A^2)$

Thus $A^{-1} = \frac{1}{2}(5I - 4A + A^2)$

$$= \frac{1}{2}\left(\begin{bmatrix} 5 & 0 & 0 \\ 0 & 5 & 0 \\ 0 & 0 & 0 \end{bmatrix} - \begin{bmatrix} 8 & -4 & 4 \\ 4 & 0 & -4 \\ -4 & 4 & 8 \end{bmatrix} + \begin{bmatrix} 4 & -3 & -3 \\ 3 & -2 & -3 \\ -3 & 3 & 4 \end{bmatrix}\right)$$

$$= \begin{bmatrix} \frac{1}{2} & \frac{1}{2} & \frac{1}{2} \\ -\frac{1}{2} & \frac{3}{2} & \frac{1}{2} \\ \frac{1}{2} & -\frac{1}{2} & \frac{1}{2} \end{bmatrix}$$

Example 2. *Obtain the eigen values and eigen vectors of the symmetric matrix.*

$$A = \begin{bmatrix} 0 & 1 & 1 \\ 1 & 0 & 1 \\ 1 & 1 & 0 \end{bmatrix}$$

Sol. $|A - \lambda I| = 0$ gives $\begin{vmatrix} -\lambda & 1 & 1 \\ 1 & -\lambda & 1 \\ 1 & 1 & -\lambda \end{vmatrix} = 0$

$\Rightarrow \quad (\lambda + 1)^2(\lambda - 2) = 0$

Hence eigen values are $\lambda = -1, -1, 2.$

Case I. $\lambda = 2$

The corresponding eigen vector is given by

$$(A - 2I)\,X = 0$$

$$\Rightarrow \quad -2x + y + z = 0$$
$$x - 2y + z = 0$$
$$x + y - 2x = 0$$

Solving, we get $x = y = z = k$, say.

Thus $X_1 = [k, k, k]'$

Case II. $\lambda = 1, -1$

The corresponding eigen vector is given by

$(A - (-1)I)\,X = 0$ *i.e.,* $(A + I)\,X = 0$

$\Rightarrow \quad x = y = z = 0$ (All three equations are equivalent)

The trial solutions are

$$[0, 1, -1] \text{ and } [1 + k, -1, -k]$$

Since these are orthogonal, we have

$$1 + k + k = 0 \Rightarrow k = -\tfrac{1}{2}$$

Hence $X_2 = \left[\frac{1}{\sqrt{2}}, 0, \frac{-1}{\sqrt{2}}\right]$, $X_3 = \left[\frac{1}{\sqrt{6}}, \frac{-2}{\sqrt{6}}, \frac{1}{\sqrt{6}}\right]$

Example 3. *Find a set of three orthonormal eigen vectors for the matrix*

$$A = \begin{bmatrix} 3 & 0 & 0 \\ 0 & 4 & \sqrt{3} \\ 0 & \sqrt{3} & 6 \end{bmatrix}$$

Sol. We have $\varphi(\lambda) = |A - \lambda I| = \begin{vmatrix} 3-\lambda & 0 & 0 \\ 0 & 4-\lambda & \sqrt{3} \\ 0 & \sqrt{3} & 6-\lambda \end{vmatrix}$

$$= (3-\lambda)^2 (\lambda - 7)$$

Thus the eigen values are 3, 3, 7.

Case I. $\lambda = 7$.

$$(A - 7I)X = 0$$

$$\Rightarrow \quad -4x_1 = 0$$

$$3x_2 + \sqrt{3}\,x_3 = 0$$

$$\sqrt{3}\,x_2 - x_3 = 0$$

Hence $\quad x_1 = 0, x_2 = \frac{1}{\sqrt{3}} x_3,$

Let $\quad x_1^2 + x_2^2 + x_3^2 =$ so that

$$x_1 = 0, x_2 = \frac{1}{2}, \; x_3 = \frac{\sqrt{3}}{2}.$$

Thus the eigen vector of unit length with eigen value 7 is

$$\left[0, \frac{1}{2}, \frac{\sqrt{3}}{3}\right].$$

Case II. $\lambda = 3$ (twice)

$$(A - 3I)X = 0$$

$$\Rightarrow \quad x_2 + \sqrt{3}\,x_3 = 0$$

$$\sqrt{3}\,x_2 + 3x_3 = 0$$

i.e., $\quad x_3 = -\frac{1}{\sqrt{3}} x_2$

Thus x_1 is arbitrary, $x_2, x_3 \left(= -\frac{1}{\sqrt{3}} x_2 \right)$

If $x_1^2 + x_2^2 + x_3^2 = 1$, then $x_1^2 + x_2^2 + \frac{1}{3} x_2^2 = 1$

Take $x_1 = 0$ so that $x_2 = \frac{\sqrt{3}}{2}, \; x_3 = -\frac{1}{2}.$

Thus the eigen vector of unit length with eigen value 3 is

$$\left[0, \frac{\sqrt{3}}{2}, \frac{-1}{2}\right]$$

We now find another eigen vector with eigen value 3 which is orthogonal to the above eigen vector,

$$\frac{\sqrt{3}}{2} x_2 - x_3 = 0$$

$$\Rightarrow \qquad x_3 = \sqrt{3}\, x_2.$$

Thus we have $x_2 = 0, x_3 = 0$. But $x_1^2 + x_2^2 + x_3^2 = 1$ so that $x_1 = \pm 1$. Thus the other eigen vector with eigen value 3 is

$$[1, 0, 0]'.$$

13·3. Quadratic Form. The expression

$$Q(x_1, x_2, ..., x_n) = \sum_{i, j = 1}^{n} a_{ij} x_i x_j \text{ with } a_{ji} = a_{ij} \qquad ...(1)$$

is called a quadratic form in the n variables $x_1, x_2, ..., x_n$. Clearly $Q(x_1, x_2, ..., x_n)$ is a homogeneous polynomial of the second degree in the n variables $x_1, x_2, ..., x_n$. The expression (1) can be put in the matrix notation as

$$Q = Q(x_1, x_2, ..., x_n) = X' AX$$

where $\quad A = (a_{ij}), X' = [x_1, x_2, ..., x_n],$

since $\quad Q = \sum_{i=1}^{n} x_i \sum_{j=1}^{n} a_{ij} x_j = \sum_{i=1}^{n} x_i (AX)_i$

and is said to be the matrix associated with the quadratic form.

Further, the expression (1) is said to be *positive definite* of Q and is always positive for all real values of X and is zero only when $X = 0$.

If $A' = A$, then the expression (1) can be expressed in the product form as

$$Q = (AX, X)$$

The eigen values of A are called eigen values of the quadratic form $Q(x_1, x_2, ..., x_n)$. It should be noted that

(1) All the eigen values of a real quadratic form are real.

(2) If the quadratic form of a matrix is postive definite, then the eigen values of the matrix are positive.

(3) The eigen values of a positive definite Hermitian form are positive.

13·4. Limit and Convergence of Sequence of Matrices. Let A be a square matrix of order n and let $\lambda_1, \lambda_2, ... \lambda_n$ be n distinct eigen values of A. Then

$$\lim_{n \to \infty} A^n = 0$$

if, and only if, each $|\lambda_i| < 1$.

Ex. *If* $A = \begin{bmatrix} 0{\cdot}9 & 0{\cdot}2 \\ 0{\cdot}3 & 0{\cdot}4 \end{bmatrix}$, *show that*

$$\lim_{n \to \infty} A^n = \begin{pmatrix} \frac{6}{7} & \frac{2}{7} \\ \frac{3}{7} & \frac{1}{7} \end{pmatrix}.$$

Norm of a matrix.

The norm of a square matrix A is a non-negative real number $\| A \|$ fulfilling the following conditions :

(1) $\| A \| > 0$ if $A \neq 0$, $\| A = 0$. if, and only if, $a = 0$

(2) $\| \alpha A \| = | \alpha | \| A \|$, α being scalar

(3) $\| A + B \| \leq \| A \| + \| B |$,

(4) $\| AB | \leq \| A \| B \|$.

It should be noted that eigen values of a matrix cannot exceed the norm of the matrix.

13·5. Reduction of a Matrix to the tridiagonal form. Let $A = (a_{ij})$ with $a_{13} = a_{14} = \cdots = a_{1k-1} = 0$ and $a_{1k} \neq 0$ for $3 \leq k \leq n$.

Set $\quad s = r_{12}(r_{12}{}^2 + \cdots + r_{1k}{}^2)^{-\frac{1}{2}}$

and $\quad t = r_{1k}(r_{12}{}^2 + \cdots + r_{1k}{}^2)^{-\frac{1}{2}}$.

so that $\quad s^2 + t^2 = 1$

and $\quad sr_{1k} - tr_{12} = 0$.

$$\text{choose } A_{\alpha,\beta} = \begin{matrix} & (\alpha, \alpha) & & (\alpha, \beta) \\ & s & \cdots & -t \\ & \vdots & 1 & \vdots \\ & (\beta, \alpha) & & (\beta, \beta) \\ & t & \cdots & s \end{matrix}$$

The matrix $A_{\alpha,\beta}$

(*i*) s occurs in (α, α) and (β, β) positions

(*ii*) t occurs in (β, α) and $-t$ in (α, β) positions

(*iii*) 1 occurs in the rest of the leading diagonal positions and 0 elsewhere.

We want the 'reduced' matrix in the tri-diagonal form having not all entries equal to zero with other entries (equal to zero) in the leading diagonal as well as in one diagonal just above the leading diagonal and one diagonal just below. This can be achieved if we post-multiply $A_{\alpha,\beta}$ by $A_{2,3} A_{2,4} \ldots A_{2,n}$ and pre-multiply by $A'_{2,3} A'_{2,4} \ldots A'_{2,n}$.

Example. *Reduce to the tri-diagonal form the matrix*

$$A=\begin{bmatrix} 1 & 2 & 2 \\ 2 & 1 & 2 \\ 2 & 2 & 1 \end{bmatrix}.$$

Sol. Since $a_{13}=2$, we choose $s-2(4+4)^{-\frac{1}{2}}=2(8)^{-\frac{1}{2}}$,

$$t=2(4+4)^{-\frac{1}{2}}=2(8)^{-\frac{1}{2}}$$

so that $$s^2+t^2=1.$$

Set $$A_{2,3}=\begin{bmatrix} 1 & 0 & 0 \\ 0 & s & t \\ 0 & t & s \end{bmatrix}=\begin{bmatrix} 1 & 0 & 0 \\ 0 & \frac{2}{\sqrt{8}} & -\frac{2}{\sqrt{8}} \\ 0 & \frac{2}{\sqrt{8}} & \frac{2}{\sqrt{8}} \end{bmatrix}$$

$\therefore$ $$A\,A_{2,3}=\begin{bmatrix} 1 & 2 & 2 \\ 2 & 1 & 2 \\ 2 & 2 & 1 \end{bmatrix}\begin{bmatrix} 1 & 0 & 0 \\ 0 & \frac{2}{\sqrt{8}} & \frac{-2}{\sqrt{8}} \\ 0 & \frac{2}{\sqrt{8}} & \frac{2}{\sqrt{8}} \end{bmatrix}$$

$$=\begin{pmatrix} 1 & \frac{8}{\sqrt{8}} & 0 \\ 2 & \frac{4}{\sqrt{8}} & \frac{2}{\sqrt{8}} \\ 2 & \frac{5}{\sqrt{8}} & \frac{-2}{\sqrt{8}} \end{pmatrix}$$

Hence $A'_{2,3}\,A\,A_{2,3}$

$$=\begin{pmatrix} 1 & 0 & 0 \\ 0 & \frac{2}{\sqrt{8}} & \frac{2}{\sqrt{8}} \\ 0 & \frac{-2}{\sqrt{8}} & \frac{2}{\sqrt{8}} \end{pmatrix}\times\begin{pmatrix} 1 & \frac{8}{\sqrt{8}} & 0 \\ 2 & \frac{4}{\sqrt{8}} & \frac{2}{\sqrt{8}} \\ 2 & \frac{5}{\sqrt{8}} & \frac{-2}{\sqrt{8}} \end{pmatrix}=\text{etc.}$$

Consequently A is reduced to the tridiagonal form.

VARIOUS METHODS FOR DETERMINING EIGEN VALUES AND EIGEN VECTORS

13·6. (*a*) Iteration Method. A standard procedure for computing approximate values of the eigen values of an m-rowed square matrix $A=(a_{ij})$ is the iteration method for eigen values.

In the method we start from any non-zero vector x_0 with n components and compute successively

$$x_1 = Ax_0, x_2 = Ax_1, ..., x_s = Ax_{s-1}.$$

For simplifying notation, we denote x_{s-1} by x and x_s by y so that

$$y = Ax$$

We now state a theorem, without proof which gives an approximation and error bounds.

Theorem. Let A be an n-rowed real symmetric matrix. Let $x \neq 0$ be any real vector with n components. Furthermore, let

$$y = Ax, m_0 = x'x, m_1 = x'y, m_2 = y'y$$

Then the quotient $\quad q = m_1/m_0 \quad$ **(Rayleigh quotient)**

is an approximation for eigenvalue λ of A and if we set $q = \lambda + \varepsilon$ that ε is the error of q_1 then

$$|\varepsilon| \leq \sqrt{\frac{m_1}{m_0} - q^2}$$

Example 1. *Let* $A = \begin{pmatrix} 8 & 1 & 1 \\ 1 & 5 & 2 \\ 1 & 2 & 5 \end{pmatrix}$. *Choose* $x_0 = \begin{pmatrix} 1 \\ 1 \\ 1 \end{pmatrix}$

Sol. Then we obtain successively

$$x_1 = \begin{pmatrix} 10 \\ 8 \\ 8 \end{pmatrix}, x_2 = \begin{pmatrix} 96 \\ 66 \\ 66 \end{pmatrix}, x_3 = \begin{pmatrix} 900 \\ 558 \\ 558 \end{pmatrix}, x_4 = \begin{pmatrix} 8316 \\ 4806 \\ 4806 \end{pmatrix}$$

Let $A = (a_{ij})$ be an n-rowed square matrix and let $\lambda_1, \lambda_2, ..., \lambda_n$ be its eigen values. Then

$$\sum_{k=1}^{n} |\lambda_k| \leq \sum_{i=1}^{n} \sum_{j=1}^{n} |a_{ij}|^2 \qquad \textbf{[Schur's inequality]}$$

The equality sign holds iff A is such that

$$A^{-1} A = AA^{-1}$$

Matrices which satisfy the above conditions are normal matrix. It can be easily proved that the Hermitian, skew-Hermitian, unitary, skew-symmetric and orthogoal matrices are normal. Let λ_n be any eigen value of the matrix A. Then $|\lambda_m|^2 <$ the sum of the left side of Schur, inequality and by taking square roots we obtain,

$$|\lambda_m| \leq \sqrt{\sum_{i=1}^{n} \sum_{j=1}^{n} \left| \begin{matrix} a \\ ij \end{matrix} \right|^2}$$

Example 2. *For the matrix*

$$A = \begin{pmatrix} 26 & -2 & 2 \\ 2 & 21 & 4 \\ 4 & 2 & 28 \end{pmatrix}$$

determine the eigen value of λ.

Sol. $30^2 + 25^2 + 20^2 = 1925 < 1936 \ (= 44^2)$

In fact, A is not normal.

$$\therefore \qquad q = m_1/m_0 = 8{\cdot}967 \quad \text{and}$$

$$\varepsilon \leq \sqrt{\frac{m}{m_0} - q^2} = 0{\cdot}311.$$

This shows that $q = 8{\cdot}967$ is an approximation for an eigen value which must lie between 8·656 and 9·278.

In fact, the eigen value of $\lambda = 9$.

Ex. *Using* $x_0' = (3\ 1\ 2)$ *and applying iteration to*

$$A = \begin{pmatrix} -2 & -1 & 4 \\ 2 & 1 & -2 \\ -1 & -1 & 3 \end{pmatrix}$$

show that $A^2 x_0 = x_0$ *and draw conclusion from this result.*

Ans. A has the eigen value 1 or -1 or both,

(*b*) **Power Series Method for finding dominant latent root and Eigen Vectors**. This method is advatageous as it enables us to determine the Eigen Values as well as the Eigen Vectors simultaneously of the given non-singular matrix *A*.

Theorem. If square matrix *A* has a single *dominant latent* roots, then these multipliers tend to this root and the normalized vectors tend to the corresponding Eigen Vectors.

Proof. Let $\lambda_i, i = 1, 2, .., n$ be n distinct eigen values such that $|\lambda_1| \geq |\lambda_2| \geq \cdots \geq |\lambda_n|$ so that λ_1 is the dominant latent root and further let c_i be the corresponding n linearly independent eigen values.

Let $v^{(i)}$ be any arbitrary vector which is normalized. Then we define $\mu^{(i+1)}$ as given by

$$A\, v^{(i)} = \mu^{(i+1)}\, v^{(i+1)}$$

where $v^{(i+1)}$ is normalized.

Now clearly $v^{(c)} = \sum_{i=1}^{n} \alpha_i\, c_i$

Also since λ_i is the eigen value, we have

$$A\, c_i = \lambda_i\, c_i,\ i = 1, 2, \ldots, n$$

$$\therefore \quad v^{(n)} = A v^{(n-1)} = A^2 v^{(n-2)} = \cdots = A^n\, v^{(0)}$$

$$\Rightarrow \quad v^{(n)} = \sum_{i=1}^{n} \alpha_i\, \lambda_i^n\, c_i = \alpha_1\, \lambda_1^n\, c_1 + \alpha_2\, \lambda_2^n\, c_2 + \cdots + \alpha_n\, \lambda_n^n\, c_n$$

$$\therefore \quad v^{(n)} \simeq \alpha_1\, \lambda_1^n\, c_1 \simeq \text{constant} \times c_1.$$

as λ_1 is the largest eigen value and n is sufficiently large so that $(\lambda_1/\lambda)^n \to 0$.

Again $\quad A v^{(i)} = \mu^{(i+1)}\, v^{(i+1)}$

$$\Rightarrow \quad \mu^{(i+2)}\, v^{(i+1)} = A\, v^{(i)} = A^{(i)}\, v^{(0)} \simeq \alpha_1 \lambda_1^{\,i} c_1$$

Also $\quad \mu^{(i+2)}\, v^{(i+2)} = A\, v^{(i+1)} = A^{(i+1)} = A^{(i+1)}\, v^{(0)} \simeq \alpha_1 \lambda_1^{i+1} c_1$

$\therefore$ Component of $v^{(n)}$ = (Component of v^{n-1}) λ_1

Thus the result is established.

Note. The eigen vector $x = \begin{bmatrix} x_1 \\ x_2 \\ x_3 \end{bmatrix}$ or $[x_1\ x_2\ x_3]'$ is called *normalized* or *orthonormal vector if* $x_1^2 + x_2^2 + x_3^2 = 1$

Example. *Determine dominant latent root and corresponding eigen vector of the following matrix.*

$$A = \begin{bmatrix} 1 & 6 & 1 \\ 1 & 2 & 0 \\ 0 & 0 & 3 \end{bmatrix}, \text{ taking } v_0 = \begin{bmatrix} 1 \\ 0 \\ 0 \end{bmatrix}$$

Ans. $\lambda_1 = 4$ and $v_1 = \begin{bmatrix} 1 \\ -5 \\ 0 \end{bmatrix}$

(*c*) **Rayleigh Quotient-Method.** Let A be any real, symmetric matrix having a dominant characteristic root λ_0 with x_0 as the corresponding characteristic vector.

Then the Rayleigh quotient for the matrix A is defined by

$$R\,(x) = \frac{x' A x}{x' x}$$

It can be shown that x_0 is a solution corresponding to λ_0, that is

$$R(x_0) = \lambda_n.$$

Rule for finding the Characteristic or Eigen Value or Latent Root Vector.

Let η_0 be an initial approximation to eigen vector.

Assume that $R(x_0) = \lambda_0$

Then $A\eta_0 - R(\eta_0)\eta_0 = A\,\eta_0 - \lambda_0\,\eta_0$

$$\simeq 0, \text{ say}$$

Then $\eta_0 \neq 0$, then we solve the equation

$$A\,\eta = R(\eta_0)\,\eta_0$$

and get the approximate solution η_1.

The process is repeated till the desired result is obtained.

Example. *Find the dominant latent root and the corresponding eigen vector of the following matrix.*

$$A = \begin{bmatrix} 1 & 6 & 1 \\ 1 & 2 & 0 \\ 0 & 0 & 3 \end{bmatrix}, \text{ taking initial eigen vector } Y_0 = \begin{bmatrix} 1 \\ 0 \\ 0 \end{bmatrix}$$

(M.Sc. Rohilkhand, 1988)

[**Hint.** $Y_1 = Ay_0 = \begin{bmatrix} 1 & 6 & 1 \\ 1 & 2 & 0 \\ 0 & 0 & 3 \end{bmatrix}\begin{bmatrix} 1 \\ 0 \\ 0 \end{bmatrix} = \begin{bmatrix} 1 \\ 1 \\ 0 \end{bmatrix} = 1\begin{bmatrix} 1 \\ 1 \\ 0 \end{bmatrix} = k_1 z_1, k_1 = 1$

$$Y_2 = Az_1 = \begin{bmatrix} 1 & 6 & 1 \\ 1 & 2 & 0 \\ 0 & 0 & 3 \end{bmatrix}\begin{bmatrix} 1 \\ 1 \\ 0 \end{bmatrix} = \begin{bmatrix} 7 \\ 3 \\ 0 \end{bmatrix} = 7\begin{bmatrix} 1 \\ \cdot 4 \\ 0 \end{bmatrix} = k_2 z_2, k_2 = 7$$

$$Y_3 = Az_3 = \begin{bmatrix} 1 & 6 & 1 \\ 1 & 2 & 0 \\ 0 & 0 & 3 \end{bmatrix}\begin{bmatrix} 1 \\ 4 \\ 0 \end{bmatrix} = \begin{bmatrix} 3\cdot 4 \\ 1\cdot 8 \\ 0 \end{bmatrix} = 3\cdot 4\begin{bmatrix} 1 \\ \cdot 52 \\ 0 \end{bmatrix}$$

$$= k_3 z_3, k_3 = 3\cdot 4$$

$$Y_4 = Az_3 = \begin{bmatrix} 1 & 6 & 1 \\ 1 & 2 & 0 \\ 0 & 0 & 3 \end{bmatrix}\begin{bmatrix} 1 \\ \cdot 52 \\ 0 \end{bmatrix} = \begin{bmatrix} 4\cdot 12 \\ 2\cdot 04 \\ 0 \end{bmatrix} = 4\cdot 12\begin{bmatrix} 1 \\ \cdot 49 \\ 0 \end{bmatrix}$$

$$= k_4 z_4, k_4 = 4\cdot 12$$

$$Y_5 = Az_4 = \begin{bmatrix} 1 & 6 & 1 \\ 1 & 2 & 0 \\ 0 & 0 & 3 \end{bmatrix}\begin{bmatrix} 1 \\ \cdot 49 \\ 0 \end{bmatrix} = \begin{bmatrix} 3\cdot 94 \\ 1\cdot 98 \\ 0 \end{bmatrix} = 3\cdot 94\begin{bmatrix} 1 \\ \cdot 50 \\ 0 \end{bmatrix}$$

$$= k_5 z_4, k_5 = 3\cdot 94$$

$$Y_6 = Az_5 = \begin{bmatrix} 1 & 6 & 1 \\ 1 & 2 & 0 \\ 0 & 0 & 3 \end{bmatrix} \begin{bmatrix} 1 \\ \cdot 50 \\ 0 \end{bmatrix} = \begin{bmatrix} 4 \\ 2 \\ 0 \end{bmatrix} = 4 \begin{bmatrix} 1 \\ \cdot 5 \\ 0 \end{bmatrix} = k_6 z_5, \; k_6 = 4$$

Hence the largest eigen value is 4 and $[1 \cdot 50]'$ is the corresponding eigen vector.

Ex. 1. *Determine the largest eigen value and eigen vector of the matrix,*

(*i*) $\begin{bmatrix} 1 & 3 & -1 \\ 3 & 2 & 4 \\ -1 & 4 & 10 \end{bmatrix}$ (*ii*) $\begin{bmatrix} 5 & 0 & 1 \\ 0 & -2 & 0 \\ 1 & 0 & 5 \end{bmatrix}$

Ans. (*ii*) 2, 4, 6

Ex. 2. *Determine the* largest eigen value and eigen vector of the matrix

$$A = \begin{bmatrix} 1 & 3 & -1 \\ 3 & 2 & 4 \\ -1 & 4 & 10 \end{bmatrix}.$$

Obtain the eigen vectors corresponding to the eigen values $-1, 2$ of the following matrix.

$$A = \begin{bmatrix} 8 & -12 & 5 \\ 15 & -25 & 11 \\ 24 & -42 & 19 \end{bmatrix}$$

Take $\eta_0 = \begin{bmatrix} 1 \\ 1 \\ 1 \end{bmatrix}$ as the initial eigen vector.

Ex 3. *Find the eigen values and eigen vectors of the matrix A.*

(*i*) $\begin{bmatrix} 5 & 0 & 1 \\ 0 & -2 & 0 \\ 1 & 0 & 5 \end{bmatrix}$ (*ii*) $\begin{bmatrix} 5 & -6 & -6 \\ -1 & 4 & 2 \\ 3 & -6 & -4 \end{bmatrix}$

Ans. (*i*) eigen values : $-2, 4, 6$

eigen vector corresponding to $\lambda = -2$ is $\begin{bmatrix} 0 \\ 1 \\ 0 \end{bmatrix}$

(*ii*) eigen values : 1, 2, 2

eigen vector corresponding to $\lambda = 1$ and $\begin{bmatrix} 3 \\ -1 \\ 3 \end{bmatrix}$

eigen vector corresponding to $\lambda = 2$ are $\begin{bmatrix} 2 \\ 1 \\ 0 \end{bmatrix}$ and $\begin{bmatrix} 2 \\ 0 \\ 1 \end{bmatrix}$

Ex. 4. *Determine the largest eigen value and the corresponding eigen vector of the matrices :*

(*i*) $\begin{bmatrix} 10 & -2 & 1 \\ -2 & 10 & -2 \\ 1 & -2 & 10 \end{bmatrix}$ (*ii*) $\begin{bmatrix} 1 & 3 & -1 \\ 3 & 2 & 4 \\ -1 & 4 & 10 \end{bmatrix}$

Ans. (*i*) $\lambda = 9, \begin{bmatrix} 1 \\ 0 \\ -1 \end{bmatrix}$ (*ii*) $11{\cdot}66, \begin{bmatrix} 0{\cdot}025 \\ 0{\cdot}422 \\ 1{\cdot}0 \end{bmatrix}$

Ex. 5. *Choose the correct alternative.*

The matrix A is singular. Then

(*i*) $\lambda = 0$ *can not be an eigen value of A.*

(*ii*) $\lambda = 1$ *is an eigen value of A.*

(*iii*) $\lambda = 0$ *is an eigen value of A.*

(*iv*) $\lambda = 0, 1$ *are eigen value of A.*

Ans. (*iii*) $\lambda = 0$ is an eigen value of A.

Ex. 6. Find the dominant eigen value of (i) $A = \begin{bmatrix} 1 & 2 \\ 3 & 4 \end{bmatrix}$

(*ii*) $A = \begin{bmatrix} 1 & 6 & 1 \\ 1 & 2 & 0 \\ 0 & 0 & 3 \end{bmatrix}$ by power method and hence find the other eizen values.

Ans. A = 5.3723, eigh vector $x_1 = (0.4574)\ y_1 + y_2$ = Trace of A = 1 + 4 = 5 in Second eigen value = $\lambda = -0.3723$

(*b*) Dominant eigen value = 4. Then eigen value and 3.1.

14

NUMERICAL ROOTS OF POLYNOMIALS AND TRANSCENDENTAL EQUATIONS IN ONE VARIABLE

14·1. Introduction. Polynomial. An expression of the forr

$$P_n(x) = a_0x^n + a_1x^{n-1} + a_2x^{n-2} + \cdots + a_{n-1}x + a_n$$

where $a_0, a_1, ..., a_n$ are constants, independent of x, and n is a positive integer, is called a polynomial in x of *degree n*, provided $a_0 \neq 0$. Any value of x which makes $P_n(x)$ equal to zero is called a *zero* of the polynomial $P_n(x)$. Every polynomial $P_n(x)$ of degree n has exactly n zeros. Geometrically, the zero of the polynomial equation $P_n(x)$ is the value of x where the graph of $P_n(x)$ crosses the axis of x.

Numerical methods for so bring algebrair equations can be divided into two types, direct and interative, the former are those which yield the exact solution in a finite number of elementary arithmetic operations. The latter method involves the assumption of an initial approximation and finding a suitable algorithm approach to successively better approximation

14·2·1. Synthetic division by a linear expression. Consider the division of the polynomial $P_n(x)$ by the linear expression $x - h$. Let $Q(x)$ be the quotient and R the remainder when divided by $(x - h)$ where

$$Q(x) = b_0x^{n-1} + b_1x^{n-2} + \cdots + b_{n-1},\ b_0 \neq 0. \text{ Thus}$$

$$a_0x^n + a_1x^{n-1} + \cdots + a_n \equiv (x-h)(b_0x^{n-1} + b_1x^{n-2} + \cdots + b_{n-1}) + R$$

Hence, on comparing the coefficients of various powers of x,

$$\left.\begin{aligned} a_0 &= b_0, \\ a_1 &= b_1 - b_0h, \\ a_2 &= b_2 - b_1h, \\ &\cdots\cdots \\ a_{n-1} &= b_{n-1} - b_{n-2}h, \\ a_n &= R - b_{n-1}h, \end{aligned}\right\} \text{ i.e., } \quad \begin{aligned} b_0 &= a_0, \\ b_1 &= a_1 - b_0h, \\ b_2 &= a_2 - b_1h, \\ &\cdots\cdots \\ b_{n-1} &= a_{n-1} + b_{n-2}h, \\ R &= a_n + b_{n-1}h. \end{aligned}$$

Conveniently the following scheme, known as Synthetic Division (Hornor's scheme), is used :—

	a_0	a_1	a_2	...	a_{n-2}	a_{n-1}	a_n
h	a_0	b_0h	b_1h	...	$b_{n-3}h$	$b_{n-2}h$	$b_{n-1}h$
	$b_0(\equiv a_0)$	b_1	b_2	...	b_{n-2}	b_{n-1}	$R = P_n(h)$

Every term in the second row is obtained by multiplication with h of the preceding term in the third row, and every term in the third row is obtained by addition of the terms above.

The remainder

$$\begin{aligned} R &= a_n + b_{n-1}h \\ &= a_n + (a_{n-1} + b_{n-2}h)h \\ &= a_n + a_{n-1}h + (a_{n-2} + b_{n-3}h)h^2 = \cdots \\ &= a_n + a_{n-1}h + a_{n-2}h + \cdots + a_1h^{n-1} + a_0h^n \\ &= P_n(h) \end{aligned}$$

Cor. 1. If the divisor is $x + h$, then we write $-h$ for h and proceed as before.

Cor. 2. If the divisor is $ax - b$, then since

$$ax - b = a\left(x - \frac{b}{a}\right)$$

we find the quotient and the remainder of $P_n(x)$ when divided by $\left(x - \frac{b}{a}\right)$ and then divide the quotient by a.

Ex. *Divide* $x^4 - 4x^3 + 6x^2 - 4x + 1$ *by* $3x - 2$.

Ans. $Q(x) = \frac{1}{3}x^3 - \frac{10}{9}x^2 + \frac{34}{27}x - \frac{40}{81}$, $R = \frac{1}{81}$.

14·2·2. Synthetic division by a quadratic expression. Consider the division of $2x^5 - 3x^4 + 4x^3 - 5x^2 + 6x - 7$ by $x^2 - x + 2$. By long division, we have

$$(x^2 - x + 2)\; 2x^5 - 3x^4 + 4x^3 - 5x^2 + 6x - 7\; (2x^3 - x^2 - x - 4) = \text{quotient}$$

$$\begin{array}{l} \underline{2x^5 - 2x^4 + 4x^3} \\ \quad -x^4 - 5x^2 \\ \quad \underline{-x^4 + x^3 - 2x^2} \\ \quad -x^3 - 3x^2 + 6x \\ \quad \underline{-x^3 + x^2 - 2x} \\ \qquad -4x^2 + 8x - 7 \\ \qquad \underline{-4x^2 + 4x - 8} \\ \qquad\quad 4x + 1 = \text{remainder.} \end{array}$$

The long division can be condensed in the following scheme :

	2	−3	4	−5	6	−7
1		2	−1	−1	−4	
−2			−4	2	2	8
	2	−1	−1	−4	4	1

quotient $= 2x^3 - x^2 - x - 4$

remainder $= 4x + 1$.

Generally, let us divide $a_0x^n + a_1x^{n-1} + a_2x^{n-2} + \cdots + a_{n-1}x + a_n$ by $x^2 - px - q$.

We have

	a_0	a_1	a_2	.....	$\ldots a_{n-1}$	a_n
p		pa_0	ps_1	.....		
q			qs_0			
	$a_0 = s_0$	$a_1 + pa_0 = s_1$	$a_2 + ps_1 + qs_0 = s_2$			

Ex. *Divide* $2x^5 - 3x^4 + 4x^3 - 5x^2 + 6x + 2$ *by* $2x^2 - x + 1$.

Ans. $Q(x) = x^3 - x^2 + x - \dfrac{3}{2}$, $R = \dfrac{7}{2}x + \dfrac{7}{2}$.

14·2·3. Synthetic division for evaluation of $P_n(x_0)$ and Derivatives of a Polynomial.

Consider the polynomial

$$P_n(x) = a_0x^n + a_1x^{n-1} + a_2x^{n-2} + \cdots + a_{n-1}x + a_n,\ a_0 \neq 0 \qquad \ldots(1)$$

where $a_0, a_1, \ldots, a_n$ are defined by the polynomials

$$\left.\begin{aligned} P_0(x) &= a_0 \\ P_1(x) &= xP_0(x) + a_1 \\ P_2(x) &= xP_1(x) + a_2 \\ &\ldots\ldots\ldots\ldots\ldots\ldots \\ P_{n-1}(x) &= xP_{n-2}(x) + a_{n-1} \\ P_n(x) &= xP_{n-1}(x) + a_n \end{aligned}\right\} \qquad \ldots(2)$$

Now $P_n(x)$ of equation (2) is the same as $P_n(x)$ of equation (1) as can be shown by multiplying by $x^n, x^{n-1}, \ldots$ the first, second, ... equations of (2) and then adding.

Differentiating equations (2) we obtain

$$\left.\begin{aligned} P_0'(x) &= 0 \\ P_1'(x) &= P_0(x) \\ P_2'(x) &= xP_1'(x) + P_1(x) \\ &\ldots\ldots\ldots\ldots\ldots\ldots \\ P_n'(x) &= xP_{n-1}'(x) + P_{n-1}'(x) \end{aligned}\right\} \qquad \ldots(3)$$

Comparing equations (3) with equations (2), we find that $P_1'(x_2)$, $P_2'(x)$, ..., $P_n'(x)$ can be obtained from $P_0(x), P_1(x), ..., P_{n-1}(x)$ in exactly the same way as $P_0(x), P_1(x), ..., P_{n-1}(x)$ can be obtained from $a_0, a_1, ..., a_{n-1}$.

x	a_0	a_1	a_2	$...a_{n-1}$	a_n
		xP_0	xP_1	$...xP_{n-2}$	xP_{n-1}
	P_0	P_1	P_2	$...P_{n-1}$	P_n
x		xP_1'	xP_2'	$...xP_{n-2}'$	
	P_1'	P_2'	P_3'	$...P_n'$	

where $P_n' \equiv P_n'(x) = \dfrac{d}{dx} P_n(x)$,

Cor. $P_n'(x) = a_0 nx^{n-1} + \cdots + 2a_{n-2}x + a_{n-1}$

$P_n''(x) = a_0 n(n-1)\, x^{n-2} + \cdots + 2a_{n-2}$

Thus $a_{n-1} = P_n'(0)$, $a_{n-2} = \dfrac{1}{2\,!} P_n''(0)$, $a_{n-3} = \dfrac{1}{3\,!} P_n'''(0)$, etc.

Example 1. *For the polynomial* $P(x) = x^3 - 3x^2 + 4x - 5$, *find* $P(2)$, $P'(2)$, $P''(2)$, *and* $P'''(2)$.

Sol.

					Thus
2	1	-3	4	-5	
		2	-2	4	
2	1	-1	2	$-1 = P(2)$	$P(2) = -1$
		2	2		
2	1	1	4	$= P'(2)$	$P'(2) = 4$
		2			
2	1	3	$= \dfrac{1}{2\,!} P''(2)$		$P''(2) = 6$
		2			
2		5	$= \dfrac{1}{3\,!} P'''(2)$		$P'''(2) = 30$

14·3. Theorems for finding Nature and Location of the roots of polynomial equations. Consider the general polynomial equation with real coefficients

$$P_n(x) = a_0 x^n + a_1 x^{n-1} + \cdots + a_{n-1}x + a_n, \quad a_0 \neq 0 \qquad ...(1)$$

We state below general theorems relating to the real roots of (1).

I. *Descartes Rule of Signs.* The number of positive roots of $P_n(x) = 0$ where the coefficients a_i are real, cannot exceed the number of variation or changes in sign of the coefficients in the polynomial $P_n(x)$

and the number of negative roots cannot exceed the number of changes in sign of the coefficients in $P_n(-x)$.

II. The largest root of equation (1) may be approximated by the root of the linear equation $a_0x + a_1 = 0$ or by the root larger in absolute value of the quadratic equation

$$a_0x^2 + a_1x + a_2 = 0$$

These approximations are accurate only when the largest root of equation (1) is much larger in absolute value than all the others.

The smallest root of equation (1) may be approximated by the root of the equation

$$a_{n-1}x + a_n = 0$$

or by the root smaller in absolute value of the quadratic equation

$$a_{n-2}x^2 + a_{n-1}x + a_n = 0.$$

III. If $P(x)$ is a real function continuous between $x = a$ and $x = b$ and $P(a)$ and $P(b)$ have opposite signs, then there are an odd number of real roots of $P_n(x) = 0$ between a and b, if $P(a)$ and $P(b)$ have the same sign, then there are an even number of real roots between a and b or there are no real roots at all between a and b.

IV. (*a*) Any polynomial equation of odd degree with real coefficients has at least one real root of sign opposite to the sign of a_0/a_n while a polynomial equation of even degree has least two real roots, one positive and one negative, provided $a_n < 0$.

(*b*) Any polynomial equation of an even degree whose last term is negative and coefficient of first term is positive, has at least two real roots, one positive and other negative.

14·4. Methods of Solving Transcendental Equations. There are various methods of solving such equations. With the help of these methods we first find a first approximation to a root of the given equation and then successively improve it.

14·4·1. Graphical Method. This method is quite useful in locating approximately the roots of an equation $f(x) = 0$. Frequently, the equation $f(x) = 0$ can be written as $f_1(x) = f_2(x)$ and the roots of $f(x) = 0$ are determined by the intersection of the graphs of the two curves $y = f_1(x)$ and $y = f_2(x)$. Sometimes we consider the intersection of the graphs $y = f(x)$ and $y = 0$. For example, to solve the equation

$$f(x) = x^n + ax + b = 0$$

we draw the graphs of $y = x^n$ and $y = -(ax + b)$ and then consider their point of intersection. In order to find the roots of the equation $e^x \sin x = 1$, we write it as $\sin x = e^{-x}$ and then determine the intersection of $y = \sin x$ with $y = e^{-x}$.

14·4·2. Bolzano or Interval Halving or Bisection Method. This method of solving a transcendental equation consists in locating the roots of the equation $f(x) = 0$ between two numbers say x_0 and x_1 such that $f(x)$ is continuous for $x_0 \le x \le x_1$ and $f(x_0)$ and $f(x_1)$ are of opposite signs so that the product $f(x_0) f(x_1) < 0$, *i.e.*, the curve crosses the x axis between x_0 and x_1. Then the desired root is approximately

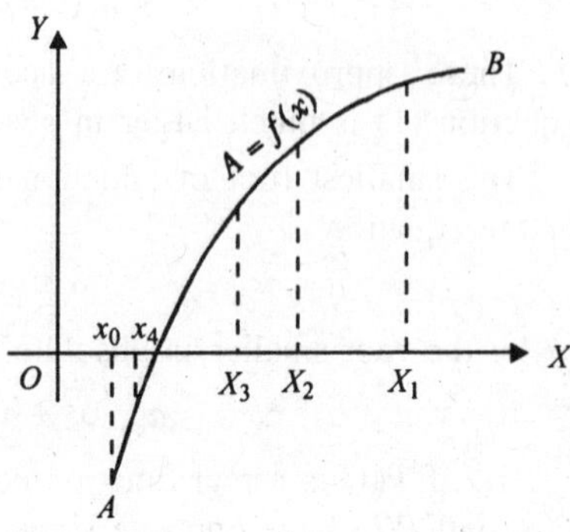

$$x_2 = \frac{x_0 + x_1}{2}$$

Now $\quad x_3 = \dfrac{x_0 + x_2}{2}$ provided that $f(x_0) f(x_2) < 0$ *i.e.*, if it lies between x_0 and x_2.

and $\quad x_3 = \dfrac{x_1 + x_2}{2}$ provided that $f(x_1) f(x_2) < 0$.

Further $\quad x_4 = \dfrac{x_0 + x_3}{2}$ provided that $f(x_0) f(x_3) < 0$, and so on.

Thus, in each iteration we either find the root with desired accuracy or we narrow the range to half the previous interval. The process is continued till it converges. After n bisections, the length of the subinteral containing h is $(b - a)/2$. If the error is $< \varepsilon_1$ then $(b - a_1)/2^n < t \Rightarrow$

$$2^n > \frac{b - a}{t} \text{ or } n > \frac{\log \dfrac{b - a}{2}}{\log 2}.$$

Example 2. *Solve $x^3 - 9x + 1 = 0$ for the root between $x = 2$ and $x = 4$ by the method of interval having or Bolzano bisection procedure.*

Sol. Here $f(x) = x^3 - 9x + 1$ so that $f(x)$ is continuous in $2 \le x \le 4$. Further, $f(2) = -9$, $f(4) = +29$ and $f(2) f(4) < 0$ so that a root lies between 2 and 4.

Now $\quad x_2 = \dfrac{x_0 + x_1}{2} = \dfrac{2 + 4}{2} = 3.$

Also $f(3) = 1$ so that $f(2) f(3) < 0$. Thus the root lies between 2 and 3

Hence $x_3 = \dfrac{x_0 + x_2}{2} = \dfrac{2 + 3}{2} = 2{\cdot}5$

Further, $f(2{\cdot}5) < 0$ so that $f(3) f(2{\cdot}5) < 0$, $\quad \therefore\ x_4 = \dfrac{3 + 2{\cdot}5}{2} = 2{\cdot}75$

Similarly $x_5 = 2{\cdot}875, x_6 = 2{\cdot}9375$ and so for higher iterations, so that the process can be continued as long as required.

14·4·3. Regula Falsi Method. (Method of False Position), **Secant Method.**

Consider the equation $f(x) = 0$ and let x_1 and x_2 be the approximations

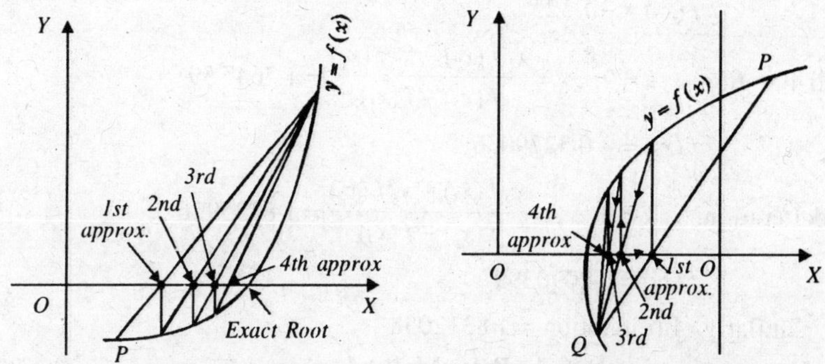

of the root of the equation where $x_1 < x_2$. Also let the corresponding ordinates $f(x_1)$ and $f(x_2)$ have opposite signs so that $f(x_1)f(x_2) < 0$.

$$x_3 = \frac{x_1 f(x_2) - x_2 f(x_1)}{f(x_2) - f(x_1)} \qquad ...(1)$$

which can be easily derived by writing down the equation of the line PQ through $P[x_1, f(x_1)]$ and $Q[x_2, f(x_2)]$ where x_3 is the abscissa of the point where this line crosses the x-axis provided that at each step $f(x_1)f(x_2) < 0$. The formula (1) can be iterated as

$$x_{n+1} = \frac{x_{n-1} f(x_n) - x_n f(x_{n-1})}{f(x_n) - f(x_{n-1})} \qquad ...(2)$$

$$= x_n - \frac{(x_n - x_{n-1}) f(x_n)}{f(x_n) - f(x_{n-1})}$$

provided that at each step $f(x_{n-1})f(x_n) < 0$.

The method with variable secant is known as *Regula falsi.*

Example 3. *Compute the real root of $x^3 - 5x + 3 = 0$ in the interval [1, 2] by the secant method by performing four iterations.*

(*B.Sc. Math Hons., Delhi 1997; Nagpur 1998*)

Sol. $f(x) = x^3 - 5x + 3.$ Interval : [1, 2]

$f(x_0) = f(1) = -1$

$f(x_1) = f(2) = +1$

Hence one root lies between 1 and 2.

using the iterative formula :

$$\therefore x_{k+1} = \frac{x_{k-1} f(x_k) - x_k f(x_{k-1})}{f(x_k) - f(x_{k-1})}$$

1st Iteration, $k = 1,\ x_2 = \dfrac{x_0 f(x_1) - x_1 f(x_0)}{f(x_1) - f(x_0)} = \dfrac{1 \times 1 - 2(-1)}{1+1} = \dfrac{3}{2} = 1{\cdot}5$

$f(x_2) = -1{\cdot}125$

2nd Iteration, $k = 2,\ x_3 = \dfrac{x_1 f(x_2) - x_2 f(x_1)}{f(x_2) - f(x_1)} = 1{\cdot}7647{\cdot}59$

$f(x_3) = -0{\cdot}3279036$

3rd Iteration, $k = 3,\ x_4 = \dfrac{x_2 f(x_3) - x_3 f(x_2)}{f(x_3) - f(x_2)} = 1{\cdot}8735976$

$f(x_4) = 0{\cdot}2090399$

Similarly, 4th iteration = 1·8312058

Convergence of Regula-Falsi Method

Let $\{x_i\}$ be a sequence obtained by a given method and let ξ be the root of the given equation. Then the method is said to be convergent iff $\lim |x_n - \xi| = 0$.

Order of convergence. If $x_n = \xi + \varepsilon_n,\ x_{n+1} = \xi + \varepsilon_{n+1}$ where ξ is a root of $f(x) = 0$, and if we are able to prove that $\varepsilon_{n+1} = k\,\varepsilon_n^p$ where k is a constant then the order of convergence of the method is ***p***.

Rate of convergence of Regula-Falsi Method

Let $$x_n = \xi + \varepsilon_n,\ x_{n+1} = \xi + \varepsilon_{n+1}$$

where ξ is a root of $f(x) = 0$ and ε_n is the error involved at the nth step. when finding out the actual root ξ. Then (2), page 265 gives

$$\xi + \varepsilon_{n+1} = \frac{(\xi + \varepsilon_n) f(\xi + \varepsilon_{n-1}) - (\xi + \varepsilon_{n-1}) f(\xi + \varepsilon_n)}{f(\xi + \varepsilon_{n-1}) - f(\xi + \varepsilon_n)}$$

$$= \xi - \frac{(\xi + \varepsilon_n) f(x_{n-1}) - (\xi + \varepsilon_{n-1}) f(x_n)}{f(x_{n-1}) - f(x_n)}$$

$$\therefore \quad \varepsilon_{n+1} = \frac{\varepsilon_{n-1} f(\xi + \varepsilon_n) - \varepsilon_n f(\xi + \varepsilon_{n-1})}{f(\xi + \varepsilon_n) - f(\xi + \varepsilon_{n-1})}$$

$$= \frac{\varepsilon_{n-1}\left[f(\xi) + \varepsilon_n f'(\xi) + \dfrac{\varepsilon_n^2}{2!} f''(\xi) + \cdots\right] - \varepsilon_n\left[f(\xi) + \varepsilon_{n-1} f'(\xi) + \dfrac{\varepsilon_{n-1}^2}{2!} f''(\varepsilon) + \cdots\right]}{[f(\xi) + \varepsilon_n f'(\xi)] - [f(\xi) + \varepsilon_{n-1} f'(\xi)]}$$

$$= \frac{1}{2}\varepsilon_{n-1}\,\varepsilon_n \frac{f''(\xi)}{f'(\xi)} + 0(\varepsilon_n^2), \qquad \text{...(3)}$$

using $f(\xi) = 0$.

We shall now find a number k such that $\varepsilon_{n+1} = c\,\varepsilon_n^k$ and therefore

$$\varepsilon_n = c\,\varepsilon_{n-1}^k$$

$$\Rightarrow \qquad \varepsilon_{n-1} = c^{-1/k}\,\varepsilon_n^{1/k}.$$

Thus from (3),

$$c\varepsilon_n^k = l.c^{1/k}\,\varepsilon_n^{1/k}\cdot\varepsilon_n \quad \text{where } l = \tfrac{1}{2}f''(\xi)/f'(\xi)$$

Hence $\qquad k = 1 + \dfrac{1}{k}$, on equating powers of ε_n

$$\Rightarrow \qquad k^2 - k - 1 = 0$$

Solving, $\qquad k = \dfrac{1 \pm \sqrt{5}}{2}$,

The positive value of k is $\frac{1}{2}(1 + \sqrt{5})$ or 1·618.

Hence $\qquad \varepsilon_{n+1} = c\varepsilon_n^{1\cdot618}$

Thus the order of convergence of regula falsi method is 1·618.

14·5. Method of Iteration. This method can be applied when the equation $f(x) = 0$ can be solved for the unknown x in terms of a function of x.

An iteration process defined by $x = g(x)$ is said to be convergent for an x_0 if the corresponding sequence $x_0, x_1, x_2, \ldots$ is convergent, that is if

$$x_1 = g(x_0)$$

Similarly we obtain a sequence of the successive approximations, $x_2, x_3, \ldots$ given by $x_2 = g(x_1), x_3 = g(x_2), \ldots, x_i = g(x_{i+1}), x_{i+1} = g(x_{i+2})$ so that each iterate is nearer to the previous iterate or these iterates move away from the actual root.

Condition under which the iteration process converges.

Case 1. Let the given equation $f(x) = 0$ be put in the form

$$x = g(x)$$

Suppose x is the exact root of this equation ...(3′)

Starting with x_0, we have

$$x_1 = g(x_0) \qquad \text{...(4)}$$

From (3) and (4),

$$x - x_1 = g(x) - g(x_0) \qquad \text{...(5)}$$

$$= (x - x_0)g'(\xi_0),\ x_0 < \xi_0 < x_1 \quad \text{by Mean value theorem.}$$

Similarly $x - x_2 = (x - x_1)g'(\xi_1),\ x_1 < \xi_1 < x_2$

$$\vdots$$

$$x - x_n = (x - x_{n-1})g'(\xi_{n-1}),\ x_{n-1} < \xi_{n-1} < x_n$$

Multiplying columnwise, we obtain

$$x - x_n = (x - x_0)g'(\xi_0)g'(\xi_1), ..., g'(\xi_{n-1})$$

Suppose that $|g'(\xi_k)| \le \varepsilon$ where ε is very small

and $$k = 0, 1, ..., n.$$

Thus $$|x - x_n| \le |x - x_0| \in^n \qquad ...(6)$$

Hence we have convergence of the iterate x_n to x if $\in < 1$ *i.e.*, if $|g'(x)| < 1$ in the interval near the value of the root, we see that the smaller the value of ε, the faster is the convergence.

Case II. Suppose we want to solve the equation $f(x) = 0$ which can be written in the form

$$g(x) = h(x)$$

We further replace this equation by the set of simultaneous equations

$$y = g(x) \qquad ...(i)$$

$$y = h(x) \qquad ...(ii)$$

Solving (*ii*) explicitly for x, we obtain

$$x = H(y) \qquad ...(iii)$$

Let x_0 be the first approximation of the root of $f(x) = 0$. Then from (*i*) y_0 is the first approximation of y and then from (*iii*), x_1 is the second approximation of x and so on for the other iterates $x_2, ..., x_n$.

Now if $$\left| \frac{d}{dx} g(x) \right|_{x=p} < \left| \frac{d}{dx} h(x) \right|_{x=p}$$

i.e., absolute value of slope of $g(x)$ is less than that of $h(x)$ at $x = p$ where p is the desired root of $f(x) = g(x) - h(x) = 0$ then if x_0 is a good approximation, then $x_n \to p$.

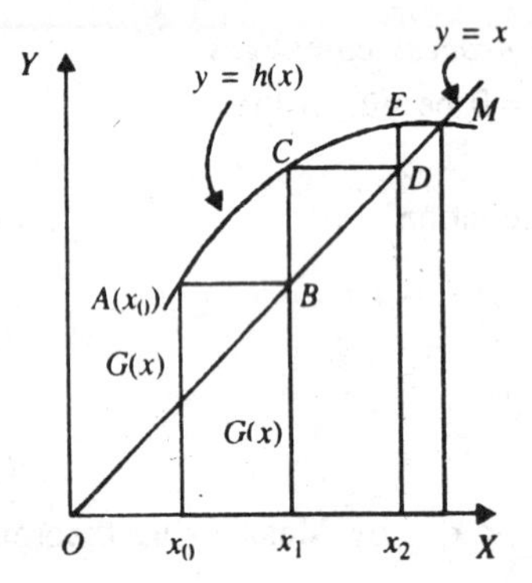

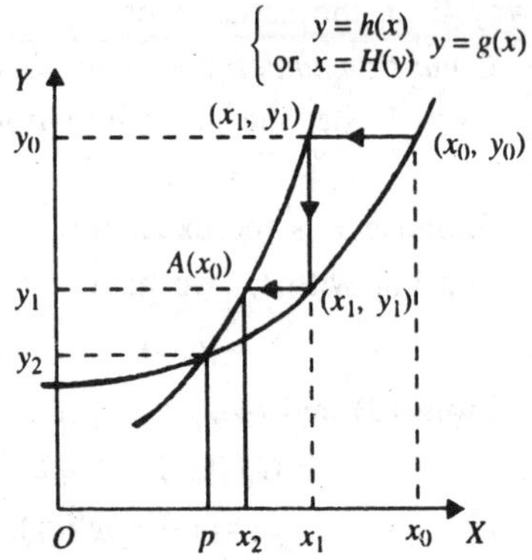

g(x) and h(x) have positive slopes.

Example 4. *Find, by iteration method, a real root of*

$$2x - \log_{10} x = 7$$

Sol. The given equation can be written as $y = g(x), y = h(x)$

i.e., $y = 2x - 7 = \log_{10} x.$

We find from the intersection of the functions $y = 2x - 7$ and $y = \log_{10} x$ that an approximate value of the root of the given equation is 3·8.

The given equation is $x = \frac{1}{2}(\log_{10} x + 7)$

Hence, First approximation $x_1 = \frac{1}{2}(\log 3{\cdot}8 + 7) = 3{\cdot}79$

Second approximation $x_2 = (\log 3{\cdot}79 + 7) = 3{\cdot}7893$

Third approximation $x_3 = \frac{1}{2}(\log 3{\cdot}7893 + 7) = 3{\cdot}7893.$

Since second and third approximations are same, we take 3·7893 as the correct value of the root of the equation to four places of decimals.

Ex. (*i*) *Find by iteration method a real root of the equation* $x^3 + x^2 = 1$.

Hint. $x = \dfrac{1}{\sqrt{1 + dx}} = g(x)$, $g'(x) = \dfrac{1}{2(1+x)^{3/2}}$ $\therefore |g'(x)| < 1$ for all x_1 particularly in(0, 1).

(*ii*) Starting with $x = {\cdot}12$, solve $x = {\cdot}21 \sin({\cdot}5 + x)$ by using iteration method.

[**Hint.** Here $g(x) = {\cdot}21 \sin({\cdot}5 + x)$, $|g'(x)| = |{\cdot}21 \cos({\cdot}5 + x)| < 1$

$\therefore$ Iteration is possible.

Now $x_1 = {\cdot}21 \sin({\cdot}5 + {\cdot}12) = {\cdot}21 \sin {\cdot}62 = {\cdot}21 \times {\cdot}5104 = {\cdot}1220$

Similarly $x_2 = {\cdot}21 \sin {\cdot}6220 = {\cdot}1224$]

14·5·1. Newton-Raphson Method for non-repeated real roots . The iteration method is helpful in finding a *real zero* of an equation. Usually, the problem is to find a recurrence relation which enables us to find out a sequence $\{x_i\}$ converging to the desired root r.

Let $y = f(x) = 0$

be the equation to the solved and let x_0 be a point near to a root of this equation. Then expanding $f(x)$ by the Taylor's series in the neighbourhood of x_0 we have

$$f(x) = f(x_0) + (x - x_0) f'(x_0) + \frac{(x - x_0)^2}{2!} f''(x_0) + \cdots = 0$$

As a first approximation x_1 we have

$$f(x_0) + (x_1 - x_0) f'(x_0) = 0$$

or $$x_1 = x_0 - \frac{f(x_0)}{f'(x_0)}, \text{ provided } f'(x_0) \neq ($$

or $$\Delta x_0 = -\frac{f(x_0)}{f'(x_0)}, \; x_1 = x_0 + \Delta x_0$$

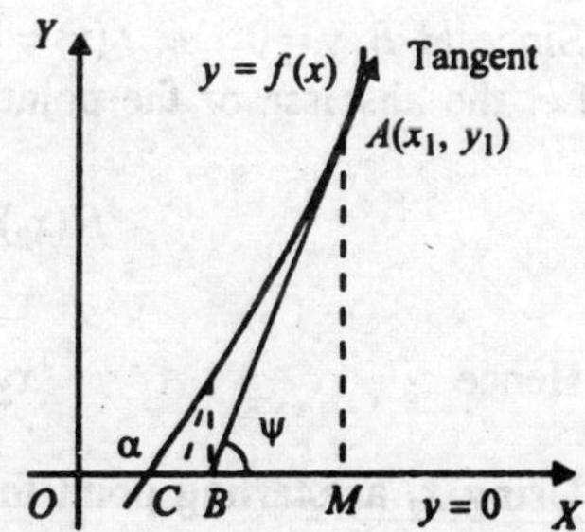

Clearly x_1 is a better approximation to the root x_0 of $f(x) = 0$. In this manner, we can get an approximation or the iteration formula

$$x_{n+1} = x_n - \frac{f(x_n)}{f'(x_n)} \qquad ...(1)$$

This is the Newton-Raphson formula having variable tangent.

Remark 1. The accuracy of Newton's method for function $f(x)$ which possess continuous first and second derivatives can be *estimated*. Thus

$$|x_2 - r| \le (x - r)^2 \cdot \frac{M}{m}$$

where $M = \max |f''(x)|$ and $m = \min |f''(x)|$ in an interval that contains the root r and the estimates x_1 and x_2.

Clearly the error decreases if $\left| (x_1 - r)^2 \frac{M}{m} \right| < 1.$

Example. *Consider the case where* $f(x) = x^2 - 5$

$\therefore$ $f'(x) = 2x$, $f''(x) = 2$, $r = \sqrt{5}$, $x_1 = 2$ *and other approximation* x_n *lies between 2 and* $\frac{9}{4}$.

Sol. Let $M = 2$, $m = 4$ so that, $|x_2 - 5| \le \frac{1}{2}(x_1 - \sqrt{5})^2$

For any two successive approximations x_{n+1} and x_n,

$$|x_{n+1} - \sqrt{5}| \le \frac{1}{2}(x_n - \sqrt{5})^2$$

This inequality shows that the number of decimal places of accuracy doubles with each successive iteration go back and forth between these two values alternately.

Remark 2. See Fig.

Geometrical interpretation of New- ton-Raphson method

We give below the *geometrical inter- pretation* of this formula.

Slope of the tangent to the curve $y = f(x)$ at the point $A(x_1, y_1)$

$$= f'(x_1)$$

Since at $B, y = 0$ *i.e.*, $f(x) = 0$, B is nearer to the desired root than is A. Let the abscissa of the point B be x_2, then from ΔABM, we have

$$f'(x_2) = \tan\psi = \frac{AM}{BM} = \frac{f(x_1)}{x_1 - x_2}.$$

Hence

$$x_2 = x_1 - \frac{f(x_1)}{f'(x_1)}.$$

Using x_2 as starting point the tangent at $[x_2, f(x_2)]$ will give

$$x_3 = x_2 - \frac{f(x_2)}{f'(x_2)}$$

The process can be repeated and the root α is approached with great rapidity. Thus we shall get the limit α approached by the sequence $\{x_n\}$ where

$$x_{n+1} = x_n - \frac{f(x_n)}{f'(x_n)}.$$

Note 1. One of the bases of Newton's method is that the tangent of the curve $y = f(x)$ at x_0 meets the x-axis at a point in between x_0 and α. The condition for the convergence of the method is that in the vicinity of α,

$$\frac{d}{dx}\left[x - \frac{f(x)}{f'(x)}\right] < 1 \quad \text{at } x = \alpha,$$

i.e., $$\left|1 - \frac{[f'(x)]^2 - f(x)f''(x)}{[f'(x)]^2}\right| < 1 \text{ at } x = \alpha$$

or $$\left|\frac{f(x)f''(x)}{[f'(x)]^2}\right| < 1 \quad \text{at } x = \alpha.$$

i.e., $$|f(\alpha)f''(\alpha)| < [f'(\alpha)]^2.$$

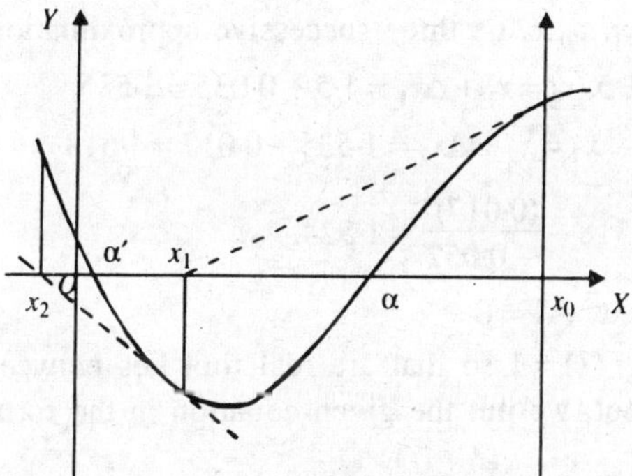

However, while the most severe limitation on the use of Newton's method is the requirement that $f'(x) \neq 0$ in the vicinity of α. Even a moderate value of $f'(x_0)$ may be more than swamped by a large value of either $f(x_0)$ or $f''(x_0)$ to produce a value x_1 that will result in a sequence that converges to a root other than the one we are interested in.

Note 2. Acceleration of convergence (Aitken's Δ^2 method)

Let $x = a$ be a root of $x = g(x)$ so that $a = g(a)$.

Let $x_1, x_2, \ldots$ be the first second, ... approximations satisfying $a = g(a)$

$\therefore \quad |a - x_{n+1}| = |g(a) - g(x_n)| \leq \in |a - x_n|, \ \in < 1$

Thus iterative method is linearly slow convergent.

Aitken has proved the following to accelerate the rate of convergence of the root of the given equation.

Now $$a - x_i \simeq \in (a - x_{i-1})$$

and $$a - x_{i+1} \simeq \in (a - x_i)$$

Dividing, $$\frac{a - x_i}{a - x_{i+1}} \simeq \in \frac{a - x_{i-1}}{a - x_i}$$

Simplifying this equation we get

$$a \simeq x_{i+1} - \frac{(x_{i+1} - x_i)^2}{x_{i+1} - 2x_i + x_{i-1}}$$

$$\simeq x_{i+1} - \frac{(\Delta x_i)^2}{\Delta^2 x_{i-1}}$$

where x_{i-1}, x_i, x_{i+1} are three successive approximations.

Thus we can obtain the new approximate value of the roots in case three successive approximations are known.

Example 5. *Solve the equation*

(*a*) *$2x = \cos x + 3$ by Aitken's Δ^2. method,*

Sol. (*a*) Let x_1, x_2, x_3 be three successive approximations.

Then if $x_1 = 1{\cdot}5, x_2 = x_1 + \Delta x_1 = 1{\cdot}5 + 0{\cdot}035 = 1{\cdot}535$,

$$x_3 = x_2 + \Delta x_2 = 1{\cdot}535 - 0{\cdot}017 = 1{\cdot}518$$

Hence $x_4 = 1{\cdot}518 - \dfrac{(0{\cdot}017)^2}{-0{\cdot}052} = 1{\cdot}524.$

(*b*) $f(x) = x^3 + x^2 - 1 = 0$

$\therefore f(0) = -1, f(1) = 1$ so that are real root lies between 0 and 1. To determine this root, we put the given equation in the form $x = g(x)$

$$\therefore x^2(x+1) = 1 \Rightarrow x = \frac{1}{\sqrt{1+x}} = g(x)$$

$\therefore g'(x) = -\frac{1}{2}(1+x)^{-3/2}$. $|g'(x)| <$ tor x lying between 0 and 1 *i.e.* $x < 1$

Hence the iterative method is applicable.

Take $x_0 = 0{\cdot}5$. We have $x_1 = g(x_0) = \dfrac{1}{\sqrt{1{\cdot}5}} = {\cdot}81649$,

$$x_2 = g(x_1) = \frac{1}{\sqrt{1{\cdot}81649}} = {\cdot}74196, \quad x_3 = g(x_2) = \frac{1}{\sqrt{1{\cdot}74196}} = {\cdot}75767$$

Similarly we can get other approximations.

Rate of convergence of Newton-Raphson formula

Let $x_n = \xi + \varepsilon_n, x_{n+1} = \xi + \varepsilon_{n+1}$ with $f(\xi) = 0$ *i.e.*, ξ is a root of

$f(x) = 0$ and ε_n is small.

Hence the equation (1), page 270 gives.

$$\varepsilon_{n+1} = \varepsilon_n - \frac{f(\xi + \varepsilon_n)}{f'(\xi + \varepsilon_n)}$$

$$= \frac{\varepsilon_n f'(\xi + \varepsilon_n) - f(\xi + \varepsilon_n)}{f'(\xi + \varepsilon_n)}$$

$$= [\varepsilon_n f'(\xi) + \varepsilon_n^2 f''(\xi) + \cdots - f(\xi) - \varepsilon_n f'(\xi) - \tfrac{1}{2}\varepsilon_n^2 f''(\xi) - \cdots]/f'(\xi + \varepsilon_n)$$

$$= \frac{1}{2}\frac{\varepsilon_n^2 f''(\xi)}{f'(\xi) + \varepsilon_n f''(\xi)} = \frac{1}{2}\varepsilon_n \frac{f''(\xi)}{f'(\xi)}\left(1 - \frac{\varepsilon_n f''(\xi)}{f'(\xi)}\right) \text{ approx.}$$

$$= \frac{1}{2}\varepsilon_n^2 \frac{f''(\xi)}{f'(\xi)} \text{ approximately.}$$

Hence the leading terms is

$$\varepsilon_{n+1} = \frac{f''(\xi)}{2f'(\xi)}\varepsilon_n^2.$$

$\therefore$ convergence is of order 2.

This is known as quadratic convergence if $\frac{1}{2}f''(\xi)/f'(\xi) > 1$. It means that the number of correct decimals is approximately doubled at every iteration, at least if the factor $f''(\xi)/2f'(\xi)$ is not too large.

Example 6. *Find the real root of $x^3 + x^2 + 3x + 4 = 0$ correct to four decimal places, using Newton-Raphson method.*

Sol. We have

$$f(x) = x^3 + x^2 + 3x + 4. \quad f(-x) = -x^3 + x^2 - 3x + 4.$$

Hence there is no real positive root and there are at most three negative roots.

Further, $\quad f(-2) = -6,\ f(-1) = +1$

Thus there is a real root of $f(x) = 0$ lying between -1 and -2.

As the initial approximation, we take $x_0 = -1$.

Also $\quad f'(x) = 3x^2 + 2x + 3,$

and $\quad f'(-1) = 4$

Hence the first approximation is by (1), page 270

$$x_1 = -1 - \frac{1}{4} = -1{\cdot}25.$$

Further, $\quad f'(x_1) = 3(-1{\cdot}25)^2 + 2(-1{\cdot}25) + 3$

$$= 5{\cdot}1875$$

Also $\quad f(x_1) = (-1{\cdot}25)^3 + (-1{\cdot}25)^2 + 3(-1{\cdot}25) + 4 = -0{\cdot}140625$

Hence second approximation is

$$x_2 = -1{\cdot}25 + \frac{0{\cdot}140625}{5{\cdot}1875} \qquad \left[x_2 = x_1 - \frac{f(x_1)}{f'(x_1)} \right]$$

$$= -1{\cdot}25 + 0{\cdot}0271 = -1{\cdot}2229 \text{ approximately}$$

Example 7. *Find by Newton's method, correct to six places of decimals, the root of the equation.* (*A.M.I.E. 1985*)

$$x \log_{10} x = 4{\cdot}772393.$$

Sol. From a table of logarithms we have, the values

$$6 \log_{10} 6 = 4{\cdot}67 \text{ and}$$

$$7 \log_{10} 7 = 5{\cdot}90,$$

so we take as the initial approximation $x_0 = 6$.

The first approximation is

$$x_1 = x_0 - \{f(x_0)/f'(x_0)\}$$

Now $\quad f(x_0) = (6 \times 0{\cdot}77815) - 4{\cdot}7772 = -0{\cdot}1083$

$$f'(x_0) = \log_{10} x_0 + \log_{10} e = 0{\cdot}778 + 0{\cdot}434 = 1{\cdot}212$$

so that $\quad 1/f'(x_9) = 0{\cdot}825$

and $\quad x_1 = 6{\cdot}089.$

The second approximation is

$$x_2 = x_1 - [f(x_1)/f'(x_1)]$$

Now $\quad f(x_1) = (6{\cdot}089 \times 0{\cdot}7845460) - 4{\cdot}7772393$

$$= -0{\cdot}0001387,$$

so that $\quad x_2 = 6{\cdot}089 + (0{\cdot}0001387 \times 0{\cdot}825)$

$$= 6{\cdot}089114,$$

which is correct to six places.

Remark 3. The formula $x_{n+1} = x_n - \dfrac{f(x_n)}{f'(x_r)}$

can be re-written as

$$x_{n+1} = \frac{x_n f'(x_n) - f(x_n)}{f'(x_n)} = \frac{N_n}{D_n}, \text{ say}$$

The computation may be arranged as follows :

n	x_n	$f(x_n)$	$f'(x_n)$	N_n	D_n	N_n/D_n
0	x_0	—	—	—	—	—
1	x_1	—	—	—	—	—
2	x_2	—	—	—	—	—
3	x_3	—	—	—	—	—

Remark 4. In the case of multiple roots, it becomes difficult for the process to converge.

Example 8. *Find the positive root of*

$$f(x) = x - 2 \sin x \qquad \text{(A.M.I.E. 1992)}$$

Sol. From a rough sketch, we conclude that the root is somewhere near $x = 2$. We also show that $f(\pi/2) = -0.429$, $f(2\pi/3) = 0.362$. Hence the root lies between $\pi/2$ and $2\pi/3$ *i.e.*, between 1·57 and 2·09 rad.

Newton-Raphson iteration formula is $x_{n+1} = x_n - \{ f(x_n)/f'(x_n) \}$

$$= \frac{-2x_n \cos x_n + 2 \sin x_n}{1 - 2 \cos x_n}$$

The first approximation is

$$x_1 = \frac{-2x_0 \cos x_0 + 2 \sin x_0}{1 - 2 \cos x_0}, \quad x_0 = 2$$

$$= \frac{3.48}{1.832}$$

$$= 1.90$$

The second approximation is

$$x_2 = \frac{-2x_1 \cos x_1 + 2 \sin x_1}{1 - 2 \cos x_1}, \quad x_1 = 1.90$$

$$= \frac{3.1211}{1.6466}$$

$$= 1.8955$$

The third approximation is

$$x_3 = \frac{-2x_2 \cos x_2 + 2 \sin x_2}{1 - 2 \cos x_2}, \quad x_2 = 1.8955$$

$$= \frac{3.1049258}{1.638056}$$

$$= 1.89549$$

14·5·2. Newton-Raphson Method for real Multiple roots and near Multiple roots. Let $x = a$ be the root of the equation $f'(x) = 0$ but not of the equation $f(x) = 0$. Then if $f(a)$ and $f''(a)$ have unlike signs, there will be two roots of $f(x) = 0$ in the neighbourhood of a, one less than a and one greater than a.

By Taylor's series expanding $f(x)$ in the neighbourhood of $x = a$, we have on neglecting terms of order higher than the second

$$f(x) = f(a) + (x - a) f'(a) + \tfrac{1}{2}(x - a)^2 f''(a)$$

In order to obtain the roots of $f(x) = 0$ near $x = a$ when $f'(a) = 0$ we solve

$$0 = f(a) + \tfrac{1}{2}(x - a)^2 f''(a)$$

Thus $\qquad (x - a)^2 = -2 f(a)/f''(a)$

or $$x = a \pm \sqrt{-2f(a)/f''(a)} \qquad ...(1)$$

The values of the roots given by this formula will be sufficiently close to the true values for the application of Newton-Raphson formula.

Suppose $x = \gamma$ is a root of multiplicity n of $f(a) = 0$ then

$$f(x) = (x - \gamma)^n g(x) \quad \text{where} \quad g(\gamma) \neq 0$$

Thus in the immediate vicinity of $x = \gamma$, we have

$$f(x) = A(x - \gamma)^n$$

so that $$f'(x) = n A(x - \gamma)^{n-1}$$

where A is the linear constant,

$$\therefore \qquad \frac{f'(x)}{f(x)} = \frac{n}{x - \gamma}$$

is the immediate neighbourhood of γ.

Hence $$\gamma = x - \frac{nf(x)}{f'(x)}$$

where x is close to γ.

Then if x_1 is in the neighbourhood of a root γ of multiplicity n, then

$$x_2 = x_1 - \frac{nf(x_1)}{f'(x_1)}$$

where x_2 is even closer approximation to γ.

Rate of Convergence of Newton's formula when there exist multiple roots.

We have $$x_{n+1} = x_n - \frac{f(x_n)}{f'(x_n)}$$

i.e., $$a + \varepsilon_{n+1} = a + \varepsilon_n - \frac{f(a + \varepsilon_n)}{f'(a + \varepsilon_n)}$$

$$\Rightarrow \qquad \varepsilon_{n1} = \varepsilon_n - \frac{f(a + \varepsilon_n)}{f'(a + \varepsilon_n)}$$

$$= \varepsilon_n \frac{f(a) + \varepsilon_n f'(a) + \frac{1}{2}\varepsilon_n^2 f''(a) + \cdots}{f'(a) + \varepsilon_n f''(a) + \frac{1}{2}\varepsilon_n^2 f'''(a) + \cdots}$$

$$= \varepsilon_n - \frac{\frac{1}{2}\varepsilon_n^2 f''(a) + \cdots}{\varepsilon_n f''(a) + \cdots} \qquad [\because f(a) = 0, f'(a) = 0]$$

$$= \varepsilon_n - \varepsilon_n \frac{\frac{1}{2} f''(a) + \cdots}{f''(a) + \cdots}, \quad f''(a) \neq 0$$

$$\simeq \frac{1}{2}\varepsilon_n$$

Hence the convergence is linear.

Example 9. *Find correct to 3 decimal places the two positive roots of*

$$2e^x - 3x^2 = 2{\cdot}5644$$

Sol. Let $f(x) = 2e^x - 3x^2 - 2{\cdot}5644$

$$f'(x) = 2e^x - 6x = 2(e^x - 3x)$$

A root of the equation

$$f'(x) = 0 \text{ is } x = 0{\cdot}6190 = a.$$

Now $e^{0{\cdot}619} = 1{\cdot}8571$

$$f''(x) = 2e^x - 6,\ f''(a) = -\,2{\cdot}2858,$$

$$f(a) = {\cdot}00317$$

$$x = 0{\cdot}6190 \pm \sqrt{\frac{-2 \times {\cdot}00317}{-2{\cdot}2858}} \qquad \left[x = a \pm \sqrt{\frac{-2f(a)}{f''(a)}}\right]$$

$$= 0{\cdot}6190 \pm \sqrt{{\cdot}000277}$$

$$= 0{\cdot}6190 \pm 0{\cdot}0166$$

$$= 0{\cdot}6356, 0{\cdot}6024.$$

Example 10. *Using synthetic division with x = 1, solve the equation*

$$f(x) = x^3 - x^2 - 1{\cdot}0001x + 0{\cdot}9999 = 0$$

Sol. We find the values of $f(1)$ and $f'(1)$.

	1	−1	−1·0001	0·9999
		1	0	− 1·0001
1	1	0	−1·0001	−0·0002 = $f(1)$
		1	1·0000	
1	1	1	−0·0001 = $f'(1)$	

Since both $f(1)$ and $f'(1)$ are small, we solve first the equation starting with $x = 1$.

	3	−2	−1·0001
1		3	1·0000
	3	1	−0·0001 = $f'(1)$
1		3	
	3	4 = $f''(1)$	

and we obtain the root

$$a = 1 - \frac{-0{\cdot}0001}{4} = 1{\cdot}000025. \qquad \left[a = 1 - \frac{f'(1)}{f''(1)}\right]$$

Now we evaluate $f(a)$ and $\frac{1}{2}f''(a)$ by synthetic division,

1·000025	1	– 1	– 1·000100	0·999900
		1·000025	0·000025	– 1·000095
		0·000025	– 1·000075	– 0·000195 = $f(a)$
1·000025		1·000025	1·000075	
	1	1·000050	0 = $f'(a)$	
1·000025		1·000025		
	1	2·000075 = $1/2 f''(a)$		

Using Eq. (1), page **286** we get

$$x = 1{\cdot}000025 \pm \sqrt{\frac{-0{\cdot}000195}{2{\cdot}000075}}$$

$$= 1{\cdot}009899, 0{\cdot}990151.$$

Order of iteration

The order of an iteration $x = f(x)$ is the order of the lowest order non-zero derivative of $f(x)$ at the solution.

Let us now consider other characteristics which influence the rate of convergence of an iteration of the form $x = f(x)$.

Let ε_n be the departure of the nth iterate from the true solution ξ so that $x_n = \xi + \varepsilon_n$. Thus ε_n in the error in the nth approximation.

Let us now expand $f(x)$ in a Taylor's series about ξ. In general

$$f(x) = f(\xi) + f'(\xi)(x-\xi) + \frac{f''(\xi)(x-\xi)^2}{2!} + \frac{f'''(\xi)(x-\xi)^3}{3!} + \cdots$$

In particular if we start with x_n for x and put $x_n - \xi = \varepsilon_n$, then

$$f(x_n) = f(\xi) + f'(\xi)\,\varepsilon_n + \frac{f''(\xi)\varepsilon_n^2}{2!} + \frac{f'''(\xi)\varepsilon_n^3}{3!} + \cdots$$

Since $f(x_n) = x_{n+1}$ and $f(\xi) = \xi$, we have

$$x_{n+1} = \xi + f'(\xi)\varepsilon_n + \frac{1}{2!} f''(\xi)\varepsilon_n^2 + \frac{1}{3!} f'''(\xi)\varepsilon_n^3 + \cdots$$

$$\Rightarrow \quad x_{n+1} - \xi = f'(\xi)\varepsilon_n + \frac{f''(\xi)}{2!}\varepsilon_n^2 + \frac{f'''(\xi)}{3!}\varepsilon_n^3 + \cdots$$

$$\Rightarrow \quad \varepsilon_{n+1} = f'(\xi)\varepsilon_n + \frac{f''(\xi)}{2!}\varepsilon_n^2 + \frac{f'''(\xi)}{3!}\xi_n^3 + \cdots$$

Thus given the $(n+1)$st iterate in terms of a series involving the nth iterate and the derivatives of $f(x)$ at the solution. If the nth iterate is a fair approximation of ξ, then ε_n should be small and the series should converge. If $f'(\xi) = 0$, then the series should converge even more rapidly. If both $f'(\xi)$ and $f''(\xi)$ are zero, convergence should be greatly accelerated.

Criteria for ending the iteration. For certain iterative method the use of $|x_{n+1} - x_n| < \varepsilon_n$ where ε_n is the nth approximation and x_{n+1} is the

$(n+1)$ st. approximation may lead to the termination of the iteration prematurely. If the iteration is converging very slowly, the change in one step may be very small. In such cases the use of $|f(x_n)| < \varepsilon_n$ may be desirable in conjunction with the previous notation.

14·5·3. Newton's formula for the determination of the *p*th roots and the reciprocal *p*th roots of a number.

Suppose we wish to find the square root of the given positive number N *i.e.* $N = x^2$.

Let $\quad f(x) = N - x^2$

then $\quad f'(x) = -2x$

so that
$$x_{i+1} = x_i - \frac{N - x_i^2}{-2x_i}, \quad \left[x_{i+1} = x_i - \frac{f(x_i)}{f'(x_i)}\right]$$
$$= \frac{1}{2}\left[x_i + \frac{N}{x_i}\right] \qquad \ldots(3)$$

This is the required iteration formula for the square root of N.

However, if we take $f(x) = x^3 - Nx, \quad N = x^2$.

so that if $f(x) = 0$, then $x = \sqrt{N}$.

Here $\quad f'(x) = 3x^2 - N,$

so that
$$x_{i+1} = x_i - \frac{f(x_i)}{f'(x_i)} = \frac{2x_i^4}{3x_i^3 - Nx_i} = \frac{2x_i^3}{2N}$$
$$= \frac{x_i(3N - x_i^2)}{2N} = \frac{1}{2}x_i\left(3 - \frac{x_i^2}{N}\right) \qquad \ldots(4)$$

which is the iteration formula.

Rate of convergence

Let $\quad p = \sqrt{N}$ so that $N = p^2$.

and write
$$x_i = p\left(\frac{1+\varepsilon_i}{1-\varepsilon_i}\right)$$
$$= p\left(1 + \frac{2\varepsilon_i}{1-\varepsilon_i}\right)$$

$$\Rightarrow \quad \frac{x_i - p}{p} = \frac{2\varepsilon_i}{1-\varepsilon_i} \quad \Rightarrow \quad \frac{1}{2}\frac{x_i - p}{p} = \frac{\varepsilon_i}{1-\varepsilon_i}$$

Now $\dfrac{x_i - p}{p}$ is known as the relative error.

Example 11. *If a root of $f(x) = 0$ lies in the interval (a, b), then find the minimum number of iterations required when the permissible error is ε.*

(B.Sc. Math Hons. Delhi, 1998)

Sol. Minimum number of iterations with $\varepsilon > 0$ as the permissible error be n where n is given by

$$\frac{b-a}{2^n} \le \varepsilon$$

$$\Rightarrow \quad \log (b-a) - n \log 2 \le \log \varepsilon$$

$$\Rightarrow \quad n \ge [\log (b-a) - \log \varepsilon]/\log 2$$

Thus the minimum $n = \left[\dfrac{\log (b-a) - \log \varepsilon}{\log 2} \right]$

where $[x]$ is the greatest integer contained in x.

Note. Thus $2\varepsilon_i (1 - \varepsilon_i)$ is the relative error in x_i and if ε_i is very small in comparison to 1.

We have

$$x_{i+1} = p\left(\frac{1+\varepsilon_{i+1}}{1-\varepsilon_{i+1}}\right) \text{by def.} \qquad ...(4')$$

Also $$x_{i+1} = \frac{1}{2} p\left[\left(\frac{1+\varepsilon_i}{1-\varepsilon_i}\right) + \left(\frac{1-\varepsilon_i}{1+\varepsilon_i}\right)\right] \text{ by (3)}$$

$$= p\left(\frac{1+\varepsilon_i^2}{1-\varepsilon_i^2}\right) \qquad ...(4'')$$

Thus $\varepsilon_{i+1} = \varepsilon_i^2$, on comparing (4′) and (4″).

If we start with $x = x_0$ such that $|\varepsilon_0| < 1$, we obtain a sequence $\varepsilon_0, \varepsilon_1, ..., \varepsilon_i ...$ which converges to zero as $i \to \infty$ and x_i converges to p.

Note. A formula for finding the **pth root** of N can be deduced as follows:

$$f(x) = x^p - N, \quad f'(x) = px^{p-1}$$

$$\therefore \quad x_{n+1} = x_n - \frac{(x_n^p - N)}{px_n^{p-1}} = \frac{(p-1)x_n^p + N}{px_n^{p-1}}.$$

$$= \frac{1}{p}\left[(p-1)x_n + \frac{N}{x_n^{p-1}}\right]$$

To find the **reciprocal** ***p*th** root of N,

Let $N^{-1/p} = x \Rightarrow N = x^{-p}; \quad f(x) = N - x^{-p}$

then $f'(x) = -px^{-p-1}$

so that $x_{i+1} = x_i (p + 1 - Nx_i^p)/p$

which is the iteration formula for finding the value of $(1/N)^{1/p}$.

For finding the **inverse** of N, the iteration formula is

$$x_{i+1} = x_i(2 - Nx_i) \qquad ...(5)$$

This can also be written as

$$1 - Nx_{n+1} = (1 - Nx_n)^2$$

and clearly shows a quadratic convergence.

Rate of Convergence

Let $$p = \frac{1}{N}$$

$$\Rightarrow \quad \frac{p - x_n}{p} = \varepsilon_n \quad \Rightarrow \quad x_n = p - p\varepsilon_n$$

we write $$x_n = p(1 - \varepsilon_n)$$

Now $$x_{n+1} = p(1 - \varepsilon_{n+1})$$

Also $$x_{n+1} = p(1 - \varepsilon_n)\left[2 - p(1 - \varepsilon_n)\frac{1}{p}\right] \text{ by (5), page 280}$$

$$= p(1 - \varepsilon_n)(1 + \varepsilon_n) = p(1 - \varepsilon_n^2)$$

$$\Rightarrow \quad p(1 - \varepsilon_{n+1}) = p(1 - \varepsilon_n^2)$$

$$\Rightarrow \quad \varepsilon_{n+1} = \varepsilon_n^2.$$

which shows a quadratic convergence.

Newton's Second Order Method. Expanding $f(x)$ into a Taylor's series about $x = x_n$ up to its third term and letting

$$x_{n+1} - x_n = h,$$

we have

$$f(x_{n+1}) = f(x_n) + hf'(x_n) + \frac{h^2}{2} f''(x_n) + \cdots = 0$$

$$\Rightarrow \quad f(x_n) + h\left[f'(x_n) + \frac{h}{2} f''(x_n)\right] = 0$$

Replacing h inside the bracket by $-f(x_n)/f'(x_n)$,

we have

$$\frac{1}{h} = \frac{-f'(x_n)}{f(x_n)} + \frac{1}{2}\frac{f''(x_n)}{f'(x_n)} \qquad \text{...(1)}$$

Example 12. *Evaluate* $\sqrt{12}$ *by applying Newton's formula.*

Sol. Since $\sqrt{9} = 3$ and $\sqrt{16} = 4$,

we take $$x_0 = 3{\cdot}5$$

Then $$x_1 = \frac{1}{2}\left(3{\cdot}5 + \frac{12}{3{\cdot}5}\right) \text{ by equation (3), Page 279}$$

$$= 3{\cdot}465$$

Further, $$x_2 = \frac{1}{2}\left(3{\cdot}465 + \frac{12}{3{\cdot}4641}\right) = 3{\cdot}4641$$

$$x_2 = \frac{1}{2}\left(3{\cdot}4641 + \frac{12}{3{\cdot}4641}\right)$$

$$= 3{\cdot}4641$$

Hence $\sqrt{12} = 3{\cdot}4641$

Example 13. *Obtain the cube root of (i) N, (ii) 12.*

Sol. (*i*) Let $f(x) = N - x^3$

then $f'(x) = -3x^2$

so that
$$x_{n+1} = x_n - \frac{N - x_n^2}{-3x_n^2}$$

$$= \frac{1}{3}\left(2x_n + \frac{N}{x_n^2}\right) \qquad ...(1)$$

(*ii*) Since $(8)^{1/3} = 2$ and $(27)^{1/3} = 3$

we shall take $x_0 = 2{\cdot}5.$

By equation (1),

First iterate
$$x_1 = \frac{1}{3}\left(2 \times 2{\cdot}5 + \frac{12}{(2{\cdot}5)^2}\right)$$

$$= \frac{1}{3}\left(5 + \frac{48}{25}\right) = 2{\cdot}3066$$

Second iterate
$$x_2 = \frac{1}{3}\left(4{\cdot}6133 + \frac{12}{(2{\cdot}3066)^2}\right)$$

$$= \frac{1}{3}(4{\cdot}6133 + 2{\cdot}2553) = 2{\cdot}2895$$

Third iterate
$$x_3 = \frac{1}{3}\left(4{\cdot}5790 + \frac{12}{(2{\cdot}2895)^2}\right)$$

$$= \frac{1}{3}(4{\cdot}5790 + 2{\cdot}2893) = 2{\cdot}2895 \text{ approx.}$$

Example 14. *Obtain the following Newton-Raphson's extended formula* (**Chebyshev formula** *of third order) for the root of $f(x) = 0$;*

$$x_1 = x_0 - \frac{f(x_0)}{f'(x_0)} - \frac{1}{2} \cdot \frac{\{f(x_9)\}^2 f''(x_0)}{\{f'(x_0)\}^3}$$

Sol. Expanding $f(x)$ in the neighbourhood of the iterate x_0 using Taylor's series, we have

$$0 = f(x) = f(x_0) + (x - x_0) f'(x_0)$$

so that
$$x - x_0 = -\frac{f(x_0)}{f'(x_0)},$$

or
$$x_1 = x_0 - \frac{f(x_0)}{f'(x_0)},$$

as a first approximation.

Again
$$0 = f(x) = f(x_0) + (x_2 - x_0) f'(x_0) + \tfrac{1}{2}(x_1 - x_0)^2 f''(x_0)$$

or
$$0=f(x_0)+(x_2-x_0)f'(x_0)+\frac{1}{2}\frac{\{f(x_0)\}^2}{\{f'(x_0)\}^3}f''(x_0)$$

Second approximation $x_2=x_0-\dfrac{f(x_0)}{f'(x_0)}-\dfrac{1}{2}\dfrac{\{f(x_0)\}^2 f''(x_0)}{\{f'(x_0)\}^3}$

Aliter. We have
$$0=f(x)=f(x_0+\Delta x_0)=f(x_0)+\Delta x_0 f'(x_0)+\tfrac{1}{2}\Delta^2 x_0 f''(x_0)+\cdots$$

Let $\Delta x_0=a_1 f_0+a_2 f_0^2+a_3 f_3^3+\cdots$ where $f_0\equiv f(x_0)$.

Hence $0=f_0+(a_1 f_0+a_2 f_0^2+\cdots)f_0'+\dfrac{1}{2\,!}(a_1 f_0+a_2 f_0^2+\cdots)^2 f_0''$
$$+\frac{1}{3\,!}(a_1 f_0+\cdots)^3 f_0'''+\cdots$$
$$=f_0(1+a_1 f_0'+\tfrac{1}{2}a_1^2 f_0'')+f_0^2(a_3 f'+\cdots) \qquad ...(1)$$

From (1), $1+a_1 f_0'=0$ so that $a_1=-\dfrac{1}{f_0'}$ ...(2)

$\therefore$ $\Delta x_\theta=a_1 f_0-\dfrac{f_0}{f_0'}$

$\therefore$ $\xi=x_0+\Delta x_0=x_0-\dfrac{f(x_0)}{f'(x_0)}$

Again from (1), $a_2 f_0'+\dfrac{1}{2}\cdot a_1^2 f_0''=0$

$\therefore$ $a_2=-\dfrac{1}{2}\cdot\dfrac{f_0''}{f_0'}\cdot\dfrac{1}{f_0'^2}=-\dfrac{1}{2}\dfrac{f_0''}{f_0'^3}$, using (2)

Hence $\xi-x_0-\dfrac{f(x_0)}{f'(x_0)}-\dfrac{1}{2}\cdot\dfrac{f''(x_0)\{f(x_0)\}^2}{\{f'(x_0)\}^3}$.

Example 15. *Show by applying formula (3) page 279 twice that the square root of $N=AB$ is given by*
$$\sqrt{N}\simeq\frac{S}{4}+\frac{N}{S}$$
where S stands for the sum of the factors A and B.

(M.A. Delhi; 1971; Delhi B.A. Hons., 1971)

Sol. We have
$$x_{i+1}=\frac{1}{2}\left(x_i+\frac{N}{x_i}\right)$$
where x_i is the approximation of $\sqrt{AB}$

and
$$x_{i+2}=\frac{1}{2}\left(x_{i+1}+\frac{N}{x_{i+1}}\right)$$
$$=\frac{1}{2}\left[\frac{1}{2}\left(x_i+\frac{N}{x_i}\right)+\frac{N}{\frac{1}{2}\left(x_i+\frac{N}{x_i}\right)}\right]$$
$$=\frac{1}{4}S+\frac{N}{S}$$
where $S=x+\frac{N}{x}$, $N=AB$

and $x=A$ since $\sqrt{AB}=A$ approx.

so that $S=A+B.$

Thus
$$\sqrt{N}\simeq\frac{S}{4}+\frac{N}{S}\cdot$$

Example 16. *Solve* $f(x)=x^3+x^2+3x+4=0$ *with the help of Chebyshev third order formula.*

Sol. In Example 6, page 273, we found that -1 is the initial approximation. Applying Chebyshev third order formula given in Example 14, page 282 together and using $x_0=-1$, $f(x_0)=1$
$$f(x_0)=3x^2+2x+3 \text{ so that } f'(x_0)=4$$
$$f''(x_0)=6x+2 \text{ so that } f''(x_0)=-4$$
Now
$$x_1=x_0-\frac{f(x_0)}{f'(x_0)}-\frac{1}{2}\frac{f''(x_0)\{f(x_0)\}^2}{\{f'(x_0)\}^3}$$
$$=-1-\frac{1}{4}-\frac{1}{2}\frac{(-4)(1)^2}{(4)^3}=-1{\cdot}21875$$
Taking
$$x=-1{\cdot}21875,\ x^2=1{\cdot}48535,\ x^3=2{\cdot}20626$$
$$f(x)=-0{\cdot}37716,\ f'(x)=5{\cdot}01855$$
$$f''(x)=-5{\cdot}30980$$
Thus second approximation is given by
$$x_2=-1{\cdot}21875-\frac{1}{2}\times\frac{-5{\cdot}30890\,(-0{\cdot}3776)^2}{(5{\cdot}01855)^3}$$
$$=-1{\cdot}21875+\frac{1}{2}\times\frac{7{\cdot}547643}{126{\cdot}408534}$$
$$=-1{\cdot}21875+{\cdot}000297=-1{\cdot}21846.$$
Similarly third and higher approximations can be obtained.

Example 17. *Solve* $x^3-12{\cdot}2\,x^2+7{\cdot}45x+42=0$*, by Newton's second order method, taking* $x_1=12$.

Sol.

1	– 12·20	7·45	42·60
	12·00	– 2·40	60·60
1	– 0·20	5·05	$\lvert 102{\cdot}60 = f(x_1)$
	12·80	141·60	
1	11·80	$\lvert 146{\cdot}65 = f'(x_1)$	
	12·00		
	$\lvert 23{\cdot}80 = \frac{1}{2} f''(x_1)$		

$$\frac{1}{h} = -\frac{146{\cdot}65}{102{\cdot}6} + \frac{23{\cdot}8}{146{\cdot}65} = -1{\cdot}27 \text{ ; by (1) } h = -7{\cdot}79, \text{ page 281}$$

$\therefore \quad x_0 = x_1 + h = 12{\cdot}00 - 0{\cdot}79 = \mathbf{11{\cdot}21}.$

14·6. Graeffe's Root-Squaring Method for solving a polynomial equation $P(x) = 0$, when the roots are real and distinct.

The method of solving an equation whose roots are near to each other is based on the fact that the given equation should be transformed into another whose roots are high powers of the roots of the given equation. Thus if the roots of the transformed equation are widely separated (Roots are said to be separated if the ratio of any root to the next larger root is negligible in comparison to unity), the roots of the given equation can be determined by taking roots of the transformed equation.

Consider the equation $a_0x^n + a_1x^{n-1} + a_2x^{n-2} + \cdots + a_n = 0, \quad a_0 \neq 0.$ Suppose that all the roots of this equation are real and different from each other. Keeping all even terms on one side and all odd terms on the other side and squaring, we have

$$(a_0x^n + a_2x^{x-2} + \cdots)^2 = (a_1x^{n-1} + a_3x^{n-3} + \cdots)^2$$

Putting $x^2 = y$ and simplifying we obtain

$$a_0^2y^n + (-a_1^2 + 2a_0a_2)y^{n-1} + (a_2^2 - 2a_1a_3 + 2a_4a_0)y^{n-2} + \cdots + (-1)^n a_n^2 = 0$$

$$\equiv b_0y^n + b_1y^{n-1} + b_2y^{n-2} + \cdots + b_n$$

so that
$$\begin{cases} b_0 = a_0^2 \\ b_1 = -a_1^2 + 2a_2\,a_0 \\ b_2 = a_2^2 - 2a_1a_3 + 2a_4a_0 \\ b_3 = -a_3^2 + 2a_0a_4 - 2a_1a_5 + 2a_6a_0 \\ \cdots\cdots\cdots\cdots\cdots\cdots\cdots\cdots \\ b_n = (-1)^n a_n^2 \end{cases}$$

$$\Rightarrow \quad (-1)^k b_k = a_k^2 - 2a_{k-1}a_{k+1} + 2a_{k-2}a_{k+2} - \cdots, \; k = 1, 2, \ldots, n$$

The above process can be repeated a number of times, say m till the roots are widely separated and the roots of the new equation

$$z^n + A_1 z^{n-1} + A_2 z^{n-2} + \cdots + A_n = 0$$

are $q_1, q_2, \ldots, q_n$ with $z = x^{2^m}$

Thus $$A_1 = -(q_1 + q_2 + \cdots + q_n)$$

$$= -q_1 \left(1 + \frac{q_2}{q_1} + \cdots + \frac{q_n}{q_1} \right) \cong -q_1$$

$$A_2 = (q_1 q_2 + q_2 q_3 + \cdots + q_{n-1} q_n)$$

$$= q_1 q_2 \left[1 + \frac{q_3}{q_1} + \cdots + \frac{q_{n-1}}{q_1} \cdot \frac{q_n}{q_2} \right]$$

$$\simeq q_1 q_2$$

$$A_3 = -(q_1 q_2 q_3 + q_1 q_3 q_4 + \cdots) = -q_1 q_2 q_3 \left(1 + \frac{q_4}{q_2} + \cdots \right)$$

$$\simeq -q_1 q_2 q_3.$$

....

$$A_n = (-1)^n q_1 q_2 \cdots q_n$$

Now, by our assumption, the roots $p_1, p_2, \ldots, p_n$ of the original equation are real and different and that

$$|p_1| > |p_2| > \cdots > |p_n|.$$

and since $q_i = q_i^{2^m}$, $i = 1, 2, \ldots, n$, we find that

$$|q_1| > |q_2| \geq \cdots \geq |q_n|$$

(The sign '>' means 'much larger than')

so that
$$\begin{cases} q_1 \simeq -A_1 \\ q_2 \simeq -A_2/A_1 \\ q_3 \simeq -A_3/A_2 \\ \ldots\ldots\ldots\ldots\ldots\ldots \\ q_n \cong -A_n/A_{n-1} \end{cases}$$

Thus knowing q's we can determine p's by taking 2^m th square root. Hence the roots of the original equation can be determined by substituting in the original equation to know whether the roots are positive or negative. The process is to be terminated when q_i is negligible compared to q_{i-1} and when, therefore, the coefficient of the equation are given in terms of the root by equation.

The coefficients b_k can be conveniently obtained from the coefficients of the original equation as follows :

Working Table

a_0	a_1	a_2	a_3.....	a_n
a_0^2	a_1^2	a_2^2	a_3^2...	a_n^2
	$-2a_0a_2$	$-2a_1a_3$	$-2a_2a_4$	
		$+2a_0a_4$	$+2a_1a_5$	
			$-2a_0a_6$	
b_0	$-b_1$	b_2	$-b_3$....	b_n

Example 18. *Use Graeff's method (squaring three times) to solve the equation*

$$x^3 - 5x^2 - 17x + 20 = 0. \qquad \textit{(B.A. Hons. Math Delhi, 1984)}$$

Sol. **Working Table**

	1	-1	-17	20
	1	25	289	400
		34	200	
First squaring	1	59	489	400
	1	348	239121	16×10^4
		-978	-47200	
Second squaring	1	2503	191921	16×10^4
	1	6265009	36933670241	256×10^8
		-383842	-800960000	
Third squaring	1	5881167	36032710241	256×10^8

Thus $\alpha_1 = \sqrt[8]{5881167} = \pm 7.017 \qquad [\because 2^3 = 8]$

$$\alpha_2 = \sqrt[8]{\frac{36032710241}{5881167}} = \pm 2.974$$

$$\alpha_3 = \sqrt[8]{\frac{256 \times 10^{-8}}{3603271041}} = \pm 3.958$$

By Descartes Rule of Signs, we find that the given equation has at most two real positive and one real negative roots. Substituting in the given equation, we find that the roots of given equation are

$$\alpha_1 = 7.017, \alpha_2 = -2.974 \text{ and } \alpha_3 = 3.958.$$

14·7. Birge-Vieta Method. We want to determine an approximate real number p of the polynomial equation

$$P_n(x) = x^n + a_1x^{n-1} + \cdots + a_{n-1}x + a_n = 0 \qquad ...(1)$$

where $a_1, ..., a_n$ are real numbers. We have, since $x = p$ is a trial factor,

$$P_n(x) = x^n + a_1x^{n-1} + \cdots + a_{n-1}x + a_n$$

$$\equiv (x-p)(x^{n-1} + b_1x^{n-2} + \cdots + b_{n-2}x + b_{n-1}) + R,$$

where $\qquad R = \text{remainder.} = P_n(p)$

Comparing the coefficients of like powers of x, we have

$$\left.\begin{array}{ll} a_1 = b_1 - p, & b_1 = a_1 + p \\ a_2 = b_2 - pb_1, & b_2 = a_2 + pb_1 \\ \dots\dots\dots & \dots\dots\dots \\ a_k = b_k - pb_{k-1}, & b_k = a_k + pb_{k-1} \\ \dots\dots\dots & \dots\dots\dots \\ a_{n-1} = b_{n-1} - pb_{n-2} & b_{n-1} = a_{n-1} + pb_{n-2} \\ a_n = R - pb_{n-1}, & R = a_n + pb_{n-1} \end{array}\right\} \quad \dots(2)$$

Introduce a quantity b_n and define the recurrence relations

$$b_k = a_k + pb_{k-1}, \quad k = 1, 2, \dots, n, b_0 = 1$$

The $\quad P_n(p) = R = b_n. \quad \dots(3)$

Now Newton-Raphson formula gives

$$p_1 = p_0 - \frac{P_n(p_0)}{P_n'(p_0)}$$

In order to determine $P_n'(p)$, we differentiate equation (2) and get

$$\frac{db_k}{dp} = b_{k-1} + p\frac{db_{k-1}}{dp}$$

and $\quad \dfrac{db_k}{dp} = c_{k-1}, c_0 = 1, 2, .., n-1 \quad \dots(4)$

$$\Rightarrow \quad c_{k-1} = b_{k-1} + pc_{k-2}. \, a_{k-1} \equiv \frac{db_k}{dp}.$$

$$\Rightarrow \quad c_k = b_k + pc_{k-1}, \; k = 1, 2, \dots, n-1$$

Thus c_k are determined from b_k in the same way as b_k are determined from a_k, and

$$p_1 = p_0 - \frac{b_n}{c_{c-1}}.$$

We have $P_n'(p) = \dfrac{dR}{dp} = \dfrac{db_n}{dp} = c_{n-1}.$

Working Table

	1	a_1	$a_2 \dots$	a_{n-2}	a_{n-1}	a_n
p		p	$pb_1 \dots$	pb_{n-3}	pb_{n-2}	pb_{n-1}
	1	b_1	$b_2 \dots$	b_{n-2}	b_{n-1}	$b_n = R = R_n(p)$
p		p	$pc_1 \dots$	pc_{n-3}	pc_{n-2}	
	1	c_1	$c_2 \dots$	c_{n-2}	$c_{n-1} = \dfrac{dR}{dp} = P_n'(p)$	

Example 19. *Use Birge-Vieta method to find a real root correct to three decimals for the equation*

$$x^3 - 5x^2 - 17x + 20 = 0.$$

Sol.

	1	−5	−17	20
1				
		1	−4	−21
	1	−4	−21	$-1 = b_n$
		1	−3	
	1	−3	$-24 = c_{n-1}$	

As a first approximation to the root near $x = 1$, we have

$$p_1 = p_0 - \frac{b_n}{c_{n-1}} = 1 - \frac{(-1)}{-24} = 0{\cdot}9583$$

	1	−5	−17	20
0·9583				
		0·9583	−3·8732	−20·0028
	1	−4·0417	−20·8732	−·0028
		0·9583	−2·9548	
	1	−3·0834	−23·8280	

As a second approximation, we have

$$p_2 = 0{\cdot}9583 - \left(\frac{-{\cdot}0028}{-28{\cdot}8280}\right) = 0{\cdot}9583 - {\cdot}000118$$

$$= 0{\cdot}9582$$

14·8. Lin-Bairstow Method for finding Complex Roots of a polynomial equation.

Consider the polynomial equation of nth degree

$$P_n(x) = a_0x^n + a_1x^{n-1} + \cdots + a_{n-1}x + a_n = 0,\ a_0 \neq 0 \qquad \ldots(1)$$

with *real* coefficients having complex roots. Since complex roots appear in *complex conjugate* pairs $a \pm ib$, a and b being real, and $i = \sqrt{-1}$ each such pair corresponds to a quadratic factor $x^2 + rx + s$ with real coefficients. The method was suggested by L. Bairstow in 1914 and modified by Hitchcock in 1944. If $P_n(x)$ is divided by a quadratic trial factor $x^2 + px + q$, one obtains as a quotient a polynomial $Q(x)$ of degree $(n - 2)$ and remainder $Rx + S$. Thus we write

$$a_0x^n + a_1x^{n-1} + \cdots + a_n \equiv (x^2 + px + q)(b_0x^{n-2} + b_1x^{n-3} + \cdots + b_{n-2}) + Rx + S$$

On comparing the coefficients of like powers of x we obtain

$$\left.\begin{aligned}
a_0 &= b_0 \\
a_1 &= b_1 + pb_0 \\
a_2 &= b_2 + pb_1 + pb_0 \\
&\cdots\cdots\cdots\cdots\cdots\cdots \\
a_k &= b_k + pb_{k-1} + qb_{k-2} \\
&\cdots\cdots\cdots\cdots\cdots\cdots \\
a_{n-2} &= b_{n-2} + pb_{n-3} + qb_{n-4} \\
a_{n-1} &= R + pb_{n-2} + qb_{n-3} \\
a_n &= S + qb_{n-2}
\end{aligned}\right\} \quad \ldots(2)$$

Setting $b_{-1} = b_{-2} = 0,$

$$b_{n-1} = R$$

and $b_n = S - pR$

the equation (2) can be rewritten in the form

$$\left.\begin{aligned}
&b_k = a_k - pb_{k-1} - qb_{k-2},\ k = 0, 1, 2, \ldots, n \\
\text{with} \quad &R = b_{n-1} \\
\text{and} \quad &S = b_n + pb_{n-1}
\end{aligned}\right\} \quad \ldots(3)$$

The coefficients b_k of the polynomial $Q(x)$ and the coefficients R and S of the remainder are functions of p and q. We want to find p^* and q^* such that $x^2 + p^*x + q^*$ is a factor of the polynomial. In other words, we want to determine p^* and q^* such that

$$R(p, q) = 0 \text{ and } S(p, q) = 0.$$

We therefore assume that p^* and q^* are near to the values p and q so that

$$\left.\begin{aligned}
p^* &= p + \Delta p \\
q^* &= q + \Delta q
\end{aligned}\right\} \quad \ldots(4)$$

where Δp and Δq are small. Thus

$$R(p + \Delta p, q + \Delta q) = 0 \text{ and } S(p + \Delta p, q + \Delta q) = 0 \quad \ldots(5)$$

Expanding in Taylor series and neglecting terms of higher powers than the first increment, we get

$$\left.\begin{aligned}
R(p, q) + \frac{\partial R}{\partial p}\Delta p + \frac{\partial R}{\partial q}\Delta q = 0 \\
S(p, q) + \frac{\partial S}{\partial p}\Delta p + \frac{\partial S}{\partial q}\Delta q = 0
\end{aligned}\right\} \quad \ldots(6)$$

These equations can be solved for Δp and Δq after substituting the values of the four partial derivatives. Thus we get the improved values of p^* and q^* and solving $x^2 + p^*x + q^* = 0$ we get the improved estimate of the roots of the equation (1).

To find the four partial derivatives, we differentiate partially equation (3).

Thus

$$\left.\begin{aligned} -\frac{\partial b_k}{\partial p} &= b_{k-1} + p\frac{\partial b_{k-1}}{\partial p} + q\frac{\partial b_{k-2}}{\partial p}; \quad \frac{\partial b_0}{\partial p} = \frac{\partial b_{-1}}{\partial p} = 0 \\ -\frac{\partial b_k}{\partial q} &= b_{k-2} + p\frac{\partial b_{k-1}}{\partial q} + q\frac{\partial b_{k-2}}{\partial q}; \quad \frac{\partial b_0}{\partial q} = \frac{\partial b_{-1}}{\partial q} = 0 \end{aligned}\right\} \quad ...(7)$$

From these we obtain

$$\frac{\partial b_k}{\partial p} = \frac{\partial b_{k+1}}{\partial q} = -c_{k-1}, \text{ say, } k = 0, 1, 2, ..., n \quad ...(8)$$

This result can be shown to hold for all k by mathematical induction since if it holds goods for all k upto $m-1$, then

$$-\frac{\partial b_m}{\partial p} = b_{m-1} + p\frac{\partial b_{m-1}}{\partial p} + q\frac{\partial b_{m-2}}{\partial p}$$

$$= b_{m-1} + p\frac{\partial b_m}{\partial q} + q\frac{\partial b_{m-1}}{\partial q} = -\frac{\partial b_{m+1}}{\partial q}$$

and thus it holds for m. It can be easily proved to hold for $k = 0, 1, 2, ...$

By virtue of Eq. (8), the Eqs. (7) can be written into a single recurrence relation

$$c_k = b_k - pc_{k-1} - qc_{k-2} \quad ...(9)$$

with $\quad c_{-1} = 0$ and $c_0 = 0 \quad k = 0, 1, 2, ..., n-1.$

Hence the c's can be computed from b's in exactly the same way as the b's from a's.

From Eq. (3) and (8),

$$R = b_{n-1}, \frac{\partial R}{\partial p} = \frac{\partial b_{n-1}}{\partial p} = -c_{n-2}, \frac{\partial R}{\partial q} = \frac{\partial b_{n-1}}{\partial q} = -c_{n-2}$$

and $$S = b_n + pb_{n-1}, \frac{\partial S}{\partial p} = -c_{n-1} - pc_{n-2} + b_{n-1},$$

$$\frac{\partial S}{\partial q} = -c_{n-2} - pc_{n-3}$$

Substituting these values in (6), we obtain equations for determining Δp and Δq,

$$\left.\begin{aligned} c_{n-2}\Delta p + c_{n-3}\,\Delta q &= b_{n-1} \\ (c_{n-1} - b_{n-1})\Delta p + c_{n-2}\Delta q &= b_n \\ \Rightarrow \quad -(pc_{n-2} + qc_{n-3})\Delta p + c_{n-2}\Delta q &= b_n \end{aligned}\right\} \quad ...(10)$$

Procedure. Compute b_k and c_k from a_k using relations (3) and (9)

$$\Delta p = -\frac{b_n c_{n-3} - b_{n-1}c_{n-2}}{c_{n-2}^2 - c_{n-3}(c_{n-1} - b_{n-1})}$$

$$\Delta q = -\frac{b_{n-1}(c_{n-1} - b_{n-1}) - b_n c_{n-2}}{c_{n-2}^2 - c_{n-3}(c_{n-1} - b_{n-1})} \quad ...(11)$$

Working Table

	1	a_1	a_2	..... a_{n-2}	a_{n-2}	a_n
$-p$		$-p$	$-pb_1$	..... $-pb_{n-2}$	$-pb_{n-2}$	
$-q$			$-q$	..... $-qb_{n-4}$	$-qb_{n-3}$	$-qb_{n-2}$
$-p$	1	b_1	b_2	..... b_{n-2}	$b_{n-1} = R$	S
$-q$		$p\,b$	$-pc_1$	..... $-pc_{n-3}$	$-pc_{n-2}$	
			$-q$	..... $-pc_{n-4}$	$-qc_{n-3}$	$b_0 = 1$
	1	c_1	c_2	..... c_{n-2}	c_{n-1}	

Step-by-Step Procedure for obtaining p and q.

1. Select the initial values for p and q. ($a_1, ..., a_n$. are n given coefficients.

2. Compute all b_k's from (3) and

3. Compute all c_k's from (9).

4. Compute Δp and Δq from (11), page 291.

5. Increase p_i and q_i by the amounts Δp and Δq so that

$$p_{i+1} = p_i + \Delta p,$$

$$q_{i+1} = q_i + \Delta q,$$

where i is the number of iterations.

6. Test for convergence. $M = |\Delta p| + |\Delta q| < \varepsilon$.

Example 20. *Solve* $x^4 - 8x^3 + 39x^2 - 62x + 50 = 0$, *using Bairstow method, taking* $p = q = 0$

Sol.

	1	−8	39	−62	50
0		0	0	0	0
0		0	0	0	0
	1	−8	39	−62	50
0		0	0	0	
0		0	0	0	
	1	−8	39	−62	

Here $b_4 = a_4 - pb_3 - qb_2 = 50$

$b_3 = -62, \qquad c_3 = -62$

$b_2 = 39, \qquad c_2 = 39$

$b_1 = -8, \qquad c_1 = -8$

$$\Delta p = -\left(\frac{b_4 c_1 - b_3 c_2}{c_2^2 - c_1(c_3 - b_3)}\right) = -\left[\frac{-(50)(8)+(62)(39)}{(39)^2 + 8(-62+62)}\right]$$

$$= -\frac{2018}{1521} = -1{\cdot}327$$

$$\Delta q = -\left(\frac{b_3(c_3 - b_3) - b_4 c_2}{c_2^2 - c_1(c_3 - b_3)}\right)$$

$$= -\left[\frac{(-62)(-62+62) - (50\times 39)}{(39)^2 + 8(-62+62)}\right]$$

$$= \frac{50\times 39}{39\times 39} = 1{\cdot}282$$

$\therefore \qquad p_1 = 0 - 1{\cdot}327 = -1{\cdot}327$ and $q_1 = 0 + 1{\cdot}282 = 1{\cdot}282$

Again

	1	– 8	39	– 62	50
1·327		1·327	– 8·855	38·119	– 37·002
– 1·282			– 1·282	8·554	
	1	– 6·673	28·863	– 15·327	12·998
1·327			– 7·094	27·186	
		1·327			
– 1·282			– 1·282	6·853	
	1	– 5·346	20·487	18·712	

Here $\quad b_4 = 50 - (-1{\cdot}32)(-15{\cdot}327) - 37{\cdot}002 = -7{\cdot}34$

$b_3 = -15{\cdot}327 \qquad c_3 = 18{\cdot}712$

$b_2 = 28{\cdot}863 \qquad c_2 = 20{\cdot}487$

$b_1 = -6{\cdot}673 \qquad c_1 = -5{\cdot}346$

$$\Delta p_1 = -\left[\frac{(-7{\cdot}341)(-5{\cdot}346) - (-15{\cdot}327)(20{\cdot}487)}{(20{\cdot}487)^2 + (5{\cdot}346)(18{\cdot}712 + 15{\cdot}327)}\right]$$

$$= -\frac{353{\cdot}248935}{601{\cdot}689663} = -0{\cdot}587$$

$$\Delta q_1 = -\left[\frac{(-15{\cdot}327)(34{\cdot}039) - (7{\cdot}34)(20{\cdot}487)}{601{\cdot}690}\right] = +1{\cdot}118$$

$\therefore \qquad p_2 = p_1 + \Delta p_1 = -1{\cdot}327 - 0{\cdot}587 = 1{\cdot}914$

$q_2 = q_1 + \Delta q_1 = 1{\cdot}282 + 1{\cdot}118 = 2{\cdot}400$

$x^2 - 1{\cdot}914\,x + 2{\cdot}400 = 0$ is one of the quadratic factors and the roots of this equation are $\dfrac{1{\cdot}914 \pm i\sqrt{2}}{2}$.

EXERCISE 14

1. (*a*) Describe a method of finding an approximate root and improving upon it of a polynomial equation.

   (*b*) Explain how to obtain with the help of graphs, the approximate value of a root of an equation $f(x) = 0$.

   (*c*) Expalin the method of iteration for finding the roots of the equation $f(x) = 0$. How can this method be improved? Show graphically that the convergence to the root is rapid if $\left|\frac{d}{dx}g(x)\right|_{x=p} < \left|\frac{d}{dx}h(x)\right|_{x=p}$ where $x = p$ is the desired root and $f(x) = g(x) - h(x)$.

2. (*a*) State the 'Location' theorem. Locate the roots of the equation $\cos x \cosh x + 1 = 0$ using graphical method.

   (*b*) By iteration, find the root up to four decimal places :

   (*i*) $e^x - 3x = 0$ (*ii*) $\tan x = x$, (*iii*) $x = \frac{1}{2} + \sin x$ near $x = 1{\cdot}5$

   (*B.Sc. Delhi, 1995*)

   (*c*) Find the smallest positive root of

   (*i*) $e^{-x} = \sin x$

   (*ii*) $\tan x + \tanh x = 0$ to five decimals. Root lies in $(1{\cdot}6, 3{\cdot}0)$

   (*A.M.I.E. 1991*)

   [**Hint.** (*a*) We have $\cos^{-1} x = -\cosh x = y$,

   (*c*) (*i*) $y = e^{-x}$ and $y = \sin x$ interest at $x = 0{\cdot}6$.]

3. (*a*) Explain the method of obtaining the derivatives of a polynomial by Synthetic Division.

   (*b*) Find the number of real roots of the polynomial equation

   $$P(x) = 4x^4 - 4x^3 + 4x^2 + k = 0.$$

   [All the roots will be real if $k < 0$ and $1 + k > 0 \Rightarrow -1 < k < 0$.]

4. (*a*) Establish Newton-Raphson's method for determining the approximate value of a root of the equation $f(x) = 0$ and show that this method converges when $|f(x)f''(x)| < [f'(x)]^2$.

   (*b*) Compute the real root to four decimals of the equation; $x^3 - 5x + 3 = 0$ in the interval [1, 2] by the Secant method.

   (*i*) $2e^{-x} = \frac{1}{x+2} + \frac{1}{x+1}$, when two roots are greater than -1.

   (*ii*) $x^6 - x^4 - x^3 - 1 = 0$

   which lies between 1 and 2 by Newton-Raphson method.

   (*I.A.S. 1990*)

   (*iii*) $x^3 - x - 4 = 0$ (*B.Sc. Hons. Math,. Delhi, 1994*)

   which lies between 1·5 and 2 by False Position method.

   (*iv*) $x^4 - x - 10 = 0$ by Newton-Raphson method, root being near $x = 2$.

   (*v*) $3x - \cos x - 1 = 0$ by Newton-Raphson method.

   (*B.Sc. Hons. Math., Delhi, 1994*)

(*vi*) $\log_e x - \cos x = 0$ by Newton-Raphson method, root lying between 1 and 2. (*I.A.S., 1995*)

(*vii*) $x^4 - 3x^3 + 2x^2 + 2x - 7 = 0$ by Chebyshev third order formula

(*viii*) $\sin x - \dfrac{x+1}{x-1} = 0$ near $x = -5\cdot5$

(*ix*) $x^3 - x - 1 = 0$ using Secant method between 1 and 2.

(*x*) $x^2 + 4\sin x = 0$ by Newton-Raphson method.

(*B.Sc., Agra, 1996, I.A.S., 1992*)

(*xi*) $x\sin x + \cos x = 0$ (*I.A.S., 1988 ; B.Sc. Hons. Math, Delhi, 1995*)

(*xii*) $x^3 - x - 4 = 0$ by Newton-Raphson method between 1 and 2 correct to three places of decimals.

(*xiii*) $x - e^{-x} = 0$ by Newton-Raphson method correct to three decimal places. (*B.Sc. Math Hons., Delhi, 1998*)

5. (*a*) The equation $x^4 - 5x^3 - 12x^3 + 76x - 79 = 0$ has two roots close to $x = 2$. Find these roots to four decimals.

(*b*) Find two roots near $x = 1$ of $2x^4 + 16x^3 + x^2 - 74x + 56 = 0$.

(*c*) Solve $2e^x - 3x^2 - 2\cdot5644 = 0$ correct to three places given that two roots are close to $0\cdot62$.

6. Evaluate by Newton-Raphson formula the following :

(*i*) $\sqrt{29}$ (*ii*) $(10)^{1/3}$ (*iii*) $\left(\frac{1}{2}\right)^{1/4}$ (*iv*) $\sqrt{18}$

(*B.Sc. Hons. Math, Delhi, 1991*)

Correct to four places of decimals. (*B.Sc. Indore, 1989*)

7. Solve :

(*a*) $x^2 + 1\cdot2x^2 - 4x - 4\cdot8 = 0$, by Birga-Vieta method, root being $x = -1$.

(*b*) $x\log_{10} x = 1\cdot2$.

(*i*) by the method of interval halving. (*ii*) by the method of false position.

8. (*a*) Use the iteration method $x_{n+1} = a - b_n^2$ to compute the positive root of the equation $x = a - bx^2$, a and b being positive. What is the condition for convergence ?

(*b*) The equation $x^2 + ax + b = 0$ has two real roots α and β. Show that the iterative method $x_{k+1} = -(a\,x_k + b)/x_k$ is convergent near $x = \alpha$ if $|\alpha| > |\beta|$

[**Hint.** The method is of the form $x_{k+1} = g(x_k)$. Also $\alpha + \beta = -a$, $\therefore\ \alpha\beta = b$

$$\therefore \quad |g'(\alpha)| = \left|\frac{b}{\alpha^2}\right| = \left|\frac{\alpha\beta}{\alpha^2}\right| = \left|\frac{\beta}{\alpha}\right| < 1 \Rightarrow |\beta| < |\alpha|.]$$

9. Solve (*i*) $x^3 - 8x^2 + 17x - 10 = 0$ (*ii*) $x^3 - 6\cdot8x^2 + 6\cdot72x - 1\cdot1 = 0$ by Graeffe's Root squaring method (squaring three times).

(*iii*) $x^3 - 6x^2 + 11x - 6 = 0$ (*iv*) $x^3 + 9x^2 + 23x + 14 = 0$

10. (*a*) Solve the equation $x^4 - 8x^3 + 39x^3 - 62x + 50 = 0$ using Bairstow's method. Start with $p = q = 0$, (*b*) Apply Bairstow's method to $x^2 - 3x^3 + 20x^2 + 44x + 54 = 0$ to find a quadratic factor close to $x^3 + 2x + 2$ and hence solve the equation.

11. Apply Newton-Raphson formula to the equation $1 - \alpha/x^2 = 0$ to give the following recurrence relation for $\sqrt{a}$.

$$x_{n+1} = x_n\left[1 - \frac{x_n^2 - a}{2a}\right]$$

12. Using Chebyshev method of third order, find a real root near $x = 2$ correct to three decimal places of the equation

$$x^3 - x^2 - x - 1 = 0$$

13. (*a*) If x_0 is an approximation to a root r of $x = g(x)$ and $g'(x_0) \neq 1$,

then $x_1 = \dfrac{g(x) - x\,g'(x_0)}{1 - g'(x_0)}$ is suitable for iteration to the root r.

[**Hint.** $x = g(x),\ f(x) = x - g(x)$

$x_1 = g(x_0),\ f'(x) = 1 - g'(x)$

$$x_1 = x - \frac{f(x)}{f'(x)} = x - \frac{x - g(x_0)}{1 - g'(x_0)} = \frac{g(x) - x\,g'(x)}{1 - g'(x)}.$$

(*b*) How would the constant α be chosen to ensure the fastest possible convergence with the iterative formula

$$x_{n+1} = \frac{\alpha\, x_n + x_2^2}{\alpha + 1} + 1\ ?$$

14. (*a*) Show that the following two sequences have both convergence of the second order with the same limit $\sqrt{a}$.

(*b*) Let $x = a$ be a simple root of the equation $f(x) = 0$. Find the root by means of iterative formula

$x_{i+1} = x_i - f(x_i)/\{f'(x_i)\}$. Show that the error of convergence is 2.

(*c*) If $x_{n+1} = \dfrac{1}{2}x_n\left(1 + \dfrac{a}{x_n^2}\right)$ and $x_{n+1} = \dfrac{1}{2}x_n\left(3 - \dfrac{x_n^2}{a}\right)$

(*B.Sc. Kurukshetra, 1987; B.A. Rohtak, 1977*)

If x_n is a suitably close approximation to $\sqrt{a}$, show that the error in the first formula for x_{n+1} is about one-third of that in the second formula and deduce that $x_{n+1} = (\frac{1}{8}x_n)\,[6 + (3a/x_n^2) - x_n^2/a]$. Give a sequence with third order convergence.

[**Hint.** Show that

$$\sqrt{a} - x_{n+1} = \frac{1}{2x_n}(\sqrt{a} - x_n)^2 \ \therefore\ \varepsilon_{i+1} = a\varepsilon_i^2]$$

[**Hint.** (*c*) $x_{n+1} - \sqrt{a} = \dfrac{1}{2}\left(x_n + \dfrac{a}{x_n}\right) - \sqrt{a} = \dfrac{1}{2x_n}(x_n - \sqrt{a})^2$

$\Rightarrow\quad \varepsilon_{n+1} = \dfrac{1}{2x_n}\varepsilon_n^2$. Hence convergence is quadratic]

Similarly $x_{n+1} - \sqrt{a} = \frac{1}{2} x_n \left(3 - \frac{x_n^2}{a}\right) - \sqrt{a}$

$$= \frac{1}{2} x_n \left(1 - \frac{x_n^2}{a}\right) + \left(x_a + \sqrt{a}\right)$$

$$= \left(x_n - \sqrt{a}\right)\left[1 - \frac{x_n}{2a}\left(x_n + \sqrt{a}\right)\right] = -\left(x_n + 2\sqrt{a}\right)\frac{(x_n - \sqrt{a})^2}{2a}$$

$= C \,.\, \varepsilon_n^2$ Thus the convergence is quadratic.

Next show that $x_{n+1} - \sqrt{a} = \frac{x_n}{8}\left(6 + \frac{3a}{x_n^2} - \frac{x_n^2}{a}\right) - \sqrt{a}$

$= -(x_n + 3\sqrt{a})(x_n - \sqrt{a})^2 / 8x_n a$

Thus $\varepsilon_{n+1} = \text{const.}\ \varepsilon_n^3$. Hence the result.

15. (In continuation of Q. 14).

(*i*) Use this method to show that $\sqrt{5} = 2{\cdot}236068$.

(*ii*) Prove that, if $\lim x_n$ exists, then this limit is $\sqrt{a}$.

(*iii*) Prove that if $a \le x_n \le 1$, then $a \le x_{r+1} \le 1$.

[**Hint.** Consider $x_{n+1} - 1$ and $x_{n+1} - a$]

(*iv*) Prove that, if $0 < a < 1$, then the iteration does converge.

16. Obtain the iterative formulae :

$$x_{n+1} = \frac{1}{3}\left(2x_n + \frac{a}{x_n^2}\right),\quad x_{n+1} = \frac{1}{5}\left(4x_n + \frac{a}{x_n^4}\right),\quad x_{n+1} = \frac{1}{2}\left(x_n + \frac{1}{ax_n}\right)$$

for $\sqrt[3]{a}$, $\sqrt[5]{a}$ and $1/\sqrt{a}$ respectively.

17. Use the following iteration formula for $1/\sqrt{a}$ to find $1/\sqrt{5}$ to four decimals :

$$x_{n+1} = (3 - ax_n^2)x_n/2$$

Show that the convergence is of second order.

18. Show that the modified Newton-Raphson method

$$x_{n+1} = x_n - \frac{2f(x_n)}{f'(x_n)}$$

gives a quadratic convergence when the equation $f(x) = 0$ has a pair of double roots in the neighbourhood of $x = x_n$. (*M.Sc. Meerut, 1971*)

[**Hint.** Here $\varepsilon_{n+1} = \varepsilon_n - \frac{2f(a+\varepsilon_n)}{f'(a+\varepsilon_n)}$

$$= \varepsilon_n - 2\frac{[f(a) + \varepsilon_n f'(a) + \frac{1}{2}\varepsilon_n^2 f''(a) + \frac{1}{3!}\varepsilon_n^3 f'''(a) + \cdots]}{f'(a) + \varepsilon_n f''(a) + \frac{1}{2}\varepsilon_n^2 f'''(a) +}$$

$$= \varepsilon_n - \frac{2\varepsilon_n\left[\frac{1}{2}f''(a) + \frac{1}{6}f'''(a) + \cdots\right]}{f''(a) + \frac{1}{3}\varepsilon_n f'''(a) + \cdots]} \quad \because f(a) = 0,\ f'(a) = 0$$

$$= \frac{1}{6}\varepsilon_n^2 \frac{f'''(a)}{f''(a)} + \cdots \qquad \simeq \frac{1}{6}\varepsilon_n^2 \frac{f'''(a)}{f''(a)} = \text{Const.}\ \varepsilon_n^2\,.$$

Hence the convergence is quadratic.]

19. By applying Newton-Raphson formula to $f(x) = 1 - \frac{1}{ax}$ obtain the recurrence formula $x_{i+1} = x_i(2 - ax_i)$

for the iterative determination of the reciprocal of a. Show that if ε_i denotes the error in x_i, there follows $\varepsilon_{i+1} = a\,\varepsilon_i^2$.

$$\left[\textbf{Hint.}\ x_{i+1} = x_i - \frac{\left(1 - \frac{1}{ax_i}\right)}{\frac{1}{ax_i^2}} = \text{etc.}\right.$$

$$\left.\text{Let } \alpha = \frac{1}{a},\ x_i = \alpha - \varepsilon_i, x_{i+1} = \alpha - \varepsilon_{i+1} = \alpha - \varepsilon_i\,[\alpha - a(\alpha - \varepsilon_i)] = \text{etc.}\right]$$

20. Show by starting from the rough approximation $x = \pi$ as a root of the equation $\sin x = \alpha x$ where α is small, that a much better approximation can be obtained as $x = (1 - \alpha + \alpha^2)\pi$.

[**Hint.** Let $f(x) = \sin x - \alpha x$

$\therefore\ \frac{f(\pi)}{f'(\pi)} = \alpha\pi(1 + \alpha)^{-1} = \alpha\pi(1 - \alpha)$ approx. Hence x = etc.]

21. In the study of Fraunhofer diffraction, we encounter the equation

$$\left(\frac{\sin x}{x}\right)^2 = \frac{1}{2}$$

Solve for x in radians.

$\left[\textbf{Hint.}\ \sin x = \frac{1}{\sqrt{2}}x.\ \text{We find from graph, } x_0 = 1{\cdot}5\ ;\right]$

22. By applying Newton-Raphson method to $f(x) = 1 - a/x^n$,

prove that $x_{k+1} = \frac{1}{n}\left[(n+1)x_k - \frac{x_k^{n+1}}{a}\right]$

if $\varepsilon_k = \alpha - x_k$ and $\alpha = a^{1/n}$, show that approximately

$$\varepsilon_{k+1} \simeq \frac{n+1}{2\alpha}\,\varepsilon_k^2.$$

23. Determine α_1 and α_2 so that the order of the iterative method $x_{i+1} = x_i - \alpha_2 g_1(x_i) - \alpha_2 g_2(x_i)$, where $g_1(x_i) = f(x_i)/f'(x_i)$, $g_2(x_i) = f(x_i)/f'(x_i) + \beta g_1(x_i)$, $\beta \neq 0$ for finding a root of the equation $f(x) = 0$ becomes as high as possible.

[**Hint.** Let $x_i = \xi + \varepsilon_i$ and $f(\xi) = 0$ in $g_1(x_i)$ and $g_2(x_i)$. Then show that

$$g_1(\varepsilon_i + \xi) = \varepsilon_i - \tfrac{1}{2} f''(\xi)\,\varepsilon_i^2 + 0(\varepsilon^3),\ g_2(\xi + \varepsilon_i) = \varepsilon_i + \tfrac{1}{2}\frac{f''(\xi)}{f'(\xi)}$$
$$[1 - 2(1 + \beta)]\,\varepsilon_i^2 + 0(\varepsilon_i^3).]$$

15

SIMULTANEOUS LINEAR ALGEBRAIC EQUATIONS

15·1. Introduction. A large number of methods of solving the system of linear equations and a variety of computers are available to solve such equations. We give below a few direct as well as indirect or ilterative methods for the solution of systems of linear equations.

Direct Methods

15·2·1. Method of Determinants. Consider the following set of n linear equations in n unknowns $x_1, x_2, ..., x_n$;

$$\left.\begin{array}{l} a_{11}x_1 + a_{12}x_2 + \cdots + a_{1n}x_n = b_1 \\ a_{21}x_1 + a_{22}x_2 + \cdots + a_{2n}x_n = b_2 \\ \cdots\cdots\cdots\cdots\cdots\cdots\cdots\cdots\cdots \\ a_{n1}x_2 + a_{n2}x_2 + \cdots + a_{nn}x_n = b_n \end{array}\right\} \text{ or } AX = B \quad ...(15\cdot1)$$

where A is a n-square non-singular matrix $(a_{ij}), i, j, = 1, 2, ..., n$ $X = (x_1, x_2....x_n)'$, $B = (b_1\ b_2...b_n)'$ and the n^2 coefficient a_{ij} and the n b_i are given. Let the determinant of the coefficients

$$D = |\, a_{ij} \,| \neq 0, j \,.\, i = 1, 2, ..., n$$

which is the condition that the solution should be unique.

Then, by Cramer's rule, the solution of equation (15·1) is given by

$$x_j = \frac{D_j}{D}, j = 1, 2\ , ..., n \quad ...(15\cdot2)$$

where $$D_j = \begin{vmatrix} a_{11}\ a_{12}.....a_{1s-1}\ b_1\ a_{1s+1}.... & a_{1n} \\ a_{21}\ a_{22}.....a_{2s-1}\ b_2\ a_{2s+1}, & a_{2n} \\ \cdots\cdots\cdots\cdots\cdots\cdots\cdots\cdots\cdots & \\ a_{n1}\ a_{n2}......a_{n2,\,s-1}\ b_n\ a_{ns+1}.... & a_{nn} \end{vmatrix}$$

$$= |\, a_{ij} \,|,\ s = 1, 2, ..., n.$$

From (15·2), we find that the solution of equations (15·1) requires the evaluation of $n+1$ determinants each of order n. A method due to Chio in which each element of the determinant to be evaluated is expressed by means of a determinant of order 2 and is

$$D=\frac{1}{a_{11}^{n-2}}\begin{vmatrix} \begin{vmatrix} a_{11} & a_{12} \\ a_{21} & a_{22} \end{vmatrix} & \begin{vmatrix} a_{11} & a_{13} \\ a_{21} & a_{23} \end{vmatrix} & \cdots & \begin{vmatrix} a_{11} & a_{1n} \\ a_{21} & a_{2n} \end{vmatrix} \\ \begin{vmatrix} a_{11} & a_{12} \\ a_{31} & a_{32} \end{vmatrix} & \begin{vmatrix} a_{11} & a_{13} \\ a_{31} & a_{33} \end{vmatrix} & \cdots & \begin{vmatrix} a_{11} & a_{1n} \\ a_{31} & a_{3n} \end{vmatrix} \\ \cdots & \cdots & \cdots & \cdots \\ \begin{vmatrix} a_{11} & a_{12} \\ a_{n1} & a_{n2} \end{vmatrix} & \begin{vmatrix} a_{11} & a_{13} \\ a_{n1} & a_{n3} \end{vmatrix} & \cdots & \begin{vmatrix} a_{11} & a_{1n} \\ a_{n1} & a_{nn} \end{vmatrix} \end{vmatrix}$$

Note. The use of determinants for the solution of linear equations becomes excessively cumberous if the number of equations exceeds four.

Example 1. *Solve the following system of equations by determinants :*

Eqs.	x_1	x_2	x_3	x_4	b
I	*3*	2	*1*	0	*8·70*
II	2	3	2	*1*	*11·25*
III	*1*	2	3	2	*11·20*
IV	*0*	*1*	2	3	*9·45*

Sol. We have $n=4$.

$$D=\begin{vmatrix} 3 & 2 & 1 & 0 \\ 2 & 3 & 2 & 1 \\ 1 & 2 & 3 & 2 \\ 0 & 1 & 2 & 3 \end{vmatrix}=\frac{1}{3^2}\begin{vmatrix} \begin{vmatrix} 3 & 2 \\ 2 & 3 \end{vmatrix} & \begin{vmatrix} 3 & 1 \\ 2 & 2 \end{vmatrix} & \begin{vmatrix} 3 & 0 \\ 2 & 1 \end{vmatrix} \\ \begin{vmatrix} 3 & 2 \\ 1 & 2 \end{vmatrix} & \begin{vmatrix} 3 & 1 \\ 1 & 3 \end{vmatrix} & \begin{vmatrix} 3 & 0 \\ 1 & 2 \end{vmatrix} \\ \begin{vmatrix} 3 & 2 \\ 0 & 1 \end{vmatrix} & \begin{vmatrix} 3 & 1 \\ 0 & 2 \end{vmatrix} & \begin{vmatrix} 3 & 0 \\ 0 & 3 \end{vmatrix} \end{vmatrix}$$

$$=\frac{1}{9}\begin{vmatrix} 5 & 4 & 3 \\ 4 & 8 & 6 \\ 3 & 6 & 9 \end{vmatrix}=\frac{1}{9}\cdot\frac{1}{5^1}\begin{vmatrix} \begin{vmatrix} 5 & 4 \\ 4 & 8 \end{vmatrix} & \begin{vmatrix} 5 & 3 \\ 4 & 6 \end{vmatrix} \\ \begin{vmatrix} 5 & 4 \\ 3 & 6 \end{vmatrix} & \begin{vmatrix} 5 & 3 \\ 3 & 9 \end{vmatrix} \end{vmatrix}$$

$$=\frac{1}{45}\begin{vmatrix} 24 & 18 \\ 18 & 36 \end{vmatrix}=\frac{540}{45}=12.$$

$$D_1 = \begin{vmatrix} 2 & 1 & 0 & 8\cdot70 \\ 3 & 2 & 1 & 11\cdot25 \\ 2 & 3 & 2 & 11\cdot20 \\ 1 & 2 & 3 & 9\cdot45 \end{vmatrix} = 21$$

Similarly, $D_2 = 15\cdot6,\ D_3 = 9,\ D_4 = 25\cdot8$

$$\therefore \qquad x_1 = \frac{D_1}{D} = \frac{21}{12} = \frac{21}{12} = 1\cdot75,\ x_2 = \frac{D_2}{D} = \frac{15\cdot6}{12} = 1\cdot30,$$

$$x_3 = \frac{D_3}{D} = \frac{9}{12} = 0\cdot75,\ x_4 = \frac{D_4}{D} = \frac{25\cdot8}{12} = 2\cdot15.$$

15·2·2. Cholesky's Method

Suppose A is a symmetric square matrix. Let us decompose A as $A = TT'$ where T is a lower triangular matrix and T' is its transpose.

Now $A^{-1} = (TT')^{-1} = (T')^{-1}T^{-1} = (T^{-1})'T^{-1}$

Thus A^{-1} can be determined if T^{-1} is known.

Consider A to be a matrix of order 3.

$$\text{Now } A = TT' = \begin{bmatrix} a_{11} & a_{12} & a_{13} \\ a_{21} & a_{22} & a_{23} \\ a_{31} & a_{32} & a_{33} \end{bmatrix} = \begin{bmatrix} t_{11} & 0 & 0 \\ t_{21} & t_{22} & 0 \\ t_{31} & t_{32} & t_{33} \end{bmatrix}.$$

$$\times \begin{bmatrix} t_{11} & t_{21} & t_{31} \\ 0 & t_{22} & t_{32} \\ 0 & 0 & t_{33} \end{bmatrix} = \begin{bmatrix} t^2_{11} & t_{11}\ t_{22} & t_{11}\ t_{21} \\ t^2_{21} + t^2_{22} & t_{21}\ t_{31} + t_{32}\ t_{33} & 0 \\ t^2_{31} + t^2_{32} + t^2_{33} & 0 & 0 \end{bmatrix}$$

where t's are to be determined in terms of a'^s.

$\therefore\ t_{11} = \sqrt{a_{11}},\ t_{11}\ t_{21} = a_{12} \Rightarrow t_{21} = a_{12}/t_{11},\ t_{11}/t_{31} = a_{13} \Rightarrow t_{31} = a_{13}/t_{11}.$

$t^2_{21} + t^2_{22} = a_{22} \Rightarrow t_{22} = \sqrt{a_{22} - t^2_{21}},$

$t_{21}\ t_{31} + t_{23}\ t_{33} = a_{23} \quad \Rightarrow \quad t_{32} = (a_{23} - t_{21}\ t_{31})/t_{22}$

$t^2_{31} + t^2_{32} + t^2_{33} = a_{35} \quad \Rightarrow \quad t_{33} = \sqrt{a_{23} - t^2_{31} - t^2_{32}}$

Thus T is determined completely.

Example. Find the inverse of A using Choleske's method where

$$A = \begin{bmatrix} 1 & 2 & 6 \\ 2 & 5 & 15 \\ 6 & 15 & 40 \end{bmatrix}$$ *(Agra 1990)*

Solution. Clearly A is a symmetric matrix. Let $A = TT'$ where T is a lower triangular matrix. Also $A^{-1} = (T^{-1})'\ T^{-1}$

Substituting the values of a_{11}, a_{12}, etc., we find

$$t_{11} = 12,\ t_{21} = 2,\ t_{31} = 6,\ t_{22} = 1,\ t_{32} = 3,\ t_{33} = 1$$

$$\therefore \qquad T = \begin{bmatrix} 1 & 0 & 0 \\ 2 & 1 & 0 \\ 6 & 3 & 1 \end{bmatrix}$$

Now we want to find T^{-1}.

Let $T^{-1} = X$ where X is a lower triangular matrix.

$\therefore \quad TX = I.$

Proceeding as in Crout's method,

we find $\quad x_{11} = 1, x_{22} = 1, x_{33} = 1, x_{33} = 1, x_{21} = -2, x_{31} = 0, x_{32} = -3$

$$\therefore \qquad T^{-1} = X = \begin{bmatrix} 1 & 0 & 0 \\ -2 & 1 & 0 \\ 0 & -3 & 1 \end{bmatrix}$$

$$(T^{-1})' = \begin{bmatrix} 1 & -2 & 0 \\ 0 & 1 & -3 \\ 0 & 0 & 1 \end{bmatrix}$$

$$\therefore \qquad A^{-1} = (T^{-1})' \, T^{-1} = \begin{bmatrix} -1 & -2 & 0 \\ 0 & 1 & -3 \\ 0 & 0 & 1 \end{bmatrix} \begin{bmatrix} 1 & 0 & 0 \\ -2 & 1 & 0 \\ 0 & -3 & 1 \end{bmatrix}$$

$$= \begin{bmatrix} 5 & -2 & 0 \\ -2 & 10 & -3 \\ 0 & -3 & 1 \end{bmatrix}$$

15·2·3. Gauss's method for the solution of a system of linear equations by successive elimination. Gauss method of solving the given equations is shortest of all the methods known. It due to German mathematicion consists in the elimination of one unknown at a time from the set of equations. Thus, for example, from evaluation of the first equations (15·1), we have

$$x_1 = \left[b_1 - \sum_{i=2}^{n} a_{1i} \cdot x_i \right] / a_{11}, \; a_{11} \neq 0 \qquad \text{...(15·3)}$$

Eliminating x_1 from the remaining $n-1$ equations with the help of (15·3) is

$$a_{ji} \left[b_1 - \sum_{i=1}^{n} a_{1i} x_i \right] \Big/ a_{11} + \sum_{i=2}^{n} a_{ji} x_i = b_j, j = 2, 3, \ldots n$$

or $$\sum_{i=2}^{n} (a_{ji} - a_{j1} \, a_{1i}/a_{11}) \, x_i = b_j - a_{j1} b_1 / a_{11}, j = 2, 3, \ldots, n$$

From first of these $n-1$ equations we find x_2 and then eliminate x_2 from the remaining $(n-2)$ equations. By successive elimination, we arrive at a single equation in the unknown x_n which can be solved and then x_{n-1} can be obtained by back substitution for x_n in the second equation from the last equation. In this manner, we find x_1 when the elimination is completed, the system assumes the form

$$a_{11}x_1 + a_{12}x_2 + \cdots + a_{1n}x_n = a_{1,\,n+1}$$

$$a_{22}^{(1)} x_2 + \cdots + a_{2n}^{(1)} x_n = a_{2,\,n+1}^{(1)}$$

$$\cdots\cdots\cdots\cdots\cdots\cdots\cdots\cdots\cdots\cdots$$

$$a_{nn}^{(n-1)} x_n = a_{n,\,n+1}^{(n-1)}$$

which is the triangular form and where the coefficients are given by

$$a_{in}^{(k)} = a_{ij}^{(k-1)} - \left[a_{ik}^{(k-1)} a_{kj}^{(k-1)} / a_{kk}^{(k-1)} \right]$$

$$k = 1, 2, \ldots, n-1, j = k+1, \ldots, n+1, i = k+1, \ldots, n$$

and $\quad a_{ij}^{(0)} = a_{ij}$

The solution is given by the formula

$$x_i = \frac{1}{a_{ii}^{(i-1)}} \left[a_{i,\,n+1}^{(i-1)} - \sum_{j=i+1}^{n} a_{ij}^{(n-1)} x_j \right], \quad i = n, n-1, \ldots, 1.$$

Notes 1. Care must be taken in the order of elimination of the unknown when they are of widely different magnitudes. In such cases, it is proper to start with the elimination of the unknown with the smallest coefficient and then proceed in order of increasing magnitude.

2. If $a_{11} = 0$, then we can interchange the variables.

3. We shall put the working in the tabular form which will be self-explanatory except for the numbers r and the column S. The numbers r are the ratios of the encircled coefficients of the boxed coefficients and S is the sum of all the coefficients and the constant of each row and is operated upon as any other number of the same row.

Example 2. *By Gauss-elimination method, solve*

$$5x_1 - x_2 - 2x_3 = 142,$$

$$x_1 - 3x_2 - x_3 = -30,$$

$$2x_1 - x_2 - 3x_3 = 5.$$

Gauss's Scheme

Sol. Explanation. The column S is used as a check, it contains the sum of all the coefficients and the constants of each row and is operated upon as any other number of the same row.

The numbers r are ratios of coefficients obtained as follows :

r_2 is the ratio of the encircled coefficient 1 of x_1 in row 2 to the boxed coefficient 5 of x_1 in row 1, etc.

Starting from the last equation, we have

$$\left.\begin{array}{lll} \text{Row 9 :} & \dfrac{-29}{14}x_3 = \dfrac{-275}{7} & \therefore\ x_3 = 19 \text{ approx.} \\ \text{Row 4 :} & -\dfrac{14}{5}x_2 - \dfrac{3}{5}x_3 = \dfrac{-292}{5} & \therefore\ x_2 = 16{\cdot}7 \\ \text{Row 1 :} & 5x_1 - x_2 - 2x_3 = 142 & \therefore\ x_1 = 39{\cdot}2 \end{array}\right\} \quad \textbf{Ans.}$$

Row	*r (ratio of coeffs).*	x_1	x_2	x_3	*C*	*S (sum)*	*Explanation*
(1)		$\boxed{5}$	-1	-2	142	144	(1)
2	$r_2 = \frac{1}{5}$	(1)	-3	-1	-30	-33	(II)
3		-1	$\frac{1}{5}$	$\frac{2}{5}$	$\frac{-142}{5}$	$\frac{-144}{5}$	$-r_2 \times (1)$
(4)		0	$\frac{-14}{5}$	$\frac{-3}{5}$	$\frac{-292}{2}$	$\frac{-309}{5}$	(2) + (3)
5	$r_3 = \frac{2}{5}$	(2)	-1	-3	5	3	(III)
6		-2	$\frac{2}{5}$	$\frac{4}{5}$	$\frac{-284}{5}$	$\frac{-288}{5}$	$-r_3 \times (1)$
7	$-r_3 = \frac{3}{14}$	0	$\left(\frac{-3}{5}\right)$	$\frac{-11}{5}$	$\frac{-259}{5}$	$\frac{-273}{5}$	(5) + (6)
8			$\frac{3}{5}$	$\frac{9}{70}$	$\frac{438}{35}$	$\frac{927}{70}$	$-r_3' \times (4)$
(9)			0	$\boxed{\frac{-29}{14}}$	$\frac{-275}{7}$	$\frac{-579}{14}$	(7) + (8)
10							

Example 3. *Solve the following equations by Gauss's Elimination method.*

$$x - 2y + 3z + 4t = \frac{9}{2},$$

$$3x - y + 2z + 5t = \frac{19}{2},$$

$$2x + 4y - 5z + t = 15,$$

$$4x + 2y - z + 3t = 12.$$

From rwo (14), we have

$$z + t = 2 \quad \therefore \quad z = -1$$

From row (4), we have

$$5y - 7z - 7t = -4 \quad \therefore \quad y = 2$$

From row (1), we have

$$x - 2y + 3z + 4t = \frac{9}{2} \quad \therefore \quad x = -\frac{1}{2}$$

Hence the solution is

$$x = -\frac{1}{2}, \; y = 2, \; z = -1, t = 3$$

From 16th row, we have

$$-20\,t = -60 \;\therefore\; t = 3$$

Gauss' Scheme

S. No.	*r (ratio)*	*x*	*y*	*z*	*t*	*c*	*Sum*	*Explanation*
(1)		$\boxed{1}$	−2	3	4	$\frac{9}{2}$	$\frac{21}{2}$	I
2	$r_1 = 3$	(3)	−1	2	5	$\frac{19}{2}$	$\frac{43}{2}$	II
3		−3	4	−9	−12	$-\frac{27}{2}$	$-\frac{63}{2}$	$-r_1 \times (1)$
(4)		0	$\boxed{5}$	−7	−7	−4	−13	(2) + (3)
5	$r_2 = 2$	(2)	4	−5	1	15	17	III
6		−2	4	−6	−8	−9	−21	$-r_2 \times (1)$
(7)		0	(8)	−11	−7	6	−4	(5) + (6)
8	$r_1 = 4$	(4)	2	−1	3	12	20	II
9		−4	8	−12	−16	−18	−12	$-r_3 \times (1)$
10		0	(10)	−13	−13	−6	−22	(8) + (9)
11	$r_4 = \frac{4}{5}$		−0	$\frac{55}{5}$	$\frac{56}{5}$	$\frac{33}{5}$	$\frac{104}{5}$	$r_4 \times (4)$
(12)			0	$\boxed{\frac{1}{5}}$	$\frac{21}{5}$	$\frac{62}{5}$	$\frac{84}{5}$	(7) + (11)
13	$r_3 = 2$		−10	14	14	8	26	$-r_5 \times (4)$
14			0	(1)	1	2	4	(10) + (13)
15	$r_6 = 5$			−1	−21	−62	−64	$-r_3 \times (12)$
16				0	−20	−60	−00	(14) + (15)

Example 4. *Solve by Gauss elimination method;*

$$10x - 7y + 3z + 5w = 6$$
$$-6x + 8y - z - 4w = 5$$
$$3x + y + 4z + 11w = 2$$
$$5x - 9y - 2z + 4w = 7$$

Sol. First we eliminate x, using the first equation *viz.*,

$$x - 0{\cdot}7y + 0{\cdot}3z + 0{\cdot}5w = 0{\cdot}6$$

then
$$3{\cdot}8y + 0{\cdot}8z - w = 8{\cdot}6$$
$$3{\cdot}1y + 3{\cdot}1z + 9{\cdot}5w = 0{\cdot}2$$
$$-3{\cdot}5y - 3{\cdot}5z + 1{\cdot}5w = 4{\cdot}0$$

Since the numerical largest y coefficient belongs to the fourth equation, we permutate the second and fourth equations. After that,

$$x - 0{\cdot}7y + 0{\cdot}3z + 0{\cdot}5w = 0{\cdot}6$$
$$y + 0{\cdot}63636z - 0{\cdot}27273w = -0{\cdot}72727$$
$$-1{\cdot}61818z - 0{\cdot}3636w = 11{\cdot}36364$$
$$1{\cdot}12727z + 10{\cdot}34545w = 2{\cdot}45455$$

Now z is also eliminated, which gives

$$x - 0{\cdot}7y + 0{\cdot}3z + 0{\cdot}5w = 0{\cdot}6$$
$$y + 0{\cdot}63636z - 0{\cdot}27273w = -0{\cdot}72727$$
$$z - 0{\cdot}02247w = -7{\cdot}02247$$
$$10{\cdot}37079w = 10{\cdot}37079$$

The final solution is now easily obtained which is

$$w = 1,\ z = -7,\ y = 4 \text{ and } x = 5.$$

Note. Solve this example by Gauss's Sheme.

15·2·4. Method of Inversion of the matrix of the coefficients.

The equations (15·1) can be represented in the matrix form

$$AX = B \qquad \text{...(15·4)}$$

where A is the matrix (a_{ij}) of the coefficients or of the system in (15·1), X and B are column matrices. $[x_1, x_2, ..., x_n]$ and $[b_1, b_2, ..., b_n]$ of the unknowns and of the constants respectively.

It is proved in 'Theory of Matrices', that equations (15·1) will be consistent if the matrices A and AB are of the same rank. Suppose that A is non-singular so that A^{-1} exists, then the equation (15·4) possesses a unique solution

$$X = A^{-1} B \qquad \text{...(15·5)}$$

Example 5. *By the method of matrices, solve the following equations for x, y, z and w.*

$$x - 3y + z = a,$$
$$2x + y - w = b,$$

$$3x - 2y - z - 2w = c,$$
$$4x - y + 3w = d.$$

Sol. The given equations can be expressed in the matrix form

$$AX = B$$

or
$$\begin{pmatrix} 1 & -3 & 1 & 0 \\ 2 & 1 & 0 & -1 \\ 3 & -2 & -1 & -2 \\ 4 & -1 & 0 & 3 \end{pmatrix} \begin{pmatrix} x \\ y \\ z \\ w \end{pmatrix} = \begin{pmatrix} a \\ b \\ c \\ d \end{pmatrix}, \ A' = \begin{pmatrix} 1 & 2 & 3 & 4 \\ -3 & 1 & -2 & -1 \\ 1 & 0 & -1 & 0 \\ 0 & -1 & -2 & 3 \end{pmatrix}$$

Now
$$\begin{vmatrix} 1 & -3 & 1 & 0 \\ 2 & 1 & 0 & -1 \\ 3 & -2 & -1 & -2 \\ 4 & -1 & 0 & 3 \end{vmatrix} = -70, \begin{vmatrix} 1 & 0 & -1 \\ -2 & -1 & -2 \\ -1 & 0 & 3 \end{vmatrix} = -2, \begin{vmatrix} -3 & -2 & -1 \\ 1 & -1 & 0 \\ 0 & -2 & 3 \end{vmatrix} = 17.$$

$$\therefore \quad A^{-1} = \frac{1}{70} \begin{vmatrix} 2 & 17 & 2 & 7 \\ -10 & 20 & -10 & 0 \\ 38 & 43 & -32 & -7 \\ -6 & -16 & -6 & 14 \end{vmatrix}$$

or
$$\begin{pmatrix} x \\ y \\ z \\ w \end{pmatrix} = \frac{1}{70} \begin{vmatrix} 2 & 17 & 2 & 7 \\ -10 & 20 & -10 & 0 \\ 38 & 43 & -32 & -7 \\ -6 & -16 & -6 & 14 \end{vmatrix} \begin{pmatrix} a \\ b \\ c \\ d \end{pmatrix}$$

$$x = \frac{1}{70}(2a + 17b + 2c + 7d),$$

$$y = -\frac{1}{7}(-a + 2b - c),$$

$$z = \frac{1}{70}(38a + 43b - 32c - 7d),$$

$$w = -\frac{1}{35}(3a + 8b + 3c - 7d).$$

15·2·5. Gauss-Jordan Reduction Method of solving $AX = B$.

Here we make use of the same technique upto the first derived system of Gauss Elimination Method. After this we write the following new system :

$$a_{11}x_1 + 0 + a_{13}^{(2)} x_3 + a_{14}^{(2)} x_4 + \cdots + a_{1n}^{(2)} x_n = a_{1,n+1}^{(2)}$$
$$a_{22}^{(1)} x_2 + a_{23}^{(1)} x_3 + \cdots \quad \cdots + a_{2n}^{(1)} x_n = a_{2,n+1}^{(2)}$$
$$a_{33}^{(2)} x_3 + \cdots \quad \cdots + a_{3n}^{(2)} x_n = a_{3,n+1}^{(2)}$$

in which the element in the first row and the second column is also reduced to zero. The remaining coefficients in the first row are given by the previous formula

$$a_{ij}^{(2)} = a_{ij}^{(1)} - \frac{a_{ij}^{(1)}}{a_{ij}^{(1)}} \cdot a_{ij}^{(1)}, \; i = 1, j = 3, 4, ..., n$$

and $$a_{2,n+1}^{(2)} = a_{2,n+1}^{(1)}$$

Continuing the process, we get

$$a_{11}x_1 = a_{1,n+1}^{(n-1)}$$

$$a_{22}x_2 = a_{2,n+1}^{(n-1)}$$

$$a_{33}x_3 = a_{3,n+1}^{(n-1)}$$

....

$$a_{nn}^{(n-1)} x_n = a_{n,n+1}^{(n-1)}$$

where $a_{ii}^{(i-1)}$ are determined from the previous equation

$$a_{ij}^{(k)} = a_{ij}^{(k-1)} - \frac{a_{ij}^{(k-1)}}{a_{kk}^{(k-1)}} a_{kj}^{(k-1)}$$

and $$a_{i\,n+1}^{(k)} = \begin{cases} a_{i\,n+1}^{(k-1)} - \dfrac{a_{ik}^{(k-1}}{a_{kk}^{(k-1)}} a_{kn+1}^{(k-1)} & \text{if } i = 1, 2, ..., k-1, k+1, ..., n \\ a_{i\,n+1}^{(k-1)} & \text{if } i = k \text{ and } k = 1, 2, ..., n-1 \end{cases}$$

The solution is given by

$$x_i = \frac{a_{i\,n+1}^{(n-1)}}{a_{ii}^{(i-1)}}, i = 1, 2, ..., n \qquad ...(i)$$

Example 6. *Solve by Gauss-Jordan Reduction Method :*

$$10x_1 + x_2 + x_3 = 12$$

$$x_1 + 10x_2 + x_3 = 12$$

$$x_1 + x_2 + 10x_3 = 12.$$

Sol. Here $$M = \begin{pmatrix} 10 & 1 & 1 & 12 \\ 1 & 10 & 1 & 12 \\ 1 & 1 & 10 & 12 \end{pmatrix}$$

$$= \begin{pmatrix} 1 & \frac{1}{10} & \frac{1}{10} & \frac{12}{10} \\ 1 & 10 & 1 & 12 \\ 1 & 1 & 10 & 12 \end{pmatrix} \text{ by } R_1 = \frac{1}{10} R_1$$

$$= \begin{pmatrix} 1 & \frac{1}{10} & \frac{1}{10} & \frac{12}{10} \\ 0 & \frac{99}{10} & \frac{9}{10} & \frac{108}{10} \\ 0 & \frac{9}{10} & \frac{99}{10} & \frac{108}{10} \end{pmatrix} \text{ by } \begin{array}{l} R_2 = R_2 - R_1 \\ R_3 = R_3 - R_1 \end{array}$$

$$= \begin{pmatrix} 1 & \frac{1}{10} & \frac{1}{10} & \frac{12}{10} \\ 0 & 1 & \frac{1}{11} & \frac{108}{99} \\ 0 & \frac{9}{10} & \frac{99}{10} & \frac{108}{99} \end{pmatrix} \text{ by } R_2 = \frac{10}{99} R_2$$

$$= \begin{pmatrix} 1 & 0 & \frac{1}{11} & \frac{108}{99} \\ 0 & 1 & \frac{1}{11} & \frac{108}{99} \\ 0 & 0 & \frac{108}{11} & \frac{108}{11} \end{pmatrix} \text{ by } \begin{array}{l} R_1 = R_1 - \frac{1}{10} R_2 \\ R_3 = R_3 - \frac{9}{10} R_2 \end{array}$$

$$= \begin{pmatrix} 1 & 0 & \frac{1}{11} & \frac{108}{99} \\ 0 & 1 & \frac{1}{11} & \frac{109}{99} \\ 0 & 0 & 1 & 1 \end{pmatrix} \text{ by } R_3 = \frac{11}{108} R_3$$

$$= \begin{pmatrix} 1 & 0 & 0 & 1 \\ 0 & 1 & 0 & 1 \\ 0 & 0 & 1 & 1 \end{pmatrix} \text{ by } \begin{array}{l} R_1 = R_1 - \frac{1}{11} R_2 \\ R_2 = R_2 - \frac{1}{11} R_3 \end{array}$$

Hence the solution set is (111, 1)

i.e., $x_1 = 1,\ x_2 = 1,\ x_3 = 1.$

Hence using (1′), $x_1 = 1, \quad x_2 = 1, \quad x_3 = 1.$

15·2·6. Crout Method. This method of finding the numerical solution of a set of linear equations has an advantage over Gauss method in that it requires less computation. Further this method is adapted to high speed computation.

Let us consider the linear system of equations

$$AX = B$$

where $A = (a_{ij}),\ i = 1, 2, \ldots m\ ;\ j = 1, 2, \ldots, n$

$B = (b_1, b_2, \ldots, b_m)',\ X = (x_1, x_2, \ldots, x_m)'.$

$$M = [A : b]$$

We define a matrix, called Auxiliary matrix, *viz.*

$$M' = [A' : b']$$

from which we get the solution column X. We get a solution called x by the backward substitution method.

Rules for obtaining the Auxiliary Matrix M' $[A' : b']$.

1. The elements of the first column of M' are identical with the corresponding elements of the first column of M. The remaining elements of the first row of M' are obtained from the corresponding element of M after dividing by the first diagonal element of M'.

   Thus $$a_{11}' = a_{11}, a_{21}' = a_{21}, ..., a_{n1}' = a_{n1}$$

   $$a_{12}' = \frac{a_{12}}{a_{11}},$$

2. The elements on or below the principal diagonal of M' are obtained by subtracting from the corresponding elements of M, the inner product of its row and column in M' when all uncalculated elements are taken zero.
3. The elements on the right of the Principal diagonal of M' are calculated as in rule (2) above and then divided by the diagonal elements in its row of M'.
4. The solution will be determined in the order of $x_n, x_{n-1}, ..., x_2, x_1$ from foot to head.

The succeeding elements of b' are obtained by subtraction from the corresponding elements of b', the inner product of column b' and its row in A' and then divided by the diagonal elements of its row in A'.

Summary

The elements of M' are obtained as follows :—

$$a'_{11} = a_{11}, a'_{21} = a_{21}, a'_{12} = \frac{a_{12}}{a_{11}}, a'_{13} = \frac{a_{13}}{a_{21}}, b'_1 = \frac{b_1}{a_{11}}$$

1. $$a'_{ij} = a_{ij} - \sum_{k=1}^{j-1} a'_{ik} a'_{kj} \;.\; i \geq j,\; a'_{22} = a_{22} - a'_{21} a'_{12},$$

   $$a'_{32} = a_{32} - a'_{31}a'_{12} = a_{33} - a'_{31}a'_{13} - a'_{32}a'_{23}.$$

2. $$a'_{ij} = \frac{1}{a'_{ij}} \left(a_{ij}' - \sum_{k=1}^{i-1} a'_{ik} a'_{kj} \right), i < j$$

   $$a'_{23} = \frac{1}{a_{22}} (a_{23} - a'_{21} a'_{12})$$

3. $b_i' = \dfrac{1}{a_{ij}'}\left(b_i - \sum_{j=1}^{i-1} a_{ik}'b_k'\right)$

$b'_2 = \dfrac{1}{a'_{22}}(b_2 - a'_{21}b_1),\ b'_3 = \dfrac{1}{a'_{33}}[b_3 - a'_{31} - b'_{13} - a'_{32}b'_2]$

4. $x_i = b_i' - \sum_{k=i+1}^{n} a'_{ik}x_x$.

The values of three unknown variables are obtained from the matrix

$$\begin{bmatrix} a'_{11} & a'_{12} & a'_{13} & : & b'_1 \\ a'_{21} & a'_{22} & a'_{23} & : & b'_2 \\ a'_{31} & a'_{32} & a'_{33} & : & b'_3 \end{bmatrix}$$

Example 7. *Solve for x_1, x_2 and x_3 by Crout's method.*

$$2x_1 + 3x_2 + 2x_3 = 2,$$
$$10x_1 + 3x_2 + 4x_3 = 16,$$
$$3x_1 + 6x_2 + x_3 = -6.$$

Sol. We have $M = \begin{bmatrix} 2 & 3 & 2 & : & 2 \\ 10 & 3 & 4 & : & 16 \\ 3 & 6 & 1 & : & -6 \end{bmatrix}$

Form third row, $x_2 = b_3'$, From second row, $x_2 = b_2' - a'_{23}x_3$,

From first row, $x_1 = b'_1 - a'_{12}x_2 - a'_{13}x_3$,

$$M' = \begin{bmatrix} 2 & \frac{3}{2} & 1 & : & 1 \\ 10 & -12 & \frac{1}{2} & : & -\frac{1}{2} \\ 3 & \frac{3}{2} & -\frac{11}{4} & : & 3 \end{bmatrix}$$

Since according to Rule 1, the diagonal element is 2 and then the remaining elements of the first row of M' are $\frac{3}{2}$, 1, 1 respectively. Also the second element of the second row is given by

$$3 - 10 \times \frac{3}{2} = 3 - 15 = -12$$

The third element of the second row $= \dfrac{4 - 10 \times 1}{-12} = \dfrac{1}{2}$

fourth " " " $= \dfrac{16 - 10 \times 1}{-12} = -\dfrac{1}{2}$

second element of the third row $= 6 - 3 \times \frac{3}{2} = \frac{3}{2}$

third " " " $= 1 - \left(3 + \frac{3}{4}\right) = -\frac{11}{4}$

fourth " " " $= \dfrac{\left[-6 - \left(3 - \frac{3}{4}\right)\right]}{-\frac{11}{4}} = 3$

Hence the solution is given by

$$x_3 = 3,$$

$$x_2 = \frac{-1}{2} - 3 \times \frac{1}{2} = -2,$$

$$x_1 = 1 - [-3 + 3] = 1$$

Example 8. *Solve by Crout's method, the following system of equations :*

$$a + 3c + d = 2$$
$$4a - b + c - 2d = 1$$
$$a + 3b + 5d = -1$$
$$3a - b + 2c + d = 3$$

Sol. We have $M = \begin{bmatrix} a & 0 & 3 & 1 & : & 2 & 7 \\ 4 & -1 & 1 & -2 & : & 1 & 3 \\ 1 & 3 & 0 & 5 & : & -1 & 8 \\ 3 & -1 & 2 & 1 & : & 3 & 8 \end{bmatrix}$

$$M' = \begin{bmatrix} 1 & 0 & 3 & 1 & : & 2 \\ 4 & -1 & 11 & 6 & : & 7 \\ 1 & 3 & -36 & \frac{7}{13} & : & \frac{37}{18} \\ 3 & 1 & 4 & \frac{22}{9} & : & \frac{17}{11} \end{bmatrix}$$

Solution is given by

$$d = \frac{6}{11}, \quad c = \frac{2}{3} - \frac{7}{33} = \frac{5}{11}$$

$$b = 7 - \left(5 + \frac{36}{11}\right) = -\frac{14}{11}$$

$$a = 2 - \left(-\frac{14}{11} \times 0 + 3 \times \frac{5}{11} + 1 \times \frac{6}{11} \right) = \frac{1}{11}$$

i.e. $$a = \frac{1}{11}, \; b = -\frac{14}{11}, \; c = \frac{5}{11}, \; d = \frac{6}{11}.$$

Note. If the elements of a given matrix A are symmetrical *i.e.*, $a_{ij} = a_{ji} \; \forall$ *i.j.*, the work of computing the auxiliary matrix is cut almost to half.

For example, $$a'_{23} = \frac{1}{a'_{23}} \left[a_{23} - \frac{a_{21}\, a_{13}}{a_{11}} \right] = \frac{1}{a'_{22}} \left[a_{32} - \frac{a_{21}\, a_{13}}{a_{11}} \right], (\because a_{23} = a_{32})$$

$$= \frac{1}{a'_{22}} \left[a_{32} - a_{13} \frac{a_{12}}{a_{11}} \right] = \frac{1}{a'_{22}} (a'_{32})$$

$$\therefore \quad a'_{23} = a'_{32}.$$

Ex. *Solve* : $x_1 + x_2 + x_3 = 1,\; 3x_1 + x_2 - 3x_3 = 5,\; x_1 - 2x_2 - 5x_3 = 10.$

Hint. Here $$M = \begin{bmatrix} 1 & 1 & 1 & : & 1 \\ 3 & 1 & -3 & : & 5 \\ 1 & -2 & -5 & : & 10 \end{bmatrix} \text{ and } M' = \begin{bmatrix} 1 & 1 & 1 & : & 1 \\ 3 & -2 & -3 & : & -1 \\ 1 & -3 & 3 & : & 2 \end{bmatrix}$$

$$\therefore \quad x_3 = 2,\; x_2 = -1 - (3 \times 2) = -7,$$

$$x_1 = 1 - (1 \times -7 + 1 \times 2) = 6.$$

15·3. Indirect Methods. Indirect methods are those in which the solution is approached by successive approximation or iterative methods.

15·3·1. The Gauss-Seidel Method. A system of simultaneous linear equations is called diagonal when in each equation the coefficient of a different unknown is greater in absolute value then the sum of the absolute values of the other coefficients. The large coefficient is usually located in the main diagonal position a_{ii}. Diagonal systems have the fundamental property of being solvable by methods of successive approximation, among which Gauss—Seidel iterative process is of great advantage so far as simplicity is concerned.

We solve each equation of the system for the unknown with the largest coefficient :

$$x_1 = c_{12}\, x_2 + c_{13} x_3 + \cdots + c_{in} x_n + d_1$$

$$x_2 = c_{21} x_1 + c_{23} x_3 + \cdots + c_{2n} x_n + d_2$$

....

$$x_n = c_{n1} x_1 + c_{n2} x_2 + \cdots + c_{n,\, n-1}\, x_{n-1} + d_n$$

Substitute any initial value $x_i^{(0)}$ for the unknown on the R.H.S. after putting other unknowns to zero, and obtain the main values $x_j^{(1)}$ on the R.H.S. Again substitute the new value $x_j^{(1)}$ on the R.H.S. and obtain improved values $x_j^{(2)}$; and continue the process until $x_j^{(m)}$ is equal to $x_j^{(m+1)}$ within the required accuracy. The $x_i^{(m)}$ are the required roots of the system.

Note. The Gauss-Seidel method suffers from the slow convergence for ill-conditioned systems.

Example 9. *Solve the following system of equations by Gauss-Seidel iteration method, (iterate up to two iterations).*

$$27x + 6y - z = 85,$$

$$6x + 15y + 2z = 72,$$

$$x + y + 54z = 110.$$

(B.A., Hons. Delhi, 1982, M. Sc. Meerut, 1986)

Sol. These equations are ready for iteration. Solving each equation for the unknown, having the large coefficient, we have

$$x = \frac{1}{27}(85 - 6y + z),$$

$$y = \frac{1}{15}(72 - 6x - 2z),$$

$$z = \frac{1}{54}(110 - x - y).$$

We start the iteration by putting $y = 0, z = 0$ and we obtain

First approximation of

$$x = x^{(1)} = \frac{85}{27} = 3{\cdot}15$$

Substituting

$$x = 3{\cdot}15, z = 0$$

in the second equation we get

First approximation of

$$y = y^{(1)} = \frac{1}{15}(72 - 18{\cdot}90) = 3{\cdot}54$$

Substituting $\quad x = 3{\cdot}15, y = 3{\cdot}54$

in the last equation, we have

First approximation of

$$z = z^{(1)} = \frac{1}{54}(110 - 3{\cdot}15 - 3{\cdot}54) = 1{\cdot}91.$$

Similarly, we obtain

Second approximation of

$$x = x^{(2)} = \frac{1}{27}(85 - 21{\cdot}24 + 191) = 2{\cdot}43$$

Second approximation of

$$y = y^{(2)} = \frac{1}{15}(72 - 14{\cdot}58 - 3{\cdot}82) = 3{\cdot}57$$

Second approximation of

$$z = z^{(2)} = \frac{1}{54}(110 - 2{\cdot}43 - 3{\cdot}57) = 1{\cdot}926.$$

Proceeding in this manner, we obtain third and higher approximation of x, y and z.

We put the above results in the tabular form.

Approx: / Variable	1	2	3	...
x	3·15	2·43		
y	3·54	3·57		
z	1·91	1·926		

15·3·2. Method of Relaxation. Relaxation method is method for the solution of a system of linear algebraic equations $Ax = d$ by successive approximations. It obtains a solution for all unknowns simultaneously. It is convenient for hand computation.

This indirect method was originally developed by R.V. Southwell for application in problems of structural engineering.

Consider the system of equations

$$\begin{aligned}
a_{11}x_1 + a_{12}x_2 + \cdots + a_{1n}x_n &= d_1 \\
a_{21}x_1 + a_{22}x_2 + \cdots + a_{2n}x_n &= d_2 \qquad \ldots(1) \\
\cdots\cdots\cdots\cdots\cdots\cdots\cdots\cdots & \\
a_{n1}x_1 + a_{n2}x_2 + \cdots + a_{nn}x_n &= d_n
\end{aligned}$$

In this method, the coefficients in the main diagonal are each equal to – 1 and the left member of each equation is made zero.

Thus we have the system of equations (1) in the following form :

$$\begin{aligned}
-x_1 + c_{12}x_2 + \cdots + c_{1n}x_n + k_1 &= 0 \\
b_{12}x_1 - x_2 + \cdots + c_{2n}x_n + k_2 &= 0 \\
\cdots\cdots\cdots\cdots\cdots\cdots\cdots\cdots & \\
b_{i1}x_1 + b_{i2}x_2 + \cdots - x_i + \cdots + b_{in}x_n + k_i &= 0 \\
\cdots\cdots\cdots\cdots\cdots\cdots\cdots\cdots & \\
b_{n1}x_1 + b_{n2}x_2 + \cdots - x_n + k_n &= 0
\end{aligned}$$

We denote by R_i, called the residual, the value of the right side of the ith equation after substituting the assumed values $x_i^{(0)}$,

Thus,
$$-x_1^{(0)} + c_{12}x_2^{(0)} + \cdots + c_{1n}x_n^{(0)} + k_1 = R_1$$
$$c_{21}x_1^{(0)} - x_2^{(0)} + \cdots + c_{2n}x_n^{(0)} + k_2 = R_2$$
$$\cdots\cdots\cdots\cdots\cdots\cdots\cdots\cdots\cdots\cdots$$
$$c_{n1}x_1^{(0)} + c_{n2}x_2^{(0)} + \cdots - x_n^{(0)} + k_n = R_n$$

The relaxation method consists in changing the assumed values of the unknowns, one or more at each step until all the R_i become negligible.

For this purpose note that not a given set $x_i^{(0)}, i = 1, 2, ..., n$ say $x^{(0)}$ for $j = k$ is changed by an amount δx_k, then R_k changes by $c_{ij}\delta x_k$. Hence to reduce a given R_i say R_k to zero, we change $x_k^{(0)}$ by $\delta x_k = R_k$. In doing so, the other $R_i, i \neq k$ will also change (R_i will change by $c_{ik}R_k$) and must be reduced to zero one by one to eliminate the largest residual appearing in the system at any stage of the process.

The method of relaxation may fail since the residuals may be small without our having approached closely to a solution.

Consider the equations
$$x + 5y = 17,$$
$$1{\cdot}500x + 7{\cdot}501y = 25{\cdot}503.$$

The solution is readily found to be
$$x = 2, \quad y = 3.$$

Next the solution of equations
$$x + 5y = 17,$$
$$1{\cdot}500x + 7{\cdot}501y = 25{\cdot}500$$
is
$$x = 17, \quad y = 0.$$

Thus we find that these equations behave very badly, a change of one part in 3rd decimal in one of the coefficients is sufficient to change the solution from (2, 3) to (17, 0). Such systems of equations, where the solution is very sensitive to tiny alterations in the co-efficients are called *ill-conditions.* The two equations are one and the same but the solutions differ widely. In fact when $x = 17, y = 0$, we have $R_1 = 0, R_2 = {\cdot}002$. Certainly these residuals are small, yet the true solution is (2, 3). This illustrates a fundamental fact about ill-conditioned equations. There are values, in no way resembling the true solution, which makes the residuals very small. By large, if the equations are ill-conditioned, only the method of elimination is a foolproof method of solution.

Example 10. *Solve the following equations by Relaxation method.*
$$9x_1 - 2x_2 + x_3 = 50,$$
$$x_1 + 5x_2 - 3x_3 = 18,$$
$$-2x_1 + 2x_2 + 7x_3 = 19.$$

Sol. The working is shown in the following table :

x_1	x_2	x_2	R_1	R_2	R_3
0	0	0	– 50	– 18	– 19
5	0	0	– 5	– 13	– 29
5	0	4	– 1	– 25	– 1
5	5	4	– 11	0	9
6	5	4	– 2	1	7
6	5	3	– 3	4	0
6	4·2	3	– 1·4	0	– 1·6
6	4·2	3·2	– 1·2	– 0·6	– 0·2
6·1	4·2	3·2	– 0·3	– 0·5	– 0·4
6·1	4·3	3·2	– 0·5	– 0	– 0·2
6·15	4·3	3·2	– 0·05	0·05	– 0·3
6·15	4·3	3·24	– 0·01	– 0·07	0·02
6·15	4·31	3·24	– 0·03	– 0·02	0

Thus $(x_1, x_2, x_3) = (6{\cdot}15, 4{\cdot}31, 3{\cdot}24)$.

Block Relaxation Method

In the previous method, relaxation was applied mechanically always reducing to zero the largest residual but the method becomes particularly advantageous if no such rule is used. In block relaxation, for example, all (or a group of) the variables are changed by the same amount δ whenever convenient, changing the residual

$$R_i \log \delta R_i = \left(-1 + \sum_{\substack{j=1 \\ j \neq i}}^{n} c_{ij} \right) \delta = C_i\, \delta, \text{ say}$$

while in the *under relaxation* or *over relaxation*, the changes in the variables are taken small or large enough to obtain new residuals with equal or opposite signs respectively to that of previous residuals.

One of the most efficient ways of using Block Relaxation is to reduce to zero the total residual *i.e.* the sum of all the R_i'*s*.

Note. The solution of the system is started by an initial Block change δ.

$$\left(\text{L.H.S.} = k_i \quad \therefore \quad \text{Sum of residuals} = \sum_{i=1}^{n} R_i + \left(\sum_{i=1}^{n} C_i \right) \delta = 0 \right.$$

$$\Rightarrow \delta = -\frac{\Sigma R_i}{\Sigma C_i}$$

The rest of the solution is obtained by relaxation of the largest residual as in the case of ordinary relaxation method.

Example 11. *Solve the following system of equations by using Block Relaxation method :*

$$-10x + 2y + 2z + 4 = 0$$
$$x - 10y + 2z + 18 = 0$$
$$x + y - 10z + 45 = 0$$

Sol. Initial solution is $x = 0, y = 0, z = 0$

Working Table

x	R_1	y	R_2	z	R_3
0	4	0	18	0	45
3	– 18	3	– 21	3	– 24
3	– 14	3	– 3	3	21
0	4	0	4	2	– 20
3	– 10	3	1	5	1
– 1	10	0	– 1	0	– 1
2	0	3	0	5	0

$$C_1 = c_{11} + c_{12} + c_{12} = -6$$
$$C_2 = c_{21} + c_{22} + c_{23} = -7$$
$$C_3 = c_{31} + c_{32} + c_{33} = -8,$$
$$\therefore \sum_{i=1}^{3} C_i = -6 - 7 - 8$$
$$= -21.$$

$$\Sigma R_i = 4 + 18 + 45 = 67,$$

We start by the initial block change

$$\delta = -\frac{\Sigma R_i}{\Sigma C_i} = \frac{67}{-21} \simeq -3.$$

Now proceed as in Example 10.

Ex 12. *Solve by the Block Relaxation Method :*

$$2x - 3y + 10z = 3, -x + 4y + 2z = 20, 5x + 2y + z = -12$$

Sol. We write the given system of equations as (*M.A. Delhi 1971*)

$$5x + 2y + x + 12 = 0$$
$$-x + 4y + 2x - 20 = 0$$
$$2x - 3y + 10z - 3 = 0$$

Working Table

x	R_1	y	R_2	z	R_3
0	12	0	– 20	0	– 3
1	8	1	5	1	9
1	20	1	– 15	1	6
– 4	– 20	0	4	0	– 8
– 3	0	1	– 11	1	– 2
0	6	3	12	0	– 9
– 3	6	4	1	1	– 11
0	1	0	2	1	10

$$C_1 = 8$$
$$C_2 = 5$$
$$C_3 = 9$$
$$\Sigma C_i = 22$$
$$\Sigma R_i = 11$$
$$R_2 = \frac{\Sigma R_i}{\Sigma C_i} = \frac{11}{22}$$
$$= \frac{1}{2}$$

– 3	7	4	3	2	– 1
– 1	– 5	0	1	0	– 2
– 4	2	4	4	2	– 3
0	– 2	– 1	– 4	0	+ 3
– 4	0	3	0	2	0

Hence $x = -4,\ y = 3, z = 2.$

15·3·5. Doolittle Method. Triangular or LU Decomposition Method for Finding the Solution of a Set of Linear equations. In this method, the matrix A is expressed as the product of lower triangular matrix L and an upper triangular matrix U, that is

$$LU = A \Rightarrow A^{-1} = (Lu)^{-1} = u^{-1}L^{-1}. \quad ...(1)$$

The diagonal forms of L or U may be specified arbitrary and it is convenient to make those of L equal to unity. Thus (1) becomes

$$\begin{bmatrix} I & 0 & 0 \\ l_{21} & 1 & 0 \\ l_{31} & l_{32} & 1 \end{bmatrix} \begin{bmatrix} u_{11} & u_{12} & u_{13} \\ 0 & u_{22} & u_{23} \\ 0 & 0 & u_{33} \end{bmatrix} = \begin{bmatrix} a_{11} & a_{12} & a_{13} \\ a_{21} & a_{22} & a_{23} \\ a_{31} & a_{32} & a_{33} \end{bmatrix}$$

Hence
$$u_{11} = a_{11}$$
$$u_{12} = a_{12}$$
$$u_{13} = a_{13}$$
$$l_{21}\, u_{11} = a_{21}$$
$$l_{21}\, u_{12} + u_{21} = a_{22}$$
$$l_{21}\, u_{13} + u_{23} = a_{23}$$
$$l_{21}\, u_{11} = a_{31}$$
$$l_{31}\, u_{12} + l_{32}\, u_{22} = a_{32}$$
$$l_{31}\, u_{13} + l_{32}\, u_{33} + u_{33} = a_{33}.$$

From these equations, we can determine the unknown compoments of L and U.

Once A has been decomposed into LU, the solution may be effected by noting that $LU\,x = b$

is equivalent to $Ux = y$

and $Ly = b.$

The second equation is solved by back substitution, giving y, the first is then solved by another back substitution, giving x.

Example 13. *Solve the equations.*

$$x + 2y + 3z = 14$$
$$2x + 3y + 4z = 20$$
$$3x + 4y + z = 14$$

by the factorization method. (*M.A. Udaipur 1986*)

Sol. Here the coefficient of matrix is $\begin{bmatrix} 1 & 2 & 3 \\ 2 & 3 & 4 \\ 3 & 4 & 1 \end{bmatrix}$

which is symmetric. Thus $L = u'$ and we may determine the elements of u from

$$\begin{bmatrix} 1 & 2 & 3 \\ 2 & 3 & 4 \\ 3 & 4 & 1 \end{bmatrix} = \begin{bmatrix} u_{11} & 0 & 0 \\ u_{12} & u_{22} & 0 \\ u_{13} & u_{23} & u_{33} \end{bmatrix} \begin{bmatrix} u_{11} & u_{12} & u_{13} \\ 0 & u_{22} & u_{23} \\ 0 & 0 & u_{33} \end{bmatrix}$$

From this, we obtain

$$u_{11} = 1 \Rightarrow u_{11} = 1,\ u_{11}\, u_{12} = u_{12} = 2,\ u_{11}u_{13} = u_{13} = 3$$

$$u^2_{21} + u^2_{22} = 4 + u^2_{22} = 3 \Rightarrow u_{22} = i\ (= \sqrt{-1})$$

$$u_{12}u_{13} + u_{22}u_{23} = 6 + i\, u_{23} = 4,\ u_{23} = 2i$$

and finally $\quad u^2_{13} + u^2_{23} + u^2_{33} = 9 - 4 + u^2_{33} = 1,\ \Rightarrow\ u_{33} = 2i$

Hence $\begin{bmatrix} 1 & 0 & 0 \\ 2 & i & 0 \\ 3 & 2i & 2i \end{bmatrix} \begin{bmatrix} y_1 \\ y_2 \\ y_3 \end{bmatrix} = \begin{bmatrix} 14 \\ 20 \\ 14 \end{bmatrix}$, $\therefore y_1 = 14,\ 2y_1 + iy_2 = 20,\ 3y_1 + 2iy_i + 2iy_3 = 4$ and $\begin{bmatrix} y_1 \\ y_2 \\ y_3 \end{bmatrix} = \begin{bmatrix} 14 \\ 8i \\ 6i \end{bmatrix}$

we now solve and we find

$$\begin{bmatrix} 2 & 2 & 3 \\ 0 & i & 2i \\ 0 & 0 & 2i \end{bmatrix} \begin{bmatrix} x_1 \\ x_2 \\ x_3 \end{bmatrix} = \begin{bmatrix} 14 \\ 8i \\ 6i \end{bmatrix} \quad \therefore \quad \begin{bmatrix} x_1 \\ x_2 \\ x_3 \end{bmatrix} = \begin{bmatrix} 1 \\ 2 \\ 3 \end{bmatrix}$$

15·4. Newton-Raphson Method of Solving Simultaneous Non-linear Equations

The method of approximation may also be extended to the solution of the system of non-linear equations. Thus if

$$F(x, y) = 0, \quad G = (x, y) = 0$$

are two equations in two unknowns and if a point (x_0, y_0) close to a solution has been determined by graphical method or otherwise, a closer point (x_1, y_1) can usually be a obtained as follows :

Let $x_1 = x_0 + \Delta x,\ y_1 = y_0 + \Delta y$. Expand $F(x_1, y_1)$ and $G(x_1, y_1)$ in Taylon's series to terms of the first degree, and assure that (x_1, y_1) is a solution *i.e.* $F(x_1, y_1) = 0,\ G(x_1, y_1) = 0$. Then approximately

$$\left.\begin{aligned} 0 &= F + \frac{\partial F}{\partial x}\Delta x + \frac{\partial F}{\partial y}\Delta y, \\ 0 &= G + \frac{\partial G}{\partial x}\Delta x + \frac{\partial G}{\partial y}\Delta y, \end{aligned}\right\} \quad \text{where } F \equiv F(x_0, y_0)$$

$\frac{\partial F}{\partial x}$, etc. are the values of $F(x, y)$ at $x = x_0$, $y = y_0$.

Solving the two equations for Δx and Δy, we get the new approximations to the solution given by

$$x_1 = x_e + \Delta x = y_0 + \Delta y$$

The process is repeated until the desired accuracy is secured.

Scheme of Computation

F	$\frac{\partial F}{\partial x} = F_x$,	$\frac{\partial F}{\partial y} = F_y$
G	$\frac{\partial G}{\partial x} = G_x$,	$\frac{\partial G}{\partial y} = G_y$
D	Δx x	Δy y

$$E = F_x G_y - F_y G_x \neq 0$$

The values of F, G, $\frac{\partial F}{\partial x}$, $\frac{\partial F}{\partial y}$, etc. are calculated for $x = x_0$, $y = y_0$.

The improved value of (x_0, y_0) is then given by

$$x_1 = x_0 - \frac{F(x_0, y_0)\, G_y(x_0, y_0) - G(x_0, y_0)\, F_y(x_0, y_0)}{F(x_0, y_0)\, G_y(x_0, y_0) - F_y(x_0, y_0)\, G_x(x_0, y_0)}$$

$$y_1 = y_0 - \frac{G(x_0, y_0)\, F_x(x_0, y_0) - F(x_0, y_0)\, G_2(x_0, y_0)}{F_z(x_0, y_0)\, G_v(x_0, y_0) - F_y(x_0, y_0)\, G_r(x_0, y_0)}$$

After $(n + 1)$ th interation, (x_n, y_n) can be written down similarly.

EXERCISE 15

1. Solve the following set of linear equations by the method of determinants :

(*i*) $10x + 2y + z = 69,$
$x + 8y + 2z = -3,$
$2x - y + 20z = 76.$

2. Solve the following equations by Gauss-elimination method :

(*i*) $2x_1 + 4x_2 + x_3 = 3,$
$3x_1 + 2x_2 - 2x_3 = -2,$
$x_1 - x_2 + x_3 = 6.$

(*ii*) $2x + y + 4z = 12.$
$8x - 3y + 2z = 20,$
$4x + 11y - z = 33.$

(*iii*) $16x_2 - 19x_2 + 14x_3 - 17x_4 = 89,$
$82x_1 + 42x_2 - 32x_3 - 8x_4 = 73,$
$55x_1 - 35x_2 + 45x_3 - 25x_4 = 67,$
$7x_1 + 19x_2 + 33x_3 + 85x_4 = 111.$

3. Use matrix method to solve the following equations :

(*i*) $6x_1 + 3x_2 + 7x_3 = 4,$

$x_1 + 5x_2 + 2x_3 = 2,$

$7x_1 + 2x_2 + 10x_3 = 5.$

(*ii*) $6.0x - 2.0y + 2.0z + 4.7w = -2.3,$

$-2.0x + 6.0y + 2.0z + 2.9w = 3.6,$

$2.0x + 2.0y + 6.0z - 15.2w = 2.2,$

$4.7x + 2.9y - 15.2z + 20.6w = 1.4$

4. Solve using Crout method, the system of equations :

(*i*) $x_1 + x_2 + x_3 = 1$

$3x_1 + x_2 - 3x_3 = 5$

$x_1 - 2x_2 - 5x_3 = 10$

(*ii*) $10x - 7y + 3z + 5w = 6$

$-6x + 8y - z - 4w = 5$

$3x + y + 4z + 11w = 2$

$5x - 9y - 2z + 4w = 7$

(*iii*) $5x + 2y + z = -12,$

$-x + 4y + 2z = 20,$

$2x - 3y + 10z = 3.$

(*iv*) $2x_i + 4x_2 + x_3 = 5$

$4x_1 + 4x_3 + 3x_3 = 6$

$4x_1 + 8x_2 + x_3 = 9$

(*v*) $x + y + 2z = 7$

$3x + 2y + 4z = 13$

$4x + 3y + 2z = 8$ (*Kurukshetra, M.Sc., 1986*)

[**Hint.** $a'_{23} = \frac{1}{-1}(4 - 3 \times 2) = 2,\ b'_2 = \frac{1}{-1}(13 - 3 \times 3 = 4)$

$a'_{33} = 2 - 4 \times 2 - (-1)\,2 = -4,\ b'_3 = 3.$]

5. Apply Gauss-Seidal iteration method to solve equations.

(*i*) $10x_1 + x_2 + x_3 = 12,$

$2x_1 + 10x_2 + x_3 = 13,$

$2x_1 + 2x_2 + 10x_3 = 14.$

(*ii*) $2x_1 - x_2 + x_3 = 5,$

$x_1 + 3x_2 - 2x_3 = 7,$

$x_1 + 2x_2 + 3x_3 = 10.$

(*iii*) $27x + 6y - z = 85$

$6x + 15y + 2z = 72$

$x + y + 54z = 110$

6. Solving the following set of linear equations by relaxation method giving the results correct to three places of decimals.

(*i*) $9x_1 - 2x_2 + x_3 = 17,$
$4x_1 + 5x_2 - 2x_3 = -9,$
$x_1 - 3x_2 - 5x_3 = 4.$

(*ii*) $15x_1 + 3x_2 - 4x_3 + 2x_4 = 49.207,$
$3x_1 + 7x_2 - 5x_3 - 4x_4 = 18.024,$
$-4x_1 - 5x_2 + 16x_3 + 5x_4 = -23.871.$
$2x_1 - x_2 + 5x_3 + 19x_4 = 54.907.$

(*iii*) $2x - 13y - 3z - 49 = 0.$
$5x - 6y + 17z - 25 = 0,$
$11x + 2y - 4z + 31 = 0.$

(*iv*) $3x + 9y - 2z - 11 = 0,$
$4x + 2y + 13z - 24 = 0,$
$x - 4y + 3z + 8 = 0.$

(*v*) $-10x + 6y + 3z + 100 = 0,$
$6x - 10y + 5z + 100 = 0$
$3x + 5y - 10z + 100 = 0.$

(*vi*) $10x_1 - 2x_2 - 2x_3 = 6$
$-x_1 + 10x_2 - 2x_3 = 7$
$-x_1 - x_2 + 10x_3 = 8$

(*vii*) $2x_1 + 2x_2 + x_3 - 2x_4 = 10$
$x_1 + 2x_2 + x_3 + x_4 = 12$
$3x_1 + x_2 - 3x_3 + x_4 = 0$
$x_1 - 3x_2 + 3x_3 - x_4 = 3.$

[**Hint.** (*iii*) Treat last equation as first; first as second and second as third.

[**Hint.** (*v*) Solve the following equations by relaxation method the roots of the given equations being 100 times. Give the initial values $x = 0, y = 1, z = 1.$

$$-10x + 6y + 3z + 1 = 0,$$
$$6x - 10y + 5z + 1 = 0,$$
$$3x + 5y - 10z + 1 = 0.$$

[**Hint.** (*vi*) The given equations can be rewritten as

$$-x_1 + \cdot 2x_2 + \cdot 2x_3 + \cdot 6 = 0$$
$$0\cdot 1\, x_1 - x_2 + \cdot 2x_3 + \cdot 7 = 0$$
$$\cdot 1\, x_1 + \cdot 1\, x_2 - x_3 + \cdot 8 = 0$$

7. (*a*) Solve the following equations by Block Relaxation method :

$$-10x + 2y + 2z + 4 = 0$$
$$x - 10y + 2z + 18 = 0$$
$$x + y - 10z + 45 = 0$$

(*b*) Consider the following ill-conditional equations :

$$\left.\begin{aligned} x - y &= 1 \\ x - 1.00001\, y &= 0 \end{aligned}\right\} \qquad \left.\begin{aligned} x - y &= 1 \\ x - 0.99999\, y &= 0 \end{aligned}\right\}$$

Find their solutoin.

[**Hint.** (*a*) $B_1 = -10 + 2 + 2 = -6$, $B_2 = 1 - 10 + 2 = -7$, $B_3 = 1 + 1 - 10 = -8$, $\Sigma B_i = -21$, $\Sigma R_i = 4 + 18 + 43 = 67$.

$$\therefore \quad \delta = -\Sigma R_i / \Sigma B_i = \frac{67}{21} \simeq 3 \Bigg]$$

8. Using Gauss Jordan reduction method solve :

(*i*) $5x + 4y - 3z = 2$,
$2x + 6y + z = 25$,
$3x + y - 4z + 26$

(*ii*) $2x_1 - x_2 + x_3 = 5$,
$x_1 + 3x_2 - 2x_3 = 7$,
$x_1 + 2x_2 + 3x_3 = 10$.

9. Solve the following system of linear equations by Doolittle method of triangularization :

(*i*) $2x_1 - x_2 + x_3 = 5$,
$x_1 + 3x_2 - 2x_3 = 7$,
$x_1 + 2x_2 + 3x_3 = 10$.

(*ii*) $9x_1 - 2x_1 + x_3 = 17$,
$4x_1 + 5x_2 - 2x_3 = -9$,
$x_1 - 3x_2 - 5x_3 = 4$.

10. Solve the following linear equations by the method of triangular decomposition :

$$2x_1 + 2x_2 + 4x_3 - 2x_4 = 0$$
$$3x_1 + x_2 + 3x_3 + x_4 = 18,$$
$$x_1 + 3x_2 + 2x_3 + x_4 = 17,$$
$$x_1 + 3x_2 + 4x_3 + 2x_4 = 27.$$

11. Use Newton's method to find the root in the neighbourhood of (3.8 – 1.8) to satisfy the system

$$x^2 + y - 1 = 0, \quad y^2 + x - 7 = 0,$$

12. Solve the equations :

(*a*) $x_1 + x_2 + x_3 = 6$,
$3x_1 + (3 + \varepsilon)\, x_2 + 4x_3 = 20$,
$2x_1 + x_2 + 3x_3 = 13$

using the Gauss elimination method where ε is so small such that $1 \pm \varepsilon^2 \approx 1$. (*B.A. Math Hons. Delhi, 1996, M.Sc. Delhi 1988*)

(*b*) $3x_1 + 2x_2 + 6x_3 = 24$
$2x_1 + 4x_2 + 3x_3 = 23$
$5x_1 + 3x_2 + 4x_3 = 33$. (*B.Sc. Guwahati 1996*)

13. Perform three iterations of Gauss-Seidal iteration method on the equations.

$$2x_1 - x_2 = 7, -x_1 + 2x_2 - x_3 = 1, -x_2 + 2x_3 = 1$$

You may start with $x_1 = x_2 = x_3 = 0$. (***B.Sc. Hons. Delhi, 1989***)

14. The curves $y = 2 \sin x$ and $y = \log x + c$ touch each other in the neighbourhood of $x = 8$. Find c and the coordinate of the point of contract, correct to three places of decimal.

[**Hint.** Equate the slopes of tangent at common point $\therefore\ 2 \cos x = \frac{1}{x}$ · The solution of this is $x = 7{\cdot}793$, then $y = 2 \sin 7{\cdot}793$ = etc.]

15. Solve the equations :

$$2x + 3y + z = 9\,,\quad x + 2y + 3z = 6\,,\quad 3x + y + 2x = 8,$$

by the factorization method. *(Udaipur, 1988)*

Ans. $x = 35/18,\ y = 29/18,\ z = 5/18.$

16. Factorize the matrix.

$A = \begin{bmatrix} 5 & -2 & 1 \\ 7 & 1 & -5 \\ 3 & 7 & 4 \end{bmatrix}$ into the form $L\ U$ and solve : $5x - 2y + 2 = 4$,

$$7x + y - 5x = 6\ 3x + 7y + 4z = 10.$$

Ans. $L = \begin{bmatrix} 1 & 0 & 0 \\ 7/5 & 1 & 0 \\ 3/5 & 41/19 & 1 \end{bmatrix},\ U = \begin{bmatrix} 5 & -2 & 1 \\ 0 & 19/5 & -32/5 \\ 0 & 0 & 327/19 \end{bmatrix}$

$x = 122/169,\ y = 284/327,\ z = 46/327$

17. (*a*) By gaross elimination find A^{-1} if $A = \begin{bmatrix} 4 & 1 & 1 \\ 2 & 3 & -2 \\ 1 & -2 & 2 \end{bmatrix}$

(*b*) Find the universe of $\begin{bmatrix} 2 & 1 & -1 \\ 0 & 2 & 1 \\ 5 & 2 & -3 \end{bmatrix}$ by Gauss method

(B.E., Bharathiar, 1997)

18. Find the inverse of the following matrix by Jordan's method

$A = \begin{bmatrix} 50 & 107 & 36 \\ 25 & 54 & 20 \\ 31 & 66 & 21 \end{bmatrix}$ *(Agra, 1990, 1997)*

19. Find the inverse of the following matrix by Triangularization method

(*i*) $A = \begin{bmatrix} 2 & -1 & 2 \\ 2 & 3 & 4 \\ -1 & 1 & -1 \end{bmatrix}$ *(Agra, 1990)*

(*ii*) Solve by triangularization method

$2x - 3y + 10z = 3,\ -3x + 4y + 2z = 20,\ 5x + 2y + 2 = 12$

and hence find the inverse *(B.Sc., Agra, 1996)*

(*iii*) Find inverse of the matrix using <u decomposing method

$A = \begin{bmatrix} 3 & 1 & 2 \\ 2 & -3 & -1 \\ 1 & -2 & 1 \end{bmatrix}$ Ans. $A = \begin{bmatrix} 1/4 & 1/4 & 1/4 \\ 3/20 & 1/20 & -7/20 \\ 1/20 & 7/20 & 11/20 \end{bmatrix}$

where $<^{-1} = \begin{bmatrix} 1/3 & 0 & 0 \\ 2/11 & -3/4 & 0 \\ 11/20 & 11/20 & 11/20 \end{bmatrix},\ u^{-1} = \begin{bmatrix} 1 & -1/3 & -5/11 \\ 0 & 1 & -7/11 \\ 0 & 0 & 1 \end{bmatrix}$

16

NUMERICAL SOLUTION OF ORDINARY DIFFERENTIAL EQUATIONS

16·1. Introduction. In this chapter we shall discuss the methods of finding to any desired degree of accuracy the numerical solution of first and second order differential equations having numerical coefficients and given initial conditions. The solution is evaluated step by step through a series of equal intervals in the independent variable so that as soon as the solution has been carried to $x = x_i$ the next step consists of evaluating the change in the solution through the interval $\delta x (= h)$ of x from x_i to x_{i+1}. When the dependent variable and its derivatives occur in the first degree only and not as higher powers or products, the equation is said to be liner otherwise it is called non-linear. Thus

$\frac{d^2y}{dx^2} + zy = x \sin x$ is a linear ordinary equation whereas $(x+y)^3 \frac{dy}{dx} = 1$ is a non-linear ordinary differential equation.

The general solution of an nth order ordinary differential equation contains n arbitary constants. Inorder to determine these, we require n conditions. If these conditions are given at one point, then these conditions known as initial conditions and the given differential equation together with the initial conditions is called an initial value problem. (IVP). If the n conditions are prescribed at more than one point then these conditions are known as boundary conditions. The given differential equation together with these boundary conditions is called a boundary value problem.

Stability of numerical methods in solving Ordinary equation

$$\frac{dy}{dt} = \lambda y, \quad t \, \varepsilon \left[t_{0,} b\right], \; y(t_0) = y_0$$

Let the grid point t_j be defined by $t_1 = t_0 + jh, j = 0, 1, \ldots, n$ where $t_0 = a$ and $t_0 + Nh = b$.

Separating the variables and integrating, the exact solution of equ. (1) is

$y(t) = y(t_0)\ \exp\ [\lambda(t - t_0)]$.

If t_n and t_{n+1} are two successive solution values, then we can easily obtain

$$y_{n+1}(t) = \exp(\lambda h)\, y(t_n)\ ,\ n = 0,1,\ \ ,N-1$$

Denoting the approximation of $e^{\lambda h}$ in its expansion up to (p + 1) terms by $E(\lambda - h)$, we have

$$y_{n+1} = E(\lambda h)\, y_n$$

$\therefore$ Truncation error of the above method $= y\left(t_{n+1}\right) - y_{n+1}$

$$= 1 + \lambda h + \ldots\ + \frac{(\lambda h)^p}{p.!} + \frac{(\lambda h)^{p+1} e^{\theta \lambda h}}{(p+1)!} - (1 + \lambda h + - + \frac{(\lambda h)^p}{p1}$$

$$= \frac{(\lambda h)^{p+1}}{(p+11)} e^{\theta \lambda h}\ ,\ 0 < \theta < 1$$

Thus truncatin error is of order $p + 1$, The integer p is called the *order of the* method. A numberical method is stable if error at any stage *i.e.* $y_n - y(\varepsilon_n) = t_n$ is bounded as $n \to \infty$. It is called *absolutely stable* if

$$|\ E(\lambda h)\ | < 1$$

and is called *relatively stable* if

$$|\ E(\lambda\, h)| \leq e^{\lambda h}, \lambda > 0$$

If $y_{n+1} = y\left(t_{n+1}\right) + \varepsilon_{n+1} = E(\lambda h)\ [\ y(t_n) + \ \varepsilon_n\ ]$.

$$\varepsilon_{n+1} \left[E(\lambda h) - \lambda h_e \right] y(t_n) + E(\lambda h)\varepsilon_n\ .$$

Differential equations can be classified into Ordinary and Partial differential equations. The first category is further classified into linear and nonlinear differential equations. In this chapter, we shall deal with only the ordinary differential equations.

One-point and two-point boundary conditions. In a step-by-step process, what is more important is the nature of the boundary and other conditions to be satisfied by the solution and not the nature of the equation itself. If the conditions are only the boundary conditions given at one point of the range of integration, then they are known as one-point boundary conditions. The conditions which are specified at two-points of the range are called 'two-point' boundary conditions.

16·2. Equations of the First order

16·2·1. Euler's Method. It is one of the oldest and best known numerical method for integrating differential equations. Consider the differential equation of the first order.

$$\frac{dy}{dx} = f(x, y) \quad ...(16{\cdot}1)$$

with the initial condition $y = y_0$ when $x = x_0$

Let $y = F(x)$ be the solution of (16·1) satisfying the initial conditions. Assume that the graph is small curve. Since small curve is practically straight for a short distance from any point on it, we have

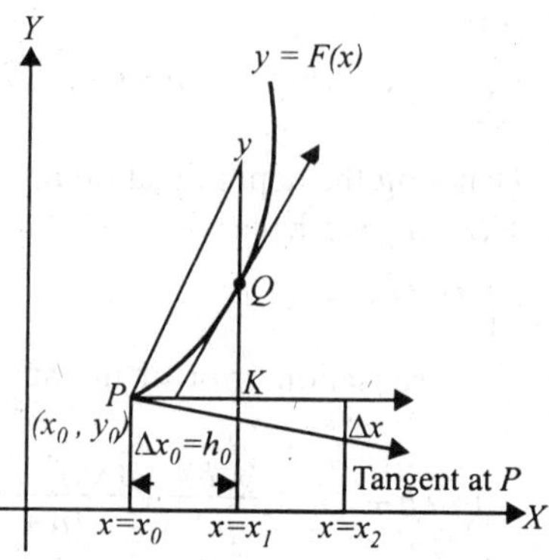

Geometrical representation of Euler's Method

$$\frac{dy}{dx} = \lim_{\Delta x \to 0} \frac{\Delta y}{\Delta x}.$$

Hence the increment Δy can be computed by the equation

$$\Delta y \simeq \frac{dy}{dx} \Delta x \text{ for small values of } \Delta x$$

$$= f(x, y)\, \Delta x = hf(x, y) \qquad \text{where } \Delta x = h.$$

Thus $\int_{y_0}^{y_1} dy = \int_{x_0}^{x_1} f(x, y)\, dx$ or $y_1 = y_0 + \int_{x_0}^{x_1} f(x, y)\, dx.$

This is the exact formula.

Replacing $f(x, y)$ by the approximation $f(x_0, y_0)$, we get

$$y_1 = y_0 + \int_{x_0}^{x_1} f(x_0, y_0)\, dx = y_0 + (x_1 - x_0) f(x_0, y_0) \text{ for } x = x_1$$

Thus $y_1 = y_0 + hf(x_0, y_0) + R$, for $x = x_1$ ...(16·2)

where R is error due to approximation and is known as truncation error.

$$y_1 = y_0 + h\left(\frac{dy}{dx}\right)_0 = y_0 + hf(x_0, y_0)$$

Next, $y_2 = y_1 + hf(x_1, y_1)$, $\quad y_3 = y_2 + hf(x_2, y_2)$,

....

$$y_{n+1} = y_n + hf(x_n, y_n). \quad ...(16{\cdot}2')$$

Thus starting from x_0 when $y = y_0$ we construct a table of y for given steps of h in x.

Drawbacks of the Method

In case h is small, this method is too slow. However if h is large this method is inaccurate.

Notes 1. If PN is drawn parallel to the tangent at Q. Then Δy computed by using slope at Q is represented by KN. Also

$$\frac{KN}{PK} = \left(\frac{dy}{dx}\right)_0, \quad KN = h\left(\frac{dy}{dx}\right)_0 = \Delta y$$

If we take the average of

$$\left(\frac{dy}{dx}\right)_0 \text{ and } \left(\frac{dy}{dx}\right)_1$$

then we have

$$\Delta y = \frac{\left(\frac{dy}{dx}\right)_0 + \left(\frac{dy}{dx}\right)_1}{2} \times h = \frac{1}{2}\,[KM + KM + MN]$$

$$= \frac{1}{2}\,(2KM + MN) = KM + \frac{1}{2}\,NM$$

which is very close to its true value.

2. The attainable accuracy is limited by the length of the step h.

3. The method is slow and has limited accuracy.

Example 1. *Given*

$$\frac{dy}{dx} = \frac{y - x}{y + x}.$$

with the boundary condition $y = 1$ for $x = 0$, find approximately for $x = 0{\cdot}1$ by Euler's method (Five Steps).

Sol. We have, using equation (16·2) and

$$\Delta x = h = \frac{0{\cdot}1 - 0}{5} = 0{\cdot}02$$

For the first step

$$\Delta y = \frac{1 - 0}{1 + 0}\,\Delta x = 0{\cdot}02$$

$$\therefore \quad y = y_0 + \Delta y = 1 + 0{\cdot}02 = 1{\cdot}02 \text{ at } x = 0{\cdot}02$$

For the second step

$$\Delta y = \frac{1{\cdot}02 - 0{\cdot}02}{1{\cdot}02 + 0{\cdot}02}\,h = 0{\cdot}0192$$

$$\therefore \quad y = 1{\cdot}02 + 0{\cdot}0192 = 1{\cdot}0392 \text{ at } x = 0{\cdot}04$$

For the third step

$$\Delta y = \frac{1{\cdot}0392 - 0{\cdot}04}{1{\cdot}0392 + 0{\cdot}04}\,h = 0{\cdot}0185$$

$$\therefore \quad y = 1{\cdot}0392 + 0{\cdot}0185 = 1{\cdot}0577 \text{ at } x = 0{\cdot}06$$

For the fourth step

$$\Delta y = \frac{1{\cdot}0577 - 0{\cdot}06}{1{\cdot}0577 + 0{\cdot}06}\,h = 0{\cdot}0179$$

$$\therefore \quad y = 1{\cdot}0577 + 0{\cdot}0179 = 1{\cdot}0756 \text{ at } x = 0{\cdot}08$$

$$\Delta y = \frac{1{\cdot}0756 - 0{\cdot}08}{1{\cdot}0756 + 0{\cdot}08} h = 0{\cdot}0172$$

$\therefore$ $$y = 1{\cdot}0758 + 0{\cdot}0172 = 1{\cdot}0928 \text{ at } x = 0{\cdot}10.$$

16·2·2. Euler's Modified Method. By taking y' as a new variable, the equation from (16·1) can be written as

$$\frac{dy}{dx} = y'$$

and $$y' = f(x, y).$$

Integrating the first equation from $x_i (= x_0 + ih)$ to x_{i+1}, we have

$$\int_{x_i}^{x_{i+1}} \frac{dy}{dx} dx = \int_{x_i}^{x_{i+1}} y' \, dx$$

or $$y_{i+1} = y_i + \int_{x_i}^{x_{i+1}} y' \, dx \qquad ...(16{\cdot}3)$$

with $$y_i' = f(x_i, y_i)$$

Using the trapezoidal rule, we obtain

$$y_{j+1} \simeq y_i + \tfrac{1}{2} h(y_i' + y_{i+1}')$$

$$= y_i + \tfrac{1}{2} h[f(x_i, y_i) + f(x_{i+1}, y_{i+1})] + R \qquad ...(16{\cdot}4)$$

where R is the error due to an approximation of the integral.

The successive approximation for y_{i+1} may be obtained by the method of iteration explained in §4·2·2, the process being repeated until no significant change occurs in y_{i+1}. This method of obtaining successive approximations is known as modified method of Euler. First y_i is obtained by unmodified method given n (16·2) and then substituted in (16·4).

Notes 1. The equation (16·2) can be obtained from equation (16·3) by using the trapezoidal rule.

2. Equation (16·4) differs from equation (16·2) in that $f(x_i, y_i)$ is replaced by the mean or average of $f(x_i, y_i)$ and $f(x_{i+1}, y_{i+1})$,

3. Euler's modified method has an error of order h^3 since on comparing (16·2), (16·4) and $f(x_1, y_1) = f(x_0, y_0) + hf'(x_0, y_0) + \frac{h^2}{12} f''(x_0, y_0) + R$ we find $R = \frac{h^3}{6} f''(x_0, y_0) - \frac{h^3}{4} f'''(x_0, y_0) = \frac{-h^3}{12} \times \text{constant}$.

Example 2. *Using Euler's modified method, obtain a solution of the equation*

$$\frac{dy}{dx} = x + |\sqrt{y}| = f(x, y),$$

with boundary condition $y = 1$ at $x = 0$ for the range $0 \le x \le 0{\cdot}6$ in steps of $0{\cdot}2$.

Sol. We have $\dfrac{dy}{dx} = x + |\sqrt{y}|$

with $\quad x_0 = 0,\ y_0 = 1,\ 0 \le x \le 0{\cdot}6,\ h = 0{\cdot}2$

By Euler's method (for $x = 0{\cdot}2$)

$$y_1 = y_0 + hf(x_0, y_0)$$
$$= 1 + 0{\cdot}2[1 + 0] = 1{\cdot}2$$

Now we apply Euler's modified method to get more correct approximation. Let $y_1', y_1'', \ldots,$ be first, second, ..., approximations to y_1.

$$y_1' = 1 + \tfrac{1}{2}(0{\cdot}2).[(1 + 0) + 0{\cdot}2 + \sqrt{1{\cdot}2}]$$
$$= 1 + 0{\cdot}1[1{\cdot}2 + 1{\cdot}095]$$
$$= 1 + 0{\cdot}1[2{\cdot}295] = 1 + 0{\cdot}2295$$
$$= 1{\cdot}2295$$
$$y_1'' = 1 + \tfrac{1}{2}(0{\cdot}2)[(1 + 0) + (0{\cdot}2 + \sqrt{1{\cdot}2295})]$$
$$= 1 + 0{\cdot}1[1{\cdot}2 + 1{\cdot}109]$$
$$= 1 + 0{\cdot}1[2{\cdot}309]$$
$$= 1{\cdot}2309$$
$$y_1''' = 1 + 0{\cdot}1[(1 + 0) + 0{\cdot}2 + \sqrt{1{\cdot}2309}]$$
$$= 1 + 0{\cdot}1[2{\cdot}309]$$
$$= 1{\cdot}2309$$

Now we find out y for $x = 0{\cdot}4$ by Euler's method

$$y_2 = 1{\cdot}2309 + 0{\cdot}2[0{\cdot}2 + \sqrt{1{\cdot}2309}]$$
$$= 1{\cdot}2309 + 0{\cdot}2[0{\cdot}2 + 1{\cdot}109]$$
$$= 1{\cdot}2309 + 0{\cdot}2[1{\cdot}309]$$
$$= 1{\cdot}2309 + 0{\cdot}2618 = 1{\cdot}4927$$

Now we apply Euler's modified method for more correct approximations.

$$y_2' = 1{\cdot}2309 + \tfrac{1}{2}(0{\cdot}2)[0{\cdot}2 + \sqrt{1{\cdot}2309} + 0{\cdot}4 + \sqrt{1{\cdot}4927}]$$
$$= 1{\cdot}2309 + 0{\cdot}1[0{\cdot}2 + 1{\cdot}109 + 0{\cdot}4 + 1{\cdot}222]$$
$$= 1{\cdot}2309 + 0{\cdot}1[2{\cdot}931]$$
$$= 1{\cdot}2309 + 0{\cdot}2931$$
$$= 1{\cdot}5240$$
$$y_2'' = 1{\cdot}2309 + 0{\cdot}1[0{\cdot}6 + \sqrt{1{\cdot}230{\cdot}9} + \sqrt{1{\cdot}524}]$$
$$= 1{\cdot}2309 + 0{\cdot}1[1{\cdot}709 + 1{\cdot}235]$$
$$= 1{\cdot}2309 + 0{\cdot}1[2{\cdot}944]$$

$$= 1{\cdot}2309 + 0{\cdot}2944$$
$$= 1{\cdot}5253$$
$$y_2''' = 1{\cdot}2309 + 0{\cdot}1[0{\cdot}6 + \sqrt{1{\cdot}2309} + \sqrt{1{\cdot}5253}]$$
$$= 1{\cdot}2309 + 0{\cdot}1[1{\cdot}709 + 1{\cdot}235]$$
$$= 1{\cdot}5253$$

Now we apply Euler's method to find the value of $y(= y_3)$ for $x = 0{\cdot}6$

$$y_3 = 1{\cdot}5253 + 0{\cdot}2[0{\cdot}4 + \sqrt{1{\cdot}5253}]$$
$$= 1{\cdot}5253 + 0{\cdot}2[0{\cdot}4 + 1{\cdot}235]$$
$$= 1{\cdot}5253 + 0{\cdot}2[1{\cdot}635]$$
$$= 1{\cdot}5253 + 0{\cdot}3270$$
$$= 1{\cdot}8523$$

For the better approximation we apply Euler's modified method.

$$y_3' = 1{\cdot}5253 + \tfrac{1}{2}(0{\cdot}2)[0{\cdot}4 + \sqrt{1{\cdot}5253} + 0{\cdot}6 + \sqrt{1{\cdot}8523}]$$
$$= 1{\cdot}5253 + 0{\cdot}1[1{\cdot}0 + 1{\cdot}235 + 1{\cdot}361]$$
$$= 1{\cdot}5253 + 0{\cdot}1[3{\cdot}596]$$
$$= 1{\cdot}5253 + 0{\cdot}3596$$
$$= 1{\cdot}8849$$
$$y_3'' = 1{\cdot}5253 + 0{\cdot}1[2{\cdot}235 + \sqrt{1{\cdot}8849}]$$
$$= 1{\cdot}5253 + 0{\cdot}1[1{\cdot}235 + 1{\cdot}373]$$
$$= 1{\cdot}5253 + 0{\cdot}1[3{\cdot}608]$$
$$= 1{\cdot}5253 + 0{\cdot}3608$$
$$= 1{\cdot}8851.$$

Now if we find out y_3''' we shall see $y_3''' = 1{\cdot}8851$.

Thus when $x = 0{\cdot}6$, $y = 1{\cdot}8851$ approx.

16·2·3. Picard's Method of Successive Approximations

Consider the differential equation

$$\frac{dy}{dx} = f(x, y) \qquad \text{...(16·5)}$$

with the initial condition $y = y_0$ at $x = x_0$.

Integrating eq. (16·5) from x_0 at x with respect to x, we have

$$\int_{y_0}^{y} dy = \int_{x_0}^{x} f(x, y)\, dx$$

$$\Rightarrow \qquad y = y_0 + \int_{x_0}^{x} f(x, y)\, dx. \qquad \text{...(16·6)}$$

In order to obtain a solution of (16·5), we solve equation (16·6) by successive approximations. Thus if $u(x)$ is a first approximation of the solution of equation (16·5), then the second approximation $u_2(x)$ is obtained by substituting $u_1(x)$ in the right hand side of (16·6) Thus

$$u_2(x) = y_0 + \int_{x_0}^{x} f\{x, u_1(x)\}\, dx.$$

The third approximation is obtained by substituting $u_2(x)$ in the right hand side of (16·6) so that

$$u_3(x) = y_0 + \int_{x_0}^{x} f\{x, u_2(x)\}\, dx.$$

and so on for further approximations.

Note. The solution of (16·5) is obtained under the following assumptions :—

(i) $f(x, y)$ is a single-valued continuous function in the domain D defined by $|x - x_0| \le a, -\infty < y < \infty$

(ii) $f(x, y)$ is bounded in D *i.e.*, $|f(x, y)| < M \quad \forall\ (x, y)$ in D.

(iii) $|f(x, y)^i - f(x, y)^{i+1}| \le L\, |y^{(i)} - y^{(i+1)}|$ where L is a + ve number.

Example 3. *Given* $\dfrac{dy}{dx} = \dfrac{y-x}{y+x} = f(x, y)$ *with $y = 1$ when $x = 0$, find approximately the value of y for $x = 0{\cdot}1$ by Picard's method.*

Sol. With $y = 1$ when $x = 0$, we have $\dfrac{dy}{dx} = 1$. Further since $y = 1$ and $\dfrac{dy}{dx} = 1$ when $x = 0, y$ will be nearly 1 for small x. Hence as a first approximation, we take $u_1(x) = 1$.

Thus to the solution of given equation, the second approximation is given by

$$u_2(x) = y_0 + \int_{x_0}^{x} f[x, u_1(x)]\, dx$$

or $$u_2(x) = y_0 + \int_{0}^{x} \frac{1-x}{1+x}\, dx = 1 + [2 \log (1 + x) - x]_0^x$$

$$= 1 + 2 \log (1 + x) - x$$

Thus when $x = 0{\cdot}1$, second approximation

$$= 1 + 2 \log 1{\cdot}1 - 0{\cdot}1$$

$$= 0{\cdot}9828.$$

Third approximation is given by

$$u_3(x) = y_0 + \int_{x_0}^{x} f\{x, u_2(x_1)\}\, dx$$

or

$$u_3(x) = 1 + \int_0^x \frac{1 + 2\log(1+x) - 2x}{1 + 2\log(1+x)}\, dx$$

$$= 1 + x - 2\int_0^x \frac{x}{1 + 2\log(1+x)}\, dx$$

$$= 1 + x - 2\int_0^t \frac{e^{2t}}{1+2t}\, dt + 2\int_0^t \frac{e^t}{1+2t}\, dt$$

where $t = \log(1+x)$ and $t = 0$ when $x = 0$.

Which can not be integrated. Hence in this example, only second approximation which is 0·9828 can be obtained.

Example 4. *Use Picard method of successive approximations to*

$$\frac{dy}{dx} = x + y$$

with boundary conditions $y = 1$ when $x = 0$, (compute three iterations only). Check the results with the exact values.

Sol. The first approximation is given by $u_1(x) = 1$

For the second approximation,

$$u_2(x) = 1 + \int_0^x f(x, 1)\, dx$$

$$= 1 + \int_0^x (x+1)\, dx = \tfrac{1}{2}x^2 + x + 1.$$

For the third approximation, we have

$$u_3(x) = 1 + \int_0^x \left\{ x + \left(\tfrac{1}{2}x^2 + x + 1 \right) \right\} dx$$

$$= \frac{1}{6}x^3 + x^2 + x + 1.$$

For the fourth approximation, we get

$$u_4(x) = 1 + \int_0^x \left\{ x + \left(\tfrac{1}{6}x^3 + x^2 + x + 1 \right) \right\} dx$$

$$= \frac{1}{24}x^4 + \frac{1}{3}x^3 + x^2 + x + 1.$$

Thus when $x = 0{\cdot}1$, we have

$$u_4(x) = \frac{1}{24}(0{\cdot}1)^4 + \frac{1}{3}(0{\cdot}1)^3 + (0{\cdot}1)^2 + 0{\cdot}1 + 1$$

$$= 1{\cdot}11030.$$

and when $x = 0{\cdot}2$, we have

$$u_4(x) = \frac{1}{24}(0{\cdot}2)^4 + \frac{1}{3}(0{\cdot}2)^3 + (0{\cdot}2)^2 + 0{\cdot}2 + 1$$

$$= 1{\cdot}2427.$$

The procedure can be continued to obtain better approximations.

For *or analytic solutions* of the given differential equation, we have

Solution. $ye^{-x} = \int xe^{-x}\,dx + c \qquad [I.F. = e^{-x}]$

$$= -xe^{-x} - e^{-x} + c$$

or $y = x - 1 + ce^x$

with $y = 1$ when $x = 0$ *i.e.*, $c = 2$

Thus $y = 2e^x - x - 1$ is the analytic solution of $\frac{dy}{dx} = x + y$.

When $x = 0{\cdot}1,\ y = 3e^{0{\cdot}1} - 0{\cdot}1 - 1 = 1{\cdot}11034,$

$$x = 0{\cdot}2,\ x = 2e^{0{\cdot}2} - 0{\cdot}2 - 1 = 1{\cdot}2428.$$

Thus the approximation obtained for $x = 0{\cdot}1$ is correct to four places of decimals while for $x = 0{\cdot}2$, it is correct to three places of decimals.

(*b*) **Picard Method for solving first order simultaneous equations.**

We shall illustrate the method of solving

$$\frac{dy}{dx} = f(x, y, z) \text{ and } \frac{dz}{dx} = g(x, y, z)$$

by an example.

Example 5. *Approximate y and z by using Picard's method for the particular solution of*

$$\frac{dy}{dx} = x + z, \quad \frac{dz}{dx} = x - y^2$$

satisfying $y = 2,\ z = 1$ *when* $x = 0$.

(*B.A. Maths. Delhi, 1989*)

Sol. Here $f(x, y, z) = x + z,$

$$g(x, y, z) = x - y^2$$

Also $x_0 = 0,\ y_0 = 2,\ z_0 = 1$

For the first approximations,

$$y_1 = y_0 + \int_0^x f(x, y_0, z_0)\, dx$$

$$= 2 + \int_0^x (x + 1)\, dx = 2 + x + \tfrac{1}{2} x^2$$

$$z_1 = z_0 + \int_0^x g(x, y_0, z_0)\, dx$$

$$= 1 + \int_0^x (x - 4)\, dx = 1 - 4x + \tfrac{1}{2} x^2$$

For the second approximations,

$$y_2 = y_0 + \int_0^x f(x, y_1, z_1)\, dx$$

$$= 2 + \int_0^x (x + z_1)\, dx = 2 + \int_0^x \left(x + 1 - 4x + \frac{x^2}{2} \right) dx$$

$$= 2 + x - \frac{3}{2} x^2 + \frac{1}{6} x^3$$

$$z_2 = z_0 + \int_0^x g(x, y_1, z_1)\, dx$$

$$= 1 + \int_0^x (x - y_1^2)\, dx$$

$$= 1 + \int_0^x \left[x - \left(2 + x + \frac{1}{2} x^2 \right)^2 \right] dx$$

$$= 1 - 4x - \frac{3}{2} x^2 - x^3 - \frac{1}{4} x^4 - \frac{1}{20} x^5$$

Similarly third approximations can be obtained for y and z respectively.

16·2·4. Method of starting the solution.

(*a*) **By Taylor's series.** Suppose we wish to obtain the solution of the first order differential equation

$$\frac{dy}{dx} = f(x, y) \qquad \text{...(16·7)}$$

with the boundary condition $y = y_0$ when $x = x_0$, where $f(x, y)$ may be differentiated as required.

Let the solution of equation (16·7) be $y = \phi(x)$ with $\phi(x_0) \neq 0$.

Expanding $\phi(x)$ as a Taylor's series near the point x_0 we have

$$y = y_0 + (x - x_0)y_0' + \frac{1}{2!}(x - x_0)^2 y_0'' + \frac{1}{3!}(x - x_0)^3 y_0''' + \cdots$$

$$= y_0 + hy_0' + \frac{1}{2!}h^2 y_0'' + \frac{1}{3!}h^3 y_0''' + \cdots \qquad ...(16·8)$$

which is the convergent series in x in the range $x_0 \le x < x_n$, say, with

$$y_0'' = \phi''(x_0) = \left(\frac{d^2y}{dx^2}\right)_{x = x_0}$$

The constants y_0', y_0'', y_0''' ... can be calculated in turn from (16·7) which can be re-written as

$$y' = f(x, y) = g_1(x, y).$$

Successive differentiation of this equation gives as

$$y'' = \frac{\partial g_1}{\partial x} + \frac{\partial g_1}{\partial y} y' = g_2(x, y, y')$$

$$y''' = \frac{\partial g_2}{\partial x} y' + \frac{\partial g_2}{\partial y'} y'' = g_3(x, y, y', y'')$$

....

$$y^{(n)} = \frac{\partial g_{n-1}}{\partial x} + \Sigma \frac{\partial g_{n-1}}{\partial y^{n-2}} y^{(n-1)} = g_n(x, y, y' \ldots y^{(n-1)})$$

Putting $x = x_0$ and $y = y_0$ in these results, we obtain the constants $y_0^{(n)}$, $n = 1, 2, 3, \ldots$ in terms of $x_0, y_0, y_0', \ldots, y_0^{(n-1)}$ Substituting the values of the constant $y_0^{(n)}$ in eq. (16·8), we obtain the solution of eq. (16·1).

N.B. $x - x_0 = ih$, then equation (16·8) becomes

$$y_i = y_0 + (ih)y_0' + \frac{1}{2!}(ih)^2 y_0'' + \frac{1}{3!}(ih)^3 y_0''' + \cdots \qquad ...(16·9)$$

Example 6. *Given* $y' \equiv \frac{dy}{dx} = y^2 + 1$ *with boundary condition* $y = 0$ *at* $x = 0$ *in the range* $0 \le x \le 1$, *obtain* y *as a series in powers of* x. *Check your result with the exact value.* (*B.Sc. Nagpur 1998*)

Sol. We have

$$y' = y^2 + 1, \qquad \therefore \quad y'(0) = 1$$

$$y'' = 2yy', \qquad y''(0) = 0$$

$$y''' = 2yy'' + 2y'^2, \qquad y'''(0) = 2$$

$$y^{iv} = 2y'y'' + 2yy''' + 4y'y'', \qquad y^{iv}(0) = 0$$

Hence substituting the values of the constants in equation (16·8) we have

$$y = 0 + x \cdot 1 + \frac{x^2}{2!} \cdot 0 + \frac{x^3}{3!} \cdot 2 + \frac{x^4}{4!} \cdot 0 + \cdots$$

$$\Rightarrow \qquad y = x + \frac{x^3}{3} + \cdots$$

Let us take $h = \Delta x = 0{\cdot}2$. Then

$$y_1 = y(0{\cdot}2) = 0{\cdot}2 + \tfrac{1}{2}(0{\cdot}008) = 0{\cdot}203 \text{ approx.}$$

$$y_1' = y'(0{\cdot}2)^2 = (0{\cdot}203)^2 + 1 = 1{\cdot}0412.$$

Exact solution is $y = \tan x$

Table

x :	0	0·2	0·4	0·6	0·8	1
y :	0	0·203	0·422	0·683	1·027	1·546

Correct value 0·557 when $x = 1$

Error 0·011 when $x = 1$.

Example 7. *Solve $y' = x + y$ numerically. Start with $x = 1$ $y = 0$ and carry to $x = 1{\cdot}2$ with $h = 0{\cdot}1$. Compare the final result with the value of the explicit solution ($e^{0.2} = 1{\cdot}221$).*

Determine the number of terms so as to obtain an accuracy of 10^{-10}, $1 \le x \le 10$

Sol. We have $x_0 = 1$, $y_0 = 0$ so that

$$y' = x + y, \qquad y_0' = x_0 + y_0 = 1$$

$$y'' = 1 + y', \qquad y_0'' = 1 + y_0 = 2$$

$$y''' = y'', \qquad y_0''' = y_0'' = 2$$

$$y^{iv} = y''', \qquad y_0^{iv} = y_0''' = 2$$

$$y^{v} = y^{iv}, \qquad y_0^{v} = y_0^{iv} = 2.$$

$$\therefore \qquad y_z = 1 + x + x^2 + \frac{2x^2}{3!} + \cdots + \frac{2x^n}{n!}$$

$$\therefore \qquad \frac{2}{(n+1)!} < 5 \times 10^{-11} \text{ which gives } n = 15.$$

Substituting the values of the constants in eq. (16·9), we obtain

$$y_1 = 0 + (0{\cdot}1) \times 1 + \frac{(0{\cdot}1)^2}{2!} \times 2 + \frac{(0{\cdot}1)^3}{3!} \times 2 + \frac{(0{\cdot}1)^4}{4!} \times 2 + \frac{(0{\cdot}1)^5}{5!} \times 2 + \cdots$$

or $\qquad y = 0{\cdot}11033847$, $x = 1 + {\cdot}1 = 1{\cdot}1$

When $y_0' = 1{\cdot}1$, $y_0 = 0{\cdot}1103384$ so that

$$x_0 = 1{\cdot}21,\ y_0'' = 2{\cdot}21,\ y_0''' = 2{\cdot}21,\ y_0^{iv} = 2{\cdot}21$$

$$\therefore \quad y = 0{\cdot}11 + (0{\cdot}1)(2{\cdot}21) + \frac{(0{\cdot}1)^2}{2}(2{\cdot}21) + \frac{(0{\cdot}1)^3}{6}\cdot(2{\cdot}21) + \frac{(0{\cdot}1)^4}{24}(2{\cdot}21)$$

$$= 0{\cdot}232$$

Again when $x_0 = 1{\cdot}2$ $y_0 = 0{\cdot}232$.

The exact solution of $y' = x + y$ is

$$y = -x - 1 + 2e^{x-1}$$

When $x = 1{\cdot}2,$

$$y = -1{\cdot}2 - 1 + 2(1{\cdot}221)$$
$$= 0{\cdot}242$$

System of Differential Equations solvable by Taylor's series.

We shall illustrate by solving the following :

Example 8. *Find a solution of the set of simultaneous equations.*

$$\left.\begin{aligned}\frac{dx}{dt} &= x + y + t\\ \frac{dy}{dx} &= 2x - t\end{aligned}\right\} \quad ...(1)$$

subject to the initial conditions $x = 0, y = 1$ at $t = 1$ as the power series in t for x and y using Taylor' method.

Sol. By Taylor's series, we have

$$x = x_0 + x_0^{(1)}t + \frac{1}{2\,!}x_0^{(2)}t^2 + \frac{1}{3\,!}x_0^{(3)}t^3 + \cdots$$

$$y = y_0 + y_0^{(1)}t + \frac{1}{2\,!}y_0^{(2)}t^2 + \frac{1}{3\,!}y_0^{(3)}t^3 + \cdots$$

where $x^{(1)} = \dfrac{dx}{dt},\ y^{(1)} \equiv \dfrac{dy}{dt}.$

Differentiating (1) twice and putting $x = 0,\ y = 1$ and $t = 1$, we obtain

$$x_0^{(1)} = 2,\ y_0^{(1)} = -1,\ x_0^{(2)} = x_0^{(1)} + y_0^{(1)} + 1 = 2$$
$$y_0^{(2)} = 2x_0^{(1)} - 1 = 3,$$
$$x_0^{(3)} = x_0^{(2)} + y_0^{(2)} = 5,\ y_0^{(3)} = 2x_0^{(2)} = 4.$$

Thus, $x = 2t + t^2 + \dfrac{5}{6}t^3 \ldots$

and $y = 1 - t + \dfrac{3}{2}t^2 + \dfrac{2}{3}t^3 \ldots$

(*b*) **By Milne's series.** Suppose the higher derivatives of $y' = f(x, y)$ ...(16·7) can be written down. Let the boundary condition be $y = y_0$ at

$x = x_0$ so that we have by starting process the numerical values for y_0, y_1, y_2, y_3, and for y'_0, y'_1, y'_2, y'_3. By a step-by-step process, it is possible for us to write down the values of all subsequent y_i and y'_i with the help of suitable numerical integration formulae. We have, on integrating (16·7) from x_{n-3} to x_{n+1}.

$$y_{n+1} - y_{n-3} = \int_{x_{n-3}}^{x_{n+1}} y'\, dx$$

$$= \frac{4h}{3}(2y'_{n-2} - y'_{n-1} + 2y'_n) + \frac{14}{45} h^5 y^{(5)}(\xi),\ x_{n-3} < \xi < x_{n+1}$$

Taking h to be so small that the remainder term is small, we have

$$y_{n+1} = y_{n-3} + \frac{4h}{3}(2y'_{n-2} - y'_{n-1} + 2y'_n) \qquad \text{...(16·10)}$$

This formula, known as a *predictor*, enables us to obtain the approximate value of y_{n-1} and then y'_{n+1} can be obtained from (16·7). A better value of y_{n+1} can be obtained as follows by making use of Simpson's one-third rule

$$y_{n+1} - y_{n-2} = \frac{h}{3}[y'_n + 4y'_{n-1} + y'_{n-2}] = \frac{h}{90}\Delta^4 y'_n$$

$$\text{Error term} = -\frac{1}{90} y^{(0)}(\xi),\ x_{n-1} < \xi < x_{n+1} \qquad \text{...(16·11)}$$

The remainder term is usually smaller than the previous remainder term and hence this formula is a better approximation to y_{n+1} and y_{n+1} given by formula (16·10). This formula is known as a *corrector*.

After the values of y_{n+1} and y'_{n+1} have been obtained, the values of the next pair *viz.* y_{n+2} and y'_{n+2} can be obtained by repetition of the process.

N.B. 1. If the four prior values of y are not given in any question, they should be obtained by using Taylor's series, Euler's or Runge-Kutta method.

N.B. 1. Let the new value of y_{n+1} obtained from Eq. (16·11) be denoted by $\bar{y}_{n+1}$ so that

$$\bar{y}'_{n+1} - y'_{n+1} = f(x_{n+1}, \bar{y}_{n+1}) - f(x_{n+1}, y_{n+1})$$

$$= \delta f_y(x_{n+1}, \eta_{n+1}),\quad y_{n+1} < \eta_{n+1} < \bar{y}_{n+1}$$

where $\delta = \bar{y}_{n+1} - y_{n+1}$.

Thus from Eq. (16·11), the change in y'_{n+1} would cause a change of $\frac{1}{3} h\, \delta f_y(x_{n+1}, \eta_{n+1})$ in y_{n+1}. Thus the first correction to the value of

y_{n+1} is multiplied by the *convergence factor* $\theta = \frac{1}{3} hf_y(x_{n+1}, \eta_{n+1})$ in obtaining the second correction to the value of y_{n+1}.

N.B. 2. We find

$$\delta = \frac{14}{45} h^5 y^{(5)}(\xi) - \left\{ \frac{-h^5}{90} y^{(5)}(\xi') \right\}$$

$$= \frac{h^5}{90} [y^{(5)}(\xi') + 28y^{(5)}(\xi)].$$

Assume that remainder form in (16·11) does not changes too rapidly from step to step, $y^{(5)}(\xi) \equiv y^{(5)}(\xi')$ and therefore, remainder term in (16·11) $\simeq \frac{-1}{29} \delta. \simeq \frac{-1}{29} (\bar{y}_{n+1} - y_{n+1})$

Example 9. *Solve numerically the equation*

$$y' = x + y$$

with the initial conditions

$$x_0 = 0,\ y_0 = 1,$$

by Milne Method from (i) $x = 0$ to $x = 0{\cdot}4$ and (ii) $x = {\cdot}20$ to ${\cdot}30$.

Sol. (*i*) Here $\quad y' = x + y,\ y'' = 1 + y',\ y''' = y'',$

$$y^{iv} = y''',\ y^{v} = y^{iv}$$

Hence $\quad y_0' = x_0 + y_0 = 1,\ y_0'' = 1 + y_0' = 2,\ y_0''' = y_0'' = 2,$

$$y_0^{iv} = y_0''' = 2,\ y_0^{v} = y_0^{iv} = 2.$$

Now taking $h = 0{\cdot}1$ and substituting these values for $i = 1, 2, 3$ respectively in eq. (16·9) we get

$$y_1 = 1{\cdot}1103,\ y_2 = 1{\cdot}2428,\ y_3 = 1{\cdot}3997,$$

Hence $\quad y_0' = x_0 + y_0 = 1.$

$$y_1' = x_1 + y_1 = 0{\cdot}1 + 1{\cdot}1103 = 1{\cdot}2103,$$

$$y_2' = x_2 + y_2 = 0{\cdot}2 + 1{\cdot}2428 = 1{\cdot}4428$$

$$y_3' = x_3 + y_3 = 0{\cdot}3 + 1{\cdot}3997 = 1{\cdot}6997,$$

$$y_4' = y_0 + \frac{4}{3} h(2y_1' - y_2' + 2y_3') \quad \text{by predictor formula}$$

$$= 1 + \frac{4}{3} \times 0{\cdot}1(2 \times 1{\cdot}2103 - 1{\cdot}4428 + 2 \times 1{\cdot}6997)$$

$$= 1{\cdot}583627$$

$$y_4 = y_2 + \frac{1}{3} h(y_2' + 4y_3' + y_4') \quad \text{by corrector formula}$$

$$= 1{\cdot}2428 + \frac{1}{3} \times 0{\cdot}1(1{\cdot}4428 + 4 \times 1{\cdot}6997 + 1{\cdot}5836)$$

$$= 1{\cdot}5703 \quad \text{when } x = 0{\cdot}4.$$

(*ii*) We tabulate the values of y and y' corresponding to $x = 0{\cdot}05, 0{\cdot}10$ and $0{\cdot}15$, $h = {\cdot}05$

x	y	y'
0·00	1·0000	1·0000
0·05	1·0526	1·1026
0·10	1·1104	1·2104
0·15	1·1737	1·3237

$$\therefore \quad \bar{y}_4 = y_0 + \frac{4h}{3}(2y_1' - y_2' + 2y_3') \quad \text{by predictor formula.}$$

$$= 1 + \frac{4}{3}({\cdot}05)[2{\cdot}2052 - 1{\cdot}2104 + 2{\cdot}6474]$$

$$= 1{\cdot}2428$$

which is corrected by

$$y_4 = y_2 + \frac{h}{3}(y_2' + 4y_3' + y_4')$$

$$= 1{\cdot}1104 + \tfrac{1}{3}({\cdot}05)(1{\cdot}2104 + 5{\cdot}2948 + 1{\cdot}4428)$$

$$= 1{\cdot}4428$$

Hence $y_4 = y_{\cdot 20} = 1{\cdot}2428$ and $y_4' = 1{\cdot}4428$

Again $\bar{y}_5 = y_1 + \frac{4h}{3}(2y_2' - y_3' + 2y_4')$

$$= 1{\cdot}0526 + \frac{4({\cdot}05)}{3}(2{\cdot}4208 - 1{\cdot}3237 + 2{\cdot}8856)$$

$$= 1{\cdot}3181$$

$$y_5 = y_3 + \frac{h}{3}[y_3' + 4y_4' + y_5']$$

$$= 1{\cdot}1737 + \tfrac{1}{3}({\cdot}05)[1{\cdot}3237 + 5{\cdot}7712 + 1{\cdot}5681]$$

$$= 1{\cdot}3181$$

Hence $y_5 = y_{\cdot 15} = 1{\cdot}3181$ and $y_5' = 1{\cdot}5681$

Again $y_6 = y_2 + \frac{4h}{3}[2y_3' - y_4' + 2y_5']$

$$= 1{\cdot}1104 + \frac{4({\cdot}05)}{3}[2{\cdot}6474 - 1{\cdot}4428 + 3{\cdot}1362]$$

$$= 1{\cdot}3997$$

$$\bar{y}_6 = y_4 + \frac{h}{3}[y_4' + 4y_5' + \bar{y}_6']$$

$$= 1{\cdot}2428 + \frac{0{\cdot}05}{3}[1{\cdot}4428 + 6{\cdot}2724 + 1{\cdot}3997]$$

$$= 1{\cdot}3997$$

Hence $y_6 = 1{\cdot}3997$, $y_6' = 1{\cdot}6997$

The result can be put in tabular form

x	y	y'
·20	1·2428	1·4428
·25	1·3181	1·5681
·30	1·3997	1·6997

16·2·5. Runge-Kutta (German Mathematicians) Method

Consider the first order differential equation

$$y' \equiv \frac{dy}{dx} = f(x, y) \qquad \text{...(16·12)}$$

with the initial condition

$$y = y_0 \text{ for } x = x_0.$$

Differentiating (16·12) successively with respect to x, we have

$$y'' = f_x + f_y f,$$

where $$f_x = \frac{\partial f}{\partial x}, f_y = \frac{\partial f}{\partial y}, f \equiv f(x, y).$$

$$y''' = (f_{xx} + f_{xy} f + f_y f_x) + (f_{xy} + f_{yy} f + f_y^2) f$$

$$= f_{xx} + 2f_{xy} f + f_{yy} f^2 + (f_x + f_y f) f_y$$

and so on.

Putting $x = x_0$, $y = y_0$,

we obtain the values of the constants y_0', y_0'', y_0''', etc.

Substituting these values in the Taylor's series (16·8), we have

$$\Delta y = hf_0 + \frac{1}{2\,!} h^2 (f_x + f_y f)_0$$

$$+ \frac{1}{3\,!} h^3 [f_{xx} + 2f_{xy} f + f_{yy} f^2 + (f_x + f_y f) f_y]_0 + \cdots \qquad \text{...(16·13)}$$

where $\Delta y = y - y_0$.

Let the increments in y be defined by the equation

$$\Delta y = a\Delta' y + b\Delta'' y + c\Delta''' y \qquad \text{...(16·14)}$$

where $$\Delta' y = f(x_0, y_0)\, \Delta x = hf_0,$$

$$\Delta'' y = hf(x_0 + mh, y_0 + m\Delta' y),$$

and $$\Delta''' y = hf[x_0 + nh, y_0 + \rho\Delta'' y + (n - \rho)\Delta' y]$$

where m, n and ρ are constants.

Using Taylor's series for two independent variables about the point (x_0, y_0), *viz.*,

$$f(x_0+h, y_0+k) = f(x_0, y_0) + \left(h\frac{\partial}{\partial x} + k\frac{\partial}{\partial y}\right) f(x_0, y_0) + \frac{1}{2!}\left(h\frac{\partial}{\partial x} + k\frac{\partial}{\partial y}\right)^2 f(x_0, y_0) + \cdots,$$

$$\Delta' y = hf_0,$$

$$\Delta'' y = h\{ f_0 + mh(f_x)_0 + m\Delta' y(f_y)_0 + \tfrac{1}{2}[(mh)^2(f_{xx})_0 + 2m^2 h\Delta' y(f_{xy})_0 + (m\Delta' y)^2(f_{yy})_0] + \cdots \quad \text{...(16·15)}$$

$$\Delta''' y = h\{ f_0 + nh(f_x)_0 + [\rho\Delta'' y + (n-\rho)\Delta' y](f_y)_0 + \tfrac{1}{2}\{(nh)^2(f_{xx})_0 + 2nh[\rho\Delta'' y + (\lambda-\rho)\Delta' y](f_{xy})_0 + [\rho\Delta'' y + (n-\rho)\Delta' y]^2(f_{yy})_0\} + \cdots \quad \text{...(16·16)}$$

Substituting the values of $\Delta' y$ in (16·15) and the value of $\Delta'' y$ in (16·16), we obtain up to terms in h^3,

$$\Delta' y = hf_0$$

$$\Delta'' y = hf_0 + m(f_x + f_y f)_0 h^2 + \tfrac{1}{2} m^2(f_{xx} + 2f_{xy} f + f_{yy} f^2)h^3$$

$$\Delta''' y = hf_0 + n(f_x + f_y f)_0 h^2 + \tfrac{1}{2}[n^2(f_{xx} + 2f_{xy} f + f_{yy} f^2) + 2m\rho(f_x + f_y f) f_y]h^3$$

Substituting these values of $\Delta' y$, $\Delta'' y$ and $\Delta''' y$ in (16·14) and simplifying we obtain the value of Δy. Also from (16·14), we get the value of Δy. Equating the values of Δy and comparing the coefficients of h, h^2 and h^3, we have,

$$a + b + c = 1,$$

$$bm + cn = \tfrac{1}{2},$$

$$bm^2 + cn^2 = \tfrac{1}{3},$$

$$c\rho m = \frac{1}{6}.$$

Thus we can choose m and n and determine the values of the four constants a, b, c and ρ. We find

$$\rho = \frac{n(m-n)}{m(3m-2)}, \quad a = \frac{6mn - 3(m+n) + 2}{6mn},$$

$$b = \frac{2-3n}{6m(m-n)}, \quad c = \frac{3m-2}{6n(m-n)}$$

Thus we can determine the value of Δy and hence find as a first approximation the value of y corresponding to the value $x_0 + h$ of x.

Runge-Kutta's Second order rule. For solving

$$y' = \frac{dy}{dx} = f(x, y) \equiv f$$

We have $k_1 = hf(x, y)$

$$k_2 = hf\left(x + \frac{h}{2b}, y + \frac{hf}{2b}\right)$$

and $\Delta y = (1-b)k_1 + bk_2$

We can given any values to the constants remembering that c cannot be zero and a and b cannot both be zero,

Range-Kutta's third-order Rule;

$$\Delta y = \tfrac{1}{6}(\Delta' y + 4\Delta'' y + \Delta''' y)$$

where $\Delta' y = hf(x, y),$

$$\Delta'' y = hf\left(x + \tfrac{1}{2}h, y + \tfrac{1}{2}\Delta' y\right),$$

$$\Delta''' y = hf(x + h, y + 2\Delta'' y - \Delta' y).$$

N.B. 1. If the error can be allowed up to the order h^5 instead of h^4, then we have

$$\Delta y = a\Delta' y + b\Delta'' y + c\Delta''' y + d\Delta^{iv} y$$

where $\Delta' y = hf(x, y)$

$$\Delta'' y = hf(x + mh, y + m\Delta' y)$$

$$\Delta''' y = h[f(x + \lambda h), y + \rho\Delta'' y + (n - \rho)\Delta' y)]$$

$$\Delta^{iv} y = y[f(x + \mu h, y + \rho\Delta''' y + \tau\Delta'' y + (\mu - \rho - \gamma)\Delta' y]$$

From this we can on giving particular values to the constants, obtain **Runge-Kutta fourth order** Rule :

$$\Delta y = \tfrac{1}{6}(\Delta' y + 2\Delta'' y + 2\Delta''' y + \Delta^{iv} y)$$

where $\Delta' y = hf(x, y).$

$$\Delta'' y = hf\left(x + \tfrac{1}{2}h, y + \tfrac{1}{2}\Delta' y\right).$$

$$\Delta''' y = hf\left(x + \tfrac{1}{2}h, y + \tfrac{1}{2}\Delta'' y\right),$$

$$\Delta^{iv} y = hf(x + h, y + \Delta''' y).$$

Thus $y_n = y_{n-1} + \Delta y_{n-1}, n = 1.$

2. In case $f(x, y)$ is a function of x alone

i.e., $f(x, y) = f(x),$

then equation (16·14) reduces to simpson's one third rule *viz.*,

$$\Delta y = \tfrac{1}{6}h\left[f(x) + 4f\left(x + \tfrac{1}{2}h\right) + f(x + h)\right]$$

where $$\Delta y = \int_x^{x+h} f(x)\, dx$$

Example 10. *Use Runge-Kutta fourth order method to solve $y' = xy$ for $x = 1{\cdot}4$. Initially $x = 1$, $y = 2$ (take $h = 0{\cdot}2$).*

Taking $h = 0{\cdot}2,$

and $f(x, y) = xy,$

Sol. We have $$\Delta y = \tfrac{1}{6}\left(\Delta' y + 2\Delta'' y + 2\Delta''' y + \Delta^{iv} y \right) \quad \text{...(1)}$$

$$\Delta' y = hf(x, y) = 0{\cdot}2(1 \times 2) = 0{\cdot}4$$

$$\Delta'' y = 0{\cdot}2\left(x + \tfrac{1}{2} h \right)\left(y + \tfrac{1}{2} \Delta' y \right)$$

$$= 0{\cdot}2(1{\cdot}1)(2{\cdot}2) = 0{\cdot}484$$

$$\Delta''' y = 0{\cdot}2\left(x + \tfrac{1}{2} h \right)\left(y + \tfrac{1}{2} \Delta'' y \right)$$

$$= 0{\cdot}2(1{\cdot}1)(2{\cdot}242) = 0{\cdot}49324$$

$$\Delta^{iv} y = 0{\cdot}2(x + h)(y + \Delta''' y)$$

$$= 0{\cdot}2(1{\cdot}2)(2{\cdot}49324) = 0{\cdot}5983776$$

$\therefore$ From (1), $$\Delta y = \tfrac{1}{6}(0{\cdot}4 + 2 \times 0{\cdot}484 + 2 \times 0{\cdot}49324 + 0{\cdot}5983776)$$

$$= 0{\cdot}4921429$$

Here $$x_1 = x_0 + h = 1 + 0{\cdot}2 = 1{\cdot}2,$$

$$y_1 = y_0 + \Delta y = 2 + 0{\cdot}4921429$$

$$= 2{\cdot}4921429.$$

Then for the second interval we have

$$\Delta' y = 0{\cdot}2(1{\cdot}2)(2{\cdot}4921429) = 0{\cdot}49842858$$

$$\Delta'' y = 0{\cdot}2(1{\cdot}2)(2{\cdot}24921499) = 0{\cdot}49482714$$

$$\Delta''' y = 0{\cdot}2(1{\cdot}2)(2{\cdot}24741357) = 0{\cdot}49443098$$

$$\Delta^{iv} y = 0{\cdot}2(1{\cdot}2)(2{\cdot}24721549) = 0{\cdot}53933177$$

$$\Delta y = 0{\cdot}50271276$$

and $$x_2 = x_0 + 2h = 1{\cdot}4,$$

$$y_2 = y_1 + \Delta y = 2{\cdot}99485866.$$

Example 11. *Prove that the solution for the equation $\dfrac{dy}{dx} = y$, $y(0) = 1$ yields $y_m = \left(1 + h + \tfrac{1}{2} h^2 \right)^m$, using second order Runge-Kutta method.*

Sol. Here $f(x, y) = y$ so $k_1 = hy$

and $$k_2 = h\left(y + \frac{hf}{2b} \right) = h\left(y + \frac{hy}{2b} \right)$$

$$\Delta y = (1-b)\,hy + bh\left(y + \frac{hy}{2b}\right)$$

$$= hy + \tfrac{1}{2}h^2 y = hy(x) + \tfrac{1}{2}h^2 y(x)$$

$$\therefore \quad \Delta y = y(x+h) - y(x) = hy(x) + \tfrac{1}{2}h^2 y(x)$$

$$\therefore \quad y(x+h) = y(x) + hy(x) + \tfrac{1}{2}h^2 y(x) = y(x)\left[1 + h + \tfrac{1}{2}h^2\right]$$

we take $y(x)$ as y_{m-1} and $y(x+h)$ as y_m

$$y_m = y_{m-1}\left(1 + h + \tfrac{1}{2}h^2\right), \; y_{m-1} = y_{m-2}\left(1 + h + \tfrac{1}{2}h^2\right)$$

$$\therefore \quad y_m = \left(1 + h + \frac{1}{2}h^2\right)\cdot\left(1 + h + \frac{1}{2}h^2\right) = y_{m-2}\left(1 + h + \frac{1}{2}h^2\right)^2$$

By induction, $y_m = y_0\left(1 + h + \tfrac{1}{2}h^2\right)^m$

$$= \left(1 + h + \tfrac{1}{2}h^2\right)^m. \qquad \because \; y_0 = y(0) = 1$$

Example 12. *Use Runge-Kutta formula of third order for solving the differential equation $y' = f(x, y) = -y$, with $y(0) = 1$, h = length of one step.*

Sol. We have $y_{n+1} = y_n + \dfrac{1}{6}(k_1 + 4k_2 + k_3)$

with $\quad k_1 = hf(x_n, y_n) = -hy_0$

$$k_2 = hf\left(x_n + \frac{h}{2}, y_n + \frac{1}{2}k_1\right) = -h\left(y_0 - \frac{1}{2}hy_0\right) = \left(-h + \frac{1}{2}h^2\right)y_0$$

$$k_3 = hf(x_n + h, y_n - k_1 + 2k_2) = -h(y_0 + hy_0) + 2\left(-h + \frac{1}{2}h^2 y_0\right)$$

$$= (-h + h^2 - h^3)y_0$$

$$y_1 = y_0 + \frac{1}{6}\left[-hy_0 + 4\left(-h + \frac{1}{2}h^2\right)y_0 + (-h + h^2 - h^3)\,y_0\right]$$

$$= \left(1 - h + \frac{1}{2}h^2 - \frac{1}{6}h^3\right)y_0$$

$$= \left(1 - h + \frac{1}{2}h^2 - \frac{1}{6}h^3\right), \quad y_0 = 1]$$

16·2·6. Adams-Bashford Method. Substituting in the Taylor's expansion, the value of the derivatives in terms of backward differences or integrating Newton-Gregory Backward formula applied to y_n' between the limits 0 to 1 for u where $u = (x - x_n)/h$ we have

$$y_{n+1} = y_n + h\left[y_n' + \frac{1}{2}\nabla y_n' + \frac{5}{12}\nabla^2 y_n' + \frac{3}{8}\nabla^3 y_n' + \frac{251}{720}\nabla^4 y_n' + \cdots\right]$$

$$\left.\begin{aligned} y_{n+1}, p &= y_n + \frac{h}{24}(55y'_n - 59y'_{n-1} + 37y'_{n-2} - 9y'_{n-2}) \\ y_{n+1}, e &= y_n + \frac{h}{24}(9y'_{n+1} + 19y'_n - 5y'_{n-1} + y'_{n-2}) \end{aligned}\right] \text{Adams-Bashford predictors} \quad ...(16\cdot17)$$

Given the first order equation

$$y' = f(x, y)$$

with the boundary value $y = y_0$ at $x = x_0$, we can find the starting values $y_0, y_1, ..., y_4$ and the derivatives $y'_0, y'_1, ..., y'_4$ by any of the methods discussed in §15·2. We then form a difference table. Putting $n = 4$ in eq. (16·17) and making use of the difference table we obtain the value of y_5 and then y'_5 from equation (16·18). This procedure may be repeated to find y_6.

Table

x	y	y'				
x_0	y_0	y'_0				
			$\nabla y'_1$			
x_1	y_1	y'_1		$\nabla^2 y'_2$		
			$\nabla y'_2$		$\nabla^3 y'_3$	
x_2	y_2	y'_2		$\nabla^3 y'_3$		$\nabla^4 y'_4$
			$\nabla y'_3$		$\nabla^3 y'_4$	
x_3	y_3	y'_3		$\nabla^3 y'_4$		
			$\nabla y'_4$			
x_4	y_4	y'_4				

N.B. 1. The formula (16·17) is called a-*predictor.* The truncation error in this formula is $h\nabla^5 y_n$.

2. An improved value of y_{n+1} may be obtained by the following formula known as a corrector.

$$y_{n+1} = y_n + h\left[y'_{n+1} - \frac{1}{2}\nabla y''_{n+1} - \frac{1}{12}\nabla^2 y'_{n+1} - \frac{1}{24}\nabla^3 y'_{n+1} - \frac{19}{720}\nabla^4 y'_{n+1}\right] - \frac{3h^6}{160} y^6(\xi), \quad x_{n-3} < \xi < x_{n+1}$$

16·3. General Approach to Predictor and Corrector.

We want to find a formula which uses the information of the function y and the derivative y'_x at the past three points together with one more old value of the derivative.

We first predict a value for y_{n+1} and then use a different formula to correct the value. We may employ the correction similarly again to correct the value of y_{n+1}. This process, can be iterated as many times as we wish.

The most general predictor formula is

$$y_{n+1} = A_0 y_n + A_1 y_{n-1} + A_2 y_{n-2} + h[B_0 y'_n + B_1 y'_{n-1} + B_2 y'_{n-2} + B_3 y'_{n-3}] \quad ...(1)$$

which consists of seven constants. Suppose this formula is exact for polynomial up to degree 4 *i.e.*, for the polynomials $y = 1, y = x, y = x^2$ and $y = x^4$. This, with $h = 1$, we have

$$1 = A_0 + A_1 + A_2,$$
$$1 = A_1(-1) + A_2(-2) + B_0 + B_1 + B_2 + B_3,$$
$$1 = A_1 + 4A_2 - 2B_1 - 4B_2 - 6B_3$$
$$1 = -A_1 - 8A_2 + 3B_1 + 12B_2 + 27B_3$$
$$1 = A_1 + 16A_2 - 4B_1 - 32B_2 - 108B_3$$

From these five equations, we can determine the values of any five constants in terms of the remaining two variables. Solving we get

$$A_0 = 1 - A_1 - A_2,$$
$$B_0 = \frac{1}{24}(55 + 9A_1 + 8A_2),$$
$$B_1 = \frac{1}{24}(32A_2 + 19A_1 - 59),$$
$$B_2 = \frac{1}{24}(8A_2 - 5A_1 + 37),$$
$$B_3 = \frac{1}{24}(A_1 - 9).$$

Taking $A_1 = 0 = A_2$, we get

$$A_0 = 1,\ B_0 = \frac{55}{24},\ B_1 = \frac{-59}{24},\ B_2 = \frac{37}{24},\ B_3 = \frac{-9}{24}$$

Substituting these values in (1), we get

Hence $y_{n+1} = y_n + \frac{h}{24}[55y'_n - 59y'_{n-1} + 37y'_{n-2} - 9y'_{n-3}]$ ...(2)

This formula is known as Adam-Bashforth type predictors.

Corrector formula

The most general corrector formula is

$$y_{n+1} = a_0 y_n + a_1 y_{n-1} + a_2 y_{n-2} + h(l_{-1} y'_{n+1} + l_0 y'_n + l_1 y'_{n-1} + l_2 y'_{n-2}) \quad ....(3)$$

Obtaining the values of the five constants in terms of the remaining constants a_1 and a_2, we get, on putting $h = 1$ and successively taking $y(x) = 1, x, x^2, x^3, x^4$.

$$a_0 = 1 - a_1 - a_2,\ l_{-1} = \frac{1}{24}(9 - a_1),\ l_0 = \frac{1}{24}(19 + 13a_1 + 8a_2)$$

$$l_1 = \frac{1}{24}(-5 + 13a_1 + 13a_2),\ l_2 = \frac{1}{24}(1 - a_1 - 8a_2)$$

Letting $a_1 = 0 = a_2$, we have

$$a_0 = 1,\ l_{-1} = \frac{9}{24},\ l_0 = \frac{19}{24},\ l_1 = -\frac{5}{24} \text{ and } l_2 = \frac{1}{24}$$

Hence the corrector formula is

$$y_{n+1} = y_n + \frac{h}{24}(9y'_{n+1} + 19y'_n - 5y'_{n-1} + y'_{n-2}) \qquad ...(4)$$

Example 14. *Using Adam-Bashforth predictor-corrector equations, evaluate* $y(1{\cdot}4)$ *if* y *satisfies* $\frac{dy}{dx} + \frac{y}{x} = \frac{1}{x^2} \Rightarrow y' = \frac{1 - xy}{x^2}$ *and* $y(1) = 1,\ y(1{\cdot}1) = 0{\cdot}996,\ y(1{\cdot}2) = 0{\cdot}986,\ y(1{\cdot}3) = 0{\cdot}972$

Sol. We have $x_0 = 1,\ y_0 = 1,\ y_1 = {\cdot}996,\ y_2 = {\cdot}986,\ y_3 = {\cdot}972,\ h = {\cdot}1$

$$x_1 = 1{\cdot}1,\ x_2 = 1{\cdot}2,\ x_3 = 1{\cdot}3$$

Using Adam's predictor formula and $n = 3$, we have

$$y_{u,p} = y_3 + \frac{h}{24}(55y'_3 - 59y'_2 + 37y'_1 - 9y'_0) \qquad ...(1')$$

$$y'_0 = \left(\frac{1 - xy}{x^2}\right)_{x_0, y_0} = \frac{1 - 1 \times 1}{1} = 0,$$

$$y'_1 = \frac{1 - 1{\cdot}1 \times {\cdot}996}{(1{\cdot}1)^2} = \frac{-{\cdot}0956}{1{\cdot}21} = -{\cdot}7902$$

$$y'_2 = \frac{1 - 1{\cdot}2 \times {\cdot}986}{(1{\cdot}2)^2} = \frac{-{\cdot}1832}{1{\cdot}44} = -{\cdot}1273$$

$$y'_3 = \frac{1 - 1{\cdot}3 \times {\cdot}972}{(1{\cdot}3)^2} = \frac{-{\cdot}2636}{1{\cdot}69} = -{\cdot}156$$

Putting these values in (1′), we get

$$y(u, p) = {\cdot}972 + \frac{1}{24}(55 \times -{\cdot}156 - 59 \times -{\cdot}1273 + 37 \times -{\cdot}7902 - 9 \times 0)$$

$$= {\cdot}972 + \frac{1}{24}(-8{\cdot}58 + 7{\cdot}5107 - 29{\cdot}2374) = {\cdot}972 - {\cdot}1263 = {\cdot}8457$$

Putting $n = 3$ in Adam's corrector formula we get,

$$y_{4,0} = y_3 + \frac{h}{24}(19y'_4 + 19y'_3 - 5y'_2 + y'_1) \qquad ...(2')$$

Taking y_u as $y_{u,p} = {\cdot}8457$, we have

$$y'_4 = \left(\frac{1 - xy}{x^2}\right) = \frac{1 - 1{\cdot}4 \times {\cdot}8457}{(1{\cdot}4)^2} = -\frac{{\cdot}18398}{1{\cdot}96} = -{\cdot}09307$$

$$\therefore \quad y_{4,p} = \cdot0972 + \frac{\cdot1}{24}(9 \times -\cdot09387 + 19 \times -\cdot156$$
$$-5 \times -\cdot1237 - \cdot7902)$$

$$= \cdot972 + \frac{\cdot1}{24}(-\cdot84483 - 2\cdot964 - \cdot8365 - \cdot7902) = \cdot972 - \cdot02266$$

$$= \cdot94934$$

$\therefore$ $y(1\cdot4) = \cdot942$ correct to 3 decimal places.

Ex. *For the solution of the first order differential equation $y' = f(x, y)$, obtain the predictor formula of the form*

$$y_{n+1} = a_1 y_n + h(a_2 y'_n + a_3 y'_{n-1} + a_4 y'_{n-2})$$

and its corresponding corrector formula.

[**Hint.** Assume that $y_n = 1,\ x, x^2, x^3$ respectively and determine the constants a_1, a_2, a_3, a_4.

The correct formula will be,

$$y_{n+1} = a_0 y_n + h_i(l_{-1} y'_{n+1} + l_0 y'_n + l_1 y'_{n-1})$$

The values of the constants are found to be

$$a_0 = 1,\ l_{-1} = \frac{5}{12},\ l_0 = \frac{8}{12},\ l_1 = -\frac{1}{12}.$$]

16·4. Differential Equations of Second Order. Any differential equation of the second order can be reduced to a system of first order equations by the introduction of auxiliary variables. Thus the second order equation

$$\frac{d^2y}{dx^2} + P\frac{dy}{dx} + Qy = f(x)$$

can be reduced to two first order equations by putting $y' = dy/dx = z$. The resulting equations are

$$z = y' = \frac{dy}{dx},\quad \frac{dz}{dx} = y'' = \frac{dy'}{dx} = f(x) - Py' - Qy.$$

These first order equations can be solved by the previously discussed methods.

Example 15. *Obtain the numerical solution of*

$$\frac{d^2y}{dx^2} - 3\frac{dy}{dx} + 2y = x\,e^{3x},\ y(0) = 4\cdot25,\ y'(0) = 6 \text{ at } x = 0\cdot1$$

Sol. We represent the given equation as a set of simultaneous equations

by assuming $\dfrac{dy}{dx} = z$ ...(*i*)

Thus the given equation has the form

$$\frac{dz}{dx} - 3z + 2y = x + e^{3x} \qquad ...(ii)$$

Finally we write the equations (*i*) and (*ii*) as follows for numerical solution

$$\frac{dy}{dx} = z = f(x, y, z) \qquad ...(iii)$$

$$\frac{dz}{dx} = x + e^{3x} + 3z - 2y = g(x, y, z) \qquad ...(iv)$$

Now $f(x, y, z) = z$

$\therefore\ f_0 = f(x_0, y_0, z_0)$

$$f_1 = f\left(x_0 + \frac{h}{2}, y_0 + \frac{h}{2} f_0, z_0 + \frac{h}{2} g_0\right)$$

$$= f(0{\cdot}05, 4{\cdot}55, 6{\cdot}525)$$

$$= 6{\cdot}525$$

$$f_2 = f\left(x_0 + \frac{h}{2}, y_0 + \frac{h}{2} f_1, z_0 + \frac{h}{2} g_1\right)$$

$$= f(0{\cdot}05, 4{\cdot}5763, 6{\cdot}5843)$$

$$= 6{\cdot}5843$$

$$f_3 = f(x_0 + h, y_0 + hf_2, z_0 + hg_2)$$

$$= f(0{\cdot}1, 4{\cdot}9084, 7{\cdot}1812)$$

$$= 7{\cdot}1812$$

$$g(x, y, z) = x + e^{3x} - 2y + 3z$$

$$g_0 = g(x_1, y_0, z_0) = g(0, 4{\cdot}25)$$

$$= 0 + e^0 - 2(4{\cdot}25) + 3(6) = 10{\cdot}5$$

$$g_1 = g\left(x_0 + \frac{h}{2}, y_0 + \frac{h}{2} f_0, z_0 + \frac{h}{2} g_0\right)$$

$$= g(0{\cdot}05, 4{\cdot}55, 6{\cdot}525)$$

$$= 0{\cdot}05 + e^{0{\cdot}15} = g(4{\cdot}55) + 3(6{\cdot}525)$$

$$= 11{\cdot}6868$$

$$g_1 = g\left(x_0 + \frac{h}{2}, y_0 + \frac{h}{2} f_1, z_0 + \frac{h}{2} g_1\right)$$

$$= g(0{\cdot}1, 4{\cdot}9034, 7{\cdot}1812)$$

$$= 0{\cdot}1 + e^{0{\cdot}3} - 2(4{\cdot}9084) + 3(7{\cdot}1812)$$

$$= 13{\cdot}1767$$

Hence $y(0{\cdot}1) = y_0 + \frac{h}{6}\left[f_0 + 2f_1 + 2f_2 + f_5\right]$

$$= 4{\cdot}25 + \frac{0{\cdot}1}{6}[6 + 2(6{\cdot}525) + 2(6{\cdot}5843) + 7{\cdot}1812]$$

$$= 4{\cdot}9067$$

Verification. Analytical solution of the equation is

$$y = e^{2x} + 2e^x + \frac{1}{2} e^{3x} + \frac{1}{2} x + \frac{3}{4}$$

$$\therefore \qquad y(0{\cdot}1) = e^{0{\cdot}2} + 2e^{0{\cdot}1} + \frac{1}{2} e^{0{\cdot}3} + \frac{0{\cdot}1}{2} + \frac{3}{4} = 4{\cdot}9067$$

16·4·1. Numerov's Method of solving equations of Second order with missing first order term.

Let differential equations be of the form

$$y'' = \phi(x, y) \text{ or } \phi = y^{(2)} \qquad ...(1)$$

Numerov has suggested the method of solving (1) by using the method of *undetermined coefficients*. Consider the equidistant values of x say $x_1, x_2, \ldots$ and the corresponding values of y say $y_1, y_2, \ldots,$ Suppose y_{k-1} and y_k are known. Consequent by $\phi_{k-1} = \phi(x_{k-1}, y_{k-1})$ and $\phi_k = \phi(x_k, y_k)$

are also known. Hence for finding the solution (1), we assume a corrector type formula of the form

$$y_{k+1} = Ay_k + By_{k-1} + h^2(C\phi_{k+1} + D\phi_k + E'\phi_{k-1}) + R \quad ...(2)$$

where A, B, C, D, E' are the constants to be determined and R is the error term.

Now $$y_{k+1} = Ey_k = y_k + h\, y_k^{(1)} + \frac{h^2}{2!} y_k^{(2)} + \frac{h^3}{3!} y_k^{(3)} + \frac{h^4}{4!} y_k^{(4)} + \frac{h^5}{5!} y_k^{(5)} + \frac{h^6}{6!} y_k^{(6)}, \quad ...(3)$$

$$y_{k-1} = E^{-1} y_k = y_k - h\, y_k^{(1)} + \frac{h^2}{2!} y_k^{(2)} + \frac{h^3}{3!} y_k^{(3)} + \frac{h^4}{4!} y_k^{(4)} + \frac{h^5}{5!} y_k^{(5)} + \frac{h^6}{6!} y_k^{(6)} \quad ...(4)$$

and $$\phi_{k+1} = \phi_k + h\, \phi_k^{(1)} + \frac{h^2}{2!} \phi_k^{(2)} + \frac{h^3}{3!} \phi_k^{(3)} + \frac{h^4}{4!} \phi_k^{(4)} + \cdots$$

where $y_k^{(p)}$ stands for the value of pth derivative of y at $x = x_k$, $\phi^{(1)} = y^{(1)}$, etc., E is the operator $1 + \Delta$.

We determine the unknown constants in such a way that (2) is consistent with (1) up to the fourth order terms. Thus we find

$$A = 2, B = 1, C = 1/12, D = 5/6, E' = 1/12.$$

$$\therefore \quad y_{k+1} = 2y_k - y_{k-1} + \frac{h^2}{12}(\phi_{k+1} + 10\phi_k + \phi_{k-1}) + R$$

Thus formula is known as Numerov's formula.

16·5. Chebyshev Polynomials. Chebyshev Polynomials are usually the most convenient functions for the approximation of a continuous function when the desired interval of approximation is finite. Chebychev Polynomials of the *first kind* $T_n(x)$ is defined as

$$T_n(x) = \cos(n \cos^{-1} x), n = 1, 2, ..., T_0(x) = 1$$

$$\Rightarrow \quad T_n(\cos\theta) - \cos(n\theta) \text{ where } x = \cos\theta.$$

and $\cos(n\theta)$ is a polynomial of degree n in $\cos\theta$,

If we put $x = \cos\theta$ in the identity

$$\cos(n+1)\theta + \cos(n-1)\theta = 2\cos\theta \cos n\theta,$$

a three term recurrence formula for Chebychev Polynomials which is

$$T_{n+1}(x) + T_{n-1}(x) = 2xT_n(x)$$

$$\Rightarrow \quad T_{n+1}(x) - 2xT_n(x) + T_{n-1}(x) = 0 \quad ...(1)$$

This helps in generating the Chebychev Polynomials.

Further, $\quad T_0(x) = 1, \; T_1 = x$

From (1), for $n = 1$ $\quad T_2 = 2x^2 + 1 = 0 \quad \Rightarrow T_2(x) = 2x^2 - 1,$

For $n = 2$, $\quad T_3 - 2x(2x^2 - 1) + x = 0 \quad \Rightarrow T_3(x) = 4x^3 - 3x,$

For $n = 3$, $T_n(x) = 8x^4 - 8x^2 + 1$, $T_5(x) = 16x^5 - 20x^3 + 5x$.

Also $T_n(x) = \dfrac{n}{\sqrt{1-x^2}} \sin(n \cos^{-1} x) = n \dfrac{\sin n\theta}{\sin\theta} = nu_{n-1}(x)$.

$= 0$ for $m \neq n$;

$= \frac{1}{2}(m+1)$ for $m = n \neq 0$, $m + 1$, for $m = n = 0$

Further, we note that

(*i*) $T_n(x)$ are alternately odd and even polynomials.

(*ii*) The leading coefficient is 2^{n-1}.

Chebyshev Polynomials of *second kind* $u_n(x)$ is defined as

$$u_n(x) = \frac{\sin[(n+1)\cos^{-1} x]}{\sin(\cos^{-1} x)} = \frac{\sin(n+1)\theta}{\sin\theta} \quad \text{where } x = \cos\theta.$$

$$\therefore \quad u_0(x) = 1, \; y_1(x) = \frac{\sin 2\theta}{\sin\theta} = 2x,$$

$$u_2(x) = \frac{\sin 3\theta}{\sin\theta} = 3 - 4(1 - x^2) = 4x^2 - 1,$$

The recurrence relation for second kind is

$$u_{n+1}(x) - 2x\, u_n(x) + u_{n-1}(x) = 0.$$

Properties of Chebychev Polynomials.

1. To represent $u_{n-1}(x)$ in terms of $T_n(x)$ and $T_{n+1}(x)$ *i.e.*, to prove

$$(1 - x^2)u_{n-1}(x) = x\, T_n(x) - T_{n+1}(x)$$

Proof. We have $u_{n-1}(x) = \dfrac{\sin n\theta}{\sin\theta} \times \dfrac{\sin\theta}{\sin\theta}$

$$\therefore \quad (1 - x^2)u_{n-1}(x) = \sin n\theta \sin\theta - \cos n\theta\cos\theta + \cos n\theta \cos\theta$$

$$= x\, T_n(x) - \cos(n+1)\theta = x\, T_n(x) - T_{n+1}(x)$$

2. Prove that the polynomials $T_n(x)$ satisfy the differential equation

$$(1 - x^2)\, y'' - xy' + n^2 y = 0$$

Proof. $y = T_n(x) = \cos n\theta$

where $x = \cos\theta \Rightarrow \dfrac{dx}{d\theta} = -\sin\theta$

$$(1 - x^2)\, T_n'(x) - x\{x\, T_n(x) - T_{n+1}(x)\}$$

$$\Rightarrow \quad \frac{dy}{dx} = y' = \frac{n \sin n\theta}{\sin\theta}; \quad y'' = \frac{d}{d\theta}\left[\left(\frac{n\sin n\theta}{\sin\theta}\right)\right] \cdot \left(-\frac{1}{\sin\theta}\right)$$

$$= n\left[\frac{n\cos n\theta}{\sin\theta} - \frac{\sin\theta}{\sin^2\theta}\cos\theta\right]\left[\frac{-1}{\sin\theta}\right]$$

$$\Rightarrow \quad (1-x^2)\,y'' = -\,n^2 \cos n\,\theta + \frac{n \sin n\,\theta}{\sin \theta} \cos \theta = -\,n^2 y + y'x.$$

$$\Rightarrow \quad (1-x^2)\,y'' - xy' + n^2 y = 0$$

3. Prove that $u_n(x)$ satisfies :

$$(1-x^2)\,y'' - 3xy' + n(n+2)\,y = 0$$

Proof. $$y' = \left[\frac{d}{d\theta}\left(\frac{\sin(n+1)\,\theta}{\sin\theta}\right)\right] \times \frac{1}{-\sin\theta}$$

$$= \left(\frac{(n+1)\cos(n+1)\theta \sin\theta - \sin(n+1)\theta\cos\theta}{\sin^2\theta}\right) \times \frac{-1}{\sin\theta}$$

$$\Rightarrow \quad (1-x^2)y' = \frac{\sin(n+1)\theta\cos\theta}{\sin\theta} - (n+1)\cos(n+1)\,\theta$$

$$= u_n(x) - (n+1)\,T_{n+1}(x)$$

Diff. *w.r.t.* x, we get

$$(1-x^2)\,y'' - 2xy' = x\,u_n'(x) + u_n(x) - (n+1)\,T_{n+1}'(x)$$

$$\Rightarrow \quad (1-x^2)\,y'' - 3xy' = y - (n+1)\,T_{n+1}'(x) \qquad \ldots(1)$$

Also $T_n'(x) = nu_{n-1}(x) = y - (n+1)(n+1)y = [1-(n+1)^2]y$

$$= -(n^2+2n)y$$

$\therefore$ (1) reduces to $(1-x^2)\,y'' - 3xy' + n(n+2)y = 0$

4. To Prove $I_{n,m} = \int_{-1}^{1} \frac{T_n(x)\,T_m(x)\,dx}{\sqrt{1-x^2}} = 0$ if $m \neq n$, (Orthogonal Property)

Proof. Put $x = \cos\theta$, then we have

$$\text{L.H.S.} = \int_0^{\pi} T_n(\cos\theta)\,T_m(\cos\theta)\,d\theta = \frac{1}{2}\int_0^{\pi} 2\cos n\,\theta\cos m\,\theta\,d\theta$$

$$= \frac{1}{2}\int_0^{\pi}\Big(\cos(n+m)\,\theta + \cos(n-m)\,\theta\Big)d\theta$$

$$= \frac{1}{2}\left(\frac{\sin(n+m)\theta}{n+m} + \frac{\sin(n-m)\theta}{n-m}\right)_0^{\pi}$$

Hence $$\int_{-1}^{+1} \frac{T_n(x)\,T_m(x)\,dx}{\sqrt{(1-x^2)}} = 0, \; m \neq n$$

T_n and T_m are said to be orthogonal over the interval $[-1, 1]$ with weighting $\dfrac{1}{\sqrt{1-x^2}}$.

5. To prove. $\sin^{-1} x = C_0 + C_1 T_1(x) + C_2 T_2(x) + \cdots$

$$= \frac{4}{\pi}\left(\frac{31x}{45} - \frac{16x^3}{45} + \frac{16}{25} dx^5 \ldots\right)$$

Proof. $\int_{-1}^{+1} \frac{\sin^{-1} x\, T_0(x)\, dx}{\sqrt{1-x^2}} = C_0 \int_{-1}^{+1} \frac{T_0(x)\, T_0(x)\, dx}{\sqrt{1-x^2}} = \pi C_0$

$$\therefore \quad C_0 = \frac{1}{\pi}\left(\frac{\sin^{-1} x}{\sqrt{1-x^2}}\right)_{-1}^{+1} = 0$$

To find C_k, $k = 1, 2, 3, 4, \ldots$

$$\int_{-1}^{+1} T_k \frac{\sin^{-1} x}{\sqrt{1-x^2}} dx = C_k \frac{\pi}{2} \Rightarrow C_k = \frac{2}{\pi} \int_0^{\pi} \cos k\theta \left(\frac{\pi}{2} - \theta\right) d\theta \text{ etc.}$$

$$C_3 = C_4 = 0, \; C_1 = \frac{4}{\pi} \; C_3 = \frac{4}{25\pi}$$

Hence the result.

6. To prove : $T_n(x) = u_x(x) - u_{n-1}(x)$

(Relation between $T_n(x)$, $u_n(x)$ and $u_{n-1}(x)$)

Proof. We have

$$T_n(x) = \cos x\theta = \frac{\cos x\theta \sin\theta}{\sin\theta} = \frac{\sin(x+1)\theta}{\sin\theta} - \frac{\sin x\theta \cos\theta}{\sin\theta}$$
$$= u_n(x) - u_{n-1}(x)$$

EXERCISE 16

1. Use Euler's method and an interval of $h = 0.1$ to find a solution of the differential equation

   (*a*) $y' \equiv \frac{dy}{dx} = x^2 + y^2$

   with boundary condition $y = 0$ for $x = 0$ in the range $0 \le x \le 1$ when $x = 0{\cdot}5$

   (*b*) $y' = xy$ with initial conditions $x_0 = 0, y_0 = 1$ at $x = 0{\cdot}4$

   (*c*) $y' = 1 + y^2$ with initial conditions $x_0 = 0, y_0 = 0$ from $x = 0$ to $x = 1$.

2. Using Euler's modified method, solve numerically the differential equation

   (*i*) $y' \equiv \frac{dy}{dx} = x + y$

   with the inital conditions $x_0 = 0, 1$ in steps of $0{\cdot}2$ at $x = 1$

   (*ii*) $y' = xy$ with the initial condition $x_0 = 0, y_0 = 1$ and $h = {\cdot}1$.

3. (*a*) Obtain by Euler's modified method five consecutive starting values for the numerical solution of

$$\frac{dy}{dx} = \log_{10} \frac{x}{y}$$

with $x_0 = 20, y_0 = 5$, and with $h = 0{\cdot}2$.

(*b*) Use Euler's modified method with one step to find the value of y at $x = 0{\cdot}1$ where

$$\frac{dy}{dx} = x^2 + y$$

and $y = 0{\cdot}94$ when $x = 0$. (*Nagpur Univ., 1998*)

4. (*a*) Solve numerically the differential equation

$$\frac{dy}{dx} = 2x - y$$

with $x_0 = 1, y_0 = 3$ by using Picard's method of successive approximations. Compare the results with those obtained from the exact analytical solution

$$y = 2x + 3e^{1-x}\ x - 2$$

(*b*) Integrate, using Picard's method, the differential equation

$$\frac{dy}{dx} = 1 + xy$$

with $x_0 = 0, y_0 = 1$,

(*c*) Apply Picard's method to find the third approximation of the solution of

(*i*) $\frac{dy}{dx} = x + y^2$ where $y = 0$ when $x = 1$

(*ii*) $\frac{dy}{dx} = y - x^2, y(0) = 1$. Hence find the value of $y(0.1)$ (*B.R. Nov. 1995*

(*d*) Solve : $\frac{d^2 y}{dx^2} = x^3 \left(\frac{dy}{dx} + y \right)$ where $y = 1, \frac{dy}{dx} = \frac{1}{2}$ when $x = 0$

[**Hint.** Put $\frac{dy}{dx} = z$ and then $\frac{dz}{dx} = x^3 (z + y)$, with $y_0 = 1, z_0 = \frac{1}{2}$, and $x_0 = 0$]. For the first approx., $y_1 = y_0 + \int_0^x f(x_0, y_0, z_0)\, dx$

$$= 1 + \int_0^x z_0\, dx = 1 + \tfrac{1}{2} x, \text{ etc.}$$

(*e*) Solve using Picard's method the simultaneous equations

$$\frac{dy}{dx} = x, \frac{dz}{dx} = x^3 (y + z)$$

with the initaial condition $y = 1, z = \frac{1}{2}$ when $x = 0 = f(x, y)$.

(*f*) Find the third approximation of the solution of the equation

$$\frac{dy}{dx} = 2 - \frac{y}{x}$$

by Picard method, where $y = 2$ when $x = 1$

5. If $y' = 1 + xy$ and $y = 2$ when $x = 0$, use Taylor's series method to obtain and check the value of y for $x = 0{\cdot}4$.

6. (*a*) Solve $y' = y \sin x + \cos x$

subject to $x = 0,\ y = 0$ with Taylor's series method.

(*b*) Solve $y' = x - y^2$ by Taylor's series expansion for $x = 0{\cdot}2\ (0{\cdot}2)\ 0{\cdot}6$. Initial values $x = 0,\ y = 1$,

7. Use Taylor series method to start the numerical solution of

(*a*) $$\frac{dy}{dx} = x^2 + y^2$$

with $x_0 = 1, y_0 = 0$ at $x = 1{\cdot}3$

(*b*) $xy'' + y' + xy = 0$ where $y(0) = 1, y'(0) = 0$ for $x = 0$ to $x = 1{\cdot}5$.

(*c*) Find a solution of the set of simultaneous equations

$$\frac{dx}{dt} = xy + 2t, \frac{dy}{dt} = 2t, = 2ty + x$$

with the inital conditions $x = 1, y = -1,\ t = 0$

[**Hint.** (*c*) Assume $x(t) = x_0 + tx_0' + \frac{t^2}{2!} x_0'' + \frac{t^3}{3!} x_0''' + \cdots$

and $y(t) = y_0 + ty_0' + \frac{t^2}{2!} y_0'' + \frac{t^3}{3!} y_0''' + \cdots$

Also differentiate the given equations w.r.t. t twice and then put $x = 1, y = -1, t = 0$]

8. (*a*) Obtain the power series for x and y which satisfy

$$\frac{dx}{dt} = x + y + t, \quad \frac{d^2y}{dt^2} = x - t$$

subject to initial conditions that $x = 0, y = 1, \frac{dy}{dt} = -1$ at $t = 0$.

[**Hint.** Let $\frac{dy}{dt} = z$. Then the given equations reduce to a new system of first order equations $\frac{dx}{dt} = x + y + t, \frac{dz}{dt} = x - t$

and $\frac{dy}{dt} = z$ Now proceed as in Q. 7 (*c*).]

(*b*) Obtain by Taylor's seris five consecutive starting values for the numerical solution of

$$\frac{dy}{dx} = 2y + 3e^x$$

With $x_0 = 0, y_0 = 0$. Check the values.

(*c*) Obtain Taylor's series of solving the differential equation

$$\frac{dy}{dx} = 1 - 2xy, y(0) = 0$$

If error in x (x)obtained from four terms of the series only is to be less than 10^{-8}, find x. Determine the number of terms in the series to obtain results correct to 10^{-10} for $x \leq 1$. (See Ex. 7, page 337)

9. (*a*) Solve numerically at $x = 0{\cdot}4, 0{\cdot}5$ by Milne's predictor-corrector method given their value at the four points

$$x = 0, 0{\cdot}1, 0{\cdot}2, 0{\cdot}3 \quad \frac{dy}{dx} = 2e^z - y,$$

With $\quad y_0 = 2, y_1 = 2{\cdot}010, y_2 = 2{\cdot}040, y_3 = 2{\cdot}090.$

(*b*) Find $y(2)$ if $y(x)$ is the solution of $\frac{dy}{dx} = \frac{1}{2}(x + y)$ assuming

$y(0) = 2, y(0{\cdot}5) = 2{\cdot}636, y(1{\cdot}0) = 3{\cdot}595$ and $y(1{\cdot}5) = 4{\cdot}968$

(*B.Sc Agra, 1995, 99*)

10. (*a*) Solve the differential equation $y' = x^3 - y^2 - 2$, using Milne's method for $x = 0{\cdot}3$ $(0{\cdot}3\ (0{\cdot}1)\ (0{\cdot}6)$. Initial values: $x = 0, y = 1$. The values of y for $x = -0{\cdot}1, 0{\cdot}1$ and $0{\cdot}2$ are to be computed by series expansion.

(*b*) Part of a numerical solution of the differential equation

$$\frac{dy}{dx} = 0{\cdot}2\,x + 0{\cdot}1\,y$$

is shown in the following the table. Use Milne's method to find the next entry in the table

x : 0·00 0·05 0·10 0·15

(*c*) Describe the error propagation in Milne's method.

11. Using Runge-Kutta method calculate the value of y if

(*a*) $y' = 1/(x + y)$ for $x = 0{\cdot}5\ (0{\cdot}5)\ 2$.

Initial Value : $x_0 = 0,\ y_0 = 1$

(*b*) $y' = x + y$. (*i*) Start with $x = 1,\ y = 0$ and carry to $x = 1{\cdot}4$ with $h = 0.1$.

(*ii*) Start with $x = 0,\ y = 1$ and carry to second increment of y with $h = 0$.

(*c*) $\frac{dy}{dx} = 0{\cdot}31 + 0{\cdot}25\ y + 0{\cdot}3x^2$ with $y = 0{\cdot}72$ when $x = 0$ for $x = 1$ with $h = 0{\cdot}5$.

(*d*) $\frac{dy}{dx} = y - x^2,\ y(0.6) = 1.7379$, using mtheod of fourth order.

(*B.R. April, 1991*)

(*e*) $\frac{dy}{dx} = \frac{y^2 - x^2}{y^2 + x^2}$ with $y_0 = 1$ when $x_0 = 0$ and $h = 0{\cdot}2$ at $x = 0.2$

(*f*) $\frac{dy}{dx} = xy/(x^2 + y^2)$ with $x_0 = 0{\cdot}1223,\ y_0 = 0{\cdot}05$.

(*g*) $10\frac{dy}{dx} = x^2 + y^2, y(0) = 1$ for the interval $0 \angle x \angle 0{\cdot}4$ with $h = 0$

(*h*) $\dfrac{d^2y}{dx^2} - y^3 = 10, y(0) = 10, y'(0) = 5$ for $x = 0{\cdot}1$

[**Hint.** $y'' = y^3 = f(x, y, y'), x_0 = 0, y_0 = 10, y'_{(0)} = 5, k_1 = hf(x_0, y_0, y_0')$

$$= hy_0^3 = 100, \; k_2 = 108, \; k_3 = 207{\cdot}2$$

$$\Delta y = h\left[y_0' + \frac{1}{6}(k_1 + k_2 + k_3)\right] = 7{\cdot}42$$

$$\therefore \quad y_{(1)} = 17{\cdot}42$$

Show (*i*) what the value of $y(0{\cdot}1)$ by fourth-order method for the initial value problem $y = -2xy^2, y(i = j)$ and $h = 0{\cdot}1$ is $0{\cdot}9901$

(*k*) Obtain the numerical solution of the simultaneous equations

$$\left.\begin{array}{r}(D+2)\,x + 3y = 0\\ 3\,x + (D+2)\,y = 2\,e^{2t}\end{array}\right\} D = \frac{d}{dt}$$

with condition at $t = 0, x = 1, y = 2$ at the point $t = 0{\cdot}1$ by (*i*) Taylor series method (*ii*) by Runge-Kutta method

[**Hint.** Rewrite the given equation as

$$\frac{dx}{dt} = -2x - 3y = f(t, x, y)$$

$$\frac{dy}{dt} = 2e^{2t} - 3x - 2y = g(t, x, y)$$

when $t = 0, x_0 = 1, y_0 = 2, h = 0{\cdot}1$

$$f_0 = f(t_0, x_0, y_0) = f(0, 1, 2) = -2 - 6 = -8$$

$$f_1 = f\left[t_0 + \frac{h}{2}, x_0 + \frac{h}{2}f_0, y_0 + \frac{h}{2}g_0\right] = f(0{\cdot}05, 0{\cdot}6, 1{\cdot}75)$$

$$= -2(0{\cdot}01 - 3(1{\cdot}75) = -6{\cdot}45$$

$$f_2 = f\left[t_0 + \frac{h}{2}, x_0 + \frac{h}{2}f_1, y_1 + \frac{h}{2}g_1\right] = f(0{\cdot}05, 0{\cdot}6775, 1{\cdot}8455)$$

$$= -2(0{\cdot}6775) - 3(1{\cdot}18455) = -6{\cdot}8915$$

$$f_3 = f\left(t_0 + \frac{h}{x_0} + \frac{hf}{2}, y_0 + hg_2\right) = f(0{\cdot}1, 0{\cdot}3109, 1{\cdot}6993)$$

$$= -2(0{\cdot}3109) - 3(1{\cdot}6993) = -3{\cdot}5697$$

$$g_0 = (t_0, x_0, y_0) = g(0, 1, 2) = -5$$

$$g_1 = g\left(t_0 + \frac{h}{2}, x_0 + \frac{h}{2}f_0, y_2 + \frac{h}{2}g_0\right)$$

$$= g(0{\cdot}05, 0{\cdot}6, 1{\cdot}75)$$

$$= 2e^{0{\cdot}1} - 3(0{\cdot}6) - 2(1{\cdot}75) = -3{\cdot}0897$$

$$g_2 = g\left[t_0 + \frac{h}{2}, x_0 + \frac{h}{2}f_1, y_0 + \frac{h}{2}g_1\right]$$

$$= g(0{\cdot}05, 0{\cdot}6755, 1{\cdot}8455)$$

$$= 2e^{0{\cdot}1} - 3(0{\cdot}6755) - 2(1{\cdot}8455)$$

$$= -3{\cdot}5072$$

$$g_3 = g(t_0 + h, x_0 + hf_2, y_0 + h\,g_0)$$

$$= g(0{\cdot}1, 0{\cdot}3109, 1{\cdot}6493)$$

$$= 2e^{0{\cdot}1} - 3(0{\cdot}3109) - 2(1{\cdot}6493)$$
$$= -\,2{\cdot}0210$$

Now $x(0{\cdot}1) = x_0 + \frac{h}{6}\left[f_0 + 2f_1 + 2f_2 + f_3\right)$

$$= 1 + \frac{1}{6}(0{\cdot}1)\,[-\,8 + 2(-\,6{\cdot}45) + 2(-\,6{\cdot}8915) - 5{\cdot}5697) = 0{\cdot}3291$$

and $y(0{\cdot}1) = y_0 + \frac{h}{6}\left[g_0 + 2g_1 + 2g_2 + g_3\right] = 2 + \frac{0{\cdot}1}{6}[-\,5 + 2(-\,3{\cdot}0897)$

$$+\,2(-\,3{\cdot}5072) - 0{\cdot}0210] = 1{\cdot}6631$$

Thus the solutions are $x(0{\cdot}1) = 0{\cdot}3291, y(0{\cdot}1) = 1{\cdot}6631]$

13. (*a*) For the differential equation $\frac{dy}{dx} = y - x^2, y(0) = 1$, starting values are given as $y(0{\cdot}2) = 1{\cdot}2186$, $y(0{\cdot}4) = 1{\cdot}4682$ and $y(0{\cdot}6) = 1{\cdot}7379$ using Milne's predictor corrector formulae, advance the solution to $x = 0{\cdot}8$ and compare it with the analytical solution (carry to four decimals).

(b) Using Adam's method find $y(0.4)$ given (*I.A.S. 1995*)

$\frac{dx}{dx} = \frac{1}{2}xy, y(0) = 1, y\ (0.1) = 1.01, (0.2) = 1.022, Y(0.3) = 1.023$

(*M.Sc. 1992*)

(*c*) *Given that* $\frac{dy}{dx} = y - \frac{2x}{y}$, $y(0) = 1, y(0{\cdot}1) = 1{\cdot}0954$

$y(0{\cdot}2) = 1{\cdot}1832, y(0{\cdot}3) = 1{\cdot}2649$, find $y\ (0{\cdot}5)$ by Adam's predictor corrector formula. (*I.A.S. 1991*)

(*d*) Given $y' = x^2 - y, y(0) = 1$ and the starting values $y(0{\cdot}1) = {\cdot}90516$, $y(0{\cdot}2) = {\cdot}82127, y(0{\cdot}3) = {\cdot}74918$,

obtain the values of $y\ (0{\cdot}4)$ and $y\ (0{\cdot}5)$. Using Adams-Bashforth predictor-corrector method.

[**Hint** (*a*) Consider the predictor formula $y_{n+1} = A_0y_n + A_1y_{n-1} + A_2y_{n-2} + h(B_0\,y_n + B_1y_{n-1})$. Here A_0, A_1, A_2, B_0, B_1 must satisfy the following equation, if the formula is to be exact for $y = 1, y_1 = x_1, y = x^2, y = x^3$ and $y = x^4$. The resulting formula is $y_{n+1} = A_0y_n + A_1y_{n-1} + A_2g_{n-2} + 6h\ (y_n' + y_{n-1}')$. Now consider the corrector formula $y_{n+1} = B_0y_n + B_1y_{n-1} + h(l_{-1}\,y_{n+1}' + l_0y_n{}' + l_1y_{n-1}')$. We get equations in five unknowns and thus infinitely many solutions.]

14. (*a*) Prove the following relation between $T_n\,(x)$ and $u_n\,(x)$.

$u_{n-j-1}\,(x) = T_i\,(x)\,u_{n-1} - T_n(x)u_{j-1}\,(x).$

(*b*) Prove that $\sum_{k=1}^{n} \frac{T_j(t_k)}{n\,(t_k - x_n)} = 0,\ j = 0$

$$= u_{j-1}\,(x_n), 0 < j < n$$

where, $T_n\,(t_k) = 0,\ u_{n-1}\,(x_n) = 0$

15. (*a*) Prove the following results for the Chebychev polynomial of the first and second kind :

(*i*) $T_m[T_n(x)] = T_n[T_m(x)] = T_{mn}(x)$

(*ii*) $T_m(x)\,T_n(x) = \frac{1}{2}[T_{m+n}(x) + T_{m-n}(x)]$

(*iii*) $\int T_m(x)\,dx = \frac{1}{2}\left(\frac{T_{n+1}(x)}{n+1} - \frac{T_{n-1}(x)}{n-1}\right),\ n \geq 1$

(*iv*) $\int_{-1}^{+1} \frac{T_m(x)\,T_n(x)}{\sqrt{1-x^2}}\,dx = \begin{cases} \pi \text{ if } m = n = 0 \\ \pi/2 \text{ if } m = n \neq 0 \\ 0 \text{ if } m \neq n \end{cases}$

(*v*) $\int_{-1}^{+1} \sqrt{1-x^2}\; u_m(x)\,u_n(x)\,dx = \begin{cases} \pi/2 \text{ if } m = n \\ 0 \text{ if } m \neq n \end{cases}$

(*b*) Show that for | x | < 1, the Chebychev polynomial satisfy $|T_2(x)| < |$

16. Prove that $T_n(x) = \cos n\theta$ where $x = \cos\theta$ has n simple zeros

$$x_k = \cos\frac{2k-1}{2x}\pi,\quad k = 1, 2, ..., n$$

which are located inside [– 1, 1]

(*ii*) Prove that $T_n'(x_k) = \frac{n}{\sqrt{1-x_k^2}} \sin(2k-1)\pi/2 \neq 0$,

$k = 1, 2, ..., n$, has n simple roots.

17. Find the third approximation of the solution of the equation

$$\frac{dy}{dx} = z,\quad \frac{dz}{dx} = x^3(y+z).$$

by Picard's method, where $y = 1, z = \frac{1}{2}$ when $x = 0$

[**Hint.** $y_n = y_0 + \int_{x_0}^{x} z_{n-1}\,dt$ and $z_n = z_0 + \int_{x_0}^{x} t^2\,(y_{n-1} + z_{n-1})\,dt$

Take $x_0 = 0, y_0 = 1, z_0 = \frac{1}{2}$

$\therefore\ y_1 = y_0 + \int_{x_0}^{x} z_0\,dt = 1 + \frac{1}{2}x,\ z_1 = z_0 + \int_{x_0}^{x} t^3\,(y_0 + z_0)\,dt = \frac{1}{2} + \frac{3}{8}x^4$

Second approximation

$$y_2 = 1 + \frac{1}{2}x + \frac{3}{40}x^5,\ z_2 = \frac{1}{2} + \frac{3}{8}x^4 + \frac{1}{16}x^5 + \frac{3}{64}x^8]$$

(*b*) Find the third approximation of

$$\frac{d^3y}{dx^3} = x^2\frac{dy}{dx} + x^4 y$$

where $y = 5$ and $\frac{dy}{dx} = 1$, when $x = 0$

Take $\dfrac{dy}{dx} = z$. Then $\dfrac{dz}{dx} = x^2 z + x^4 y$ and $x_0 = 0,\ y_0 = 5$ and $z_0 = 1$]

18. Assuming

$$y_{n+1} = a_1 y_n + a_2 y_{n-1} + h(b_0 y'_{n+1} + b_1 y'_n + b_2 y'_{n-1}) \qquad ...(1)$$

and that $y(t)$ is a polynomial of degree 4, deduce Milne's formula

$$y_{n+1} = y_{n-1} + \frac{h}{3}(y'_{n+1} + 4y'_n + y'_{n-1})$$

[**Hint.** To simplify the algebra, use the substitution $s = (t - t_n)/h$ and $y(s) = 1, s, s^2, s^3$ and s^4 respectively in (1) to get the following equations:

$1 = a_1 + a_2,\ 1 = -a_2 + b_0 + b_1 + b_2,\ 1 = a_2 + 2b_0 - 2b_2,\ 1 = -a_2 + 3b_0 + 3b_2,$

$1 = a_2 + 4b_0 - 4b_2$

Solving; we find $a_1 = 0, a_2 = 1, b_0 = \frac{1}{3}, b_1 = \frac{4}{5}, b_2 = \frac{1}{3}$]

17

ERRORS

17·1. Nature or Types of Errors. The checking procedures are adopted to eliminate all mistakes whether these are human *i.e.*, due to the computer or technical *i.e.*, due to the calculating device. Even when all such mistakes have been eliminated from a computation, the solution is not generally exact on account of inherent errors of various types. They are classified as follows:

1. **Truncation errors.** These are caused by the use of a closed form, such as the first few terms of an infinite series to express a quantity defined by limiting process. For example such errors appear when a definite integral is computed by Simpson's rule or when a differential equation is solved by some difference method. This error is not machine-caused but is due to the method.

2. **Round-off errors.** These errors are unavoidable in most of the calculations since some of the quantities in the calculations will be non-terminating decimals and for practical reasons only certain number of figures will be carried in a calculation. These are due to the fact that in computational work we have to deal with approximations. If a number is rounded of with decimal place a change in the number of $5 \times 10^{-(n+1)}$ may occur and may increase or decrease this also put error on the number.

3. **Propagation or Inherited or Inherent Errors.** These errors are due to the approximate nature of the Applied formula used in the solution. It is caused by the use of previous points calculated by the computer which already are errors owing to the two errors above since we are already off the solution curve, we cannot expect any new points we compute to be as the correct solution curve.

4. **Absolute, Relative and Percentage Errors**. The absolute error of number, measurement, or calculation is the numerical difference between the true value of the quantity and its approximate value as given or obtained by measurement or calculation. The relative error is the absolute error divided by the true value of the quantity. The percentage error is

100 times the relative error. If a be the true value of the quantity and a' its approximate value, then

Absolute error $= a - a'$

Relative error $= \dfrac{a - a'}{a}$ or rather $\left| \dfrac{a'}{a} - 1 \right|$

Percentage error $= 100 \times \dfrac{a - a'}{a}$.

The relative and percentage errors are independent of the units while absolute error is expressed in terms of units used.

If The first significant figure of a number is k, and the number is correct to n significant figures, then the relative error in this number is less than $1/(k.\, 10^{n-1})$.

If the relative error in an approximate number is less than $1/[(k+1)\, 10^{n-1}]$, the number is correct to n significant figures.

If the relative error of any number is not greater than $1/(2 \cdot 10^n)$, the number is certainly correct to n significant figures.

For example, if the true value is 0·0002314 and the approximate value is 0·0002315, the relative error

$$\frac{\cdot 0000001}{\cdot 0002314} = \frac{1}{2314} < \frac{1}{2000} = \frac{1}{2 \times 10^3} = .5 \times 10^{-3}$$

17·2. General method of obtaining an error of a function

Let $$w = f(n_1, n_2, \ldots, n_m)$$

be a function of independent numbers $n_1, n_2, \ldots, n_m$ and let Δw be the error in w corresponding to the errors $\Delta n_1, \Delta n_2, \ldots, \Delta n_m$ in these numbers respectively.

Thus $$w + \Delta w = f(n_1 + \Delta n_1, n_2 + \Delta n_2, \ldots, n_m + \Delta n_m)$$

In order to obtain an expression for Δw, we write Taylor's expansion of the right side. Then

$$w + \Delta w = f(n_1, n_2, \ldots, n_m) + \left(\Delta n_1 \frac{\partial f}{\partial n_1} + \Delta n_2 \frac{\partial f}{\partial n_2} + \cdots + \Delta n_m \frac{\partial f}{\partial n_m} \right)$$
$$+ \frac{1}{2!} \left[(\Delta n_1)^2 \frac{\partial^2 f}{\partial n_1^2} + \cdots + (\Delta n_m)^2 \frac{\partial^2 f}{\partial n_m^2} + 2 \Delta n_1 \Delta n_2 \frac{\partial^2 f}{\partial n_1 \partial n_2} + \cdots \right] + \cdots$$

Since errors $\Delta n_1, \Delta n_2, \ldots, \Delta n_m$ are assumed to be so small that their squares and products and highest powers can be neglected, we have

$$w + \Delta w = w + \Delta n_1 \frac{\partial f}{\partial n_1} + \Delta n_2 \frac{\partial f}{\partial n_2} + \cdots + \Delta n_m \frac{\partial f}{\partial n_m}$$

$$\Rightarrow \quad \Delta w = \Delta n_1 \frac{\partial w}{\partial n_1} + \Delta n_2 \frac{\partial w}{\partial n_2} + \cdots + \Delta n_m \frac{\partial w}{\partial n_m}$$

This is the general form which enables us to compute the error of a function of m independent numbers.

$$\text{Relative error} = \frac{\Delta w}{w}$$

$$= \frac{\partial w}{\partial n_1} \cdot \frac{\Delta n_1}{w} + \frac{\partial w}{\partial n_2} \cdot \frac{\Delta n_2}{w} + \cdots + \frac{\partial w}{\partial n_m} \cdot \frac{\Delta n_m}{w}.$$

Example 1. *If $w = f(x, y, z) = 3x^2 y/z$, find the relative maximum error when $\Delta x = \Delta y = \Delta z = \cdot 001$ and $x = y = z = 1$.*

Sol. Here $\quad w = 3x^2 y/z$

$$\therefore \quad \frac{\partial w}{\partial x} = \frac{6xy}{z}, \frac{\partial w}{\partial y} = \frac{3x^2}{z}, \frac{\partial w}{\partial z} = -\frac{3x^2 y}{z^2}$$

$$\therefore \quad \Delta w = \frac{6xy}{z} \Delta x + \frac{3x^2}{z} \Delta y - \frac{3x^2 y}{z^2} \Delta z$$

Taking the absolute value of both sides since the errors Δx, Δy or Δz may be positive or negative, we have

$$| \Delta w | \le \left| \frac{6xy}{z} \Delta x \right| + \left| \frac{3x^2}{z} \Delta y \right| + \left| \frac{3x^2 y}{z^2} \Delta z \right|$$

$$\therefore \quad | \Delta w | \simeq \left| \frac{6xy}{z} \Delta x \right| + \left| \frac{3x^2}{z} \Delta y \right| + \left| \frac{3x^2 y}{z^2} \Delta z \right|$$

$$\simeq (6 + 3 + 3) \cdot 001 = \cdot 012. \quad \text{Also } w = \left(\frac{3x^2 y}{z} \right) = 3$$

when $x = y = z = 1$

$$\text{Hence } \frac{\Delta w_{\max}}{w} = \frac{\cdot 012}{3} = \cdot 004$$

Example 2. *Find the relative error in the following function*

$w = k. n_1^{a_1} n_2^{a_2} \ldots n_m^{a_m}$. k and a's are constants.

Sol. Here $w = k. n_1^{a_1} n_2^{a_2} \ldots n_m^{am}$

$$\Rightarrow \quad \log w = \log k + a_1 \log n_1 + a_2 \log n_2 + \cdots + a_m \log n_m$$

$$\text{Differentiating,} \quad \frac{\Delta w}{w} = 0 + a_1 \frac{\Delta n_1}{n_1} + a_2 \frac{\Delta n_2}{n_2} + \cdots + a_m \frac{\Delta n_m}{n_m}$$

where $\Delta n_1, \Delta n_2, \ldots, \Delta n_m$ are the errors positive or negative in $n_1, n_2, \ldots, n_m$ respectively.

Hence relative error

$$= a_1 \frac{\Delta n_1}{n_1} + a_2 \frac{\Delta n_2}{n_2} + \cdots + a_m \frac{\Delta n_m}{n_m}$$

$$\le a_1 \left| \frac{\Delta n_1}{n_1} \right| + a_2 \left| \frac{\Delta n_2}{n_2} \right| + \cdots + a_m \left| \frac{\Delta n_m}{n_m} \right|$$

Cor. 1. $w = k.\, n_1^{a_1} n_2^{a_2} n_3^{a_3}$

$$\Rightarrow \log w = \log k + a_1 \log n_1 + a_2 \log n_2 + a_3 \log n_3$$

$$\therefore \quad \frac{\Delta w}{w} = 0 + a_1 \frac{\Delta n_1}{n_1} + a_2 \frac{\Delta n_2}{n_2} + a_3 \frac{\Delta n_3}{n_3}$$

Relative error

$$\le a_1 \left| \frac{\Delta n_1}{n_1} \right| + a_2 \left| \frac{\Delta n_2}{n_2} \right| + a_3 \left| \frac{\Delta n_3}{n_3} \right|$$

Cor. 2. Put $k = 1, a_1 = 1, a_2 = 1, ..., a_m = 1$ so that

$$w = n_1 n_2, ..., n_m$$

then $$\frac{\Delta w}{w} = \frac{\Delta n_1}{n_1} + \frac{\Delta n_2}{n_2} + \cdots + \frac{\Delta n_m}{n_m}$$

$\therefore$ Relative error

$$\le \left| \frac{\Delta n_1}{n_1} \right| + \left| \frac{\Delta n_2}{n_2} \right| + \cdots + \left| \frac{\Delta n_m}{n_m} \right|$$

Thus the relative error of a product of n approximate numbers is therefore equal to the algebraic sum of the relative errors of the separate numbers.

Cor. 3. Put $k = 1, a_1 = 1, a_2 = -1, a_3 = 0 = \cdots = -a_m$ so that

$$w = \frac{n_1}{n_2}$$

and Relative error $$= \frac{\Delta n_1}{n_1} - \frac{\Delta n_2}{n_2}$$

Thus the relative error of a quotient is equal to the algebraic sum of the relative errors of divisor and dividend but the error is the sum of the errors.

Alternatively. We have

$$\Delta w = \frac{n_1 + \Delta n_1}{n_2 + \Delta n_2} - \frac{n_1}{n_2} = \frac{n_2 \Delta n_1 - n_1 \Delta n_2}{n_2(n_2 + \Delta n_2)} = \frac{n_1 n_2}{n_2} \frac{\left[\frac{\Delta n_1}{n_1} - \frac{\Delta n_2}{n_3} \right]}{(n_2 + \Delta n_2)}$$

$$= n_1 \frac{\left(\frac{\Delta n_1}{n_1} - \frac{\Delta n_2}{n_2} \right)}{n_2 + \Delta n_2}$$

Let M be the max $\left(\left|\dfrac{\Delta n_1}{n_1}\right|, \left|\dfrac{\Delta n_2}{n_2}\right|\right)$

$$\therefore \qquad \Delta w \le \frac{n_1(M+M)}{n_2 - Mn_2} = \frac{2n_1M}{n_2(1-M)}, \quad \Delta w \le \frac{2n_1M}{n_2(1-M)}$$

17·2·1. Rounding off Numbers. We know that

$$\pi = 3{\cdot}1415926$$
$$\sqrt{2} = 1{\cdot}4142134$$
$$e = 2{\cdot}71828182$$

The figures on the right are approximating to the true value, of the numbers on the left.

Suppose it is desired to round off a number correct to n·significant figures. Then we observe one of the following rules :

(1) Discard all digits to the right of the nth place. If the discarded number is less than half a unit in the nth place, leave the nth digit unchanged.

(2) Discard all digits to the right of the nth place and if the discarded number is greater than half a unit in the nth place, add 1 to the nth digit.

(3) Discard all digits to the right of the nth place and if the discarded number is *exactly* half a unit in the nth place leave the nth digit unchanged if it is an even number but increase by 1 if it is odd.

Note. If a number is correct to n significant figures, then the absolute error cannot be greater than half a unit in the nth place. For example, if the number 3·1416 is correct to five figures, then

absolute error $0{\cdot}0001 \times \dfrac{1}{2} = 0{\cdot}00005$.

The relative error in a number correct to n significant figures is obtained by :

Theorem. *If the first significant figure of a number is k and the number is correct to n significant figures, then the relative error is less than* $1/(k \times 10^{n-1})$.

We shall not prove the theorem but shall illustrate it with a few examples.

Example 3. *Find the relative error in the number 834·12 correct to five significant figures.*

Sol. Here $\quad k = 8, n = 5$, Absolute error $= \frac{1}{2}\, 10^{-2} = 0{\cdot}005$

Relative error $\le \dfrac{0{\cdot}005}{864{\cdot}32 - 0{\cdot}005} = \dfrac{5}{864320 - 5}$

$$\left[\text{Rel. Error} = \frac{\text{Absolute error}}{\text{Number} - \text{Absolute Error}}\right].$$

$$= \frac{5}{10 \times 86432 - 5}$$

$$= \frac{1}{2 \times 86432 - 1}$$

$$= \frac{1}{2\left(86432 - \frac{1}{2}\right)} < \frac{1}{2 \times 80000} = \frac{1}{2 \times 8 \times 10^4}$$

$$< \frac{1}{8 \times 10^4} = \frac{1}{8} \times 10^{-4}$$

Thus the theorem is verified for the number 834·12.

17·3. Accuracy in the determination of arguments from a Tabulated Function.

Whenever we need the value of the logarithm of a number or the trigonometric ratio of an angle, we consult the log Tables which gives us the values of the desired function. However, if there is an error in the tabulated function, the argument determined from this function will be incorrect at least to some degree. We shall now investigate the accuracy of the argument whose value is required.

Let x be the argument and y, the entry of function so that we have

$$y = f(x)$$

Differentiating, $\Delta y = f'(x)\Delta x$ to the first approximation

$$\Rightarrow \quad \Delta x = \Delta y / f'(x) = \Delta f(x)/f'(x). \qquad ...(1)$$

This formula gives the (small) error in the argument x when $\Delta f(x)$ is the error in the computed function while values are tabulated.

Example 4. *Find the error in the argument x when y, the tabulated function, is*

(*i*) $y = f(x) = log_{10} x$

(*ii*) $y = f(x) = tan\, x$

(*iii*) $y = f(x) = log_{10} sin\, x$

Sol. We have

(*i*) $f(x) = \log_{10} x = \log_{10} e \times \log_e x$

$$\therefore \quad f'(x) = 0{\cdot}4343 \times \frac{1}{x}$$

$$\text{Now} \quad \Delta x = \frac{\Delta f(x)}{f'(x)} = \frac{x}{0{\cdot}4343} \cdot \Delta f(x)$$

$$= 2{\cdot}3026\, x\, \Delta f(x).$$

(*ii*) $f(x) = \tan x$

$$f'(x) = \sec^2 x$$

$$\therefore \qquad \Delta x = \frac{\Delta f(x)}{f'(x)} = \frac{\Delta f(x)}{\sec^2 x} \text{ radians}$$

$$= 206264{\cdot}8 \cos^2 x \, \Delta f(x) \text{ seconds}$$

(*iii*) $f(x) = \log \sin x$

$$= \log_{10} e \times \log_e \sin x \quad f'(x) = 0{\cdot}4343 \times \frac{\cos x}{\sin x}$$

$$\therefore \qquad \Delta x = \frac{\Delta f(x)}{f'(x)}.$$

$$= 2{\cdot}3026 \tan x \, \Delta f(x) \text{ radians.}$$

Ex. *Find the error in* Δx *when*

(*i*) $f(x) = \log_{10} \tan x$ (*ii*) $f(x) = \sin x^0$ (*iii*) $f(x) = 10^x$,

17·4. Number of terms to be retained while solving $\frac{dy}{dx} = f(x)$ by Taylor's series method.

We have $y(x) = y(x_n) + (x - x_n)y(x_n) + \frac{1}{2!}(x - x_n)^2 y(x_n) + \cdots$

$$+ \frac{1}{p!}(x - x_n)^p y(x_n) + \frac{(x - x_n)^{p+1}}{(p+1)!} y^{(p+1)}(\xi_0), \; x_n < \xi_0 < x.$$

Let $\qquad x_n = x_0 + nh, \; n = 0, 1, \ldots, N \qquad$ ...(1)

$$x_N = x_0 + Nh = p$$

and $\qquad x = x_{n+1}$ in (1) above, we get

$$y(x_{x+1}) = y(x_n) + hy'(x_n) + \frac{h^2}{2!} y''(x_n) + \cdots$$

$$+ \frac{h^p}{p!} y^{(p)}(x_n) + \frac{h^{p+1}}{(p+1)!} y^{(p+1)}(\xi_0)$$

$$\Rightarrow \qquad y_{n+1} = y, \; hk_p(x_n, y_n, h) \; ; \; n = 0, 1, 2,$$

where $hk_p(x_n, y_n, h) = hy'(x_n) + \frac{h^2}{2!} y''(x_n) + \cdots + \frac{h_p}{p!} y^{(p)}(x_n)$ and $y(n+1)$ is approximated by y_{n+1}, omitting the last term.

This is called Taylor's series method of order p.

For $p = 1$, $y_{n+1} = y_n + hy' = y_n + hf(x_n, y_n)$, $n = 0, 1, \ldots$ which is Euler's method.

Note. The number of terms to be included in (2) is fixed up by the permissible error. If this error is ε and the series is curtailed at the term $y^{(p)}(x_n)$, then

$$h^{p+1} y^{(p+1)}(\xi_0) < (p+1)! \, \varepsilon$$

or $$| h^{p+1} f^p(\xi) | < (p+1)! \, \varepsilon.$$

Thus we determine p in case h is given while if p is given we find an upper bound on h.

Example 5. *Given* $y' = x + y$, $y(0) = 1$, $x \in (0, 1)$, *solve for* y *by Taylor's series method. Determine the number of terms to be retained so as to have an accuracy of* 10^{-10}.

Sol. We have $\quad y' = x + y, y'' = 1 + y', y''' = y'', \ldots,$

$$y^{(p+1)} = y^{(p)},\ p = 2, 3, \ldots$$

$$\therefore \qquad y'(0) = 1, y''(0) = 2, \ldots, y^{(p)}(0) = 2,$$

Hence $$y(x) = 1 + x + x^2 + \frac{3}{5} \cdots + \frac{2x^p}{p!} + \cdots$$

In order to obtain the results which will be accurate upto 10^{-10} for $x \le 1$, we have $\dfrac{2}{(p+1)!} < \cdot 5 \times 10^{-10}$

This gives $p \simeq 15$.

Hence about 15 terms are required to obtain the accuracy when $x \le 1$.

EXERCISE 17

1. Show that the relation error in approximation formula in solved example 14, chapter 3, is $\dfrac{1}{8}\left(\dfrac{A-B}{A+B}\right)^3$
2. Show that the approximate formula for the maximum relative error of (*i*) a quotient $\dfrac{a}{b}$ is $\dfrac{\varepsilon}{a} + \dfrac{\eta}{b} \cdot$ (*ii*) the square root of b_r is $\dfrac{x_r}{2b_r}$ where a and b are positive numbers whose rounded values a_2 and b_2 lie in the range $a \pm \varepsilon, b \pm \eta$ respectively, ε and η are positive numbers small as compared with a and b.
3. (*a*) Round the numbers 0·01201 and – 239·85 upto three significant digit.

   (*b*) If $\pi = 3{\cdot}14$ in place of 3·14156, find the absolute and relative errors.

   (*Cal. Uni. 1995*)
4. N_1 and N_2 are two numbers each rounded off to a significant digit.

   (*a*) What will be the relative error in the product $N_1 N_2$?

   (*b*) Show that $w(N_1 \mid N_2) \le \frac{5}{4}\sqrt{10}$ where w denotes the max. absolute error in the sth significant digit.

   (*c*) Hence or otherwise, show that for $m = 1, 2, \ldots$

   $$w(N_1^{m}) \le 10^{n - \frac{1}{m}} [(1 + 5 \times 10^{-n})^{m}] - 1$$
5. A table of values of the function $f(x) = \log_{10} \sin x$ for $0{\cdot}01 \le x \le 1{\cdot}0$ is be constructed at equal intervals. Find the length of the interval of x and the number of significant digits to be kept in the tabulated values so that if a linear interpolation formula is used for any x_0 lying between two successive tabulated values the max. absolute error will not exceed 5×10^{-5}

6. If random samples of n items are drawn from a finite population of N items and variance σ^2, the sampling variance of the means of such samples is $\frac{\sigma^2}{n} \cdot \frac{N-n}{N-1}$. Find the approximate relative error involved in using $\frac{\sigma}{\sqrt{n}}$ instead of $\frac{\sigma}{\sqrt{n}} \sqrt{\frac{N-n}{N-1}}$ as the standard error.

$$\left[\textbf{Hint.}\text{ Relative error} = \left(\frac{\sigma}{\sqrt{n}} - \frac{\sigma}{\sqrt{n}}\sqrt{\frac{N-m}{N-1}}\right) \Big/ \frac{\sigma}{\sqrt{n}}\sqrt{\frac{N-m}{N-1}} = \sqrt{\frac{N-m}{N-n}}\right]$$

7. Discuss the propagation of an error committed in recording the values of a function in its successive differences.

8. Let $\sin x$ be approximated by first three non-zero terms of its Maclaurin expansion.

(*a*) For any x find a bound on the magnitude of the truncation error.

(*b*) If $|x| < 1$ is correctly rounded to 10 decimal digits, find the bound in the error in the correctly rounded value of x^2.

9. Neglecting round off error, how many terms of the Maclaurin expansion for $\sin x$ are required to obtain a maximum error of less than 10^{-7} in the range $\left(0, \frac{\pi}{2}\right)$? In the range $[0, \pi]$?

10. Suppose you have a table of $\sin x$ at an interval $h = \cdot 1$. How many tabular points would have to be used in interpolating in this table to assure a truncation error of less than (*a*) 10^{-2} (*b*) 10^{-4}?

11. Obtain Taylor's series solution of

$$\frac{dy}{dx} = 1 - 2xy, \; y_0 = 0$$

If the error in y_n, obtained from four terms of the series only, is to be less than 10^{-6} after rounding, find it. Determine the number of terms in series to obtain the result correct to 10^{-10} for $x \leq 1$.

12. Obtain a second degree polynomial approximation to $f(x) = (1+x)^{1/2}$, $x \in [0, 0{\cdot}1]$ using the Taylor's Series expansion about $x = 0$. Use the expansion to approximate $f(\cdot 05)$ and bound the truncation error.

13. Calculate $\sqrt{e}$ with an accuracy to within $0{\cdot}00001$.

[**Hint.** $e^{1/2} = 1 + \frac{1}{2} + \frac{2}{2!}\left(\frac{1}{2}\right)^2 + \cdots + \frac{1}{n!\,2^n} + R_n$

where $R_n = \frac{1}{n!\,2^n} \cdot \frac{\frac{2}{n}}{n + \frac{1}{2}} \leq \frac{1}{2^n\, n!\,(2n+1)} < 0{\cdot}00001$

which is the case when $n = 6$ as then $R_6 < \frac{1}{2^6\, 6!\,(13)} < \frac{1}{10000}$

In general, $e^x = 1 + \frac{x}{1!} + \frac{x^2}{2!} + \cdots + \frac{x^n}{n!}$, $0 < x < n!$

where $R_n = \frac{x^{n+1}}{(n+1)!} + \frac{x^{n+2}}{(n+2)!} = \frac{x^n}{n!}\left[\frac{x}{n+1} + \frac{x^2}{(n+1)(n+2)}\right]$

$$\leq \frac{x^n}{n!}\left[\frac{x}{n+1} + \left(\frac{x}{n+1}\right)^2\right] \leq \frac{x^n}{n!}\frac{\frac{x}{n+1}}{1-\frac{x}{n+1}} < \frac{x}{n!}\cdot\frac{x}{(n+1-x)}$$

14. Find the number of terms so that the sum of the series for e^x at $x = 1$ is correct to 8 decimal places.

Hint. $$e^x = 1 + x + \frac{x^2}{2!} + \cdots + \frac{x^{n-1}}{(n-1)!} + \frac{x^n}{n!}e^{\theta}, \quad 0 < \theta < x.$$

where $R_n = \frac{x^n}{n!}e^{\theta}$

Maximum relative error $= \left[\left(\frac{x^n}{n!}e^{\theta}\right)_{\theta = x} / e^x\right]_{x=1} = (x^n/n!)_{x=1} = \frac{1}{n!}$

$\therefore \quad \frac{1}{n!} < \frac{1}{2} \cdot 10^{-8}$ (Given)

This gives $n = 12$.

18

NUMERICAL SOLUTION OF PARTIAL DIFFERENTIAL EQUATIONS

18·1. Introduction. Partial differential equations and systems of such equations appear in the description of physical processes, for example, in hydrodynamics, the theory of elasticity, the theory of electromagnetism (Maxwell's equations), and quantum mechanics. The solutions of the equations describe possible physical reactions that have to be fixed through boundary conditions, which may be of quite a different character. We shall confine ourselves to second order partial differential equations as such equations are found in the applications.

18·2. Classification of the second order partial differential equations.

The general linear equation in the second derivatives is of the form

$$A\frac{\partial^2 u}{\partial x^2} + 2B\frac{\partial^2 u}{\partial x\,\partial y} + C\frac{\partial^2 u}{\partial y^2} + D\frac{\partial u}{\partial x} + E\frac{\partial u}{\partial y} + Fu = G \qquad ...(1)$$

where $A, B, C, D, E, F,$ and G are functions of $x, y, \frac{\partial u}{\partial x}$ and $\frac{\partial u}{\partial y}$.

It is the convention in partial differential equations, to denote

$$p = \frac{\partial u}{\partial x}\,;\, q = \frac{\partial u}{\partial y}\,;\, r = \frac{\partial^2 u}{\partial x^2}\,;\, s = \frac{\partial^2 u}{\partial x\,\partial y}\,;\, t = \frac{\partial^2 u}{\partial y^2}.$$

Also, we have

$$du = \frac{\partial u}{\partial x}dx + \frac{\partial u}{\partial y}dy = pdx + qdy\,;\;\; d \equiv \left(\frac{\partial}{\partial x}dx + \frac{\partial}{\partial y}dy\right)$$

$$dp = d\left(\frac{\partial u}{\partial x}\right) = \left(\frac{\partial}{\partial x}dx + \frac{\partial}{\partial y}dy\right)\left(\frac{\partial u}{\partial x}\right)$$

$$= \frac{\partial^2 u}{\partial x^2}dx + \frac{\partial^2 u}{\partial y\,\partial x}dy$$

$$= rdx + sdy, \qquad \text{assuming } \frac{\partial^2 u}{\partial x \partial y} = \frac{\partial^2 u}{\partial y \partial x}$$

Similarly $dq = sdx + tdy$

Equation (1) can now be written in the form

$$Ar + 2Bs + Ct + Dp + Eq + Fu = G$$

The equation is called

hyperbolic if $B^2 - AC > 0$

parabolic if $B^2 - AC = 0$

elliptic if $B^2 - AC < 0$

Example 1. *Classify the partial differential equations :*

(*i*) $y^2 \frac{\partial^2 u}{\partial x^2} - \frac{\partial^2 u}{\partial y^2} + 4u = 0$

(*ii*) $x^2 \frac{\partial u^2}{\partial x^2} + (1 - y^2) \frac{\partial^2 u}{\partial y^2} = 0, - \infty < x < \infty, - 1 < y < 1.$

(*iii*) $\frac{\partial u}{\partial t} = \frac{\partial^2 u}{\partial x^2}$

Sol. (*i*) Here $A = y^2$, $C = -1, B = 0$

$\therefore \; B^2 - AC = 0 - y^2 (-1) = y^2 > 0$

Hence the given equation is hyperbolic.

(*ii*) Here $A = x^2, B = 0, C = 1 - y^2$

$\therefore \; B^2 - AC = 0 - x^2(1 - y^2) = x^2 (y^2 - 1) < 0 \quad (\because \; y^2 < 1)$

Hence the given equation is elliptic.

(*iii*) Here we have $\frac{\partial^2 u}{\partial x^2} - \frac{\partial u}{\partial t} = 0$

$\therefore \; A = 1, B = 0, C = 0$

Hence $\quad B^2 - AC = 0 - 0 = 0$

so that the given equation is parabolic.

18·2·1. Hyperbolic equation

Consider the wave equation in one dimension,

$$\frac{\partial^2 u}{\partial x^2} = \frac{1}{c^2} \frac{\partial^2 u}{\partial t^2} \qquad \text{...(1)}$$

where c is the propagation velocity.

The general solution of equation (1) can be easily written down

$$u(x, t) = u = f(x + ct) + g(x - ct)$$

The solution contains two arbitrary functions f and g. We now give the physical meaning. Let u be interpreted as the superposition of two waves travelling in opposite directions.

Let us take $f(z) = -\frac{1}{2}\cos \pi z$ and $g(z) = \frac{1}{2}\cos \pi z$

so that
$$u = f(x+ct) + g(x-ct)$$
$$= -\frac{1}{2}\cos \pi (x+ct) + \frac{1}{2}\cos \pi (x-ct)$$
$$= \sin \pi x \sin \pi ct$$

This describes a standing wave with $u(0, t) = u(1,t) = 0$ and $u(x, 0) = 0$. Choosing the initial conditions as

$$u(x, 0) = f(x), \quad \frac{\partial}{\partial t} u(x,o) = g(x),$$

so that the general solution (1) is

$$u(x,t) = \frac{1}{2}[f(x+ct) + f(x-ct)] + \frac{1}{2c}\int_{x-ct}^{x+ct} g(\xi)d\xi$$

Cor. Putting $y = ct$ in (1), this equation reduces to

$$\frac{\partial^2 u}{\partial x^2} - \frac{\partial^2 u}{\partial y^2} = 0$$

with the solution $u = f(x+y) + g(x-y)$.

18·3. Derivatives in terms of Differences

We have already defined the operators on page 108.

we have
$$\delta y_i = y_{i+\frac{1}{2}} - y_{i-\frac{1}{2}}, \quad \mu y_i = \frac{1}{2}\left(y_{i+\frac{1}{2}} + y_{i-\frac{1}{2}}\right)$$

$\mu\delta\, y_i = hDy_i$, neglecting D^2 and higher derivatives

$$\Rightarrow \quad hDy_i = \frac{1}{2}\left(y_{i+1} - y_{i-1}\right) \text{ approx.}$$

Further
$$\partial^2 y_i = y_{i+1} - 2y_i + y_{i-1}$$
$$= h^2D^2 y_i \text{ approx.}$$

Thus symbolically, $h^2D^2 = \partial^2$ or $D^2 = \frac{1}{h^2}\partial^2$ ...(1)

$$\Rightarrow \quad h^2D^2 y_i = y_{i+1} - 2y_i + y_{i-1},$$
$$h^3D^3 y_i = \nabla^3 y_i = y_i - 3y_{i-1} + 3y_{i-2} - y_{i-3}$$
$$\left(1 + \frac{\delta^2}{4}\right) y_i = y_i + \frac{\delta^2}{4} y_i = y_i + \frac{1}{4}\left(y_{i+1} - 2y_i + y_{i-1}\right)$$
$$= \frac{1}{4}\left(y_{i+1} + 2y_i + y_{i-1}\right) = \mu^2 y_i$$

Also
$$\nabla y_i = y_i - y_{i-1} = hDy_i$$
$$\nabla^2 y_i = \nabla\left(y_i - y_{i-1}\right) = y_i - 2y_{i-1} + y_{i-2} = h^2D^2 y_i$$
$$\nabla^3 y_i = y_i - 3y_{i-1} + 3y_{i-2} - y_{i-3} = h^3D^3 y_i$$

Now $y(x+h) = y(x) + hDy(x) + \frac{h^2}{2} D^2y(x) + \cdots = e^{hD} y(x)$

$\therefore \quad y(x-h) = e^{-hD}y(x)$

Again $\quad \nabla y_i = y(x) - y(x-h) = y(x) - e^{-hD} y(x) = (1 - e^{-hD}) y_i$

$$\therefore \quad \nabla = hD - \frac{1}{2}h^2D^2 + \frac{1}{6}h^3D^3 -, \ldots..$$

$$\nabla^2 = h^2D^2 - h^3D^3 + \cdots$$

Further $\quad hDy_i = \Delta y_i = y_{i+1} - y_i$

$$h^2D^2y_i = \Delta^2 y_i = y_{i+2} - 2y_{i+1} + y_i$$

$$h^3D^3y_i = \Delta^3 y_i = y_{i+3} - 3y_{i+2} + 3y_{i+1} - y_i$$

Example 2. *Find the solution of the differential equation* $y'' + x^2y = 0$ *with the boundary conditions* $y(0) = 0$, $y(1) = 1$ *at the points 0·25, 0·50 and 0·75 by approximating the differential equation with a second-order difference equation.*

Sol. We have

	$i = 0$	$i = 1$	$i = 2$	$i = 3$	
x	0	0·25	0·50	0·75	1
y	y_0	y_1	y_2	y_3	y_4

From (1), page 376, we have $D^2 = \frac{1}{h^2} \partial^2$,

$$h = \frac{1}{4} = 0{\cdot}25, \quad y'' = \frac{1}{h^2}\left(y_{i+1} - 2y_i + y_{i-1}\right)$$

Hence the differential equation for $x = x_i$ is $D^2y_i + x_i^2 y_i = 0$

$$\therefore \quad 16\left(y_{i+1} - 2y_i + y_{i-1}\right) + x_i^2 y_i = 0$$

Given $y_0 = 0$, $y_4 = 1$. Also put $i = 1$, $x_1 = \frac{1}{4}$ as $x = {\cdot}25$

$$\therefore \quad 16y_2 - 32y_1 + 16y_0 + \frac{1}{16} y_1 = 0 \qquad \ldots(1)$$

When $\quad i = 2, x = \frac{1}{2}, h = \frac{1}{4}$, we have

$$16y_3 - 32y_2 + 16y_1 + \frac{1}{4} y_2 = 0 \qquad \ldots(2)$$

When $\quad i = 3, x = \frac{3}{4}, h = \frac{3}{4}$, we have

$$16y_4 - 32y_3 + 16y_2 + \frac{9}{16} y_3 = 0 \qquad \ldots(3)$$

From the equations (1), (2) and (3) for y_1, y_2, y_3, we get

$$512y_1 - 256y_2 = 0$$

$$64y_1 - 127y_2 + 64y_3 = 0$$

$$256\, y_2 - 503\, y_3 + 256\, y_4 = 0$$

$\therefore$ Solving, we get $y_1 = \cdot 7748,\ y_2 = \cdot 5223,\ y_3 = \cdot 2617$

Example 3. *Determine the values of y at the pivotal points of the interval [0, 1] if y salisfies the boundary value problem.*

$$y''' + 2y = 12x^2 + 2$$

with $$y(0) = 0,\ y(1) = y'(1) = 0$$

Case (*a*). *Use n = 2 and approximate y′ by averaged central differences and third derivative by unsymmetrical expressions.*

Case (*b*). *Use n = 3 and approximate y‴ at* $x = \frac{1}{3}$ *by unsymmetrical expressions and by averaged central differences at* $x = \frac{2}{3}$.

Sol. (*a*) $n = 2$ means that [0, 1] is divided into two equal parts *i.e.*, $h = \frac{1}{2}$.

We have $y''' + 2y = 12x^2 + 2$ *i.e.*, $D^3 y_i + 2y_i = 12x_i^2 + 2$ ...(1)

We know that $D^3 y_i = \frac{1}{h^3}\left(y_{i+2} - 3y_{i+1} + 3y_i - y_{i-1}\right)$

If $h = \frac{1}{2}$, the given equation reduces to

$$8\left(y_{i+2} - 3y_{i+1} + 3y_i - y_{i-1}\right) + 2y_i = 12x_i^2 + 2 \qquad ...(2)$$

when $x_1 = \frac{1}{2}, i = 0,\ 8\, y_3 - 24y_2 + 26y_i - 8y_0 = 12 \times \frac{1}{4} + 2 = 5$ ...(3)

$y'(1) = 0, y_2' = 0, i = 2, y_{i+1} - y_{i-1} = 0$ at $i = 2, y_3 = y_1$ as $\mu\,\delta = hD$

$\therefore$ From (3), $8y_1 - 24y_2 + 26y_1 - 8 \times 0 = 5$

$$\Rightarrow \quad 34y_1 = 5 \quad \therefore \quad y_1 = \frac{5}{34} = 0{\cdot}147$$

(*b*) If $h = \frac{1}{3}$, equation (1) becomes

$$27\left(y_{i+2} - 3y_{i+1} + 3y_i - y_{i-1}\right) + 2y_i = 12x_i^2 + 2$$

When i = 1, $27y_3 - 81y_2 + 81y_1 - 27y_0 + 2y_1 = \frac{10}{3}$

$$\Rightarrow \quad -81y_2 + 83y_1 = \frac{10}{3} \quad .... (i) \quad (\because y_0 = y_3 = 0)$$

$$x = \frac{2}{3},\quad D^3 y_i = \frac{1}{h^3}\mu\delta^3 y_i = \frac{1}{2h^3}\left(y_{i+2} - 2y_{i+1} + 2y_{i-1} - y_{i-2}\right)$$

$$i = 2,\ 27(y_4 - 2y_3 + 2y_1 - y_0) + 4y_2 = 2\,(12\cdot\frac{4}{9} + 2) = \frac{44}{3}$$

$$y_1 + \frac{31}{54} y_2 = \frac{22}{81}$$

$$y_2 = \frac{9336}{62523} = \cdot 149$$

$$y_1 = 46 \cdot \frac{007}{249} = \cdot 1855$$

Example 4. *Determine the values of y at pivotal points of [0, 1] if y satisfies*

$$y^{iv} + 81y = 81x^2$$

with the boundary value $y(0) = 0, y(1) = 0, y''(0) = y''(1) = 0$
use $n = 3$ and symmetrical expressions for the derivatives.

Sol. $y^{iv} = D^4 y_i = \frac{\delta^4}{h^4} = \frac{1}{h^4}\left[y_{i+2} - 4y_{i+1} + 6y_i - 4y_{i-1} + y_{i-2}\right]$

$i = 0$	$i = 1$	$i = 2$	$i = 3$
0	1/3	2/3	3/3 = 1
y_0	y_1	y_2	y_3

$$y(0) = y_0 = 0, \; y(1) = 0 \text{ i.e., } y(3) = 0$$

$$D^2 y_i = \frac{1}{h^2}\left[y_{i+1} - 2y_i + y_{i-1}\right] = 0 \text{ at } i = 0,3$$

When $i = 0$, $\quad y_i - 2y_o + y_{-1} = 0 \quad \therefore \quad y_1 = y_{-1}$

when $i = 1$,

$$81\left(y_3 - 4y_2 + 6y_1 - 4y_o + y_{-1}\right) + 81y_1 + 81\left(\frac{1}{9}\right)$$

$$\Rightarrow \quad 81\,(y_4 - 4y_3 + 6y_2 - 4y_1 + y_o) + 81y_2 = 81\left(\frac{4}{9}\right)$$

$$6y_1 - 4y_2 = \frac{1}{9} \qquad i = 3, y_4 = -y_2$$

$$-4y_1 + 6y_2 = \frac{4}{9}$$

$\therefore \quad y_1 = .1222, \; y_2 = \cdot 1555$

Example 5. *Determine the values of y at pivotal points of (0, 1) satisfying* $y^{iv} + 81y = f(x)$
with boundary value $y(0) = 0, \quad y'(0) = 0 \quad , \quad y''(1) = 0 \quad y'''(1) = 0$

Use n = 3 and symmetrial approximation for the derivatives, when $f(x) = 729\,x^2$.

Sol. The given difference equation can be witten as follows :

$$D^4 y_i + 81y_i = f(x)$$

$$\frac{1}{h^4}\left[y_{i+2}-4y_{i+1}+6y_i-4y_{i-1}+y_{i-2}\right]+81y_1=f(x_i)$$

When $i=1,\ h=\frac{1}{3},\ \left[y_3-4y_2+6y_1-4y_o+y_{-1}\right]+y_1=\left[\frac{9}{9},\ 1\right]$

When $i=2,\ h=\frac{2}{3},\ y_4-4y_3+6y_2-4y_1+y_o+y_2=[4, 2]$

and $\qquad y_5-4y_4+6y_3-4y_2+y_1+y_3=[9,3]$

By given conditions,

$$y'(0)=0=y_0'=y^1-y_{-1}=0 \Rightarrow y_{-1}=y_1$$

$$y''(1)=y_3'' \Rightarrow y_4-2y_3+y_2=0 \Rightarrow y_4=2y_3-y_2$$

$$y'''(1)=y_3''' \Rightarrow y_5-2y_4+2y_2-y_1=0$$

$$\Rightarrow \quad y_5=2y_4-2y_2+y_1$$

$$=4y_3-2y_2-2y_2+y_1$$

$$=4y_3-4y_2+y_1$$

We have the system of equations

$$8y_1-4y_2+y_3=1, 1 \qquad ...(1)$$

$$-4y_1+6y_2-2y_3=4, 2 \qquad ...(2)$$

Substituting the value of y_3 from (1) in (2) and (3), we get

$$2y_1-4y_2+3y_3=9, 3 \qquad ...(3)$$

$$-4y_1+6y_2-2(1-8y_1+4y_2)=4, 2$$

$$2y_1-4y_2+3(1-8y_1+4y_2)=9, 3$$

$$\therefore \quad 12y_1-2y_2=6, 4 \quad \therefore\ 11y_1=4y_2$$

$$-22y_1+8y_2=6, 0$$

$$12y_1-\frac{22}{4}y_1=4 \Rightarrow \frac{26}{4}y_1=4$$

$$y_1=\frac{16}{26}=\cdot 6154$$

$$y_2=1\cdot 6923$$

$$y_3=2\cdot 8462$$

Eigen Value Problem

Z_{ij} is the value of the function z at $x=x_i,\ y=y_j$

$$D_x=\frac{\partial}{\partial x},\ D_y=\frac{\partial}{\partial y},\ D_{xy}=\frac{\partial^2}{\partial x\,\partial y}$$

$$D_x Z_{i,j}=\frac{\mu\delta}{h}(Z_{i,j}),\ =\frac{1}{2h}(Z_{i+1,j}-Z_{i-1,j})$$

Where h and k are the spacing between pivotal points along x-axis and y-axis respectively.

$$D_{xx} Z_{i,j} = \frac{\delta^2}{h^2} = \frac{1}{h^2}(Z_{i+1,j} - 2Z_{i,j} - Z_{i-1,j})$$

$$D_y Z_{i,j} = \frac{h\delta}{k} Z_{i,j} = \frac{1}{2k}\left[Z_{i,j+1} - Z_{i,j-1}\right]$$

$$D_{yy} Z_{i,j} = \frac{\delta^2}{k^2}(Z_{i,j}) = \frac{1}{k^2}\left[Z_{i,j+1} - 2Z_{i,j} + Z_{i,j-1}\right]$$

$$\nabla^2 = \frac{\partial^2}{\partial x^2} + \frac{\partial^2}{\partial y^2}$$

$$\nabla^2 Z_{i,j} = \frac{1}{h^2}[Z_{1+1,j} - 2Z_{i,j} + Z_{i-1,j} - 2Z_{i,j} + Z_{i,j-1}]$$

Letting $h = k$

$h^2\nabla^2 =$

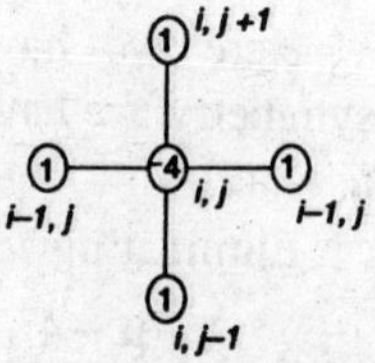

(1) $i, j+1$
(1) $i-1, j$ — (–4) i, j — (1) $i-1, j$
(1) $i, j-1$

Example 6. *The equation* $\frac{\partial^2 u}{\partial x^2} + \frac{\partial^2 u}{\partial y^2} + \lambda u = 0$ *is considered inside the region with corners (0, 0), (3, 0) (3,3) and (0,3), u = 0 on the boundary, use mesh sizes h = 1 and h = ¾ find the smallest value of λ.*

Sol. Case I. $h = 1$. We have

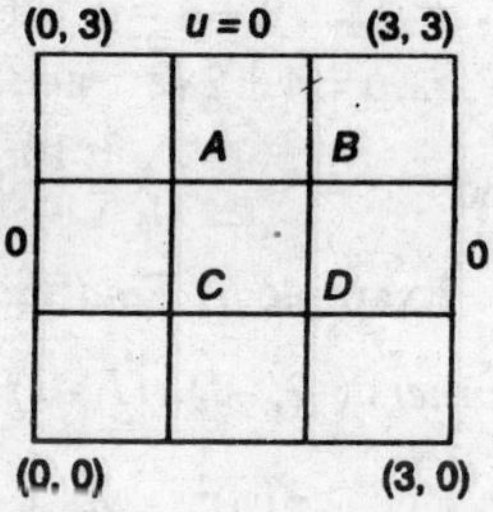

$$-4A + C + B + \lambda A = 0 \qquad ...(1)$$
$$-4B + A + D + \lambda B = 0 \qquad ...(2)$$
$$-4C + A + D + \lambda C = 0 \qquad ...(3)$$
$$-4D + C + B + \lambda D = 0 \qquad ...(4)$$

(0, 3) u = 0 (3, 3)
A B
0 C D 0
(0, 0) (3, 0)

From eq. (1) and (4),

$$-4A + C + B + \lambda A = -4D + C + B + \lambda D$$
$$\Rightarrow B + C = (4 - \lambda) A = 4(\quad \lambda) D$$
$$\Rightarrow A = D$$

From (2) and (3), $-A + D = (4 - \lambda) B = (4 - \lambda) B = (4 - \lambda) C$

$$\Rightarrow B = C$$

$$-4A + 2B + \lambda A = 0 \Rightarrow \frac{2B}{A} = (4 - \lambda)$$

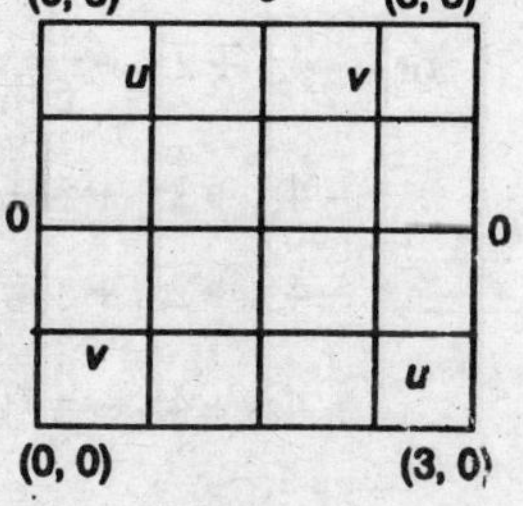

From (2), $-4B + 2A + \lambda B = 0$ $[\because D = A]$

$$(4 - \lambda) = \frac{2A}{B}$$

$$\therefore \frac{2A}{B} = \frac{2B}{A} \Rightarrow A = B = D = C$$

(0, 3) 0 (3, 3)
u v
0 0
v u
(0, 0) (3, 0)

$$-4v + 2v + \lambda v = 0$$

$$\lambda v = 2v \quad \Rightarrow \quad \lambda = 2,\ v \neq 0$$

Case II.

$$h = \frac{3}{4}, \text{ Now } \mu = \frac{9}{16}\lambda \quad (\because \mu = h^2\lambda) \quad \therefore \lambda = \frac{16}{9}\mu$$

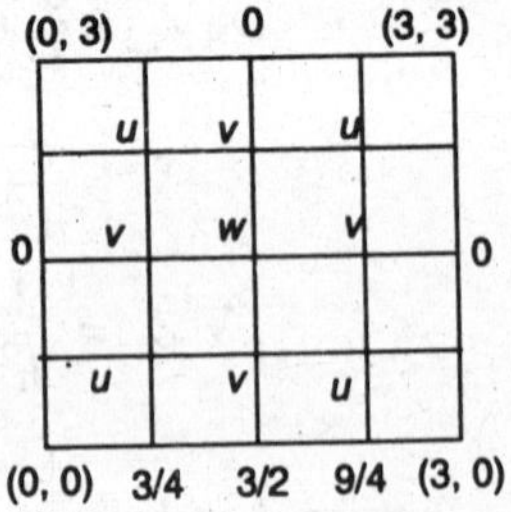

$$\left(\frac{\partial^2}{\partial x^2} + \frac{\partial^2}{\partial y^2}\right)u + h^2\lambda u = 0$$

or $\nabla^2 + h^2\lambda\, u = 0 \Rightarrow \nabla^2 + \mu u = 0$

i.e.

$$-4v + 2u + \mu v = 0$$

$$-4v + 2u + w + \mu v = 0$$

$$-4w + 4v + \mu w = 0$$

Here we have 9 pivotal points but by symmetry, we have only 3 which are devoted by u, v, w.

Eliminating u, v, w, we get

$$\begin{vmatrix} \mu - 4 & 2 & 0 \\ 2 & \mu - 4 & 1 \\ 0 & 4 & \mu - 4 \end{vmatrix} = 0$$

$$\Rightarrow \quad (\mu - 4)\,[\,(\mu - 4)^2 - 4] - 4\,[\,\mu - 4] = 0$$

$$(\mu - 4)\,[(\mu - 4)^2 - 8] = 0$$

$$\mu - 4 = 0\ , \quad (\mu - 4)^2 - 8 = 0.$$

$\therefore\ \mu = 4 \pm 2\sqrt{2}$ *i.e.*, pivotal point $= 4 - 2\sqrt{2}$ (smallest eigen value)

and $\lambda = \dfrac{16}{9}\mu = \dfrac{16}{4}(4 - 2\sqrt{2}).$

Example 7. *Find the value of Z at the pivotal points of square with corners (–1, –1), (1, –1), (1, 1), (–1, 1), z = 0 on the boundary and* $h = \frac{1}{2}$.

Sol. $\left(\dfrac{\partial^2}{\partial x^2} + \dfrac{\partial^2}{\partial y^2}\right) z = x^2 y^2.$

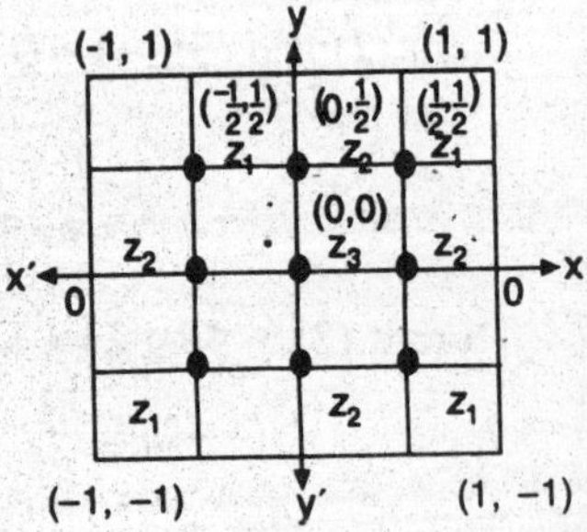

$$-4z_1 + 2z_2 - \frac{1}{4}\left(\frac{1}{4}\cdot\frac{1}{4}\right) = 0$$

or $-4z_1 + 2z_2 - \dfrac{1}{64} = 0$ ...(1)

$$-4z_2 + 2z_1 + z_3 - \frac{1}{4}(0) = 0$$

$-4z_2 + 2z_1 + z_3 = 0$ ...(2)

$$-4z_3 + 4z_2 - \frac{1}{4}(0) = 0$$

$$4z_3 = 4z_2 \quad \Rightarrow \quad z_3 = z_2 \qquad ...(3)$$

$$-3z_2 + 2z_1 = 0 \quad \Rightarrow \quad z_2 = \frac{2}{3}z_1$$

$$-4z_1 + 2 \times \frac{2}{3}z_1 - \frac{1}{64} = 0$$

$$-\frac{8}{3}z_1 = \frac{1}{64} \quad \Rightarrow \quad z_1 = \frac{-3}{512} = -\cdot 00586$$

$$\therefore \qquad z_2 = \frac{2}{3}z_1 = \frac{2}{3} \times \cdot 00586 = \cdot 00391 = z_3.$$

Example 8. *Find the smallest Eigen value for the equation*

$$\frac{\partial^2 u}{\partial x^2} + \frac{\partial^2 u}{\partial y^2} + \lambda u = 0 \qquad ...(1)$$

in a triangular domain which is bounded by the lines $x = 0$, $y = 0$, $x + y = 1$ *and it is known that* $u = 0$ *on the boundary. Use* $h = \frac{1}{4}$ *and* $\frac{1}{5}$.

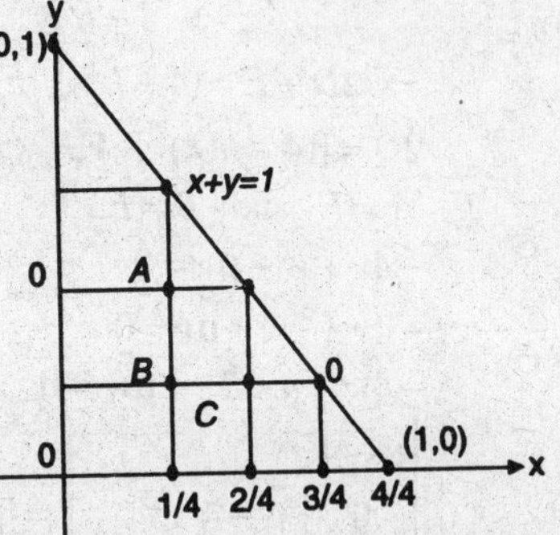

Sol. We have

$$-4A + B + h^2 \lambda A = 0 \qquad ...(1)$$

$$-4B + C + h^2\lambda B = 0 \qquad ...(2)$$

$$-4C + B + h^2\lambda C = 0 \qquad ...(3)$$

From (1) and (3),

$$B - (4 - h^2\lambda) A = (4 - h^2\lambda) C + B$$

$$\Rightarrow A = C$$

For $h = \frac{1}{4}$, At equation (1)

$$\mu = h^2\lambda = \frac{1}{16}\lambda \Rightarrow \lambda = 16\mu$$

$$-4v + u + \mu v = 0$$

$$-4v + 2u + \mu v = 0$$

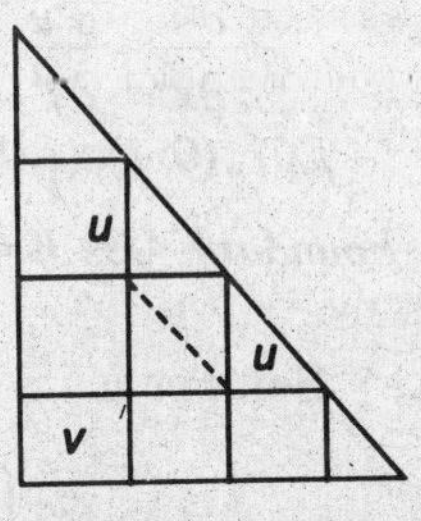

$$\begin{vmatrix} \mu - 4 & 1 \\ 2 & \mu - 4 \end{vmatrix} = 0 \Rightarrow (\mu - 4)^2 - 2 = 0$$

$\Rightarrow \mu = 4 \pm \sqrt{2} = 4 - 1{\cdot}414$, smallest eigen value.

$\therefore \lambda - 16\mu = 16 \times 2{\cdot}586 = 41{\cdot}376 \;\; = 2$ cdot 586

For $h = \frac{1}{5}$,

$$-4A + B + h^2\lambda A = 0 \qquad ...(1)$$

$$-4B + A + D + C + h^2\lambda B = 0 \qquad ...(2)$$

$$-4C + B + E + h^2 \lambda C = 0 \qquad ...(3)$$

$$-4D + B + E + h^2 \lambda D = 0 \qquad ...(4)$$

$$-4E + D + C + F + h^2 \lambda D = 0 \qquad ...(5)$$

$$-4F + E + h^2 \lambda F = 0 \qquad ...(6)$$

From eq. (3) and (4) $B + E = (4 - h^2\lambda)C$

$$= (4 - h^2\lambda)D \Rightarrow C = D$$

From eq. (1), $B = (4 - h^2\lambda) A$

From eq. (6), $E = (4 - h^2 \lambda)F$

From eq. (2), $\dfrac{2D + A}{4 - h^2\lambda} = B \quad \Rightarrow 2D + A = (4 - h^2\lambda)^2 A$

$$2D = [(4 - h^2 \lambda)^2 - 1\,]A$$

From eq. (5), $\dfrac{2D + F}{4 - h^2\lambda} = E$

$$\Rightarrow 2D + F = (4 - h^2\lambda)^2 F;$$

$$2D = [(4 - h^2\lambda)^2 - 1]F$$

$\therefore \quad A = F$ and $B = E.$

$$-4v + w + \mu v = 0$$

$$-4v + 2w + \mu v = 0$$

$$-4w + u + 2v + \mu w = 0$$

$$\begin{vmatrix} \mu - 4 & 0 & 1 \\ 0 & \mu - 4 & 2 \\ 1 & 2 & \mu - 4 \end{vmatrix} = 0 \qquad \Rightarrow (\mu - 4)\,[(\mu - 4)^2 - 4] - (\mu - 4) = 0$$

$$(\mu - u)\,[(\mu - u)^2 - 5] = 0$$

$$\mu = 4 \pm \sqrt{5}\,,\ 4\ 4 - \sqrt{5} = 4 - 2{\cdot}236 = 1{\cdot}764$$

$$\lambda = 25 \times 1{\cdot}764 = 44{\cdot}1$$

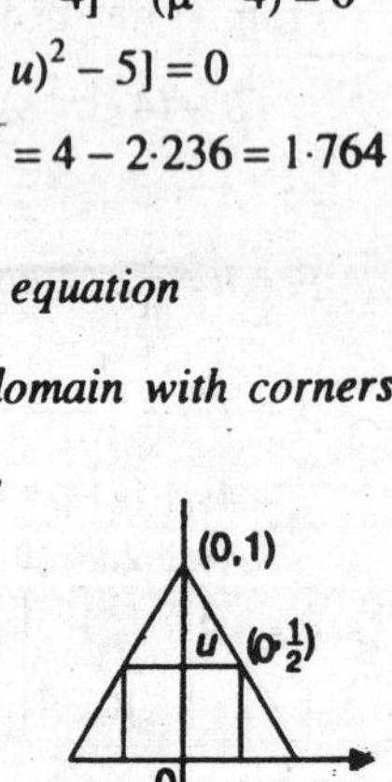

Example 9. *Find the smallest eigen value of the equation*

$\dfrac{\partial^2 u}{\partial x^2} + \dfrac{\partial^2 u}{\partial y^2} + \lambda(x^2 + y^2)\, u = 0$ *for a triangular domain with corners* $(-1,0)$, $(0, 1)$, $(1,0)$ *where we have* $u = 0$ *on the boundary. Use* $h = \dfrac{1}{2}$ *and* $h = \dfrac{1}{3}$.

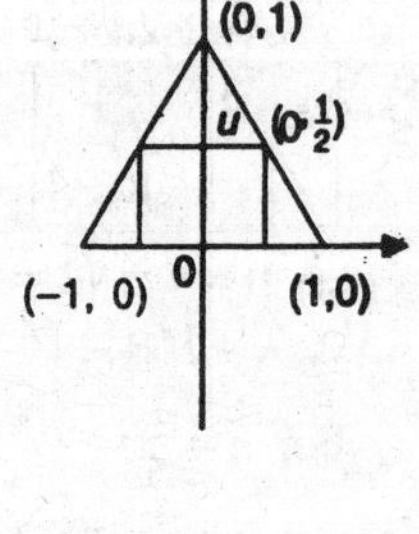

Sol. For $h = \dfrac{1}{2}$

$$-4u + h^2\lambda\left(0 + \frac{1}{4}\right)u = 0,$$

$$\Rightarrow \qquad -4u + \frac{1}{4}\lambda\left(\frac{1}{4}\right)u = 0$$

$u\,(64 - \lambda) = 0 \Rightarrow \lambda = 64 =$ smallest eigen value, $u \neq 0$

For $h = \frac{1}{3}$,

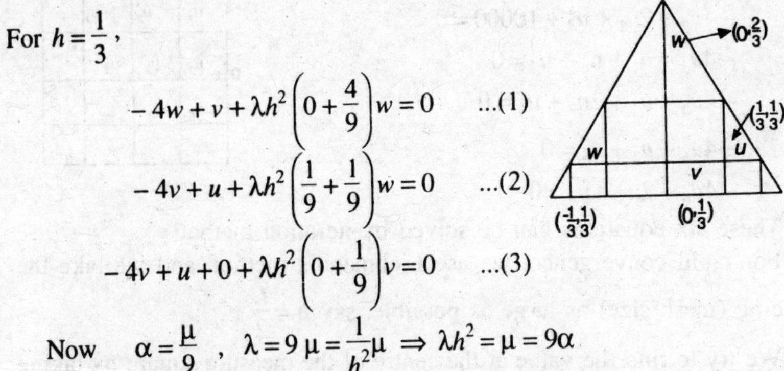

$$-4w + v + \lambda h^2 \left(0 + \frac{4}{9}\right) w = 0 \quad \text{...(1)}$$

$$-4v + u + \lambda h^2 \left(\frac{1}{9} + \frac{1}{9}\right) w = 0 \quad \text{...(2)}$$

$$-4v + u + 0 + \lambda h^2 \left(0 + \frac{1}{9}\right) v = 0 \quad \text{...(3)}$$

Now $\quad \alpha = \frac{\mu}{9}$, $\lambda = 9\,\mu = \frac{1}{h^2}\mu \Rightarrow \lambda h^2 = \mu = 9\alpha$

Eliminating w, v and u from (1), (2) and (3), we get

$$\begin{vmatrix} 2\alpha - 4 & 1 & 0 \\ 2 & \alpha - 4 & 1 \\ 0 & 1 & -4 + 4\alpha \end{vmatrix} = 0$$

$$(2\alpha - 4)\,[(\alpha - 4)(4\alpha - 4) - 1] - 1[2(4\alpha - 4)] = 0$$

$$(2\alpha - 4)\,[4\alpha^2 - 4\alpha - 16\alpha + 16 - 1] - 8\alpha + 8 = 0$$

$$(2\alpha - 4)\,[4\alpha^2 - 20\alpha + 15] - 8\alpha + 8 = 0$$

$$(\alpha - 2)\,(4\alpha^2 - 20\alpha + 15) - 4\alpha + 4 = 0$$

$$f(\alpha) = 4\alpha^3 - 28\alpha^2 + 51\alpha - 26 = 0.$$

At $\alpha = 0$, $f(\alpha)$ is –ve.

and at $\alpha = 1$, $f(\alpha)$ is +ve

Hence the root α lies between 0 and 1. We can prove that $\alpha = \cdot 89$.

$$\begin{array}{r|rrrrl} & 4 & -28 & 51 & -26 & \\ & & 3{\cdot}2 & -19{\cdot}84 & 24{\cdot}928 & \\ \cdot 8 & 4 & -24{\cdot}8 & 31{\cdot}16 & -1{\cdot}07 & f(\cdot 8) \\ & & 3{\cdot}2 & -17{\cdot}28 & - & \\ & 4 & -21{\cdot}6 & 13{\cdot}88 & - & f'(\cdot 8) \end{array}$$

18·4. Liebmann's method. Laplace equation $\nabla^2 u = \frac{\partial^2 u}{\partial x^2} + \frac{\partial^2 u}{\partial y^2} = 0$ with appropriate conditions can be solved with the method of iterations starting with any tentative value of u at the pivotal points and averaging successfully the values at the four adjoining points or corners. Thus improved values of u are obtained. For solving Laplace equation by numerical method, let us consider the problem of steady state temperature. Pivotal at points of a square grid *i.e., Lattice* of mesh size $h = \frac{l}{4}$. where l is the size of the square.

5000	10^4	10^4	10^4	5000
	u_1	u_2	u_1	
0	u_3	u_4	u_3	
	u_5	u_6	u_5	

$$-4u_1 + u_2 + u_3 + 10000 = 0$$
$$-4u_2 + 2u_1 + u_4 + 10000 = 0$$
$$-4u_3 + u_1 + u_4 + u_5 = 0$$
$$-4u_4 + u_2 + 2u_3 + u_6 = 0$$
$$-4u_5 + u_3 + u_6 = 0$$
$$-4u_6 + 2u_5 + u_4 = 0$$

These six equations can be solved by iteration method.

For rapid convergence, we use Liebmann's method and we take the spacing (mesh size) as large as possible, say $h = \frac{L}{2}$.

We try to find the value at the centre of the measure square by taking the average of four adjoining corners. $4u_4 = 10000$. *i.e.* $u_4 = \frac{1}{4}$ [10000] = 2500. We proceed to find the initial values at the centres of the small squares by averaging the values diagonally adjoining. ∇^2 is invariant.

$$\overset{0}{u_1} = \frac{1}{4}\ (5000 + 10000 + 2500]$$
$$= \frac{1}{4}(17500) = 4375$$
$$\overset{0}{u_5} = \frac{1}{4}\ [2500] = 625$$

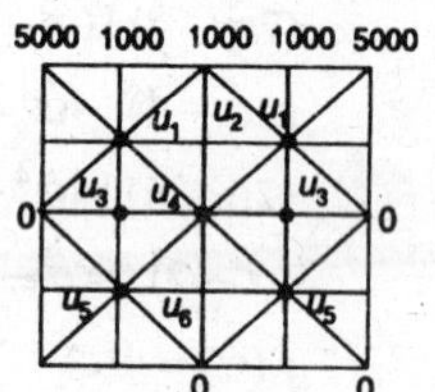

And the remaining values are obtained by averaging the values at four adjoining points.

$$\overset{0}{u_2} = \frac{1}{4}\ (4375 + 4375 + 10000 + 2500]$$
$$= \frac{1}{4}\ [21250] = 5312$$
$$\overset{0}{u_3} = \frac{1}{4}\ [4375 + 2500 + 625 + 0] = \frac{1}{4}\ [7500] = 1875$$
$$\overset{0}{u_6} = \frac{1}{4}\ [2500 + 2 \times 625 + 0] = \frac{1}{4}\ (3750) = 938$$
$$\overset{(1)}{u_1} = \frac{1}{4}[10000 + u_2^0 + u_3^0 + 0] = \frac{1}{4}\ [(10000 + 5312 + 1875]$$
$$= \frac{1}{4}(17187) = 4297$$
$$\overset{(1)}{u_2} = \frac{1}{4}\ [10000 + 2u_1^{(1)} + u_4^0] = \frac{1}{4}\ [10000 + 2 \times 4297 + 2500]$$
$$= \frac{1}{4}(92194) = 5274$$
$$\overset{(1)}{u_3} = \frac{1}{4}\left[u_5^0 + u_4^0 + u_1^{(1)}\right] = \frac{1}{4}(2500 + 625 + 4297) = \frac{1}{4}(7432) = 1856$$

$$u_4^0 = \frac{1}{4}[5274 + 1856 + 1856 + 938] = \frac{1}{4}[9924] = 2481$$

$$u_5^{(1)} = \frac{1}{4}[1856 + 938] = \frac{2794}{4} = 699$$

$$u_6^1 = \frac{1}{4}[2481 + 699 + 699] = 970.$$

Example 10. *Evaluate z at pivotal points of a square lattice of sides A = 4h = 2 and h =* $\frac{1}{2}$, $\nabla^2 z = x^2y^2$, $z = 0$ on the boundary.

Sol. Assume the origin at the centre of the square, $z_1 = \left(-\frac{1}{2}, \frac{1}{2}\right)$ is same as $z_2 = \left(\frac{1}{2}, \frac{1}{2}\right)$

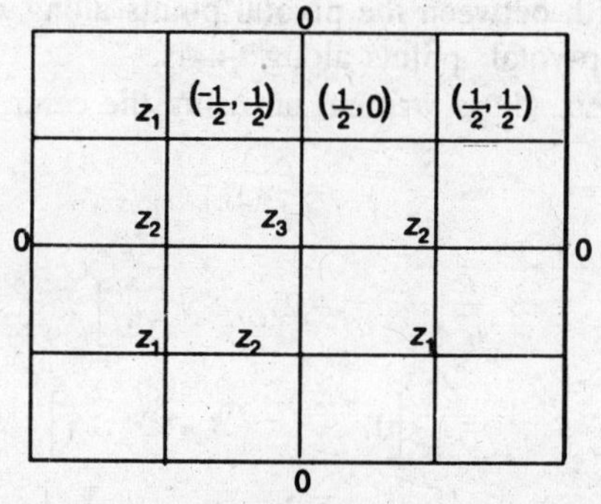

Also $\quad -4z_3^0 = 0$

$\Rightarrow \quad z_3^0 \equiv 0$

Also $-4z_1^0 + z_3^0 - h^2x^2y^2 = 0$

$\Rightarrow \quad -4z_1^0 + z_3^0 - \frac{1}{64} = 0,$

$$h = \frac{1}{2},\ x^2y^2 = \frac{1}{16}$$

$\Rightarrow \quad z_1^0 = -\cdot 0038$

Also $-4z_2^0 + 2z_1^0 + z_3^0 = 0$

$\Rightarrow \quad z_2^0 = \frac{1}{4}(2z_1^0) = -\cdot 0019$

Now $z_1^0 = \cdot 0038,$

$z_2^0 = -\cdot 0019,$

$z_3^0 = 0.$

Again, $-4z_1^1 + 2z_2^0 - \frac{1}{64} = 0$

$\Rightarrow \quad z_1^1 = \frac{1}{2}z_0^0 - \frac{1}{256} = -\cdot 0010 - \cdot 0038 = -\cdot 0048$

Also $\quad -4z_2^1 + 2z_1^1 + z_3^0 = 0$

$z_2^1 = -\cdot 0024,\ -4z_3^1 + 4z_2^1 = 0 \quad \Rightarrow \quad z_3^1 = z_2^1$

$z_1^2 = -\cdot 0052,$

$z_2^2 = -\cdot 0033,$

$z_1^3 = -\cdot 005,$

$z_2^3 = -\cdot 0035.$

Example 11. *The differential equation satisfying the deflextion $y\,(x\,,t)$ at a point x on the string at time t is*

$$a^2\frac{\partial^2 y}{\partial x^2}=\frac{\partial^2 y}{\partial t^2} \qquad ...(1)$$

and $y(x\,,t)$ *satisfies the boundary* conditions

$$y(0\,,t)=0$$
$$y\,(L\,,t)=0$$

and initial conditions $y\,(x\,,0)=f\,(x)$

$$y_t\,(x,0)=0.$$

Sol. Let $y_{i,j}$ represent the value of deflection $y\,(x,t)$ at the point $x=x_i\,,t=t_j$, consider a rectangular domain in the xt-plane with h as space length between the pivotal points along x-axis & k is the space between the pivotal points along t-axis.

Eq. (1) is written, on using the central differences, as

$$a^2\frac{\delta^2}{h^2}\cdot y_{i,j}=\frac{\delta^2}{k^2}\,y_{i,j}$$

$$\Rightarrow \frac{a^2}{h^2}\left[y_{i+1,j}-2y_{i,j}+y_{i-1,j}\right]$$

$$=\frac{1}{k^2}\left[y_{i,j+1}-2y_{i,j}+y_{i,j-1}\right]$$

$$\Rightarrow \alpha^2\left[y_{i+1,j}-2y_{i,j}+y_{i-1,j}\right]=\left[y_{i,j+1}-2y_{i,j}+y_{i,j-1}\right],$$

where $\quad \alpha^2=a^2\,k^2/h^2.$

If we know deflection at times j and $j-1$ then we can know the deflection at $j+1$.

The deflection $y(x,t)$ at the point $x=x_i$ and time $t=t_{j+1}$ is given by recurrence formula.

$$y_{i,j+1}=\alpha^2\left[y_{i+1,j}-y_{i-1,j}\right]+2(1-\alpha^2)\,y_{i,j}-y_{i,j-1}$$

in terms of deflection at the points x_{i-1}, x_i, x_{i+1} and times $t_{j-1}\,,t_j$

$$y_{i,j+1}=\alpha^2(y_{i+1,j}+y_{i-1,j})+2\,(1-\alpha^2)\,y_{i,j}-y_{i,j-1}$$

Deflection $y(x,t)$

$$x=x_i\quad,\quad t=t_{j+1}$$

in terms of the expression

$$x=x_i\,,x_{i-1},x_{i+1} \text{ and } t=t_i \text{ and } t_{j-1}$$

when $\alpha=1$

$$y_{i,j+1}=(y_{i+1,j}+y_{i-1,j})-y_{i,j-1} \qquad ....\textbf{(A)}$$

and conditions are written in the form

$y_{0,j}=0$ ....(D) where h is the spacing between the pivotal points along

$x-$ axis, $\quad h=\dfrac{L}{N}, \quad y_{N,j}=0$ ...(B)

$$y_{i,0}=f(x_i)$$

and $\quad y_t(x,0)=0$

$$\frac{1}{2k}\left[y_{i,j+1}-y_{i,j-1}\right]=0$$

when $\quad j=0, \quad y_{i,1}=y_{i,-1}$

$$y_{i,1}=\frac{1}{2}\left[y_{i+1,0}+y_{i-1,0}\right] \quad \text{...(C)}$$

From (A) putting $y=0$, eq. (A) with the starting formula (B) & (C).

Boundary conditions (D) & (E), determine the $y_{i,j}$. For knowing $y_{i,j}$ at all pivotal points and then steps $j=0$, $j=1$, $y_{i,2}$ is computed from (A) $y_{0,2}$ and $y_{N,i}$ are given by (D) & (E).

Thus, in general, starting from the time $t=t_j$ with known values of $y_{i,j}$, and $y_{i,j-1}$ the deflection $y_{i,j+1}$ is computed from the formula (A).

Example 12. *Solve the parabolic equation*

$$K\frac{\partial^2 u}{\partial x^2}=\frac{\partial u}{\partial t}, \quad 0<x<t, \quad t>0 \quad \text{...(1)}$$

with boundary conditions

$$u_x(0,t)=0 \quad \text{...(2)}$$

$$u(L,t)=a$$

and initial condition $u(x,0)=u_0$. ...(3)

Sol. To solve eq. (1) we try to reduce it in simple form by introducing a new variable

$$\bar{x}=\frac{x}{L}.$$

Now $\quad \dfrac{\partial u}{\partial x}=\dfrac{\partial u}{\partial \bar{x}}\cdot\dfrac{\partial \bar{x}}{\partial x}=\dfrac{1}{L}\dfrac{\partial u}{\partial \bar{x}}$

$$\frac{\partial^2 u}{\partial x^2}=\frac{\partial}{\partial x}\left(\frac{1}{L}\frac{\partial u}{\partial \bar{x}}\right)\frac{1}{L}=\frac{1}{L^2}\frac{\partial^2 u}{\partial \bar{x}^2}$$

$$\bar{t}=\frac{K}{L^2}\cdot t$$

$\therefore$ (1) reduces to

$$\frac{\partial^2 u}{\partial \bar{x}^2}=\frac{\partial u}{\partial t}\cdot\frac{K}{L^2}$$

$$\frac{\partial^2 u}{\partial \bar{x}^2}=\frac{\partial u}{\partial \bar{t}} \quad \text{... (4)} \qquad \text{Letting } \bar{u}=\frac{u-a}{u_0-a}$$

$$\frac{\partial \bar{u}}{\partial \bar{x}} = \frac{\partial \bar{u}}{\partial u} \cdot \frac{\partial u}{\partial \bar{x}} = \frac{1}{u_0 - a} \frac{\partial u}{\partial \bar{x}}$$

$$\frac{\partial^2 u}{\partial \bar{x}^2} = \frac{1}{u_0 - a} \cdot \frac{\partial^2 u}{\partial \bar{x}^2}$$

$$\frac{\partial \bar{u}}{\partial t} = \frac{\partial \bar{u}}{\partial u} \cdot \frac{\partial u}{\partial t} = \frac{1}{u_0 - a} \frac{\partial u}{\partial t}$$

Rewriting eq. (4), we have

$$(u_0 - a) \frac{\partial^2 \bar{u}}{\partial \bar{x}^2} = (u_0 - a) \frac{\partial \bar{u}}{\partial t} \text{ and}$$

conditions are $\bar{u}_x (0, \bar{t}) = 0,\ \bar{u}(1, \bar{t}) = 0,\ \bar{u}(\bar{x}, 0) = 1$

For $\bar{u}_x (0, t) = \dfrac{u_0 - a}{L} \bar{u}_{\bar{x}} (0, \bar{t})$

since $\dfrac{\partial u}{\partial x} = \dfrac{u_0 - a}{L} \dfrac{\partial \bar{u}}{\partial \bar{x}},\quad \bar{u}_{\bar{x}} (0, t) = 0$

Now when $x = L$, we have $\bar{x} = \dfrac{x}{L} = 1.$

Hence $\bar{u}(1, \bar{t}) = \dfrac{u(L, t) - a}{u_0 - a}$

but $u(L, t) = a \quad \therefore \quad \bar{u}(1, \bar{t}) = 0$

$\bar{u}(\bar{x}, 0) = \dfrac{u(\bar{x}, 0) - a}{u_0 - a}$

but $u(x, 0) = u_0 \quad \therefore \quad \bar{u}(\bar{x}, 0) = 1$

Thus $\bar{u}_{\bar{x}} (0, \bar{t}) = 0,\ \bar{u}(1, \bar{t}) = 0$ and $\bar{u}(\bar{x}, 0) = 1$

The given differential equation along with the boundary conditions (*b.c.*) and initial condition reduces to, after dropings bars,

$\dfrac{\partial^2 u}{\partial x^2} = \dfrac{\partial u}{\partial t}$ with boundry conditions.

$$u_x (0, t) = 0$$
$$u(1, t) = 0$$
$$u(x, 0) = 1$$

Using central difference for the 2nd derivative of the temperature for u & forward difference for the 1st derivative w.r.t. t.

Taking h along x-axis, k along t-axis.

$$\frac{\delta^2}{h^2} \left[u_{i,j} \right] = \frac{1}{k} \Delta u_{i,j}$$

where $u_{i,j}$ is the value of the temperature for u at the point $x = x_i$ and time $t = t_j$.

$$\frac{k}{h^2}\left[u_{i,j} - 2u_{i,j} + u_{i-1,j}\right] = u_{i,j+1} - u_{i,j}$$

or $$u_{i,j+1} = \alpha\left[u_{i+1,j} + u_{i-1,j}\right] + (1-2\alpha)\,u_{i,j} \qquad \text{...(A)}$$

which is the recurence formula for determining temperature for u at $x = x_i$ and time $t = t_{j+1}$ in terms of the value substituted at the points x_{i-1}, x_i, x_{i+1} and $t = t_j$

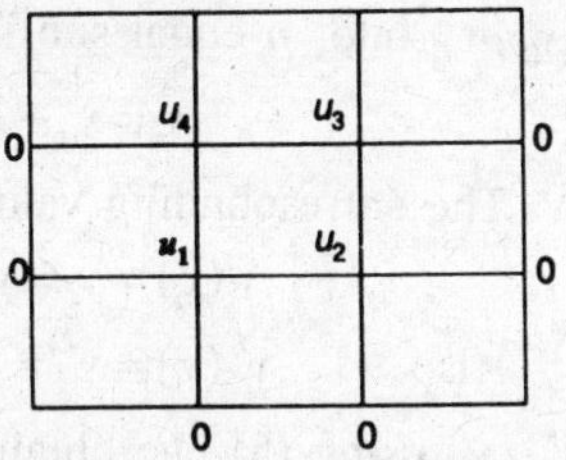

For particular case if $\alpha = \dfrac{1}{2}$

$$u_{i,j+1} = \frac{1}{2}(u_{i+1,j} + u_{i-1,j})$$

and the condition $u_x(0, t) = 0$ gives

$$\frac{\mu\delta}{h}(u_{i,j}) = \frac{1}{2h}(u_{i+1,j} - u_{i-1,j}) = 0 \text{ at } i = 0$$

$$\Rightarrow \quad u_{1,j} = u_{-1,j}$$

Formula (A) gives the value at $x = x_i$ and $t = t_{i+1}$ by taking the average of the values of u at $x = x_{i-1}$ and $x = x_{i+1}$

Replacement of Derivatives by Finite-Difference Approximations

Let the *x-y* plane be divided into rectangular regions with sides $x = ih$, $i = 0, 1, 2, \ldots$ and $y = jk$, $j = 0, 1, 2, \ldots$ where $\Delta x = h$ and $\Delta y = k$.

The points of intersection of the family of lines $x = i\,h$, $y = jk$ are called *mesh* points, *lattice* points or *grid* points.

Expanding $y(x+h)$ in Taylor's series, we have

$$y(x+h) = y(x) + hy'(x) + \frac{h^2}{2}y''(x) + \frac{h^3}{6}y'''(x) + \cdots \qquad \text{...(1)}$$

$$\therefore \quad y'(x) = \frac{y(x+h) - y(x)}{h} - \frac{h}{2}y''(x)\ldots.$$

$$\Rightarrow \quad y'(x) = \frac{y(x+h) - y(x)}{h} + 0(h), \text{ 0 being the order} \qquad \text{...(2)}$$

Similarly, we obtain

$$y(x-h) = y(x) - hy'(x) + \frac{h^2}{2}y''(x) - \frac{h^3}{6}y'''(x) + \cdots \qquad \text{...(3)}$$

$$\Rightarrow \quad y'(x) = \frac{y(x) - y(x-h)}{h} + 0(h) \qquad \text{...(4)}$$

Adding (2), and (3) and then dividing by 2, we find the better approximation

$$y'(x) = \frac{y(x+h) - y(x-h)}{2h} + 0(h^2) \qquad \text{...(5)}$$

Again, adding (1) and (3) we obtain

$$y''(x) = \frac{y(x-h) - 2y(x) + y(x+h)}{h^2} + 0(h^2) \qquad \text{...(6)}$$

Thus (5) and (6) give us the replaced values of $y'(x)$ and $y''(x)$ respectively in terms of finite–difference approximations. In case of boundary value problem with boundary conditions, we divide the range $[x_0, x_n]$ into n equal sub-intervals, each of width h so that

$$x_i = x_0 + ih, \ i = 1, 2, ..., n.$$

The corresponding values of y at this points are given by

$$y(x_i) = y_i = y(x_0 + ih), \ i = 1$$

Also $\qquad y'(x_i) = y'_i = y'(x_0 + ih)$

$\therefore$ using (5), we obtain

$$y_i' = \frac{y_{i+1} - y_{i-1}}{2h} + 0(h^2),$$

$$y_i'' = \frac{y_{i-1} - 2y_i + y_{i+1}}{h^2} + 0(h^2) \quad \text{, etc.}$$

Thus using (5) and (6) we have

$$u_x = \frac{u_{i+1,j} - u_{i-1,j}}{2h} + 0(h^2)$$

$$u_{xx} = \frac{u_{i-1,j} - 2u_{i,j} + u_{i+1,j}}{h^2} + 0(h^2)$$

where $\qquad u_{i,j} = u(ih, jk) = u(x, y)$

Similarly, $\qquad u_y = \dfrac{u_{i,j+1} - u_{i,j-1}}{2k} + 0(k^2)$

$$u_{yy} = \frac{u_{i,j-1} - 2u_{i,j} + u_{i,j+1}}{k^2} + 0(k^2)$$

Thus with the help of these results, we obtain the finite difference analogus of partial differential equations.

Consider Laplace equation in two dimensions *i.e.*, elliptic equation

$$u_{xx} + u_{yy} = 0$$

Converting with finite-difference analogue, we have

$$\frac{1}{h^2}\left[u_{i+1,j} - 2u_{i,j} + u_{i-1,j}\right] + \frac{1}{k^2}\left[u_{i,j+1} - 2u_{i,j} + u_{i,j-1}\right] = 0$$

For a square mesh, we put $k = h$, so that

$$u_{i,j} = \frac{1}{4}\left[u_{i+1,j} + u_{i-1,j} + u_{i,j+1} + u_{i,j-1}\right]$$

This formula is known as *standard* five-point formula as it uses the values of u at any point and its four neighbouring points (see Fig. 1. (a))

Sometimes, we use the following formula obtained by expanding the terms on the R.H.S by Taylor's, series:

$$u_{i,j} = \frac{1}{4}\left[u_{i-1,j-1} + u_{i+1,j-1} + u_{i+1,j+1} + u_{i-1,j+1}\right]$$

This result is known as *diagonal* five-point formula as it uses the values of u at the diagonal points (see fig. 1. (b)

Notes 1. The error in the diagonal formula is four times that of the standard formula.

2. The following result can be established.

$$u_{xy} = \frac{1}{4h^2}\begin{pmatrix}-1 & 1\\ 1 & -1\end{pmatrix} u_{i,j} = \frac{1}{4h^2}\left[u_{i-1,j} - u_{i+1,j+1} - u_{i-1,j+1} + u_{i,j-1}\right]$$

18·5. Laplace's Equation

Let us solve the equation

$$\nabla^2 = \frac{\partial^2 u}{\partial x^2} + \frac{\partial^2 u}{\partial y^2} = u_{xx} + u_{yy} = 0$$

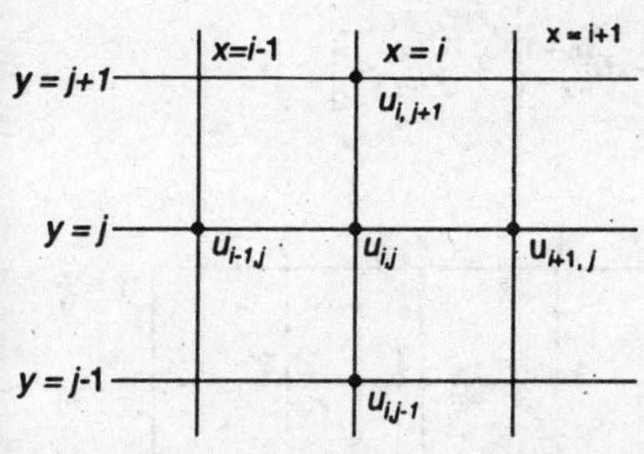

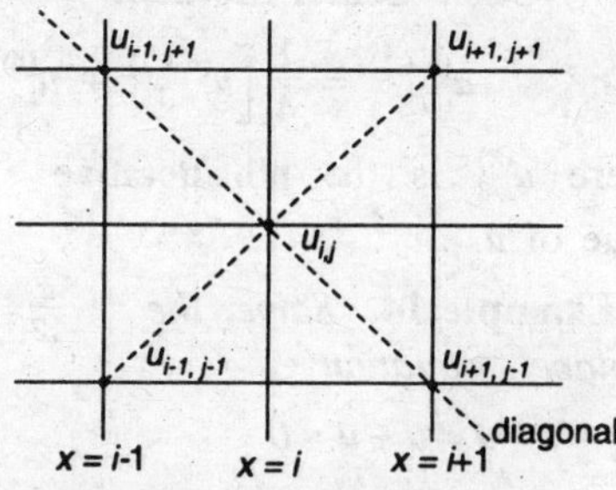

in a square region R with boundary C and divided into a network of small squares with side h. Let C_i be the values of $u\,(x,y)$ on the boundary C and the interior mesh points are shown in the following figure:

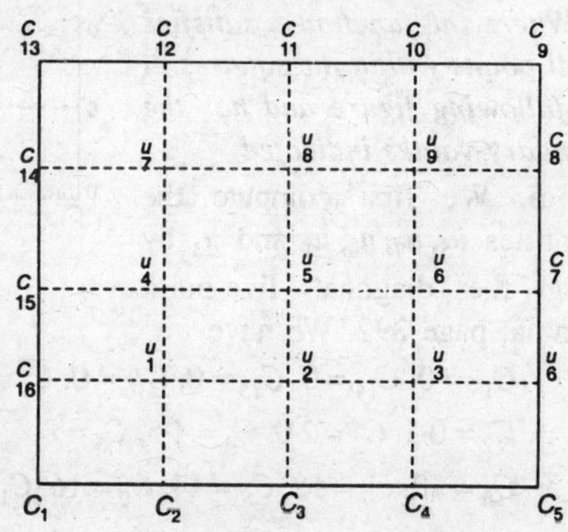

The approximate values of the function $u(x, y)$ at the interior mesh points are computed by using the diagonal two-point formula in the order u_5, u_7, u_9, u_1 and u_3.

They are

$$u_5 = \frac{1}{4}\left[C_1 + C_5 + C_9 + C_{13}\right]; \quad u_7 = \frac{1}{4}\left[C_{15} + u_5 + C_{11} + C_{13}\right]$$

$$u_9 = \frac{1}{4}\left[u_5 + C_7 + C_9 + C_{11}\right]; \quad u_1 = \frac{1}{4}\left[C_1 + C_3 + u_5 + C_{15}\right]$$

$$u_3 = \frac{1}{4}\left[C_3 + C_5 + C_7 + u_5\right].$$

The remaining quantities are then computed in the order u_8, u_4, u_6 and u_2 by the standard five-point formula.

Thus $u_8 = \frac{1}{4}\left[u_5 + u_9 + c_{11} + u_7\right]$; $\quad u_4 = \frac{1}{4}\left[u_1 + u_5 + u_7 + C_{15}\right]$,

$$u_6 = \frac{1}{4}\left[u_3 + C_7 + u_9 + u_5\right]; \quad u_2 = \frac{1}{4}\left[C_3 + u_3 + u_5 + u_1\right]$$

When all the nine u's have been computed, their values can be improved by

(*i*) **Jacobi's Method.**

$$u_{i,j}^{(n+1)} = \frac{1}{4}\left[u_{i-1,j}^{(n)} + u_{i+1,j}^{(n)} + u_{i,j-1}^{(n)} + u_{i,j+1}^{(n)}\right]$$

(*ii*) **Gauss-Seidel Method.**

$$u_{i,j}^{(n+1)} = \frac{1}{4}\left[u_{i-1,j}^{(n+1)} + u_{i+1,j}^{(n)} + u_{i,j-1}^{(n+1)} + u_{i,j+1}^{(n)}\right]$$

where $u_{i,1}^{(n)}$ is the nth iterative value of $u_{i,j}$.

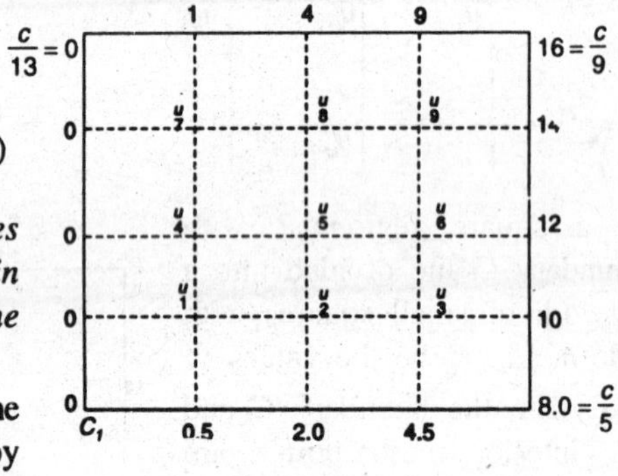

Example 14. *Solve the Laplace's equation*

$$\nabla^2 u = u_{xx} + u_{yy} = 0 \qquad \text{...(1)}$$

Where the function u satisfies at all points within the squares in the following figure and has the boundary values indicated,

Sol. We first compute the quantities u_5, u_7, u_9, u_1 and u_3 by using the diagonal five-point formula, page 392. We have

$C_{13} = 0, C_{14} = 0, C_{15} = 0, C_{16} = 0, C_1 = 0,$

$C_2 = 0{\cdot}5, C_3 = 2{\cdot}0, C_4 = 4{\cdot}5, C_5 = 8,$

$C_6 = 10, C_7 = 12, C_8 = 14, C_9 = 16, C_{10} = 9, C_{11} = 4, C_{12} = 1.$

Now $u_5 = \frac{1}{4}\left[C_1 + C_5 + C_9 + C_{13}\right] = \frac{1}{4}[0 + 8 + 16 + 0] = 6;$

$$u_7 = \frac{1}{4}\left[C_{15} + u_5 + C_{11} + C_{13}\right] = \frac{1}{4}[0 + 6 + 4 + 0] = 2{\cdot}5$$

$$u_9 = \frac{1}{4}\left[u_5 + C_7 + C_9 + C_{11}\right] = \frac{1}{4}[6 + 12 + 16 + 4] = 9{\cdot}5.$$

$$u_1 = \frac{1}{4}\left[C_1 + C_3 + u_5 + C_{15}\right] = \frac{1}{4}[0 + 2 + 6 + 0] = 2.$$

$$u_3 = \frac{1}{4}\left[C_3 + C_5 + C_7 + u_5\right] = \frac{1}{4}[2 + 8 + 12 + 6] = 7.$$

We now compute in order the quantities u_8, u_4, u_6 and u_2 by standard five-point formula, page 391.

Thus $$u_8 = \frac{1}{4}\left[u_5 + u_9 + c_{11} + u_7\right] = \frac{1}{4}[6 + 9{\cdot}5 + 4 + 2{\cdot}5] = 5{\cdot}5,$$

$$u_4 = \frac{1}{4}\left[u_1 + u_5 + u_7 + C_{15}\right] = \frac{1}{4}[2 + 6 + 2{\cdot}5 + 0] = 2{\cdot}625$$

$$u_6 = \frac{1}{4}\left[u_3 + C_7 + u_9 + u_5\right] = \frac{1}{4}[7 + 12 + 9{\cdot}5 + 6] = 8{\cdot}625,$$

$$u_2 = \frac{1}{4}[C_3 + u_3 + u_5 + u_7] = \frac{1}{4}[2 + 7 + 6 + 2] = 4{\cdot}25$$

We now compute the following quantities by Gauss-Seidel method

$$u_{i,j}^{(n+1)} = \frac{1}{4}\left[u_{i-1,j}^{(n+1)} + u_{i+1,j}^{(n)} + u_{i,j-1}^{(n)} + u_{i,j+1}^{(n)}\right]$$

Here $$u_1 = u_{(1)}^{(2)} = \frac{1}{4}\left[u_{0,1}^{(2)} + u_{2,1}^{(1)} + u_{1,0}^{(2)} + u_{1,2}^{(1)}\right]$$

$$= \frac{1}{4}\left[C_{16}^{(2)} + u_2^{(0)} + C_3^1 + u_4^{(0,1)}\right] = \frac{1}{4}[0 + 9{\cdot}38 + 0 + 18{\cdot}75]$$

$$= \frac{1}{4}(28{\cdot}13) = 7{\cdot}03$$

$$u_2 = u_{2,1} = \frac{1}{4}\left[u_{1,1}^{(2)} + u_{3,1}^{(1)} + u_{2,0}^{(2)} + u_{2,2}^{(1)}\right]$$

$$= \frac{1}{4}[7{\cdot}03 + 6{\cdot}25 + 0 + 25] = \frac{1}{4}(38{\cdot}28) = 9{\cdot}57$$

$$u_3 = u_{3,1} = \frac{1}{4}\left[u_{2,1}^{(2)} + u_{4,1}^{(1)} + u_{3,0}^{(2)} + u_{3,2}^{(1)}\right]$$

$$= \frac{1}{4}\left[u_2^{(2)} + C_6 + C_4 + u_0^{(1)}\right] = \frac{1}{4}[9{\cdot}57 + 0 + 18{\cdot}35]$$

$$= \frac{1}{4}(28{\cdot}32) = 7{\cdot}08$$

18·6. Consider the Parabolic equation *viz.*, heat conduction equation

$$\frac{\partial u}{\partial t} = \frac{\partial^2 u}{\partial x^2} \qquad ...(1)$$

Let the $x - t$ plane be divided into smaller rectangles by drawing straight lines

$$x = ih,\ i = 0, 1, 2, \ldots\ldots,\ t = jk,\ j = 0, 1, 2, \ldots\ldots$$

Replace the partial derivatives by their finite-difference **analogue** *viz.*,

$$\frac{\partial u}{\partial t} = \frac{u_{i,j+1} - u_{i,j}}{k}$$

and
$$\frac{\partial^2 u}{\partial x^2} = \frac{u_{i-1,j} - 2u_{i,j} + u_{i+1,j}}{h^2}$$

so that the equation (1) reduces to

$$\Rightarrow \quad \frac{1}{k}\left[u_{i,j+1} - u_{i,j}\right] = \frac{1}{h^2}\left[u_{i-1,j} - 2u_{i,j} + u_{i+1,j}\right]$$

where $u_{i,j+1} = u_{i,j} + r\left[u_{i-1,j} - 2u_{i,j} + u_{i+1,j}\right]$ with $r = k/h^2$ and this formula is valid only when $0 < r \le \frac{1}{2}$.

Also in order to determine the values of $u_{i,j+1}$, we need the values of $u(x, y)$ along the jth row only and is, therefore called the explicit formula.

Example 15. *Given the equation* $\frac{\partial u}{\partial t} = \frac{\partial^2 u}{\partial x^2}$ *and the boundary conditions* $u(0, t) = u(5, t) = 0$

and $u(x, 0) = x^2 (25 - x^2)$,

use the explicit method to obtain the solution for

$$x = x_i = ih,\ (i = 0, 1, \ldots.., 5;\ h = 1) \text{ and } t = jk\left(j = 0, 1, 2, \ldots.., 5,\ k = \frac{1}{2}\right)$$

Sol. We compute a few quantities u_{11}, u_{02}.....and then give below the table of the values of $u_{i,j}$ for different values of i and j and $h = 1, k = \frac{1}{2}$.

Here $h = 1, k = \frac{1}{2}$ $\therefore$ $r = \frac{k}{h^2} = \frac{1}{2}$ and $\therefore$ explicit formula reduces to

$$u_{i,j+1} = \frac{1}{2}\left(u_{i-1,j} + u_{i+1,j}\right)$$

Also $u(0, 0) = 0$, $u(1,0) = 24$, $u(2,0) = 2^2(25 - 2^2) = 84$,

$u(3, 0) = 3^2(25 - 3^2) = 144$, $u(4, 0) = 4^2(25 - 4^2) = 144$, $u(5, 0) = 0$

$$u_{1,4} = \frac{1}{2}(u_{0,3} + u_{2,3}) = \frac{1}{2}(0 + 53{\cdot}25) = 26{\cdot}625$$

$$u_{2,3} = \frac{1}{2}(u_{1,2} + u_{3,2}) = \frac{1}{2}(39 + 67{\cdot}5) = 53{\cdot}25$$

$$u_{4,4} = \frac{1}{2}(u_{3,3} + u_{5,3}) = \frac{1}{2}(49{\cdot}5 + 0) = 24{\cdot}75$$

$$u_{4,5}=\frac{1}{2}(u_{3,4}+u_{5,4})=\frac{1}{2}(43\cdot5+0)=21\cdot75$$

$$u_{1,1}=\frac{1}{2}(u_{0,0}+u_{2,0})=\frac{1}{2}(0+84)=42$$

$$u_{1,2}=\frac{1}{2}(u_{0,1}+u_{2,1})=\frac{1}{2}(0+78)=39$$

$$u_{2,2}=\frac{1}{2}(u_{1,1}+u_{2,1})=\frac{1}{2}(42+78)=60$$

$$u_{3,2}=\frac{1}{2}(u_{2,1}+u_{4,1})=\frac{1}{2}(78+57)=67\cdot5$$

$$u_{1,3}=\frac{1}{2}(u_{0,2}+u_{2,2})=\frac{1}{2}(0+60)=30$$

Table giving the values of $u_{i,j}$

$j\downarrow$ \ $i\rightarrow$	0	1	2	3	4	5
0	0·0	24·0	84·0	144·0	144·0	0·0
1	0·0	42·0	78·0	78·0	57·0	0·0
2	0·0	39·0	60·0	67·5	39·0	0·0
3	0·0	30·0	53·25	49·5	33·75	0·0
4	0·0	26·625	39·75	43·50	24·75	0·0
5	0·0	19·875	35·06	32·25	21·75	0·0

18.7Crank-Nicolson formulas. Writing the formula for jth and $(j+1)$th rows and then averaging, we get the approximate value of $\frac{\partial^2 u}{\partial x^2}$ as

$$\frac{\partial^2 u}{\partial x^2}=\frac{1}{2}\left[\left(u_{i-1,j}-2u_{i,j}+u_{i+1,j}\right)+\left(u_{i-1,j+1}-2u_{i,j+1}+u_{i+1,j+1}\right)\right]$$

The given differential equation, becomes

$$(1+r)\,u_{i,j+1}=u_{i,j}+\frac{1}{2}r\left(u_{i-1,j+1}+u_{1+1,j}+u_{i+1,j+1}+u_{i-1,j}-2u_{i,j}\right).$$

Letting $C_i=u_{i,j}+\frac{1}{2}r\left(u_{i-1,j}-2u_{i,j}+u_{i+1,j}\right)$, we get the iterative formula known as Jacobi's formula :

$$u_i^{(n+1)}=\frac{r}{2(1+r)}\left[u_{i-1}^{(n)}+u_{i+1}^{(n)}\right]+\frac{C_i}{1+r},$$

Replace $u_{i-1}^{(n)}$ by $u_{i-1}^{(n+1)}$

We get Gauss-Seidel iteration formula

$$u_i^{(n+1)} = \frac{r}{2(1+r)}\left[u_{i-1}^{(n+1)} + u_{i+1}^{(n)}\right] + \frac{C_i}{1+r}.$$

EXERCISES 18

1. Classify the following equations :

(*i*) $\frac{\partial u}{\partial t} = \frac{\partial^2 u}{\partial x^2}$ (*ii*) $\frac{\partial^2 u}{\partial x^2} + \frac{\partial^2 u}{\partial y^2} = 0$ (*iii*) $u_{xx} + 2y_{yy} = 2u_{xy}$

(*iv*) $u_{xx} + u_{yy} = 2u_{xy}$ (*v*) $u = 4u_{xx} + \left[x^2 + 4y^2\right) u_{xy} = \sin(x+y)\; u_{yy}$ (*vi*) $u_{xy} - u_x = 0$

(*M.S. Nov. '91*)

2. Solve the equation $u_{xx} + u_{yy} = 0$ in the domain of the following figure by Gauss-Seidel' method.

Ans. (*a*) first two iterates for $u(x, y)$ are

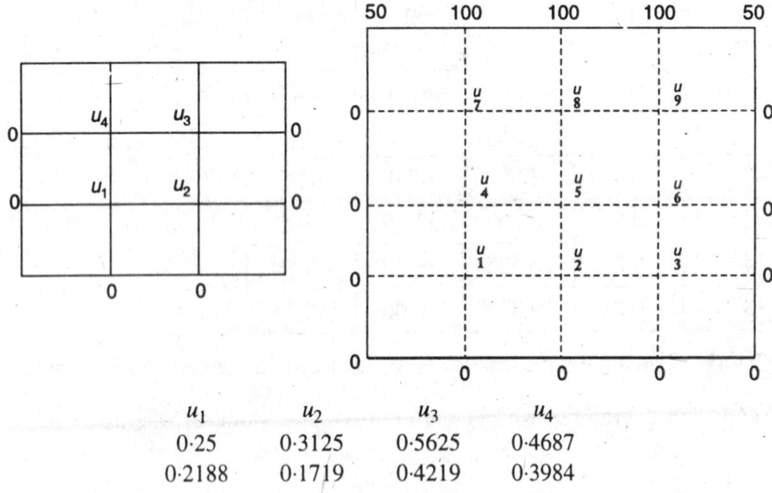

u_1	u_2	u_3	u_4
0·25	0·3125	0·5625	0·4687
0·2188	0·1719	0·4219	0·3984

(*b*) First two iterates for different quantities are:

u_1	u_2	u_3	u_4	u_5	u_6	u_7	u_8	u_9
7·03	9·57	7·08	18·94	25·10	18·98	43·02	52·97	42·99
7·13	9·83	7·20	18·81	25·15	18·84	42·94	52·77	42·90

[**Hint.** $u_5^{(1)} = \frac{1}{4}[0+0+0+50] = 25,$

$$u_7^{(1)} = \frac{1}{4}[0 + 25 + 100 + 25] = 43{\cdot}75$$

$u_1^{(1)} = 6{\cdot}25,\ u_3^{(1)} = 6{\cdot}25,\ u_8^{(1)} = 53{\cdot}12,\ u_4^{(1)} = 18{\cdot}75,\ u_0^{(1)} = 18{\cdot}75$

$$u_2^{(1)} = 9{\cdot}38$$

$$u_1^{(2)} = u_{1,1} = \frac{1}{4}\left[u_{0,1} + u_{2,1} + u_{1,0} + u_{1,2}\right] = \frac{1}{4}\left[C_{16} + u_2 + C_3 + u_4\right]$$

$$= \frac{1}{4}\ [0 + 9{\cdot}38 + 0 + 18{\cdot}75] = 28{\cdot}13$$

$$u_2 = u_{2,1} = \frac{1}{4}\left[u_{1,1}^{(2)} + u_{3,1}^{(1)} + u_{2,0}^{(2)} + u_{2,2}^{(1)}\right]$$

$$= \frac{1}{4}\ [7{\cdot}03 + 6{\cdot}25 + 0 + 25] = 9{\cdot}57$$

$$u_3 = u_{3,1}^{(2)} = \frac{1}{4}\left[u_{2,1}^{(2)} + u_{4,1}^{(1)} + u_{3,0}^{(2)} + u_{3,2}^{(1)}\right]$$

$$= \frac{1}{4}\left[u_2^{(2)} + C_0^{1\cdot 1} + C_4 + u_5^{(1)}\right]$$

$$= \frac{1}{4}\ [9{\cdot}57 + 0 + 0 + 18{\cdot}75] = 7{\cdot}08\ ;\ \text{etc.}]$$

3. Solve the equation $\frac{\partial^2 u}{\partial x^2} + \frac{\partial^2 u}{\partial y^2} = 0$ by using Gauss-Seidel method for the region bounded by the square $0 \le x \le 4$ and $0 \le y \le 4$ with $h = 10; k = 10$, the boundary conditions being $u = 0$ at $x = 0$ and $x = 8 + 2y$

$$u = \frac{1}{2}x^2 \text{ at } y = 0 \text{ and } u = x^2 \text{ when } y = 4$$

to compute the values of u at the interval mesh points.

4. Solve the parabolic equation $\frac{\partial^2 u}{\partial x^2} - \frac{2\partial u}{\partial t} = 0$

where $u(0, t) = 0$, $u(4, t) = 0$, $u(x, 0) = x(4 - x)$,

Assume $h = 1, k = 1$. Find the values of u upto $t = 4$

Ans. The values of $u_{(i,j)}$ are tabulated below:

$$u_{i,j+1} = \frac{1}{2}\left(u_{i+1,j} + u_{i-1,j}\right)$$

j \ $i \rightarrow$	0	1	2	3	4
0	0	3	4	3	0
1	0	2	3	2	0
2	0	1·50	2	1·50	0
3	0	1	1.5	1	0
4	0	·75	1	·75	0
5	0	·5	·75	·50	0

5. Solve the equation $\frac{\partial u}{\partial t} = \frac{\partial^2 u}{\partial x^2}$

subject to the initial and boundary conditions:

6. Solve the heat equattion $\frac{\partial u}{\partial t} = C^2 \frac{\partial^2 u}{\partial x^2}$, given that $u(0, t) = u(l, t) = 0$ and $u(x, 0) = x$ where l is the length of the bar.

[**Hint.** $u = XT$ where $X = A\cos kx + B\sin kx$, $T = e^{-k^2 e^2 t}$

$$X(o) = o = A,\ x(l) = o = B\sin kl \qquad k = n\pi/l$$

$$u = \Sigma B_n e^{-x^2\pi c^2 t^2/l^2} \sin\frac{n\pi x}{l}$$

$$u(x, 0) = x = \Sigma B_n \sin\frac{n\pi x}{l},\ B_n = \frac{2}{l}\int_0^l x\sin\frac{n\pi x}{l}\,dx = \frac{-2l}{x\pi}(-1)^x$$

$$u(x, t) = \frac{-2l}{\pi}\sum_{l}^{\infty}(-1)^n_{\pi} e^{-n^2 t^2 c^2 t/l^2} \sin\left(\frac{n\pi x}{l}\right).]$$

7. The function u satisfies Laplace's equation at all points within the square in the following figure and has the boundary values indicated. Compute a solution correct to two decimals.

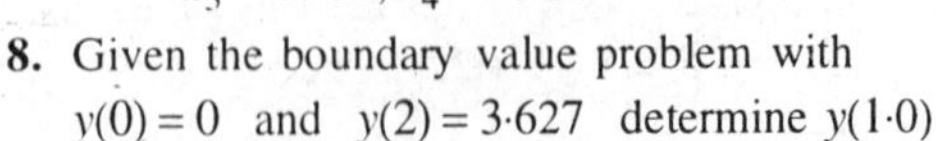

Ans. $u_1 = 26{\cdot}66$, $u_2 = 33{\cdot}33$,

$u_3 = 43{\cdot}33$, $u_4 = 46{\cdot}66$.

8. Given the boundary value problem with $y(0) = 0$ and $y(2) = 3{\cdot}627$ determine $y(1{\cdot}0)$

[**Hint.** Divide the interval [0, 2] into four equal parts so that $h = 0{\cdot}5 = \frac{1}{2}$ ·

use 317·3, the $y_{0+1} - 2u_i + u_{i-1} = h^2 d_i$

$\therefore\ 4(v_0 - 2v_1 + y_2) = y_1,\ y(y_1 - 2y_2 + y_3) = y_2$ and $u\,(y_2 - 2y_3 + y_1) = y_3$

with boundary conditions $y_2 = 0, y = 3{\cdot}627$] **Ans.** $y = 1{\cdot}813$

9. Solve : $y'' + y + 1 = 0$ with the boundary conditions $y = 0$ when $x = 0$ and $y = 0$ when $x = 1$

[**Hint.** Choose $h = \frac{1}{2}$ *i.e.*, $x = 2$] **Ans.** $y_1 = \frac{1}{4} = 0{\cdot}14283$

10. Solve the boundary value problem :

(*a*) $\dfrac{\partial u}{\partial t} + x\dfrac{\partial u}{\partial x} + y = x,\ x > 0$ whebn $t > 0$

(*b*) $\dfrac{\partial^2 u}{\partial x^2} = 0$

when $u(0, y) = y^2$ and $u(1, y) = 1$ (*Nagpur Univ., 1995*)

11. Solve : $y'' + y = 1,\ 0 \le x \le 1$

12. Solve the partial differential equation $\dfrac{\partial u}{\partial x} = \dfrac{\partial u}{\partial y}$ ·

13. In solving equation $u_t = x^2 u_{xx}$ by Crank-Nicolson method, to simplify method we take $(\Delta x)^2/\alpha^2 K$ as

(*a*) $\frac{1}{2}$ (*b*) 2 (*c*) 1 (*d*) 0.

19

FITTING OF CURVES AND CUBIC SPLINES

19·1. Introduction. *The Principle of Least Squares. Suppose we have a number of independent linear equations in n unknowns $x_1, x_2, ..., x_n$,

$$\left.\begin{array}{l} a_{11}x_1 + a_{12}x_2 + \cdots + a_{1n}x_n = b_1 \\ a_{21}x_1 + a_{22}x_2 + \cdots + a_{2n}x_n = b_2 \\ \cdots\cdots\cdots\cdots\cdots\cdots\cdots\cdots\cdots\cdots \\ a_{m1}x_1 + a_{m2}x_2 + \cdots + a_{mn}x_n = b_m \end{array}\right\} \text{ or } \sum_{j=1}^{n} a_{ij}\, x_j = b_i,\ i = 1, 2, \cdots, m \quad ...(19·1)$$

where a's and b's are constants. If $m = n$, we can, in general**, find a unique set of values satisfying the given system of equations. However, if $m > n$ that is if the number of equations is greater than the number of unknowns, no such solution exists. We, therefore, try to find those values of $x_1, x_2, ..., x_n$ which will satisfy the given system of equations as nearly as possible.

The principle of least squares asserts that these values are those which make S a minimum where

$$S = \sum_{i=1}^{m} (a_{i1}x_1 + a_{i2}x_2 + \cdots + a_{in}x_n - b_i)^2 = \sum_{i=1}^{m} E_i^2 ,$$

where $\qquad E_i = a_{i1}x_1 + a_{i2}x_2 + \cdots\cdots + a_{in}x_n - b_i$

---

* The problem of solving m observation equations for unknowns, numbering less than m seems to have engaged the attention of astronomers since about 1750. The principle of least squares as means of accomplishing this solution approximately, appears to have been used originally by Gauss in 1795 but it was named and first published in 1805 by Legendre.

** The condition for a unique solution is that the rank of the $m \times n$ matrix of coefficients *viz.*, $A = (a_{ij})$ should be equal to therank of the augmented matrix (a, b) which is an $m \times (n + 1)$ matrix obtained by adding to the n columns of A one more column of b's] and the common rank is n. If this common rank is $> n$, then we get no solution.

We shall call these values the *best* or *most plausible values* in the least square sense.

By a well-known theorem of differential calculus, the extreme values of the function $f \equiv f(x_1, x_2 \dots\dots, x_n)$ are given by $\frac{\partial f}{\partial x_1} = \frac{\partial f}{\partial x_2} = \dots = \frac{\partial f}{\partial x_n} = 0,$

provided that the partial derivatives exist.

Applying this theorem, S will have a maximum or minimum, for those values of the unknowns x_j $(j = 1, 2, \dots, n)$ which satisfy the following n equations :

$$\frac{\partial S}{\partial x_j} = 0,\ j = 1, 2, \dots, n \text{ i.e. } \sum_{i=1}^{m} a_{i1} E_i = 0,\ \sum_{i=1}^{m} a_{i2} E_i = 0, \dots, \sum_{i=1}^{m} a_{in} E_i = 0 \quad \dots(19{\cdot}2)$$

These n equations are called the *normal equations* corresponding to the equations (19·2). They can be solved like simultaneous equations for the n unknowns $x_1, x_2, \dots, x_n$. If the common rank of the matrix of coefficients and the augmented matrix is n, we can get a unique solution for $x_1, x_2, \dots, x_n$.

Note. To see the values which we obtain from the normal equations will make S a minimum, we calculate the second order partial derivatives

$$\frac{\partial^2 f}{\partial x_i^2} = 2 \sum_{i=1}^{m} a_{ij}\quad \frac{\partial^2 f}{\partial x_j \partial x_l} = 2 \sum_{i=1}^{m} a_{ij} a_{il};\ j, l = 1, 2, \dots, n, j \neq l.$$

These values make $\frac{\partial^2 f}{\partial x_1^2} y_1^2 + \dots + \frac{\partial^2 f}{\partial x_n^2} y_n^2 + 2 \frac{\partial^2 f}{\partial x_1 \partial x_2} y_1 y_2 + \dots$ ...(19·3)

positive definite and hence S is a minimum.

Alternatively. If $x_{10}, x_{20}, \dots, x_{n0}$ are the solutions of the normal equations (19·2), we can write S as

$$S = \sum_{i=1}^{m} \Big[(a_{i1}x_{10} + a_{i2}x_{20} + \dots + a_{in}x_{n0} - b_i)$$

$$+ \{a_{i1}(x_1 - x_{10}) + \dots + a_{in}(x_n - x_{n0})\}\Big]^2$$

$$= \sum_{i=1}^{m} (a_{i1}x_{10} + a_{i2}x_{20} + \dots + a_{in}x_{n0} - b_i)^2 + [2(x_1 - x_{10}) \Sigma a_{i1} E_{i0}$$

$$+ 2(x_2 - x_{20}) \Sigma a_{i2} E_{10} + \dots + 2(x_n - x_{n0}) \Sigma a_{in} E_{i0}]$$

$$+ \sum_{i=1}^{m} [a_{i1}(x_1 - x_{10}) + \dots + a_{in}(x_n - x_{n0})]^2$$

Since $x_{10}, x_{20}, \dots, x_{n0}$ are the solutions of the normal equations, the second group of terms vanishes. $\therefore \quad S \geq \sum_{i=1}^{m} (a_{i1}x_{10} + \cdots\cdots + a_{in}x_{n0} - b_i)^2.$

Hence S is a minimum when the values are obtained by solving the normal equations.

Ex. Verify that (19·3) gives a positive definite quadratic form for $n = 2, 3$.

Example 1. *Form normal equations and hence find the most plausible values of x and y from the following equations :*

$$x + y = 3{\cdot}01,\ 2x - y = 0{\cdot}03,\ x + 3y = 7{\cdot}03,\ 3x + y = 4{\cdot}97.$$

Sol. Here $S = (x + y - 3{\cdot}01)^2 + (2x - y - 0{\cdot}03)^2 + (x + 3y - 7{\cdot}03)^2 + (3x + y - 4{\cdot}97)^2$

$= f(x, y)$

The normal equations are $\dfrac{\partial S}{\partial x} = 0$ and $\dfrac{\partial S}{\partial y} = 0$

i.e., $(x + y - 3{\cdot}01) + 2\,(2x - y - 0{\cdot}03) + (x + 3y - 7{\cdot}03) + 3(3x + y - 4{\cdot}97) = 0$

and $(x + y - 3{\cdot}01) - (2x - y - 0{\cdot}03) + 3(x + 3y - 7{\cdot}03) + (3x + y - 4{\cdot}97) = 0$

i.e., $\quad 15x + 5y = 25$ and $5x + 12y = 29{\cdot}01$.

Solving these equations, we get $x = 0{\cdot}9997$, $y = 2{\cdot}0009$ approx.

Example 2. *Three independent measurements on each of the angles A, B, C of a triangle are as follows :*

	A	B	C
	39·5	*60·3*	*80·1*
	39·3	*62·2*	*80·3*
	39·6	*60·1*	*80·4*
Total	*118·4*	*182·6*	*240·8*

Obtain the best estimate of the three angles taking into account the relation that sum of the angles is equal to 180°.

Sol. Let the three observations on A be denoted by x_1, x_2, x_3, on B by y_1, y_2, y_3, and on C by z_1, z_2, z_3 respectively. Let the best estimate of A, B and C be θ_1, θ_2 and $180 - \theta_1 - \theta_2$. Then by the principle of least squares, we want E to be minimum where

$$E = \sum_{i=1}^{3} (x_i - \theta_1)^2 + \sum_{i=1}^{3} (y_i - \theta_2)^2 + \sum_{i=1}^{3} (z_i - 180 + \theta_1 + \theta_2)^2$$

Thus the equations to determine the best values of θ_1, θ_2 are given by

$$0 = \frac{\partial E}{\partial \theta_1} = -\Sigma\,(x_i - \theta_1) + \Sigma\,(z_i - 180 + \theta_1 + \theta_2),$$

and $$0 = \frac{\partial E}{\partial \theta_2} = -\Sigma\,(y_i - \theta_2) + \Sigma\,(z_i - 180 + \theta_1 + \theta_2).$$

These give $\Sigma x + 3\theta_1 + \Sigma z - 540 + 3\theta_i + 3\theta_2 = 0$

or $$6\theta_1 + 3\theta_2 - 417{\cdot}6 = 0 \quad ...(1)$$

and $$\Sigma x + 3\theta_2 + \Sigma z - 540 + 3\theta_1 + 3\theta_2 = 0$$

or $$3\theta_1 + 6\theta_1 - 481{\cdot}8 = 0 \quad ...(2)$$

Solving (1) and (2) for θ_1 and θ_2, we get

$$\hat{\theta}_1 = 39{\cdot}27,\ \hat{\theta}_2 + 60{\cdot}66,\ \hat{\theta}_3 = 180 - \hat{\theta}_1 - \hat{\theta}_2 = 80{\cdot}07$$

19·2. Fitting of Curves. The fitting of a curve to a given data is very important both from the point of view of theoretical and practical Statistics. In theoretical Statistics, we shall find that the study of regression and correlation (whether total, partial or multiple) can be regarded as fitting of linear curves (st. lines, planes, hyper-planes) to the given bivariate or multivariate frequency or probability distributions. In practical Statistics, it enables us to get a close functional relation between x and y. The relation is, in general, expressed by a polynomial, but other types of relationship, algebraic, exponential or logarithmic can also be fitted by using the principle of least squares.

19·2·1. Fitting of Algebraic Polynomials or Parabolic Curves. Let it be required to fit the curve whose equation is

$$Y = a_0 + a_1 x + a_2 x^2 + \cdots + a_p x^p, \quad ...(19{\cdot}4)$$

the coefficients $a_0, a_1 \cdots\cdots, a_p$, being constants and $a_p \neq 0$, to a number $n\ (n > p + 1)$ of given points or values (x_1, y_1) ; (x_2, y_2).......; (x_n, y_n) of the two variables so that we have to determine the $(p + 1)$ constants a_0, a_1,........, a_p for best fit. The curve whose equation is (19·4), is called the *parabola of degree p*. Substituting x_i for x in (19·4), we get

$$Y_i = a_0 + a_1 x_i + a_2 x_i^2 + \cdots + a_p x_i^p \ (i = 1, 2, ..., n).$$

The quantity Y_i is generally called the *expected value* of y corresponding to $x = x_i$, contrasted with the value y_i which is called the *observed value* of y corresponding to the same value x_i of x. These two values are, in general, different as the observed points do not necessarily lie on a parabolic curve. The difference $y_i - Y_i$ is called the residual or *error of estimate* corresponding to $x = x_i$ and, according to the *principle of least squares*, we choose the constants a_0, a_1, ...,a_p so that sum of squares of residuals is minimum.

Now $$S = \sum_{i=1}^{n} (y_i - Y_i)^2 = \sum_{i=1}^{n} (y_i - a_0 - a_1 x_i - \cdots - a_p x_i^p)^2. \quad ...(19{\cdot}5)$$

Equating to zero the partial derivatives of (19·5) with respect to $a_1, a_2,...,a_p$, we get $(p + 1)$ normal equations. On simplification, these equations reduce to

$$\left.\begin{aligned}\Sigma y &= na_0 + a_1\Sigma x + \cdots\cdots + a_p\Sigma x^p \\ \Sigma xy &= a_0\Sigma x + a_1\Sigma x^2 + \cdots\cdots + a_p\Sigma x^{p+1} \\ \Sigma x^2 y &= a_0\Sigma x^2 + a_1\Sigma x^3 + \cdots\cdots + a_p\Sigma x^{p+2} \\ &\cdots\cdots\cdots\cdots\cdots\cdots\cdots\cdots\cdots\cdots \\ \Sigma x^p y &= a_0\Sigma x^p + a_1\Sigma x^{p+1} + \cdots\cdots + a_p\Sigma x^{2p}\end{aligned}\right\} \qquad \ldots(19{\cdot}6)$$

where Σ everywhere stands for $\sum_{i=1}^{n}$.

These equations are called the *normal equations* for fitting the parabola of degree p. They are $(p + 1)$ in number and when solved as simultaneous equations give the values of the $(p + 1)$ unknowns $a_0, a_1 \ldots, a_p$.

In particular, when $p = 1$, the normal equations for fitting the straight line $Y = a_0 + a_1 x$ of best fit to the given set of points (x_i, y_i), $i = 1, 2, \ldots, n$ are

$$\left.\begin{aligned}\Sigma y &= a_0 n + a_1\Sigma x \\ \Sigma xy &= a_0\Sigma x + a_1\Sigma x^2\end{aligned}\right\} \qquad \ldots(19{\cdot}7)$$

Thus the best fitting linear function for the points (x_i, y_i), $i = 1, 2, \ldots, n$ may be expressed in the form

$$\begin{vmatrix} x & y & 1 \\ \Sigma x & \Sigma y & n \\ \Sigma x^2 & \Sigma xy & \Sigma x \end{vmatrix} = 0.$$

When $p = 2$, the normal equations for fitting the second degree parabola $Y = a_0 + a_1 x + a_2 x^2$ of best fit to the set of given points (x_i, y_i) are

$$\left.\begin{aligned}\Sigma y &= a_0 n + a_1\Sigma x + a_2\Sigma x^2 \\ \Sigma xy &= a_0\Sigma x + a_1\Sigma x^2 + a_2\Sigma x^3 \\ \Sigma x^2 y &= a_0\Sigma x^2 + a_1\Sigma x^3 + a_2\Sigma x^4\end{aligned}\right\} \qquad \ldots$$

Note. *Change of origin.* When the values x_i $(i = 1, 2, \ldots\ldots, n)$ correspond to equal increment h and n is *odd* say $= 2m + 1$, the equations (19·6) can be simplified by taking the origin of x at the middle of the value and the common increment h as unit of measurement. Thus if a be the middle value, $u = (x - a)/h$ takes the values $-m \ldots\ldots, -1, 0, 1, \ldots\ldots, m$. Hence, we get $\Sigma u = 0 = \Sigma u^3 = \ldots\ldots$

If n is *even*, say $2m$ we take the origin of x at the mean of the middle pair of values and $h/2$ as the new unit *i.e.*

$$u = \frac{x - (\text{mean of two suitable or middle terms})}{\frac{1}{2}h}$$

The values of x then become $-(2m - 1), -(2m - 3), \ldots\ldots, -3, -1, 1$ $3, \ldots\ldots, (2m - 3), (2m - 1)$, and again $\Sigma u = 0 = \Sigma u^3 = \ldots\ldots$

Example 3. *The weights of a calf taken at weekly intervals are given below. Fit a straight line using the method of least squares, and calculate the average rate of growth per week.*

Age (x)	*1*	*2*	*3*	*4*	*5*	*6*	*7*	*8*	*9*	*10*
Weight (y)	*52·5*	*58·7*	*65·0*	*70·2*	*75·4*	*81·1*	*87·2*	*95·5*	*102·2*	*106·4*

Sol. Since n is even, we take the origin of x at the mean of the middle pair of values, and $\frac{1}{2}$ as the new unit that is we take $u = \dfrac{x - 5·5}{\frac{1}{2}} = 2x - 11.$

The values of u, then are $-9, -7, -5, -3, -1, 1, 3, 5, 7, 9$

We find that $\Sigma u = 0$, $\Sigma u^2 = 330$, $\Sigma y = 796·2$ and $\Sigma uy = 1016·8$

The normal equations for fitting the straight line $y = a + bu$ are

$$\Sigma y = na + b\Sigma u, \; \Sigma uy = a\Sigma u + b\Sigma u^2.$$

On substituting the values of Σy, etc., we have

$$796·2 = 10a, \; 1016·8 = 330b \; ; \quad \therefore \quad a = 79·62, \; b = 3·08 \text{ approx.}$$

The regression equation is, therefore,

$$y = 79·62 + 3·08u = 79·62 + 3·08\,(2x - 11) \quad \text{or} \quad y = 45·74 + 6·16x$$

The average rate of growth per week is 6·16 units.

Example 4. *Fit a second degree parabola to the following data, taking x as the independent variable.*

x :	*1*	*2*	*3*	*4*	*5*	*6*	*7*	*8*	*9*
y :	*2*	*6*	*7*	*8*	*10*	*11*	*11*	*10*	*9*

Sol. Introduce two new variables u and v connected by the relations $u = x - 5$, $v = y - 7$. Since n is odd and the values of x are equidistant, we take the origin for the x series at the middle value 5. This makes $\Sigma u = 0 = \Sigma u^3$. Let us fit the curve $v = a_0 + a_1 u + a_2 u^2$.

The calculations of Σuv, $\Sigma u^2 v$, etc., are shown in the following table :

	u	v	uv	u^2	$u^2 v$	u^3	u^4
	−4	−5	20	16	−80		256
	−3	−1	3	9	−9		81
	−2	0	0	4	0		16
	−1	1	−1	1	1		1
	0	3	0	0	0		0
	1	4	4	1	4		1
	2	4	8	4	16		16
	3	3	9	9	27		81
	4	2	8	16	32		256
Total	0	11	51	60	−9	0	708

Substituting the values of Σu, etc., in the normal equations (19·8), we get

$$11 = 9a_0 + 0 + 60a_2 \,;\, 51 = 0 + 60a_1 + 0 \,;\, -9 = 60a_0 + 0 + 708a_2.$$

Solving these simultaneous equations we obtain

$$a_0 = 3{\cdot}0,\ a_1 = 0{\cdot}85,\ a_2 = -0{\cdot}27.$$

Hence the equation of the curve is $v = 3 + 0{\cdot}85u - {\cdot}027u^2$

or $y - 7 = 3 + 0{\cdot}85(x-5) - 0{\cdot}27(x-5)^2$ or $y = -1 + 3{\cdot}55x - 0{\cdot}27x^2$

Ex. Show that if the pair (x_i, y_i), $i = 1, 2, \ldots, n$ occurs with the frequency f_i where $\Sigma f_i = N$, the normal equations, for fitting the parabola of the second degree $y = a_0 + a_1x + a_2x^2$ are

$$\Sigma f_i\, y_i = N\, a_0 + a_1\, \Sigma f_i\, x_i + a_2\, \Sigma f_i\, x_i^2,$$

$$\Sigma f_i\, x_i\, y_i = a_0\, \Sigma f_i\, x_i + a_1\, \Sigma f_i\, x_i^2 + a_2\, \Sigma f_i\, x_i^3,$$

$$\Sigma f_i\, x_i^2\, y_i = a_0\, \Sigma f_i\, x_i^2 + a_1\, \Sigma f_i\, x_i^3 + a_2\, \Sigma f_i\, x_i^4.$$

19·2·2. Fitting of Curves of the types $y = ab^x$ and $y = ax^b$.

(Curves reducible to st. lines.) Whenever it is required to fit curves of exponential types to the data, the procedure is to take common logarithms of both sides of the equation and then form normal equations to find the *best* values of the constants. For example, to fit the curve $y = ab^x$ we proceed as follows :

We have $\log y = A + Bx$ where $A = \log a$, $B = \log b \Rightarrow Y = A + Bx$

we have $\quad S = \Sigma(\log y - y)^2 = \Sigma\,(\log y - A - Bx)^2 = f(A, B)$

The normal equations are

$$0 = \frac{\partial S}{\partial A} \Rightarrow \Sigma \log y = nA + B\Sigma x \,;\, 0 = \frac{\partial S}{\partial B} \Rightarrow \Sigma x \log y = A\Sigma x + B\Sigma x^2$$

In the second case $y = ax^b$, we have $\log y = \log a + b \log x = A + bX$

In this case it may be useful to plot the data on a log— paper.

The normal equations can be solved for A and b and then by taking the antilogs, a and b can be found out.

Example 5. *Fit the exponential curve* $y = ae^{bx}$ *to the following data, e being Napierian base, 2·71828 :*

x :	0	2	4
y :	5·012	10	31·62

(I.A.S. 1993)

Sol. The equation to the curve can be written as $\log y = \log_{10} a + (b \log e)x$

or $\quad Y = A + Bx$ where $Y = \log_{10} y$, $A = \log_{10} a$, $B = b \log_{10} e$

The normal equations are $nA + B\Sigma x = \Sigma Y$; $A\Sigma x + B\Sigma x^2 = \Sigma xY$

The calculations of Σx, ΣY, Σx^2 and ΣxY are shown in the following table :

	x	y	$Y(=\log y)$	x^2	xY
	0	5·012	0·7	0	0
	2	10·00	1·0	4	2
	4	31·62	1·5	16	6
Total	6		3·2	20	8

Thus we have $3A+6B=3{\cdot}2$ and $6A+20B=8$

Solving, we get $A=0{\cdot}666$, $B=0{\cdot}2$; $\therefore$ $a=4{\cdot}642$, $b=0{\cdot}46$ approx.

Hence equation of the curve fitted to the data is $y=4{\cdot}642e^{0{\cdot}46x}$

19·2·3. Fitting of Logistic curve. $y=\dfrac{a}{1+be^{cx}}$, $c<0$

This is an S-shaped curve, As $x\to\infty$, $y\to a$,

Thus a can be estimated from the data since it is the value of y when x is reasonably large.

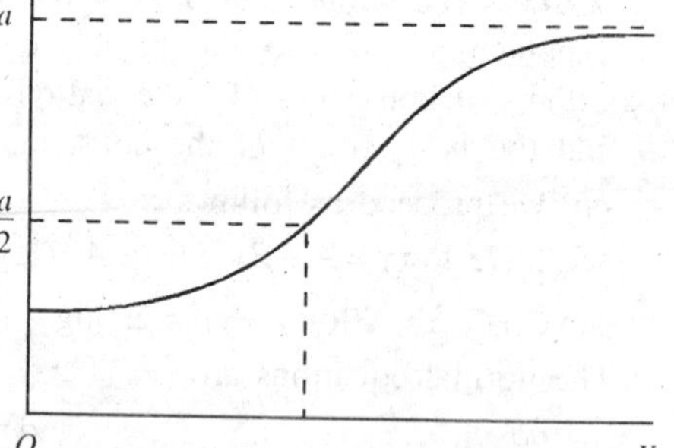

The given equations can be written as

$$\frac{1}{y}=\frac{1}{a}+\frac{b}{a}e^{cx}$$

$$\frac{1}{y}-\frac{1}{a}=\frac{b}{a}(e^c)^x \Rightarrow Y=AB^x$$

where $Y=\dfrac{1}{y}-\dfrac{1}{a}$, $A=\dfrac{b}{a}$, $B=e^c$

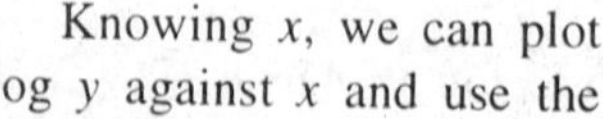

Knowing x, we can plot log y against x and use the method of least squares to estimate A and B and therefore to estimate b and c.

Such curves are widely used in the study of growth of populations.

19·3. Fitting a cubic spline. Data have been fitted to polynomials and exponential curves. These approximations are not generally proper and in Numerical Analysis techniques have been devised where splines particularly cubic splines have been fitted to data, integration, etc. It is called 'spline fitting since a curve is drawn through the given points (x_i, y_i), $i=0, 1, 2, \ldots n$ in the interval $[a, b]$ where $a=x_0<x_1<\cdots\cdots<x_x=b$ in such a way that

(*i*) it passes through the given points and (*ii*) its slope and curvature are continuous functions on $[a, b]$.

Let $S(x)$ be a cubic spline satisfying the following conditions:

(*i*) $S(x_i)=y_i$, $i=0, 1, \ldots\ldots, n$

(*ii*) $S(x)$, $S'(x)$ and $S''(x)$ are continuous on $[a, b]$, and (*iii*) $S(x)$ is a cubic polynominal in each sub-interval $\left(x_i, x_{i+1}\right)$, $i = 0, 1, \ldots, N-1$.

Since $S(x)$ is a cubic in sub-interval (x_i, x_{i+1}) of width h, $S''(x)$ must be linear and, $\therefore$, can be assumed to be of the form

$$S''(x) = \frac{1}{h}\left[(x_{1+1} - x)\, S''(x_i) + (x - x_i)\, S''(x_{i+1})\right]$$

$$= \frac{1}{h}\left[(x_{i+1} - x)\, M_i + (x - x_i)\, M_{i+1}\right] \qquad \ldots(1)$$

with $S''(x_i) = M_i$, $S'(x_i) = m_i$

Integrating (1) twice to determine $S(x)$ and using the condition (1) to determine the constants of integration, we have

$$S(x) = \frac{1}{6h}\left[(x_{i+1} - x)^3\, M_i + (x - x_i)^3\, M_{i+1}\right]$$

$$+ \frac{1}{h}(x_{i+1} - x)\left[y_i - \frac{h^2}{6} M_i\right] + \frac{1}{h}(x - x_i)\left[y_{i+1} - \frac{h^2}{6} M_{i+1}\right] \quad \ldots (2)$$

If M_i, M_{i+1} are known, then $S(x)$ as given by (2) is the required cubic spline,

Next since the condition of continuity of $S'(x)$ *i.e.*, $S'(x_i -) = S'(x_i +)$, *we easily obtain the recurrence relation.*

$$M_{i-1} + 4M_i + M_{i+1} = \frac{6}{h^2}\left(y_{i-1} - 2y_i + y_{i+1}\right),\ i = 1, 2, \ldots N-1, \qquad \ldots(3)$$

This formula gives $(n-1)$ equations in $(n+1)$ unknowns viz., $y_0, y_1, \ldots\ldots, y_n$. We therefore need two more relations in order that a unique interpolating cubic spline may be found. Sometimes, for convenience, we choose end conditions $M_o = M_n = 0$ so that in such a case the spline is known as a '*natural* cubic spline' fitted to the data.

Note. Using the condition of continuity, we can derive the recurrence *formula*: $m_{i-1} + 4m_i + m_{i+1} = \frac{3}{h}\left(y_{i+1} - y_{i-1}\right)$, $i = 1, 2, \ldots, N-1$ *where* $m_i = S'(x_i)$.

Example 6. *Fit a natural cubic spline to the following data :*

x :	2	3	4
y :	*11*	*49*	*121*

Hence compute (*i*) $y(2{\cdot}5)$ and (*ii*) $y'(2)$.

Sol. Here $n = 2$ and $M_o = M_n = 0$, $y_o = 11$, $y_1 = 49$, $y_2 = 123$, $h = 3 - 2 = 4 - 3 = 1$, $x_o = 2$, $x_1 = 3$, $x_2 = 4$.

$\therefore$ The equation (3) with $i = 1$ is

$$M_o + 4M_1 + M_2 = \frac{6}{h^2}\left(y_0 - 2y_1 + y_2\right)$$

which reduces to $4M_1 = 6\,(11 - 2 \times 49 + 123) = 6 \times 36 = 216$

$$\Rightarrow M_1 = 54.$$

For the interval $2 \le x \le 3$, the cubic spline as given by (2), page 408) is

$$S(x) = \frac{1}{6}\,[(3-x)^3 \times 0 + (x-2)^3 \times 54] + (3-x) \times 0$$

$$\left(11 - \frac{1}{6} \times 0\right) + (x-2)\,(49 - \frac{1}{6} \times 54)$$

$$= 9(x-2)^3 + 11\,(3-x) + 40\,(x-2)$$

$$= 9\,(x-2)^3 + 29x - 47$$

$$S'(x) = 27\,(x-2)^2 + 29$$

$$\therefore \quad y'(2) = S'(2) = 29$$

$$S(0{\cdot}25) = y\,(0{\cdot}25) = 9\,({\cdot}25 - 2)^3 + 29 \times {\cdot}25 - 47$$

$$= 9\,(-1{\cdot}75)^3 + \frac{29}{4} - 47 = -\,88 \text{ approx.}$$

Example 7. *Obtain the cubic spproximation for the function given in the tabular form*

x :	*0*	*1*	*2*	*3*
$f(x)$:	*1*	*2*	*33*	*244*

and $M_0 = 0, M_3 = 0$ *(I.A.S. 1997)*

Sol. We have $M_{i-1} + 4M_i + M_{i+1} = 6\,(f_{i+1} - 2f_i + f_{i-1})$, $i = 1, 2$

$$\therefore \quad M_0 + 4M_1 + M_2 = 6\,(f_2 - 2f_1 + f_0)$$

$$M_1 + 4M_2 + M_3 = 6\,(f_3 - 2f_2 + f_1)$$

using $M_0 = 0$, $M_3 = 0$ and the given functtional values, we get

$$4M_1 + M_2 = 6\,(33 - 2 \times 33 + 2) = 80$$

$$M_1 + 4M_2 = 6\,(244 - 2 \times 33 + 2) = 1080$$

Solving, $M_1 = -\,24$, $M_2 = 276$,

$$S(x) = 6h_i\left[\,(x_{i+1} - x)^3\,M_i + (x - x_i)^3\,M_{i+1}\,\right]$$

$$+ \frac{1}{h_i}\left(y_i - \frac{h_i^2}{6}\,M_i\right)\left(x_{i+1} - x\right) + \frac{1}{h_i}\left(y_{i+1} - \frac{h_i^2}{6}\,M_{i+1}\right)(x - x_i)$$

Thus the cubic spline in the corresponding interval becomes

Interval	*Corresponding cubic spline*
[0, 1]	$-\,4x^3 + 5x, t$
[1, 2]	$50x^3 - 16x^2 + 1672 - 53$
[2, 3]	$-\,46x^3 + 414x^2 - 985x + 715$

EXERCISE 19

1. Fit a natural cubic spline to the following data:

(*a*)

x :	0	$\frac{\pi}{2}$	π
y :	0	1	0

Also find $y\left(\frac{\pi}{6}\right)$

(*b*)

x :	0	$\frac{\pi}{4}$	$\frac{\pi}{2}$	$\frac{3\pi}{4}$	π
y :	0	$\frac{1}{\sqrt{2}}$	1	$\frac{1}{\sqrt{2}}$	0

Using the condition that $M_0 = M_N = 0$, we find $M_1 = -0.7440$, $M_2 = -1.053$ and $M_3 = -0{\cdot}7440$, Thus in $0 \le x \le \frac{\pi}{2}$, Cubic spline is

$S(x) = \frac{4}{\pi}\left[-0{\cdot}1240x^3 + 0{\cdot}7836x\right]$ and $S\left(\frac{\pi}{6}\right) = 0.4998$

[**Hint.** (*a*) $n = 2$, $M_o + 4M_1 + M_2 = \frac{6 \times 4}{\pi^2}[0 - 2 + 0]$ $\therefore$ $M_1 = \frac{-12}{\pi^2}$

In the interval $(0, \frac{\pi}{2})$, the spline is given by

$S(x) = \frac{2}{\pi}\left[-\frac{2x^3}{\pi^2} + \frac{3x}{2}\right]$, $S\left(\frac{\pi}{6}\right) = 0.4815$]

2. Fit the following four points by the cubic spline

i	0	1	2	3
x_i	1	2	3	4
y_i	1	5	11	8

use the end conditions $y_0'' = y_3' = 0$

Hence compute (*i*) y (1.5) and (*ii*) y' (2). (*I.A.S. 1994*)

3. The following table gives value of x and corresponding values of y.

x :	1	2	3
y :	–8	–1	18

using cubic spline interpolation, determine y (1.5).

4. The following points are given :

(1, 10), (2, 17), (3, 36), (4, 73), (5, 134)

Determine the cubic spline $S(x)$ for the interval (2,3) given that

(*i*) $S'(1) = y'(1)$ and $S'(5) = y'(5)$

and (*ii*) $S''(1) = y''(1)$ and $S''(5) = y''(5)$.

20. Miscellaneous Topics

20. Application of Method of Undetermined Coefficients in deriving Simpson's formula

20.1. *Derive Simpson's 1/3 rule using the method of unde- termined coefficients.*

$$\text{Assume} \int_{-h}^{h} y_x\, dx = a_{-1}\, y_{-1} + a_0\, y_0 + a_1\, y_1 \qquad ...(1)$$

where a_{-1}, a_0 and a_1 are the coefficients to be determined. Further assume that the formula (1) is exact, when y_x is $1, x$ or x^2. Thus putting $y_x = 1$, x or x^2 successively in (1) and integrating we obtain

$$a_{-1} + a_0 + a_1 = 2h; \qquad -a_{-1} + a_1 = 0 \quad \text{and} \quad a_{-1} + a_1 = \frac{2}{3}h$$

From last two equations we get a_{-1} and a_1 and then we obtain a_0. Substituting the following *values*

$$\therefore \quad a_{-1} = \frac{h}{3},\ a_1 = \frac{h}{3} \text{ and } a_0 = \frac{4h}{3} \text{ in (1), we obtain}$$

$$\int_{-h}^{+h} y\, dx = \frac{h}{3}(y_{-1} + 4y_o + y_1), \text{ which is Simpson's } \frac{1}{3}\text{-rule.}$$

Ex. *Using the method of undetermined coefficients, derive the formula* :

(*i*) $\displaystyle\int_0^h y_x\, dx = \frac{h}{2}(y_0 + y_1) - \frac{h^3}{24}(y''_0 + y''_1).$

(*ii*) $\displaystyle\int_0^h y_x\, dx = \frac{h}{2}\left(y_0 + y_1\right) + \frac{h^2}{12}(y'_0 - y'_1).$

[**Hint.** (*i*) Assume $\displaystyle\int_0^h y_x\, dx = a_0\, y_0 + a_1 y_1 + a_2\, y_0'' + a_3\, y_1''$

where $y_0 = 0, y_1 = h$ and $a_{0,}\, a_1, a_2$ and a_3 are the coefficients to be determined by taking $y_x = 1, x, x^2$ or x^3 successively. We have $h = a_0 + a_1, \frac{h^2}{2} = a_1\, h, \frac{h^3}{2} = a_1^2\, h + 2a_2 + 2a_3, \frac{h^4}{4} = h^3\, a_1 + 6ha_3.$]

20.2 Application of Cubic Spline in evaluating the integral

$$I = \int_{x_0}^{x_n} ydx$$

Divide the interval $[x_0, x_n]$ into n equal sub-intervals each of length h such that $x_i - x_{i-1} = h, i = 1, 2, ..., n$ *i.e.* $x_0 < x_1 < x_2 < x_{i-1} < x_i < \cdots < x_n$

Now if $S(x)$ is the cubic spline in the sub interval (x_{i-1}, x_i), we have

$$I = \int_{x_0}^{x_n} y dx = \sum_{i-1}^{n} \int_{x_{i-1}}^{x_i} S(x)\, dx$$

$$= \sum_{i=1}^{n} \int_{x_i - 1}^{x_i} \left[\frac{1}{6h} \left\{ (x_i - x)^3 M_{i-1} + (x - x_{i-1})^3 M_i \right\} \right.$$

$$\left. + \frac{1}{h} (x_i - x) \left(y_{i-1} - \frac{h^2}{6} M_{i-1} \right) + \frac{1}{h} (x - x_{i-1}) \left(y_i - \frac{h^2}{6} M_i \right) \right] dx$$

Integrating and simplifying, we have

$$I = \sum_{i=1}^{n} \left[\frac{h}{2} \left(y_{i-1} + y_i \right) - \frac{h^3}{24} \left(M_{i-1} + M_i \right) \right]$$

where $\quad M_{i-1} + 4M_i + M_{i+1} = \frac{6}{h^2} \left(y_{i-1} - 2y_i + y_{i+1} \right), i = 1, 2, \ldots\ldots, n$

Example 9. *Evaluate* $I = \int_0^1 \frac{1}{1+x} dx$, using the cubic spline method.

Sol. Take $h = 0.5, n = 2.$ so that we have the following table of values of x and $y = \frac{1}{1+x}$

TABLE

x :	0	0.5	1
y :	1	$\frac{2}{3} = 0.66$	0.50

Taking $M_0 = M_2 = 0$ we obtaining $4M_1 = 6 \times 4 \left(1 - \frac{4}{3} + \frac{1}{2} \right) = 4$.

$\therefore \quad M_1 = 1,$

$$I = \frac{1}{4} \left(1 + \frac{2}{3}\right) - \frac{1}{192} (0 + 1) + \frac{1}{4} \left(\frac{2}{3} + \frac{1}{2} \right) - \frac{1}{192} (1)$$

$$= \frac{1}{4} \times \frac{5}{3} - \frac{1}{192} - \frac{1}{4} \times \frac{7}{6} - \frac{1}{192} = \frac{37}{96} = 0.385$$

20.3 Application of Difference Equations to Integration

The finite difference equation method can be used conveniently some times in the evaluation of integrals. We give below an illustration.

Example 10. *If* $u_n = \int_0^\pi \frac{\cos n x\, dx}{5 - 3 \cos x}$, *prove that*

$$3u_{n+2} - 10u_{n+1} + 3u_n = 0, \; n \neq -1$$

Hence prove that if n is an integer, then $u_n = \frac{\pi}{4} \left(\frac{1}{3} \right)^{|n|}$

Sol. We have $\quad u_n + u_{n+1} = \int_0^\pi \frac{\cos nx + \cos (n+2)\, x\, dx}{5 - 3 \cos x}$

$$= 2\int_0^{\pi} \frac{\cos (n+1) x \cos x \, dx}{5 - 3 \cos x}$$

$$= 2\int_0^{\pi} \cos (n+1) x \left\{ -1 + \frac{5}{5 - 3 \cos x} \right\}$$

$$= \frac{10}{3}\int_0^{\pi} \frac{\cos (n+1)x}{5 - 3 \cos x} = \frac{10}{3} u_{x+1}$$

is $\qquad 3u_{n+2} - 10u_{n+1} + 3u_n = 0 \qquad ...(i)$

This is a difference equation with auxiliary equation

$$3E^2 - 10E + 3 = 0 \Rightarrow (3E-1)(E-3) = 0$$

$\therefore \qquad E = 3$ or $\frac{1}{3}$ with the solution

$$u_n = A3^n + B\left(\frac{1}{3}\right)^n \qquad (ii)$$

To determine A and B, we find u_0 and u,

$$u_0 = \int_0^{\pi} \frac{dx}{5 - 3 \cos x} = \frac{\pi}{4} \text{ on putting } \tan \frac{x}{2} = t$$

and $$u_1 = \int_0^{\pi} \frac{\cos x \, dx}{5 - 3 \cos x} = \frac{1}{3}\int_0^{\pi} \left[-1 + \frac{5}{5 - 3 \cos x} \right] dx$$

$$= -\frac{1}{3}\int_0^{\pi} dx + \frac{5}{3} u_0 = -\frac{\pi}{3} + \frac{5}{3} \cdot \frac{\pi}{4} = \frac{\pi}{12}$$

$\therefore$ From (*ii*), $u_0 = A + B = \frac{\pi}{4}$ and $u_1 = 3A + \frac{1}{3} B = \frac{\pi}{12}$

$\because \qquad A_0 = 0$ and $B = \frac{\pi}{4} \quad \therefore \quad u_n = \frac{\pi}{3}\left(\frac{1}{3}\right)^n$

Also since $u_{-n} = u_n$, we have $u_n = \frac{\pi}{4}\left(\frac{1}{3}\right)^{|n|}$.

Ex. If $I_n = \int_0^1 (\log x)^n x^a$ where $a > 0$ and n is a positive inger then show that

$(a+1) I_n + nI_{n-1} = 0$ and hence show that

$$I_n = \frac{(-1)^n n}{(a+1)^{n+1}}.$$

21. MISCELLANEOUS EXERCISES

This chapter includes questions on various topics and are more trickly and hence hint have been given

1. Choose the correct answer :

(*a*) A matrix has eigen values $-1, 2, 3$. The eigen values of $(A+I)^2$ are

(*i*) 2, 9, 16 (*ii*) 0, 4,. 9 (*iii*) 0, 9, 16 (*iv*) 1, 4, 9

(*b*) Let $f = [-10, 15] \to R$ be a unimodal function with $f(0) = 20$ and $f(5) = 10$. Then the maximum of f over $[-10, 15]$ will be in the interval

(*i*) $[-10, 0]$ (*ii*) $[-10, 5]$ (*iii*) $[0, 15]$ (*iv*) $[0, 5]$

(*c*) For the function $f(x_1, x_2) = x_1^2 + 2x_2^2$, the direction of descent in the Newton method at (0, 1) is

(*i*) $(-1, 0)$ (*ii*) $(-1, 2)$ (*iii*) $(-2, 1)$ (*iv*) $(0, -1)$

(*d*) The error equation of an iteration method to find a simple root of the equation $f(x) = 0$ is of the form $\varepsilon_{k+1} = c\varepsilon_k \varepsilon_{k-1}$ where c is independent of k. The order of method is

(*i*) 1·62 (*ii*) 2 (*iii*) 1 (*iv*) 1·5

2. Find the eigen values and eigen vectors of the matrix

$$\begin{bmatrix} 2 & -1 & 1 \\ -1 & 2 & -1 \\ 1 & -1 & 2 \end{bmatrix}$$

3. Obtain the cubic spline approximation for the function given in the tabular form

x :	0	1	2	3
$f(x)$:	2	3	10	29

[**Hint.** Spline equations are $4M_1 + M_2 = 36, M_1 + 4M_2 = 72$

$\therefore \quad M_1 = 24/5, M_2 = 84/5$].

4. Dtermine the constants w_0, w_1 and x_0 such that the method

$$\int_{-1}^{+1} f(x) = w_0 f(x_0) + w_1 f(-x_0) \qquad ...(1)$$

is of the highest possible order and find the order of the method. Use this method to evaluate $\int_0^1 \frac{dx}{1+x^2}$.

[**Hint.** Let $f(x) = ax^2 + bx + c$ then (*i*) gives

$$2a/3 + 2c = a(w_0 + w_1)\, x_0^2 + b(w_0 - w_1)\, x_0 + c(w_0 + w_1)$$

$\therefore \; w_0 - w_1 = 0, w_0 + w_1 = 2, 3x_0^2 = 1$. Hence etc.]

5. Determine the constants α, β and γ such that the method

$$y''(x_0) = \alpha y(x_0 - h) + \beta y(x_0) + \gamma y(x_0 + h)$$

is of the highest possible order. Find the error term and the order of the

is of the highest possible order. Find the error term and the order of the method. Determine the optimal value of h such that

maximum round off error + maximum truncation error = minimum.

[**Hint.** $y(x_0 + h) = y(x_0) \pm hy_0' + \frac{h^2}{2} y_0'' + \cdots$

Equate coefficients of various terms and get values of α, β and γ]

6. The equation $x^3 - 5x^2 + 4x - 3 = 0$ has one root near $x = 4$. Determine the value of k which will give the fastest convergence with the iteration formula

$$x_{n+1} = \left[3 + (k-4)\, x_n + 5x_n^2 - x_n^3 \right] / k,$$

k being an integer. Use this value of k to find x_2

(*B.A. Hons. Maths, Delhi, 1996*)

[**Hint.** Let $g(x) = \left[3 + (k-4)\, x_n + 5x_n^2 - x_n^3 \right] / k$

Show that $g'(4) = (k - 12)/k$ which should vanish for fastest convergence $\therefore\ k = 12,\ x_1 = \left(3 + 8x_0 + 5x_0^2 - x_0^3 \right.\ 12 = 4{\cdot}25$ when $x_0 = 4$

7. Given $\frac{dy}{dx} = e^x + y^2$ with $y(0) = 1$, find approximately the value of y for $x = 0{\cdot}1$ by Picard method.

[**Hint.** $y_1 = y_0 + \int_{x_0}^{x} f(x, y_0)\, dx = 1 + \int_0^x (e^x + 1)\, dx = e^x + x,$

$$y_2 = 1 + \int_0^x \{e^x + (e^x + x)^2\}\, dx = \text{etc.}$$

8. Obtain a linear polynomial approximation to the function $f(x) = x^2$ on $[0, 1]$ using the least square approximation, with $w(x) = 1$.

[**Hint.** $z = \int_0^1 [x^2 - (c_0 + c_1 x)]^2\, dx = \text{minimum}$

Put $\frac{\partial z}{\partial c_0} = 0,\ \frac{\partial z}{\partial c_1} = 0$ and find c_0 and c_1]

9. Use Runge-Kutta fourth-order method to obtain the value of $y(0{\cdot}1)$ for the initial value problem

$$y' = -\,2xy^2,\ y(0) = 1,\ \text{step size } h = 0{\cdot}1$$

[**Hint.** $k_1 = -\,(0{\cdot}1)(2 \times 0 \times 1) = 0,\ k_2 = 0{\cdot}1\,[-\,2(0{\cdot}05)1] = -\,0{\cdot}01$

$k_3 = 0{\cdot}1\,[-\,2(0{\cdot}05)(1 - 0{\cdot}005)^2] = -\,0{\cdot}0099$

$k_4 = 0{\cdot}1\,[-\,2(0{\cdot}1(1 - 0{\cdot}0099)^2] = -\,0{\cdot}0196]$

10. A root of the equation $e^x = 1 + x + \frac{x^2}{2} + \frac{x^3}{6} e^{0{\cdot}3x}$

is close to $2{\cdot}5$. Find this root correct to three decimal places, using Newton-Raphson method.

$$\left[\textbf{Hint. } x_{n+1} = x_n - \frac{f_n}{f_n'},\ f(x) = e^x - 1 - x - \frac{x^2}{2} - \frac{x^3}{6} e^{0.3x}\right.$$

$$f'(x) = e^x - 1 - x - \frac{1}{6} e^{0.3x} (0.3x^3 + 3x^2) = e^x - 1 - x - \frac{1}{2} e^{0.3x} (0.1x^3 + x^2)$$

$$x_4 = 2.5 - \frac{0.0455}{0.4130} = 2.3922,\ x_2 = 2.3922 - \frac{0.0026}{0.2778} = 2.3648$$

$$\left. x_3 = 2.3648 - \frac{0.00045}{0.5444} = 2.3641 \right]$$

11. The initial value problem

$$y' = x^2 + y^2,\ y(1) = 1$$

is given . find $y(1.1)$ using (*i*) Taylor's series method of second order and (*ii*) Runge-Kutta method of fourth-order assume that $h = 0.1$

(*B.A. Hons. Maths, Delhi, 1994*)

[**Hint.** (*i*) $y(1.1) = y(1) + hy'(1) + h^2 y''(1)$

$$= 1 + 0.1(1+1) + \frac{(0.1)^2}{2}(2 + 2 \times 1 \times 1) = 1.23$$

$$y' = x^2 + y^2,\ y' = 2x + 2yy',\ y'(1) = 2,$$

$$k_1 = hf(1, 1) = 0.2,\ k_2 = hf(1.05, 1) = 0.2325$$

$$k_3 = hf(1.05, 1.5625) = 0.23471,\ k_4 = hf(1.61, .23471) = 0.27345$$

12. The method $x_{n+1} = x_n - \dfrac{f_n}{g_n}$

$$g_n = [f(x_n + f_n) - f_n]/f_n,\ n = 0, 1, 2, \ldots$$

where $f_n = f(x_n)$ is used to find a simple root of the equation $f(x) = 0$. Determine the order of the method. Apply this method to find two iterations when finding a root of the equation $x^4 - x - 10 = 0$ starting with $x_0 = 1.8$.

[**Hint.** $f_n = f(\xi + \varepsilon_n) = \varepsilon_n f' + \frac{1}{2}\varepsilon_n^2 f'' + \cdots + f(\xi) = 0,$

$$g_n = [f(x_n) + f_n f'(x_n) + \frac{1}{2} f_n^2 f''(x_n) + \cdots - f_n]/f_n$$

$$- f_n' + \frac{1}{2} f_n f_n' \qquad [\because f(x_n) = f_n]$$

$$\left. \varepsilon_{n+1} = \frac{1}{2}\left[1 + f'(\xi)] \frac{f''(\xi)}{f'(\xi)} \varepsilon_n^2 + 0(\varepsilon_n^3) \right],\ \text{order is 2} \right]$$

13. Solve the system of equations

$$x_1 + x_2 - x_3 = 2,$$

$$2x_1 + (2 + \varepsilon) x_2 + 5x_3 = -3$$

$$3x_1 + 2x_2 - (3 + \varepsilon) x_3 = 6$$

using the Gauss-elimination method with partial pivoting where ε is small such that $0(\varepsilon^3)$ can be neglected. (*B.A. Hons. Math, Delhi, 1997*)

[**Hint.** Eliminate x_1 and write equations as $x_1 + x_2 - x_3 = 2, - x_2 - \varepsilon x_3 = 0,$ $\varepsilon x_2 + 7x_3 = -7$, as pivot in the second equation was small. Eliminate x_2 and get answer by back substitution.]

14. Solve the equation using the Gauss-elimination method where ε is so small such that $1 \pm \varepsilon^2 \simeq 1$

$$x_1 + x_2 + x_3 = 6$$
$$3x_1 + (3+\varepsilon)\, x_2 + 4x_3 = 20$$
$$2x_1 + x_2 + 3x_3 = 13$$

15. Set up the Gauss-Seidel iteration scheme in matrix form to solve the system of equations

$$\begin{bmatrix} 4 & 1 & 0 \\ 2 & 4 & 2 \\ 0 & 1 & 4 \end{bmatrix} \begin{bmatrix} x \\ y \\ z \end{bmatrix} = \begin{bmatrix} 3 \\ 0 \\ 3 \end{bmatrix}$$

Find the rate of convergence of the iteraction scheme.

[**Hint.** $D + L = \begin{bmatrix} 4 & 0 & 0 \\ 2 & 4 & 0 \\ 0 & 1 & 4 \end{bmatrix}$, $(D+L)^{-1} = \begin{bmatrix} 1/4 & 0 & 0 \\ -1/8 & 1/4 & 0 \\ 1/32 & -1/16 & 1/4 \end{bmatrix}$

$$H = -(D+L)^{-1} u^{-1} = -\frac{1}{64} \begin{bmatrix} 16 & 0 & 0 \\ -8 & 16 & 0 \\ 2 & -4 & 16 \end{bmatrix} \begin{bmatrix} 0 & 1 & 2 \\ 0 & 0 & 2 \\ 0 & 0 & 0 \end{bmatrix}$$

$$C = -(D+L)^{-1} b = \frac{1}{64} \begin{bmatrix} 16 & 0 & 0 \\ -8 & 16 & 0 \\ 2 & -4 & 16 \end{bmatrix} \begin{bmatrix} 3 \\ 0 \\ 3 \end{bmatrix} = \frac{1}{64} \begin{bmatrix} 48 \\ -24 \\ 54 \end{bmatrix}$$

Iteration matrix $H = \begin{bmatrix} 0 & -1/4 & 0 \\ 0 & 1/8 & -1/2 \\ 0 & -1/32 & 1/8 \end{bmatrix}$

Eigen values are $\lambda = 0$ and the roots of $\lambda^2 - \frac{1}{4}\lambda = 0$

$\rho(H) = 0{\cdot}25$, $v = 1{\cdot}39$; $x^{(1)} = Hx^{(0)} + C = [0{\cdot}95 \; -0{\cdot}875 \; 0{\cdot}96875]$

16. (*a*) Solve the following set of linear equations by the use of Gauss-Seidel iteration method correct to 2 decimals. Use the initial approximation as $x = 0{\cdot}8$, $y = 0{\cdot}2$ and $z = -1{\cdot}3$.

$$6x + y - z = 6$$
$$3x + 7y - 2z = 6{\cdot}5$$
$$x - 3y - 5z = 6{\cdot}25$$

(*b*) For what values of a, Gauss-Seidel method converges for the system of equations

$$\begin{bmatrix} 1 & -a \\ -a & 1 \end{bmatrix} \begin{bmatrix} x_1 \\ x_2 \end{bmatrix} = \begin{bmatrix} b_1 \\ b_2 \end{bmatrix}$$

(*B.A. Hons. Maths, Delhi, 1995*)

[**Hint.** (*b*) The Gauss-Seidel iteration method is

$$\begin{bmatrix} 1 & 0 \\ -a & 1 \end{bmatrix} x^{(k+1)} = \begin{bmatrix} 0 & a \\ 0 & 0 \end{bmatrix} x^{(k)} + \begin{bmatrix} b_1 \\ b_2 \end{bmatrix}$$

$$\Rightarrow \quad x^{(k+1)} = \begin{bmatrix} 0 & a \\ 0 & a^2 \end{bmatrix} x^{(k)} + \begin{bmatrix} 1 & 0 \\ a & 1 \end{bmatrix}\begin{bmatrix} b_1 \\ b_2 \end{bmatrix}$$

The eigen values of the Gauss-Siedel iteration matrix are given by

$$|H - \lambda I| = \begin{vmatrix} -\lambda & a \\ 0 & a^2 - \lambda \end{vmatrix} = \lambda(\lambda - a^2) = 0, \ H = \begin{bmatrix} 0 & a \\ 0 & a^2 \end{bmatrix}$$

$\therefore \ \lambda = 0, a^2.$

For $|a| < 1$, the Gauss-Seidel method converges.

17. Solve the equation $e^x - 3x = 0$ for a simple root, taking starting value $x_0 = 0{\cdot}0$ by successive iteration method correct up to 3 decimal figures.

[**Hint.** $x = \frac{1}{3} e^x = g^{(x)}$, say *(B.A. Maths. Hons. Delhi, 1995)*

$$g'(x) = \frac{1}{3} e^x = \frac{1}{3} e^0 = \frac{1}{3} < 1$$

$\therefore \quad x_{k+1} = g(x_k)$ converges

$$x_1 = \frac{1}{3} e^x = \frac{1}{3} = \cdot 333333$$

$$x_2 = \frac{1}{3} e^x = \frac{1}{2} e^{\cdot 333333} = \frac{1}{3} \times 1{\cdot}3956124 = 0{\cdot}465204.$$

Similarly $x_3, x_4, \ldots$ can be obtained

By Newton-Raphson method,

$$f(x) = e^x - 3x = 0, \ f'(x) = e^x - 3 \text{ root lying between 0 and 1}$$

Let $\quad x_0 = 0, \ f(0) = 1, f'(0) = 1 - 3 = -2,$

$$x_1 = x_0 - \frac{f_0}{f_0^1} = 0 - \frac{1}{-2} = \frac{1}{2} = 0{\cdot}5, \text{ etc.}$$

18. Find the inverse of the coefficient matrix of a system of equations given by

$$\begin{bmatrix} 1 & 1 & 1 \\ 4 & 3 & -1 \\ 3 & 5 & 3 \end{bmatrix}\begin{bmatrix} x_1 \\ x_2 \\ x_3 \end{bmatrix}\begin{bmatrix} 1 \\ 6 \\ 4 \end{bmatrix}_1$$

by using Gauss-Jordan method with partial pivorting and hence solve the system.

$$\left[\textbf{Hint.}\text{ Reduce} \begin{bmatrix} 1 & 1 & 1 \\ 4 & 3 & -1 \\ 3 & 5 & 3 \end{bmatrix}\begin{bmatrix} 1 & 0 & 0 \\ 0 & 1 & 0 \\ 0 & 0 & 1 \end{bmatrix} \sim \begin{bmatrix} 1 & 0 & 0 \\ 0 & 1 & 0 \\ 0 & 0 & 1 \end{bmatrix}\begin{bmatrix} \frac{7}{5} & \frac{1}{5} & \frac{-2}{5} \\ -\frac{3}{2} & 0 & \frac{1}{3} \\ \frac{11}{10} & \frac{-1}{5} & \frac{-1}{10} \end{bmatrix} \right.$$

$$\text{so that} \begin{bmatrix} x_1 \\ x_2 \\ x_3 \end{bmatrix} = \begin{bmatrix} \frac{7}{5} & \frac{1}{5} & \frac{-2}{5} \\ \frac{-3}{2} & 0 & \frac{1}{2} \\ \frac{11}{1} & \frac{-1}{5} & -\frac{1}{10} \end{bmatrix}\begin{bmatrix} 1 \\ 6 \\ 4 \end{bmatrix} = \begin{bmatrix} 1 \\ \frac{1}{2} \\ -\frac{1}{2} \end{bmatrix} \Bigg]$$

19. Define the following :

(*i*) Convergence of a numerical scheme;

(*ii*) Stability of a numerical method.

Establish a relation of these with consistency of the solution.

20. The method

$$y_{n+1} = y_n + \frac{1}{4}(k_1 + 3k_2)$$

$$k_1 = hf(x_n, y_n)$$

$$k_2 = hf\left(x_n + \frac{2h}{3},\ y_n + \frac{2}{3}k_1\right)$$

is used to solve the initial value problem $y' = \lambda y$, $\lambda < 0$. Show that the interval of absolute stability of the method is $(-2, 0)$.

21. Determine the convergence factor for the Gauss-Jacobi, and Gauss-Seidel methods for the system

$$\begin{bmatrix} 4 & 0 & 2 \\ 0 & 5 & 2 \\ 5 & 4 & 10 \end{bmatrix} \begin{bmatrix} x_1 \\ x_2 \\ x_3 \end{bmatrix} = \begin{bmatrix} 4 \\ -3 \\ 2 \end{bmatrix}$$

(Delhi, B.Sc. Math Hons., 1997)

We have

$$4x_1 + 0 \cdot x_2 + 2x_3 = 4,$$

$$0 \cdot x_1 + 5x_2 + 2x_3 = -3,$$

$$5x_1 + 4x_2 + 10x_3 = 2.$$

$$\mu_i = \sum_{\substack{j=1 \\ j \neq i}}^{n} \left| \frac{a_{ij}}{a_{ii}} \right| \quad \forall\ i.$$

Gauss-Jacobi method converges if $\mu_i < 1$

where
$$\mu_1 = \frac{a_{12} + a_{13}}{a_{11}} = \frac{2}{4} = 0.5$$

$$\mu_2 = \frac{a_{21} + a_{22}}{a_{22}} = \frac{2}{3} = 0.4$$

$$\mu_3 = \frac{a_{31} + a_{32}}{a_{33}} = \frac{9}{10} = 0.9$$

Gauss-Seidel method converges if $\eta_i < 1$

where
$$\eta_i = \frac{\beta_0}{i - \alpha_i},$$

where
$$\alpha_i = \sum_{j=1}^{i-1} \left| \frac{a_{ij}}{a_{ii}} \right|, \quad \beta_i = \sum_{j=i+1}^{n} \left| \frac{a_{ij}}{a_{ii}} \right|$$

and
$$\alpha_1 = \beta_{11} = 0.2$$

$$\alpha_2 = \frac{a_{21}}{a_{22}} = 0$$

$$\alpha_3 = \frac{a_{31} + a_{32}}{a_{33}} = \frac{9}{10} = .9$$

$$\beta_1 = \frac{a_{12}+a_{13}}{a_{11}} = \frac{1}{2} = 0{\cdot}5$$

$$\beta_2 = \frac{a_{23}}{a_{22}} = \frac{2}{5} = 0{\cdot}4\,, \qquad \eta_1 = \frac{\beta_1}{1-\alpha_1} = 0{\cdot}5,$$

$$\eta_2 = \frac{\beta_2}{1-\alpha_2} = 0{\cdot}4\,, \quad \eta_3 = 0$$

$$\eta = \max(\eta_1, \eta_2, \eta_3) = \max(0, 0{\cdot}5, 0{\cdot}4) = 0{\cdot}5$$

Convergence factor for Gauss-Seidel method is 0·5.

Convergence factor for Gauss-Jacobi Method is $\mu = 0{\cdot}9$.

22. (*a*) Find the Laplace transform of the function $f(t)$ where

$f(t) = \sin\left(t - \frac{2\pi}{3}\right),\ t > \frac{2\pi}{3}$; 0 for $t < \frac{2\pi}{3}$.

(*b*) Find the solution of the initial value problem

$y'' + 5y' + 4y = 0, y(0) = 0 = y'(0): f(t) = 1$ for $0 < t < 1, = 0$ otherwise.

23. (*a*) Show that in fitting the curve $y = ab^x + c$ to the three points $(x_1, y_1), (x_2, y_2)$ and (x_3, y_3) where x_1, x_2, x_3 are in G.P., the value of c is given by

$$c = \frac{y_1y_3 - y_2^2}{y_1 - 2y_2 + y_3}$$

and that the equation of the curve is reduced to the linear form $Y = A + Bx$, where

$$Y = \log_{10}(y - c), A = \log a, B = \log_{10} b.$$

(*b*) By proper transformation, convert the relation $y = a + bxy$ or $y = \frac{b}{x(x-a_1)}$ to a linear form and find the equation to fit the data

x :	–4	1	2	3
y :	4	6	10	8

[**Hint.** (*a*) c can be obtained from $x_1 = x_3 = x_2^2$.

i.e., $(y_1 - c)(y_2 - c) - (y_2 - c)^2 \Rightarrow c - \frac{y_1y_3 - y_2^2}{y_1 + y_3 - 2y_2}$ (*D.Sc. D.R., 1995*)

(*b*) Let $x = Y, \frac{1}{xy} = X$ so that $Y = a + bX$.]

24. Assuming that the follow table of values of x and y is known and that the polynomial fitted to the data is linear. Then estimate y_k for $x = x_k$ is given by

$$y_1 = y_i + \left(\frac{y_{i+1} - y_i}{x_{i+1} - x_i}\right)(x_k - x_i).$$

25. (*a*) The table below given the velocity v of a moving particle at time t seconds. Find the distance covered by the particle in 12 seconds and also the acceleration at $t = 2$ seconds

x:	0	2	4	6	8	10	12
y:	4	6	16	34	60	94	136

[**Hint.** Let S be the distance covered in t secs, then $\frac{dS}{dt} = v$ and $a = \frac{dv}{dt}$ in

$$S = \int_0^{12} v\,dt = \text{etc. and } a = \left(\frac{dv}{dt}\right)_{t=2} = \frac{1}{k}\left[\Delta v_0 - \frac{1}{2}\Delta^2 v_0 + \frac{1}{3}\Delta^3 v_0\right]$$

where $v_0 = 6$.]

26. (*a*) Evaluate the determinant given below (nth order) by forming the difference equation

$$D_n = \begin{vmatrix} 2\cos\theta & 1 & 0 & 0 & \cdots & 0 \\ 1 & 2\cos\theta & 1 & 0 & \cdots & 0 \\ 0 & 1 & 2\cos\theta & 1 & \cdots & 0 \\ \vdots & \vdots & \vdots & 1 & \cdots & \\ 0 & 0 & 0 & 1 & \cdots & 2\cos\theta \end{vmatrix}$$ *(AMIE)*

[**Hint.** $D_n = 2\cos\theta\, D_{n-1} - D_{n-2}$ so that auxiliary equation is

$$E^2 - 2E\cos\theta + 1 = 0$$

$$\Rightarrow \quad E = e^{\pm i\theta}\; D_n = A\,e^{in\theta} + B\,e^{-in\theta} = A'\cos n\theta + B'\sin n\theta]$$

(*b*) Show that n straight lines, no two of which are parallel and no three of which meet in a point, divided a plane into $\frac{1}{2}(n^2 + n + 2)$ parts.

[**Hint.** Let y_n = no. of compartments formed by n st. lines. Then

$$y_{n+1} = y_n + (n+1) \Rightarrow y_n = D^{-1}(n+1) = \text{etc.}]$$

(c) In a plane, n circles are drawn so that each circle intersects all the other cricles and no three of them meet in a point plane is divided into $(n^2 - n + 2)$ compartment.

[**Hint.** If y_n be the number of compartments when n circles are drawn, then $y_{n+1} = y_n + 2n \Rightarrow \Delta y_n = 2n$]

27. Using Taylor's series method, find $y(1{\cdot}1)$ and $y(1{\cdot}2)$ correct to four decimal places given $\frac{dy}{dx} = xy^{\frac{1}{3}}$ and $y(1) = 1$. *(M.Sc., 1991)*

$$\left[\textbf{Hint. } y_0' = 1\,(1)^{\frac{1}{3}} = 1,\; y'' = \frac{1}{3}xy^{-2/3}y' + y^{1/3} = \frac{4}{3},\; y_0''' = \frac{8}{9}\right]$$

Ans. $y_1 = y_0 = hy_0' + \frac{h^2}{2!}y_0'' + \cdots = 1{\cdot}10681,\; y_2 = y_1 + hy_1' + \frac{h^2}{2!}y_1'' + \cdots = 1{\cdot}227727$

28. Compute y at $x = 0{\cdot}25$ by Modified Euler method given

$$y' = 2xy = f(x, y),\; y(0) = 1.$$ *(B.R. Nov., 1995)*

Ans. 1·0645, Exact value = 1·0645

$$\text{[Use: } y_{n+1} = y_n + h\left[f\left(x_n + \frac{1}{2}h,\; y_n + \frac{1}{2}hf(x_n, y_n)\right)\right]$$

29. Determine the value of $y(0{\cdot}4)$ using Milne's method given $y' = xy + y^2$, $y(0) = 1$, use Taylor series to get the values of $y(0{\cdot}1), y(0{\cdot}2)$ and $y(0{\cdot}3)$. *(M.Sc., 1989)*

Ans. $y(0{\cdot}4) = 1{\cdot}83698$

$$\left[\textbf{Hint. } y_{4,p} = y_i + \frac{4h}{3}[2y_1' - y_2' + 2y_3'] = 1{\cdot}2771,\ y'_{4,p} = 2^1(1 + x_4^2)\,y_4^2 = 0{\cdot}9460\right.$$

By corrector method, $y_{4,b} = y_2 + \frac{h}{3}\left[y_2' + 4y_3' + y'_{4,p}\right]$

30. (*i*) Show that

(*a*) $u_{xy}(x_0, y_0) = \frac{1}{2h}[u(x_0 + h, y_0) - u(x_0 - h, y_0)],\ x_0 - h < \theta < x_0 + h$

where the truncation error is $-\frac{1}{6}h^2\,u_{xx}(\theta, y_0)$

(*b*) $u_{xy}(x_0, y_0) = \frac{1}{4hk}\Big[u(x_0 + h, y_0 + k) - u(x_0 - h, y_0 + k)$

$\qquad - u(x_0 + h, y_0 - h) + u(x_0 - h, y_0 - k)\Big]$.

(*ii*) Solve $u_{xx} + u_{yy} = 0$ over the square mesh of side 4 units satisfying the boundary condition $u(x, y) = x^2$ for $0 \le x \le 4$.

(*iii*) Obtain the following recurrence formula for the parabolic equation $u_{xx} = a\,u_t$

with boundary conditions $u(0, t) = T_0$, $u(l, t) = T_1$), $nh = l$ and with initial conditions $u(x, 0) = f(x)$, $0 < x < 1$

$u_{i,\,j+1} = \lambda\,u_{i+1,\,j} + (1 - 2\lambda)\,u_{i,\,j} + \lambda\,u_{i-1,\,j}$

what is the corresponding result for $\lambda = \frac{1}{2} = \frac{k}{ah^2} \Rightarrow k = \frac{1}{2}h^2$.

This result is known as Bender-Schmidt recurrence formula.

31. Answer the following question:

(*a*) Iteration method is a self correcting method.

(*b*) The order of convergence of Newton Raphson method is_____.

(*c*) Can a non-linear relation between x and y be transformed to a linear relation always?

(*d*) Which of the following formula is a particular case of Runge-Kutta formula of the second order?

(*i*) Taylor's series formula

(*ii*) Euler's modified formula

(*e*) The finite difference formula giving solution to Poisson's equation $u_{xx} + u_{yy} = -f(x, y)$ with $\Delta x = \Delta y = h$ is given by $u_{ij} = \cdots$

(*f*) If $C_n = 2^{n/2}\cos\frac{n\pi}{4}$, $S_n = 2^{n/2}\sin\frac{n\pi}{4}$, prove that $\Delta C_n = S_n$ and $\Delta S_n = C_n$.

(*g*) Using power method, calculate the dominant eigen values of:

(*i*) $A = \begin{bmatrix} 10 & 2 & 1 \\ 2 & 10 & 1 \\ 2 & 1 & 10 \end{bmatrix}$ (*ii*) $A = \begin{bmatrix} 3 & 2 & 4 \\ 2 & 0 & 2 \\ 4 & 2 & 3 \end{bmatrix}$

Ans. (*i*) 13, (1, 1, 1)′ (*ii*) 8, $(1, \frac{1}{2} 11)'$

Then $x_1 = g(x_o) = \frac{\cos(0.5)}{0.52} = 0.61694$

$x_2 = g(n_2) + \frac{\cos(0.515.94)}{0.51594} = 0.51924$

$x_3 = g(x_2) + \frac{Cu\,(0.51)}{0.5194} = 0.51655$, etc.

The root is 0.5177 correct to 4 decaides

(*iii*) Let $f(x) = x^3 - 2x - 5 + 0 \;: f(x) + -(0), (f(3) = 16 > 0$

Root lies between 2 and 3 and in closer to 2

Also $x = (2x+5)^3 + g(1)$ in $g'(x) = \frac{2}{3} \cdot \frac{1}{(2x+5)^{2/3}}$ in $| \, g'(x)/< 1 \, V^{x \, in}$

Table $x_0 = 2.0$, then $x_1 = (2x_0 + 5)^{1/3} = 2.0801$

35. Given $y' + xy' + y = 0, y(0) + 1, y'(0) = 0_1$ find the value of $y\,(0.1)$, using Range kutta method of fourth ordes for second, order differential equation.

Hint. $y + - xy' - y - Z = -x - y$ where $x = y'$.

Thus $y'' + Z' = -x2 - y2$

$\frac{dy}{dx} z = f_1\,(x_1 y_1 z)$, $\frac{dz}{dx} = -xz - y = f_2 - (x_1 y_1)$

given $y_0 = 1_1\, z_0 = y_0' = 0$

Show that $k_1 = hf_1\,(x_0, y_0, z) = (0.1) f_1\,(0, 1, 0) = (0.1)\,(0 = 0)$

$l_1 = hf_2\,(x_0, y_0, z) = -0.1$, $k_2 = hf_1\,(x_0 + \frac{1}{}h, y_0 + \frac{1}{}k, z_0 + \frac{1}{2}l_1)$

$l_2 = -0.0995$, ...

Example. Solve the differential equation $y' = (1 + y, y\,(0) = 1$, $t <\, [0,1]$ by Euler method using $h = 0.1$. If the exact value is $y\,(1) = 3.4564$, find the exact error

Solution. Euler's method is

$y_{n+1} = y_n + hy_n$, $= y_n + h\,(t_n + y_n)$, given $= (1+h)\,y_n + ht_n$

Here h = 0.1, y (0) =1

$y_1 = y_0\,(1 + 0.1) + (0.1)\,(01 = 1.1$

$y_2 = (1.1)\,(1.1) + (0.1)\,(0.1) = 1.22$

$y_3 + (1.22)\,(1.22) + (0.1)(0.1) = 1.362$

$y_4 = 1.5282, y_5 = 1.72102, y_6 = 1.943122,$

$y_7 = 2.197439, \frac{4}{8} = 2.487178, y_9 = 2.815895$

$y_{10} = 3.187485 = y\,(1)$

$\therefore$ Actual error $y(1) \_ y_{10} = 3.436564 - 3.187485 = 0.2491$.

ANSWERS

MISCELLANEOUS EXERCISES

1. (*a*) (*iii*) (0, 9, 16) (*b*) (*ii*) [– 10, 5]

(*c*) (*iv*) (0, – 1) (*d*) (*i*) 1·62

2. $|A - \lambda I| = 0 \Rightarrow \lambda^3 - 6\lambda^2 + 9\lambda - 4 = 0 \quad \therefore \quad \lambda = 1, 1, \cdot 4$

$\lambda = 1; [1\ 1\ 0]^T$ and $[-1, 0, 1]^T$ or mutiples $x_1 - x_2 + x_3 = 0$

$\lambda = 4\ [1\ -1\ 1]^T$.

3. $f(x) = \dfrac{14}{5}[(3-x)^3 - (3-x)] + 19x - 28$

4. $w_0 = w_1 = 1;\ x_0 = 1/\sqrt{3}$

(*b*) $\displaystyle\int_{x=0} \frac{dx}{1+x^2} = \frac{1}{2}\int_{t=-1}^{+1} \frac{dh}{1+\left(\frac{1+t}{2}\right)^2}$ where $2x = 1 + t$

$\displaystyle = \frac{1}{2}\int_{-1}^{+1} \frac{4dt}{4+(1+t)^2} = \frac{1}{2}\left[\frac{4}{4+\left(1+\frac{1}{\sqrt{3}}\right)^2} + \frac{4}{4+\left(1-\frac{1}{\sqrt{3}}\right)}\right]$ by the use of the formula

$\displaystyle = \frac{1}{2}\left[\frac{4}{6\cdot4880} + \frac{4}{4\cdot1786}\right] = \frac{1}{2}[1\cdot5738] = 0\cdot7869.$

5. $\alpha = \gamma = 1/h^2,\ \beta = 2/h^2$

Truncation error (T.E.) $= \dfrac{h^2}{12} y''(\xi)$, order $= 2$

Max (T.E.) $= \dfrac{h^2}{12} M_4$ and Max (Round off error) $= 4\varepsilon/h^2$

choose h such that $\dfrac{h^2}{12} M_4 + \dfrac{4\varepsilon}{h^2} =$ minimum which gives $h = \left(\dfrac{48\varepsilon}{M_4}\right)^{1/4}$

6. $x_2 = \left[3 + 8x_1 + 5x_1^2 - x_1^3\right]/12$ when $x_1 = 4\cdot25$

$\therefore\ x_1 = [3 + 3\cdot4 + 9\cdot3125 - 76\cdot5625]/12 = 4\cdot212$

7. Put $x = 0\cdot1$ after integration and get y_2.

8. $x - \dfrac{1}{6}$ is the linear polynomial.

9. $y(0\cdot1) = y_1 = 1 + \dfrac{1}{6}[0 - 0\cdot02 - 0\cdot0198 - 0\cdot0196] = 0\cdot9901$

10. $x_3 = 2\cdot3641$, value of root of the equation $= 2\cdot364$

11. $y(0\cdot1) = y_1 = 1 + \dfrac{1}{6}[k_1 + 2k_2 + 2k_3 + k_4] = 1\cdot2342$

12. $f = x^4 - x - 10, x_0 = 1\cdot8, f_0 = -1\cdot3024$

$x_0 + f_0 = 0{\cdot}4976,\ f(x_0 + f_0)t - 10{\cdot}4363,\ g_0 = 7{\cdot}0131$

$x_1 = 1{\cdot}9857,\ f_1 = 3{\cdot}5616$

$x_1 + f_1 = 5{\cdot}5473,\ f(x_1 + f_1) = 931{\cdot}4018$

$g_1' = 260{\cdot}5122,\ x_2 = 1{\cdot}9720$

13. $x_3 = \dfrac{-7}{7-\varepsilon^2},\ x_2 = \dfrac{7\varepsilon}{7-\varepsilon^2},\ x_1 = \dfrac{7-2\varepsilon^2-7\varepsilon}{7-\varepsilon^2}$

14. $x_1 = 3, x_2 = 1, x_3 = 2$

16. (*a*) $x = \frac{1}{6}(6 - y_n + z_n),\ y_{n+1} = \frac{1}{7}(6{\cdot}5 - 3x_{n+1} + 2z_n),$

$z_{n+1} = \frac{1}{3}(x_{n+1} - 3y_{n+1} - 6{\cdot}25);\ x = 0{\cdot}75, y = 0{\cdot}25, z = -1{\cdot}25.$

(*b*) $x^{(k+1)} = \begin{bmatrix} 0 & a \\ 0 & a^2 \end{bmatrix} x^{(k)} + \begin{bmatrix} 1 & 0 \\ a & 1 \end{bmatrix}\begin{bmatrix} b_1 \\ b_2 \end{bmatrix}$

$\lambda I = \begin{vmatrix} -\lambda & a \\ 0 & a^2 - \lambda \end{vmatrix} = \lambda(\lambda - a^2) = 0$ gives $\lambda = 0, a^2$

the Gauss-Seidel method converges

22. (*a*) $L[f] = \int_0^\infty e^{-st} f(t)\,dt = \int_{2\pi/3}^\infty e^{-st} \sin\left(t - \frac{2\pi}{3}\right) dt$

$= \int_0^\infty e^{-s\left(u + \frac{2\pi}{3}\right)} \sin u\, du,\ u = t - \frac{2\pi}{3}$

$= e^{-\frac{2\pi s}{3}} L[\sin u]$

$= e^{-\frac{2\pi s}{3}}/(1 + s^2).$

(*b*) $s^2 y(5) - 5{\cdot}0 - 0 + 5sy(s) - 0 + 4y = \oint_0^1 e^{-st}\,dt$

$= -\frac{1}{s}(e^{-s} - 1)$

$\therefore\ y(s) = \dfrac{1 - e^{-s}}{s(s+1)(s+4)}$ etc.

EXERCISE 1, Page 33

1. (*i*) $(a-1)a^x.$ (*ii*) $a(b^c - 1)b^{cx},$

(*iii*) $-\dfrac{x \cdot 2^x}{(x+2)\,!}$ (*iv*) $-$ cosec 2^{x+1}

(*v*) $2\sinh\frac{b}{2}\cosh\left(a + \frac{b}{2} + bx\right)\cdot$ (*vi*) $2\sinh\frac{b}{2}\sinh\left(a + \frac{b}{2} + bx\right)\cdot$

(*vii*) $\dfrac{\sin a}{\cos ax \cos a\,(x+1)}\cdot$ (*viii*) $\tan^{-1}\left(\dfrac{a}{1+a^2x+a^2x^2}\right)\cdot$

(*ix*) $\log_e\left(1+\dfrac{1}{x}\right)\cdot$ (*x*) $\log\left(1+\dfrac{\Delta f(x)}{f(x)}\right)$ (*xi*) 0

2. (*i*) $6x + 6$. (*ii*) $a(e-1)^4 e^x$. (*iii*) $-36, a_1 a_2 \ldots a_n \cdot (n\,!)$.

(*iv*) $abcd\,(10\,!)$. (*v*) $a(b^c-1)^2 b^{cx}$. (*vi*) $2(\cosh x - 1)\sin(x+h)$

(*vii*) $6x - 2$.

3. (*i*) $6x$. (*ii*) $\dfrac{6}{(1+x)^2}\cdot$ (*iii*) $6h^2(3x-h)$ (*iv*) $\dfrac{6x h^2 + 6h^3}{(x+2h)^3}$

(*v*) $2(\cos h - 1)[\sin(x+h)+1]$ (*vi*) e^x.

(*vii*) $(e^{ah}-1)^n \cdot e^{ax+b}$

5. (*b*) 755.

7. (*a*) $P(x) = \frac{1}{6}(x^3+3x^2-16x+6) = -\frac{1}{6}x^{(3)} - x^{(2)} + x^{(1)} - 1$.

(*b*) (*i*) $u_x = x^3 - 2x^2 + 7x - 5$. (*ii*) $x^2 + 2x$

(*c*) (*i*) $\dfrac{e^x}{1-e} + c$, c being a constant. (*ii*) $\frac{1}{4}x^{(4)} + 2x^{(3)} + \frac{9}{2}x^{(2)} + 12x^{(1)} + c$.

(*iii*) $ax + c$. (*iv*) $\frac{1}{3}x^{(3)} + c$.

8. (*a*) $y = 2x^{(3)} + 3x^{(2)} + 2x^{(1)} - 10$, $\Delta y = 6x^{(2)} + 6x^{(1)} + 2$.

$\Delta^2 y = 12x^{(1)} + 6$, $\Delta^3 y = 12$.

(*b*) $x^{(4)} - 6x^{(3)} + 13x^{(2)} + x^{(1)} + 9$. (*c*) $\beta^2 = \gamma^2 - 4\gamma\alpha$.

(*e*) (*i*) 65540 should be corrected to 65536.

(*ii*) 0·38945 should be corrected to 0·38942. (iii) $\dfrac{1}{6}n(n+1)(2n+1)$

20. 1·399 **22.** (*a*) 31. (*b*) 96·4

23. $l_{35} = 394$, $l_{42} = 326$, $l_{47} = 274$,

24. 7880. **25.** (*a*) 0·6239987. (*b*) 16·1.

(*c*) – 13°34′40°7″, (using Backward interpolation formula).

26. (*a*) 7th term = 30, General term = $r(r-1)$. (*b*) 15

27. 53·6 thousand 28·196 – 0·12529 (*a*) 0.

(*b*) $u_x = x^2 + x + 1$, $u_{10} = 111$ (*d*) 139.

30. $u_4 = 1$ approx.

33. (*i*) 30, (*ii*) 540, (*iii*) 240, (*iv*) 120, (*v*) 1.

43. $e\left\{1 + t + t^2 + \frac{5}{4}t^3\right\}$

EXERCISE 2, Page 69

1. (*b*) $-\dfrac{a+b}{a^2b^2}, \dfrac{ab+bc+ca}{a^2b^2c^2}; \dfrac{abc+bcd+cda+dab}{a^2b^2c^2d^2}$

5. 2·8169. **6.** 2300

7. (*a*) $(x-5)^3+17(x-5)^2+98(x-5)+194$.

(*b*) $(x-3)^3+6(x-3)^2+2(x-3)-13$

8. (*d*) $f(5)=33, f(6)=67$. **9.** 147.

10. (*a*) (*i*) x^3+x^2-x+2. (*ii*) $\frac{1}{60}(10x^3+240x^2-70x+480)$ (*iii*) -1

(*b*) 3·625 **12.** 24.

13. (*a*) $x^3-9x^2+17x+6$.

(*b*) $-18+18x+x(x-1)(-6)+x(x-1)(x-3)(-3)$

$+x(x+1)(x-3)(x-5)\cdot 6+x(x-1)(x-3)(x-5)(x-6)\cdot 1$

15. 2·09454. **16.** (*a*) 109·25 (*b*) 2·7357.

17. $(x-4)$; $(x-3)(x-4)$, $(x-3)(x-4)(x-7)$: $x(x-3)(x-4)(x-7)$.

23. $a=\cdot 46996, f(a)=1\cdot 56250$ **26.** $(-1)^n\, x_0x_1 \ldots x_n$.

EXERCISE 3, Page 87

3. (c) $\delta'^2=n^2\delta^2+\frac{n^2(n^2-1)}{12}\delta^4+\frac{n^2(n^2-1)(n^2-4)}{360}\delta^4+\cdots$

4. $y_{11}=2196$, $y_{16}=1786$. **5.** (*a*) 16·9216 (*b*) 6·06

7. (*i*) 33·41 **8.** 0·173965, 0·173965.

9. 2·5282738 by each of the formulae.

10. (*a*) 0·38891. (*b*) 0·38873. **11.** (*a*) 0·9340

13. $A_0=f(x_0)$, $4A_1=f(x_0, x_1)$, $A_2=f(x_0, x_1, x_2)$, etc.

EXERCISE 4, Page 101

1. (*a*) 2·8 **2.** (*b*) 53·7647 **3.** (a) 82.8 approx. (b) 434. 8·34.

5. 45·70 **6.** 37·2 **7.** (*a*) 3·091. (*b*) 2·1028

8. (a) (*i*) 3·5914 (*ii*) 3·092 (*b*) 1·109. (c) 2·0945. **9.** 0·66.

10. (*i*) 0·67239, (*ii*) $x=0\cdot 3015$, (*iii*) 1·5 (*iv*) 2·740

11 17·68. **12.** 0·8536 **13.** 0·267949 **15.** 4·26

EXERCISE 5, Page 115

5. (*a*) (*i*) 98008, – 1986, (*b*) 0·9872.

6. (*a*) ·002000, – 0·0000040.

7. (*a*) ·02455, – ·0003. (*b*) $f'(4)=\frac{17}{6}=2\cdot 883, f''(4)=\frac{31}{36}=0\cdot 861$.

8. 1626. **9.** (*b*) 9810 **10.** 2·644225.

11. (*a*) $f(96\cdot 92189)=43\cdot 2641$ **12.** 0·235. **14.** $f'(3)=18, f''(3)=18$.

16. 1 to 110 **17.** 0.533 cm/sec. 4 – 56 cm/sec^2 **18.** 0.6923

19. minimum = 0. at 0.2, max. = 0.25 at 1

EXERCISE 6, Page 143

2. (*c*) (*i*) 710 sq. ft. (*d*) Speed = 3087 m./sec., height = 112·75 km.

3. (*c*) 0·6932 **4.** (*b*) 1·299. (*d*) 53·87, 53·60.

5. (*a*) 0·29159. (*b*) 0·7854, the value of $\pi = 3{\cdot}1312, \frac{\pi}{4} = 7{\cdot}854 \Rightarrow \pi = 3{\cdot}1416$

(*c*) 1·13935 (*d*) 2·97049

6. (*b*) 4·05016. (*d*) 4·05098.

True value of the integral = 4·05095.

in case of (*b*) error = – 0·00011,

in case of (*d*) error = – 0·00003.

8. (*a*) 1·37040, (*b*) 1·3569.

10. 0·5236 **12.** The first formula.

14. (*a*) $0, \mp\frac{1}{\sqrt{2}}, \mp\frac{1}{\sqrt{2}}$.

(*b*) $\begin{bmatrix} x_1 = -0{\cdot}2899 \\ x_2 = 0{\cdot}5266 \end{bmatrix}$ or $\begin{bmatrix} x_1 = 0{\cdot}6899 \\ x_2 = -0{\cdot}1266. \end{bmatrix}$

(*c*) $\alpha = \pm\sqrt{0{\cdot}6}, p = \frac{5}{9}, q = \frac{8}{9}, \log 2 = 0{\cdot}694$ approx.

16. (*i*) 0·0490291 (*ii*) 0·0008333290

22. 9·688448

27. (*i*) 0·499968 (*ii*) 0·7854 (*iii*) 1·8518 (*iv*) 0·567143

28. (*ii*) ·4999. **29.** 3.854 **31.** 30.032 **32.** 0.7854 **33.** 7.7833 **35.** 1.0635 sec.

EXERCISE 7, Page 169

1. (*a*) $\frac{n(n+1)(2n+1)}{6}$. (*ii*) $\frac{n^2(n+1)^2}{4}$.

(*b*) $\frac{1}{5}n^5 + \frac{1}{2}n^4 + \frac{1}{3}n^3 - \frac{1}{30}n$, $r^4 = (r+3)^{(4)} - 6(r+2)^{(3)} + 7(r+1)^{(2)} - r$.

2. (*i*) $\frac{1}{4}(n+1)n(n-1)(n-2)$.

(*ii*) $\frac{1}{12}n(n+1)(n+2)(3n+13)$.

(*iii*) $\frac{1}{12}n(n+1)(3n^2+31n+74)$.

(*iv*) $\frac{1}{15}\{(3n-1)(3n+2)(3n+5)(3n+8)(3n+11)+880\}$.

3. (*i*) $\frac{1}{12} - \frac{1}{2(n+2)(n+3)}$, (*ii*) $\frac{n(3n+5)}{8(3n+1)(n+4)}$, (*iii*) $\frac{n}{2n+1}$, $S = \frac{1}{2}$,

5. $n(3n^2+6n+1)$.

6. $\frac{1}{8}[(n-1)(2n+1)(2n+3)(2n+5)+15]$.

7. $\frac{1}{4}n(n+1)(n+4)(n+5)$.

8. $\frac{n}{3n+1}$.

9. $\frac{5}{2} - \frac{2n+5}{2(n+1)(n+2)}$.

10. (*i*) $u_x = \frac{1}{x(x+1)(x+3)} = \frac{x+2}{x^{(4)}} = \frac{1}{(x+1)^{(3)}} + \frac{2}{x^{(4)}}$.

(ii) $\frac{29}{36}-\frac{1}{n+3}-\frac{3}{2(n+2)(n+3)}-\frac{4}{3(n+1)(n+2)(n+3)}$.

$\frac{29}{36}-\frac{6n^2+27n+29}{6(n+1)(n+2)(n+3)}$.

11. $\frac{1}{6}n(n-1)(2n^2+3n+6)$.

12. $t_x=x^3+6x^2+6x,\ S_n=n(n+1)(3n^2+23n+46)$.

13. $2^{n+1}-2+\frac{3}{2}n(n+1)$.

14. $2^{n+2}-2-\frac{n}{6}(2n^2-15n+19)$.

15. $2^{n+2}-4-\frac{1}{6}n(n+1)(2n+7)$.

16. $u_x=3^x-1+\frac{1}{2}x(x+3), \frac{1}{2}(3^{n+1}-3)+\frac{n}{6}(n^2+6n-1)$.

17. $u_x=(x^2+x)2^x,\ S_n=(n^2-n+2)2^{2n+1}-4$.

18. $u_x=\frac{2x+3}{x(x+1)}\cdot\frac{1}{3_n},\ S_n=1-\frac{1}{n+1}\cdot\frac{1}{3^n}$.

19. $(n+1)\frac{a^{n+1}}{a-1}-\frac{1}{a-1}-\frac{a^{n+2}}{(a-1)^2}\frac{a^2}{(a-1)^2}$

20. $34-(4n^2+12n+17).\frac{1}{2^{n-1}}$.

21. $1-3[(n+1)\,!]+(n+2)\,!$

23. $\frac{1}{x+n+1}$.

24. $\frac{1}{4}[(x+n)^2(x+n-1)^2-x^2(x-1)^2]$.

30. (i) $(5n+2)(n+3)^{(5)}/90$. (ii) $(2n^2+6n+1)(n+3)^{(4)}/36$

EXERCISE 8, Page 199

1. (i) Second, (ii) Four, (iii) Second, (iv) Third.

(v) Third, $u_n+c_1+nc_2+c_3\left(-\frac{1}{2}\right)^n$

2. (a) (i) $y_{n+2}-5y_{n+1}+6y_n=9$ (ii) $U_{n+2}-7U_{n+1}+12U_n=0$

(iii) $y-4y+4y=0$ (c) $(x^2+x)\Delta^2y-2x\Delta y+2y=0$

4. $u_x=c2^x-2(x+1)$ where c is the periodic function with period 1.

5. $u_x=c.2^x+3^x$.

6. (i) $u_x=\frac{1}{3}x^{(3)}+\frac{1}{2}x^{(2)}+1$. (ii) $u_x=3^x-\frac{1}{2}x$

7. $u_x=c.\frac{1}{3^x}+\frac{1}{2}x-\frac{3}{4}$.

8. (i) $u_x=c_1+\frac{x^{(2)}}{2}-x-\frac{\cos\left(x-\frac{3}{2}\right)}{2\sin\frac{1}{2}}$.

(*ii*) $u_x = \frac{c}{(x+1)(x+2)} + \frac{x}{3}$.

(*iii*) $u_x = \frac{1}{(x-1)\,!}$ (*iv*) $u_x = -\sin\left(nx - s\frac{\pi}{2}\right)$ where $n + 4h + s$,

or $nx = \frac{1}{2}c^{n+1}[(-1)^n e^{ink} - e^{-ink}]$.

9. $u_x = c.\,3^x - 2^x$.

10. $u_x = c.[(x-1)\,!]$

11. $u_x = c\prod_{x=1}^{x-1} x^2 = c(x-1)^2\,!$

12. $u_x = c.x^{-1}$

13. $u_x = c_1 3^x + c_2 5_x$.

14. $u_x = (c_1 + c_2 x)(-1)^x + c_3 2^x$.

15. (*i*) $u_x = 2^x(c_1 \cos\theta x + c_2 \sin\theta x)$, $\theta = -\pi/3$.

(*ii*) $y_n = c_1(1+k^2)^x \cos(\tan\beta/\alpha\, n + c_2)$, $\alpha = 1$, $\beta = k$

16. (*i*) $u_x = a^x\left(c_1 \cos\frac{\pi}{2}x + c_2 \sin\frac{\pi}{2}x\right) + \frac{\cos a(x-2) + a^2 \cos ax}{a^4 + 1 + 2a^2 \cos 2a}$

(*ii*) $u_x = c_1 \cos\left(\frac{x\pi}{2} + c_2\right) - \frac{x}{2}, \sin\frac{\pi}{2}x$ ...(1)

To find *P.I.*, let the trial solution be

$$v_2 = A \sin\frac{x}{2} x\, B \cos\frac{\pi}{2} x \qquad \text{...(2)}$$

Substituting in (1), we get on simplification $0 = \sin\frac{\pi}{2}x$ and hence (2) is not a solution.

Next let $v_x = Ax \sin\frac{\pi}{2}x + Bx \cos\frac{\pi}{2}x$

We get, on substitution in the given equation,

$$A = -\frac{1}{2} \text{ and } B = 0$$

(*ii*) $u_x = c_1 \cos\frac{\pi}{2}x + c_2 \sin\frac{\pi}{2}x + P.I.$ where $P.I. = \frac{1}{E^2+1}\sin\frac{\pi}{2}x$

$$= I.P. \text{ of } \frac{1}{2i}\left[\frac{1}{E-i} - \frac{1}{E+i}\right] e^{\frac{i\pi}{2}x}$$

(*iii*) $u_x = c_1 4^x + c_2 3^x + \frac{(a\cos x + b \sin x)}{a^2 + b^2}$

where $a - \cos 2 - 7\cos 1 + 12$, $b = \sin 2 - 7 \sin 1$,

$c_1 = \frac{3a - a\cos 1 - b \sin 1}{a^2 + b^2}$, $c_2 = \frac{-4a + a\cos 1 + b\sin 1}{a^2 + b^2}$.

(*iv*) $u_x = ca^x + \frac{\sin b\,(x-1) - a \sin bx}{a^2 - 2a\cos b + 1}$

17. (i) $u_x = c_1 \cos\left(\frac{\pi}{4}x\right) + c_2 \sin\left(\frac{\pi}{4}x\right) + c_3 \cos\frac{3\pi x}{4} + c_4 \sin\frac{3\pi x}{4}.$

(ii) $u_x = c_1 \cos\frac{\pi}{2}x + c_2 \sin\frac{\pi}{2}x + 2^x$ with $c_1 = 0, c_2 = -2.$

18. (i) $u_x = c_1 2^x + c_2 3^x + \frac{1}{6} 5^x.$

(ii) $u_x = x!\left\{\sum_0^{x-1} \frac{C_1(-1)^x}{(x!)} + C_2\right\}$

19. (i) $u_x = c_1 4^x + c_2(-1)^x - \frac{1}{4} \cdot 3^x.$

(ii) $u_x = c_1 + c_2 2^x - x + \frac{a^{x+1}}{(a-1)(a-2)}.$

20. $u_x = (c_1 + c_2 x)(-1)^x + \frac{1}{20} x^{(5)} + \frac{1}{4}(x-1)(-1)x.$

21. $u_x = c_1(-1)^x + c_2 8^x - 2^x(3x^2 - 5x + 2)/54.$ use §8·5·2 (i)

22. $u_x = c_1 2^x + c_2(-2)^x + 3^x.$

23. $u_x = c_1 2^x + c_2(-2^x) + \frac{x}{8} 2x.$

24. $u_x = (c_1 + c_2 x)2^x + 2^{x-3} \cdot x(x-1)$

25. $u_x = c2^x + \frac{x(x-1)(2x-1)}{3} - 2^{x-2}$

26. $u_x = c_1 \cdot 1^x + (c_2 + c_3 x)2^x + \frac{2^x}{24} x^{(3)}$

28. $u_x = (\sqrt{2})^x(c_1 \cos\theta x + c_2 \sin\theta x) + \frac{1}{2}x + \frac{3}{4}2^x$ where $\theta = \tan^{-1}\sqrt{7}.$

29. (i) $u_x = -2^x + cx!$ (ii) $u_x = \frac{x}{2}(x!)$

30. $u_x = a^{x(x-1)} p^x + \frac{qa^{2-x}}{a^{-1} - p} + c_1$

31. $u_x = \frac{v_{x+1}}{v_x}$ where

$$v_x = c_1\left(a + \frac{\sqrt{a^2 - 4b}}{2}\right)^2 + c_2\left(a - \frac{\sqrt{a^2 - 4b}}{2}\right)^x_2, \quad a^2 > 4b$$

32. $u_x = ce^{(x-1)} + \frac{e^{1/4}}{3} x^{(3)}.$ **33.** $u_x = ca^x + x^2 a^{x-1}.$

34. $(u_x = (m^2 + n^2)^{x/2}[c_1 \cos\theta x + c_2 \sin\theta\, x] + \frac{1}{n^2} \cdot m^x.$

35. (i) $u_x = \cos(c.2^x).$ (ii) $u_x = \sin(c.2^x),$ (iii) $y_x = e^{A\left(\frac{1}{2}\right)x}$

36. $u_x = c - a^x - \frac{a^{x-1}}{x+2}.$

37. $u_x = ce^{-ax(x-1)} \cdot \dfrac{e^{ax}}{e^{a-1}}, a \neq 0.$

38. (*i*) $u_x = \tan[c + B^{-1}(\tan^{-1} a^{-x})]$

41. (*ii*) $u_x^2 = a^{1/2x(x-1)} \left[\dfrac{a^{-x+1} + c}{a^{-2} - 1} \right]$

42. (*i*) $u_x = \left(\dfrac{c\Gamma x - 1}{\Gamma x + 1} \right)^{1/2}$ (*ii*) $u_x = \sqrt{a} \tan \left\{ c_1 \cos \dfrac{2\pi x}{3} + c_2 \sin \dfrac{2\pi x}{3} \right\}$

45. $u_x = (Ax^2 + Bx + C)^{1/2}.$

46. $u_x = B - \sum_{1}^{n} \left(\dfrac{1}{n + A} \right).$

47. $u_x = A \exp[Bx^3 + Cx^2 + Dx]$

49. $u_x = c_1 + c_2 x + p_x$ where

(*i*) $p_x = \dfrac{a^{x+1}}{(a-1)^2}, \quad a \neq 1$ (*ii*) $p_x = \dfrac{e^{b(x+1)}}{(e^b - 1)^2}, \quad b \neq 0.$

(*iii*) $p_x = \dfrac{-\sin cx}{2(1 - \cos c)}, \quad c \neq 2n\pi$ (*iv*) $p_x = \frac{1}{3}x^2$ (*v*) $p_x = \dfrac{1}{6}x^2.$

(*vi*) $p_x = \dfrac{(e^{b(n+1)})}{(c^b - 1)^3} [(e^b - 1)x - (e^b + 1)], \quad b \neq 0.$

50. $u_x = c_1 + c_2 t^x + p_x, \ t \neq 1, \ y_x = c_1 + c_2 t^x + q_x, \ t = 1.$

(*i*) $p_x = \dfrac{x}{1-t}, q = \frac{1}{2}x^2$ (*ii*) $p_x = \dfrac{xt}{1-t}, q_x = x^{\frac{12}{6}}.$

(*iii*) $p_x = \dfrac{kt^x}{t-1}, \quad q_x = \frac{1}{2}x^2$ (*iv*) $p_x = \sum_{r=0}^{n} \dfrac{(1 - t^{n-r})r^t}{1-t}, q_x = \frac{1}{6}x^2.$

52. $y_h = 4 \cdot 2^h + 1$. Sequence of values 5, 9, 17, 33, 65, 129, ...

EXERCISE 14, Page 304

2. (*b*) (*i*) 0·6184 (*ii*) 4·4934. (*iii*) 1·4971
(*c*) (*i*) 0·58853 (*ii*) 2·36502

3. (*b*) $x_3 = 1·77993.$

4. (*b*) (*i*) 3·3548 (*ii*) 1·4036. (*iii*) 1·7964337
(*iv*) 1·85558 – 1·2, 1·4 (*v*) 0·6071 (*vi*) 1·4246
(*viii*) – 5·5174 (*ix*) – 1·9338 (*x*) 2·7984 (*xi*) – 0·6897, 0·7701
(*xii*)1·796322(*xiii*) 0·567

5. (*a*) $x_1 = 1·8333, x_2 = 2·1667$ (*b*) $x_1 = 1·123106$ and $x_2 = 1·121320.$

6. (*i*) 5·385165 (*ii*) 2·154435 (*iii*) 0·795 (*iv*) 2·62

7. (*a*) – 1·2 (*b*) 2·72

8. $ab < 3/4$. **9.** 5·004, 2·0008, 0·9995

10. (*a*) $1 \pm i, 3 \pm 4i$ (*b*) $x^2 + 1{\cdot}9412x + 1{\cdot}9537$,

$x = -0{\cdot}9706 + 1{\cdot}0058i$. **12.** 1·83931. **13.** (*b*) $\alpha = 0{\cdot}63$ (c) 0.518, $\frac{\text{use cos }(0.52)}{0.52}$ e

21. 1·391 radians **23.** $\alpha_1 = (1 + 2\beta)/2\beta, \alpha_2 = -\frac{1}{2}\beta$, order is 3

EXERCISE 15 page 331

1. (*i*) $x = 7, y = -2, z = 3$.

2. (*i*) $x_1 = 2, x_2 = -1, x_3 = 3$. (*ii*) $x = 3, y = 2, z = 1$,

3. (*i*) $x_1 = \frac{7}{11}, x_2 = \frac{3}{11}, x_3 = 0$.

4. (*i*) $x_1 = 6, x_2 = -7, x_3 = 2$ (*ii*) $x = 5, y = 4, z = -7, w = 1$.

(*iii*) $x = \frac{8}{5}, y = -\frac{16}{5}, z = \frac{22}{15}$

(*iv*) $x_1 = \frac{1}{2}, x_2 = \frac{3}{4}, x_3 = 1$ (*v*) $x = -1, y = 2, z = 3$

5. (*i*)

x	1	2	3	4
x_1	1·20	0·9	1·00	1·00
x_2	1·06	1·00	1·00	
x_3	0·95	1·00	1·00	

(*ii*)

x	1	2	3	4	
x_1	2·5	2·5	3·33	2·86	~ 3
x_2	1·5	2·5	1·77	2·08	~ 2
x_3	1·5	0·83	1·06	0·99	~ 1

(*iii*) $x^{(3)} = 2{\cdot}426, y^{(3)} = 3{\cdot}572, z^{(3)} = 1{\cdot}926$.

6. (*i*) $x_1 = 1{\cdot}247$, $x_2 = -2{\cdot}434$, $x_3 = 0{\cdot}910$

(*ii*) $x_1 = 2{\cdot}031$, $x_2 = 2{\cdot}683$, $x_3 = -1{\cdot}118$. $x_4 = 3{\cdot}111$.

(*iii*) $x = -1861$, $y = -4{\cdot}181$, $z = 0{\cdot}542$

(*iv*) $x = -0{\cdot}55004$, $y = 1{\cdot}79126$, $z = 1{\cdot}73968$.

(*v*) $x = 175, y = 195, z = 160$. (*vi*) $x_1 = 1, x_2 = 1, x_3 = 1$

7. (*a*) $x = 2y = 3, z = 5$ (*b*) $x = 100,001, y = 100,000$

$x = {\cdot}91,999, y = -100,000$.

8. (*i*) $x_1 = -3, x_2 = -3, x_3 = 1$ approx

(*ii*) $x_1 = 1{\cdot}247, x_2 = -2{\cdot}334, x_3 = 0{\cdot}910$

14. $c = 0{\cdot}054, x = 7{\cdot}79, y = 1{\cdot}996$

EXERCISE 16 page 366

1. (*a*) 0·030 at $x = {\cdot}5$, (*b*) 1·061106 at $x = 0{\cdot}4$.

(*c*) 0, 0·2, 0·4080, 0·6413, 0·9236, 1·2942

2. (*i*) $(y)_{x=1} = 3{\cdot}2566495$. (*ii*) $y_6 = 1{\cdot}1972281$.

3. (*a*) $y_3 = \frac{1}{12}x^4 - \frac{5x^2}{6} + \frac{7x^2}{2} - \frac{35}{6}x + \frac{73}{12}$, $(y)x = 2 =$ etc,

(*b*) $y_4 = 1{\cdot}0983, h = {\cdot}1,$ Exact value $= -\ 1{\cdot}1{\cdot}66391$

4. (*a*) $y = \frac{73}{12} - \frac{-35}{6}x_1 + \frac{7}{2}x_1{}^2 - \frac{5}{6}x^3 + \frac{x^4}{12}$ (*b*) $y = x + \frac{1}{3}x^3 + \frac{1}{15}x^5$

(*c*) (*i*) $y + 1 + x + \frac{3}{2}x^2 + \frac{4}{3}x^3 + \frac{17}{12}x^4$

(*ii*) $y = 1 + x + \frac{x^2}{2} - \frac{x^3}{16} - \frac{x^4}{12}$ ....,y (0.1) + 1.1048

(*d*) $y = 1 + \frac{1}{2}x + \frac{3}{40}x^5 + \frac{1}{60}x^6 + \frac{x^9}{192}$

$z = \frac{1}{2} + \frac{3}{8}x^4 + \frac{x^5}{10} + \frac{3}{64}x^8 + \frac{7}{360}x^9 + \frac{1}{256}x^{12}$

(*e*) $y_3 = 1 + \frac{1}{2}x + \frac{3}{40}x^5 + \frac{1}{60}x^6 + \frac{1}{192}x^9$

$z_2 = \frac{1}{2} + \frac{3}{8}x^4 + \frac{1}{10}x^5 + \frac{3}{64}x^8 + \frac{7}{360}x^9 + \frac{1}{256}x^{12},$

(*f*) $2x - 2\log x = \frac{1}{3}(\log x^3)$.

5. $y = 2{\cdot}587$ at $x = 0{\cdot}4$

6. (*a*) $y = x + \frac{1}{6}x^3 + \frac{1}{120}x^5 + \cdots$

(*b*) $x = 0{\cdot}2$, $0{\cdot}4$, $0{\cdot}6$
$y = 0{\cdot}8812$, $0{\cdot}8081$, $0{\cdot}7120$.

7. (*a*) $0{\cdot}4158$ at $x = 1{\cdot}3$

(*c*) Let $z = \frac{dy}{dt}$ so that

$x^{(1)} = x + y + t$

$y^{(1)} = z$

$z^{(1)} = x - t$

$x = t + \frac{1}{2!}t^2 + \frac{1}{3!}t^3 + \cdots$

$y = 1 - t + \cdots$

$z = -1 + \frac{1}{3!}t^3 + \cdots$

(*c*) $x(t) = 1 - t + 2t^2 - \frac{8}{2}t^3 + \cdots$

and $y(t) = -1 + t - \frac{3}{2}t^2 + \frac{4}{3}t^3 + \cdots$

8. (*a*) $x = t + \frac{t^2}{2!} + \frac{t^3}{3!} + \frac{t^4}{4!} + \frac{2t^5}{5!} + \cdots,\ y = 1 - t + \frac{t^4}{4!} + \frac{t^5}{5!} + \cdots,$

$z = -1 + \frac{t^3}{3!} + \frac{t^4}{4!} + \frac{t^5}{5!} + \cdots$

(*b*) $y(1) = 13{\cdot}91,\ y(1{\cdot}1) = 17{\cdot}87$

9. (*a*) $y_4 = 2{\cdot}162,\ y_5 = 2{\cdot}256$ (*b*) $y_4 = 6{\cdot}8733$ (*c*) $1{\cdot}2772, 1{\cdot}2797$.

10.

x	0·3	0·4	0·5	·06
y	0·061493	0·45625	0·29078	0·12566

11. (*a*)

x	0·5	1·0	1·5	2·0
y	1·3571	1·5837	1·7565	1·8950

(*b*)

x	1·1	1·2	1·3	1·4
y	0·359	0·202188	0·354843	0·532377

(*c*)

x	0·5	1·0	1·5
y	0·8891	0·6674	0·5794.

(*k*) $y_{n+1} = y_n + \frac{1}{6}(k_1 + 2k_2 + 2k_3 + k_4)$ with

$$k_1 = hf, \; k_2 = hf\left(t_n + \frac{h}{2}, \; y_n + \frac{h}{2}\right),$$

$$k_3 = hf\left(t_n + \frac{h}{2}, \; y_n + \frac{k_2}{2}\right),$$

$$k_4 = hf(t_n + h, y_n + k_1)$$

$$k_1 = 0{\cdot}2, \; k_2 = 0, \; k_3 = 0{\cdot}31544, k = 0{\cdot}44371$$

$$y_1 = y(0{\cdot}2) = 2{\cdot}3138.$$

12. $y(0{\cdot}4) = 2{\cdot}20, y(0{\cdot}5) = 3{\cdot}20, y(0{\cdot}6) = 4{\cdot}49$

13. (*d*) $y(0{\cdot}4) = {\cdot}6897$. (*d*) $y = y_0 + \frac{1}{6}(k + 2k_2 + 2k_3 + k_4)$

where $k = 0.1385$, $k_4 = 0.1386$, $k_4 = 0.1967$, (e) Here $k_1 = 0.2$

$k_2 = 0.19671$, $k_3 = 0.1967$, $k_4 = 0.1891$

$y(0.2) = y_0 + Dy = 1.19598$ (*b*) $y(0.4) = 1.0410$

EXERCISE 17, Page 381

8. (*a*) $|E| < x^7/7$? (*b*) $3 \times 10^{-10}/2 + 10^{-20}/4$

9. (*a*) 6 (*b*) 9. **10.** (*a*) 3 (*b*) 3. **15.** 0·03836, 0·1421.

Consider the Gauss elimination method where we have three linear equations only :

$$\left.\begin{array}{l} a_{11}x_1 + a_{12}x_2 + a_{13}x_3 = b_1 \\ a_{21}x_1 + a_{22}x_2 + a_{23}x_3 = b_2 \\ a_{31}x_1 + a_{32}x_2 + a_{33}x_3 = b_3 \end{array}\right\} \text{ or } AX = B \text{ where } A = \begin{bmatrix} a_{11} & a_{12} & a_{13} \\ a_{21} & a_{22} & a_{23} \\ a_{31} & a_{32} & a_{33} \end{bmatrix}, B = \begin{bmatrix} b_2 \\ b_2 \\ b_3 \end{bmatrix}$$

EXERCISE 18, Page 408

(*i*) Parabolic (*ii*) Elliptic (*vi*) Hyperbolic (*vi*) Elliptic in the region outside the elliptic $\frac{x^2}{4} + \frac{y^2}{1} = 1$,

Hyperbolic region outside the ellipose $\frac{x^2}{4} + \frac{y^2}{1} = 1$,

Parabolic the region outside the ellipse $\frac{x^2}{4}+\frac{y^2}{1}=1$

We shall now put the above procedure conviently in matrix form.

Augmented matrix

$$[A \mid b] = \left[\begin{array}{ccc|c} a_{11} & a_{12} & a_{13} & b_1 \\ a_{21} & a_{22} & a_{23} & b_2 \\ a_{31} & a_{32} & a_{33} & c_3 \end{array}\right] \quad R_2 \to R_2 - \frac{a_{21}}{a_{11}} R_1, R_3 \to R_3 - \frac{a_{31}}{a_{11}} R_1$$

$$= \left[\begin{array}{ccc|c} a_{11} & a_{12} & a_{13} & b_1 \\ & a_{22}^{(1)} & a_{23}^{(1)} & b_2^{(1)} \\ & a_{32}^{(1)} & a_{23}^{(1)} & b_3^{(1)} \end{array}\right] \quad R_3 \to R_3 - \frac{a_{32}^{(1)}}{a_{22}^{(1)}} R_2$$

$$= \left[\begin{array}{ccc|c} a_{11} & a_{12} & a_{13} & b_1 \\ & a_{22}^{(1)} & a_{23}^{(1)} & b_2^{(1)} \\ & & a_{33}^{(2)} & b_3^{(2)} \end{array}\right] \quad \text{which is the desired form.}$$

Note. Here the elimination was performed in 2 stages only. Thus in case of a system of n equation given by eqn. (1), the elimination is performed in $(n-1)$ *stages.* Sometimes the process stops at the $r\,(<n)$ the stage of elimination. In such a case we have the derived system as follows:-

$$a_{11}x_1 + a_{12}x_2 + \ldots + a_{1n}x_n = b_1 \quad , \quad a_{11} \neq 0$$

$$a_{22}^{(1)}x_2 \quad + \quad + d_{2n}^{(1)}x_n = b_2^{(1)} \quad , \quad a_{22}^{(1)} \neq 0$$

$$a_{rr}^{(r-1)}x_2 + \ldots \quad + a_{rn}^{(r-1)}x_n = b_2^{(r-1)} \quad , \quad a_{rr}^{(r-1)} \neq 0$$

$$0 = b_{r+1}^{(r-1)}$$

$$\vdots \qquad \vdots$$

$$0 = b_n^{(r-1)}$$

We had assumed above that in the elimination procedure, points (diagonal elements) $a_{11}, a_{22}^{(1)}, \ldots a_{nn}^{(n-1)}$ are not all zero. Even if one of $a_{ii}^{(i-1)}\left(a_{11}^{(1)}=a_{11}\right)$ is zero but very small in magnitude as compared with the remaining elements in the i the column, division by a small number may lead to the growth of the round offers. In both these cases, we rearrange the remaining rows (with row upto n the row) so as to obtain a non vanishing point or to made it the largest element in magnitude in that column. The strategy is known as pivoting.

Example. Using the Gauss elemination method show that the system of equations

$$\begin{bmatrix} 3 & 2 & -1 & -4 \\ 1 & -1 & 3 & -1 \\ 2 & 1 & -3 & 0 \\ 0 & -1 & 8 & -5 \end{bmatrix} \begin{bmatrix} x_1 \\ x_2 \\ x_3 \\ x_4 \end{bmatrix} = \begin{bmatrix} 10 \\ -4 \\ 16 \\ 3 \end{bmatrix}$$

are inconsistent.

Solution. The augment Matrix is

$$[A1b] = \left[\begin{array}{cccc|c} 3 & 2 & -1 & -4 & 10 \\ 1 & -1 & 3 & -1 & -4 \\ 2 & 1 & -3 & 0 & 16 \\ 0 & -1 & 8 & -5 & 3 \end{array}\right]$$

$$= \left[\begin{array}{cccc|c} 3 & 2 & -1 & -4 & 10 \\ 0 & -5/3 & 10/3 & 1/3 & -22/3 \\ 0 & 3 & -9 & 2 & 24 \\ 0 & -1 & 8 & -5 & 3 \end{array}\right] \begin{array}{l} R_2 \to R_2 - \frac{1}{3} R_1 \\ = R_3 \to R_3 - 2R_2 \\ R_4 \to R_4 + \frac{1}{3} R_3 \end{array}$$

$$= \left[\begin{array}{cccc|c} 3 & 2 & -1 & -4 & 10 \\ 0 & -5/3 & 10/3 & -1/3 & -22/5 \\ 0 & 0 & -3 & 13/5 & 54/5 \\ 0 & 0 & 5 & -13/3 & 11 \end{array}\right] \begin{array}{l} R_3 \to R_3 + \frac{9}{5} R_2 \\ R_4 \to R_4 + \frac{5}{3} R_3 \end{array}$$

$$= \left[\begin{array}{cccc|c} 3 & 2 & -1 & -4 & 10 \\ 0 & -5/3 & 10/3 & -1/3 & -22/3 \\ 0 & 0 & -3 & 13/5 & 54/5 \\ 0 & 0 & 0 & -0 & 29 \end{array}\right]$$

We cannot determine x_4 from the last equation Hence the given system of equation is inconsistent.

Example. Solve the system of equation,

$$\begin{bmatrix} 1 & 2 & -3 & 1 \\ 0 & 1 & 3 & 1 \\ 2 & 3 & 1 & 1 \\ 1 & 0 & 1 & -1 \end{bmatrix} \begin{bmatrix} x_1 \\ x_2 \\ x_3 \\ x_4 \end{bmatrix} = \begin{bmatrix} -5 \\ 6 \\ 4 \\ 1 \end{bmatrix}.$$

Using Gauss elimination method.

Solution.

$$[A1b] = \left[\begin{array}{cccc|c} 1 & 2 & -3 & 1 & -5 \\ 0 & 1 & 3 & 1 & 6 \\ 2 & 3 & 1 & 1 & 4 \\ 1 & 0 & 1 & 1 & 1 \end{array}\right] \begin{array}{l} R_3 \to R_3 - 2R_1 \\ R_4 \to R_4 - R_1 \end{array}$$

$$\begin{bmatrix} 1 & 2 & -3 & 1 & -5 \\ 0 & 1 & 3 & 1 & 6 \\ 0 & -1 & 0 & -1 & 14 \\ 0 & -2 & 4 & 0 & 6 \end{bmatrix} R_3 \to R_3 - R_2$$

$$= \begin{bmatrix} 1 & 2 & -3 & 1 & 5 \\ 0 & 1 & 3 & 1 & 6 \\ 0 & 0 & 10 & 0 & -20 \\ 0 & 0 & -10 & 2 & 22 \end{bmatrix} R_4 \to R_4 - R_3$$

$$= \begin{bmatrix} 1 & 2 & -3 & 1 & -5 \\ 0 & 1 & 3 & 1 & 6 \\ 0 & 0 & 10 & 0 & -20 \\ 0 & 0 & 0 & 2 & 2 \end{bmatrix}$$

From back substitution, we have

$2x_4 = 2 \Rightarrow x_4 = 1$, $10x_3 = -20 \Rightarrow x_3 = -2,$

$x_2 \times 3x_3 + x_4 = 6 \Rightarrow x_2 + (1 - 6) + 1 = 6 \Rightarrow x_2 = 1$

Al_2 $x_1 + 2x_2 - 3x_3 + x_4 = 5 \Rightarrow x_1 - 2 + 6 + 1 = 5 \Rightarrow x_1 = 0$

Thus $x_4 = 1,\ x_3 = -2,\ x_2 = 1,\ x_1 = 0.$

INDEX

I

J

L

M

N

O

P

Q

R

S

T

U

V

W